Programming Guide For ROBOTIS MINI

By Chi N. Thai

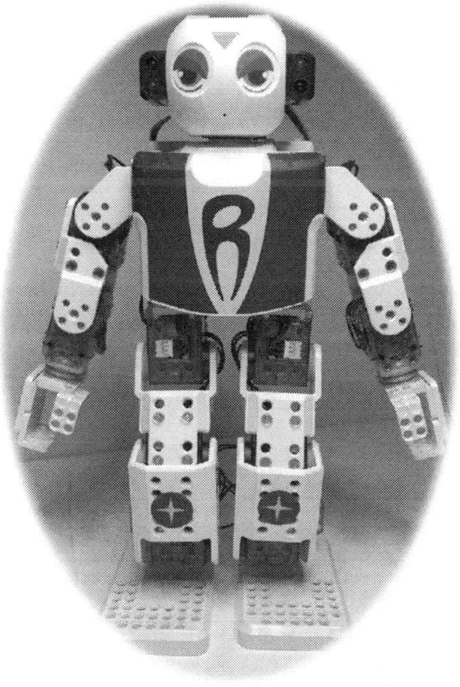

CNT Robotics LLC, Buford

2020

Author: Chi N. Thai

roboteer@comcast.net

www.cntrobotics.com

Programming Guide for ROBOTIS MINI (2020)

ISBN 978-0-9993918-3-9 (Paperback).

Library of Congress Control Number: 2020900988.

Published by CNT Robotics LLC, Buford, GA, U.S.A.

Printed by Amazon Kindle Direct Publishing.

To my parents for giving me life.

To Christine and Emily for their enduring love and support.

To Jenson and all my grand nieces and nephews.

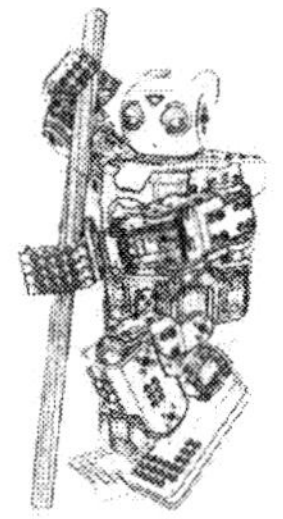

Preface

The MINI™ programmable humanoid robotic kit was released by ROBOTIS® in 2014. It is recommended for users aged 12 or older and is available from many on-line shops. It comes with a free Mobile App called MINI that allows beginner users to control and operate the MINI through its preprogrammed motions such as bowing, walking, gymnastics and dancing moves. But so far, the potential use for this kit in educational robotics is largely untapped due to lack of information and practicum sources for users to develop their own projects and to help them learn more about humanoid robotics programming. This book is designed to fill that void by using a unique approach of developing MINI projects using 2 main software tools:

1. ROBOTIS own software tools such as TASK, MOTION and R+m.PLAY700.
2. EDBOT™ software tool to allow the use of standard development environments such as MIT SCRATCH 2 and PYTHON.

This approach allows MINI users to assess the capabilities and constraints of these tools in the development of their existing or potential project ideas.

Who is this book for?

The author strives to create a book that is useful to two main groups of users:

- Existing owners of the MINI kit who are ready to go beyond the features that the MINI Mobile App provides.
- Beginners in humanoid robotics who look for a firm foundation in robotics concepts and problem-solving approaches. This group includes self-leaners as well as instructors who need a programming guide for the MINI that can be used for themselves as well as for their students, as detailed programming techniques are provided with the reasonings behind them.

What do you need?

This book will require at the minimum one MINI kit and the ROBOTIS RoboPlus software suite which is freely available at this web link (http://en.robotis.com/service/downloadpage.php?ca_id=10#). This book is using the R+TASK V.3 tool (available in desktop and mobile versions) and the R+m.PLAY700 App (available for both Android and iOS devices).

The user needs to acquire a couple of IRSS-10 NIR sensors and the DMS-80 NIR sensor for projects requiring the use of sensors, and one extra XL-320 actuator in certain projects that need to rotate the MINI's head piece.

The EDBOT software tool/license needs to be purchased separately from Robots in School Ltd. (https://shop.ed.bot/collections/products) for each MINI robot, if the user wants to practice with MIT SCRATCH 2 and PYTHON projects. The user also has the option to purchase their special MINI kit which includes one IRSS-10 NIR and one EDBOT software license.

The user also needs access to a basic USB webcam in order to practice on PYTHON projects using the Computer Vision module named "OpenCV".

Finally, Threading applications in SCRATCH and PYTHON will require the use of a second MINI robot.

What is in this book?

This book has 8 Chapters:

- Chapter 1 (25 pages) provides a systemic view of the MINI kit from its hardware components and compatible software tools to its many operating configurations. This chapter also includes the author's interpretation of ROBOTIS Dynamixel System Design Paradigm.
- Chapter 2 (7 pages) provides pre-assembly and post-assembly tips for the MINI kit.
- Chapter 3 (15 pages) focuses on the use of the ROBOTIS MANAGER tool in updating/testing/recovering the firmware for the main robot controller (OpenCM9.04-C), as well as for the actuators and sensors used in the construction of the MINI robot.
- Chapter 4 (45 pages) provides MOTION foundational concepts such as KEY-FRAME, TIME-LINE and MOTION-UNIT as well as how-to techniques to use the ROBOTIS MOTION tool in creating human-like movements for the 3-D MINI model in its Graphical Environment. It includes a Motion Creation Practicum to illustrate various MOTION components such as MOTION-UNIT, MOTION-LIST/PAGE, MOTION-GROUP and more advanced tools such as EXIT-MOTION and JOINT-GROUP. This practicum especially points out the differences between the robot's "Graphical" moves and its "Physical" moves which are constrained by the presence of the real surface that the physical robot must travel on. A custom MOTION file with the author-created MOTION examples is included with this book.
- Chapter 5 (8 pages) is a review of the features provided in the MINI App that already has extensive information provided in the ROBOTIS web-based e-Manual. It also shows how to interface this App with the author custom MOTION file.
- Chapter 6 (102 pages) provides the foundational knowledge in computer programming structures and techniques with the ROBOTIS TASK tool. It uses a Beginner's Approach based on the "Sense-Think-Act" Paradigm to provide a step-by-step procedure to go from specifications for the robot's tasks to the writing and testing of the actual TASK codes that allow the robot to perform the original user-assigned tasks. It uses a "Spiral-Teaching" approach applied through a series of projects designed to deepen and broaden the reader's understanding in several robotics areas:
    - Robot Remote Control: from Single/Multiple RC Inputs via "Message Filtering" techniques used for a single robot to more general "Message Shaping" techniques to be used between a pair of "Leader/Follower" robots in "Open-Loop" and "Closed-Loop" operations in the "Mimicking MINIs" project.
    - Autonomous sensor-based Behavior Control as well as Mixed-Control applications are showcased using several projects: RC Smart Avoider, Autonomous Random Walker, Autonomous Dowel Scanner.
    - Actuator Control: from simple Motion-Page-Play to Sensor-based Exit-Motion techniques using Motion Pages "0" and "-1", to the mixing of Motion-Page-Play with Targeted-Servo Goal-Position Control via the use of variable Joint-Offset features to provide Independent Control of Head and Shoulder joints while performing standard Motion-Play on the overall robot. The variable Torque-Level feature is used to demonstrate the use of the MINI's arms as an Adaptive Gripper to handle different-sized objects.
- Chapter 7 (39 pages) illustrates the use of the R+m.PLAY700 App to access the Smart Device features in a TASK program which allow the MINI's Hardware Controller to access the Multimedia Services of a mobile device such as a Smartphone or Tablet. Selected features are

demonstrated: Touch Areas, Text Displays, Instruments Play, Text-to-Speech, Speech Recognition and Color Tracking with Mobile Video Camera.

- Chapter 8 (114 pages) assumes the reader to have some programming experiences in using SCRATCH 2 and PYTHON. It demonstrates the use of EDBOT to obtain control of two MINI robots from a single SCRATCH 2 or PYTHON program. It showcases Remote/Autonomous/Mixed Control applications using the SCRATCH 2 "Multi-Sprite" and "Broadcast-Message" features, while it demonstrates the use of the "Threading" and "OpenCV" modules in PYTHON applications.

This book provides appropriate source codes and tutorial videos (via YouTube®) to illustrate the presented concepts. Please visit http://www.cntrobotics.com/mini-book for access options to the source codes and tutorial videos.

And please, do let the author know about your experiences in using the materials in this book, this will help the author improve the next edition and the companion resources. Please visit the author web site www.cntrobotics.com for updates and other robotics materials. Please send your comments and suggestions to roboteer@comcast.net.

Trademark Attributions

Content in this book made references to the following trademarks:

ROBOTIS product names are trademarks or registered trademarks of ROBOTIS, Inc. Used with permission by ROBOTIS Co. Ltd.

Android®, Android Studio®, YouTube®, and Chrome® are registered trademarks of Google Inc.

SCRATCH® is a registered trademark of the Scratch Team at the Massachusetts Institute of Technology.

Windows®, Visual Studio Code® are registered trademarks of Microsoft Inc.

iOS®, iPhone® and iPad® are registered trademarks of Apple Corporation.

Bluetooth® is a registered trademark of Bluetooth SIG, Inc.

Edbot® is a registered trademark of Robots in Schools Ltd.

Python® is a registered trademark of the Python Software Foundation.

JavaScript® is a registered trademark of Oracle Corporation.

Buford, GA, USA Chi N. Thai

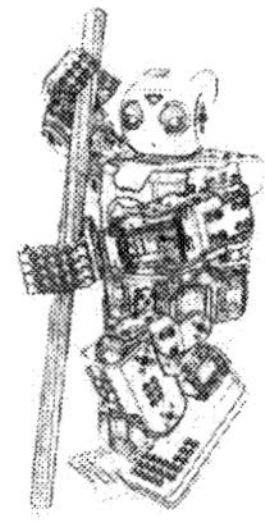

Contents

Chapter 1: Systemic View of ROBOTIS® MINI™

ROBOTIS releases the MINI kit in 2014 (http://www.robotis.us/robotis-mini/) along with the Mobile App called MINI for iOS and Android/Chromebook OS which can be used at a minimum level for entertainment purposes (https://play.google.com/store/apps/details?id=com.robotis.darwinmini, https://itunes.apple.com/us/app/robotis-mini/id948481762). But this kit is much more capable, as shown later in this book, when other software tools such as ROBOTIS R+TASK™/R+MOTION™ or EDBOT® are leveraged for the purpose of teaching robotics fundamentals to students at all ages.

The goal of Chapter 1 is to provide the reader with a systemic view of the MINI kit illustrating how its varied hardware components integrate with a host of software tools at a high conceptual level while leaving the detailed operational details for later chapters.

1.1 Components of MINI kit

In this Section 1.1, the author describes the hardware components that come with the MINI kit when purchased, along with additional hardware components that can be used to enhance the use of this robotics kit.

1.1.1 Hardware Controller: OpenCM9.04-C

The most important component of any robotics kit is of course its "brain", i.e. Hardware Controller, which is the OpenCM9.04-C. The actual microcontroller chip used for the OpenCM9.04-C is an STM32F103CB with 128 KB of Flash Memory – see back side of PCB's picture on the left in Fig. 1.1 (beneath the QC sticker). The partition of this 128 KB memory space into various firmware components will be described in Section 1.3.1.

Fig 1.1 also shows the locations of the components that the user will mostly interact with when using this Hardware Controller:

- The ON-OFF power switch.
- Two connectors for the 2-pin cables coming from the two Li-Ion battery cradles LBB-040 or LBB-041 (see Section 1.1.3).
- Four 3-pin Dynamixel™ ports where the 16 servo motors XL-320s will be connected to, in a daisy chain fashion (more details about the term "Dynamixel" are provided in Section 1.1.2). The STM32F103CB microcontroller communicates to the 16 XL-320s via Serial Communications Channel 1 (USART Ch.1) at a high speed of 1 Mbps, as lots of data throughput are needed to control 16 servo motors at the same time.

- Four General-Purpose-Input-Output (GPIO) connectors for various sensors and the LED module (see Section 1.1.6). GPIO ports are connected directly to the I/O ports of the STM32F103CB microcontroller.
- One "external" Serial Communications port where the BT-210 Bluetooth® receiver (and other 5-pin compatible serial communications devices – see Section 1.1.4) can be connected. The STM32F103CB microcontroller communicates to these devices via Serial Communications Channel 2 (USART Ch.2) at a slow speed of 57.6 Kbps, as less data throughput is required for these devices.
- One User Button to be used when recovering firmware for the OpenCM9.04-C and/or the servo motors XL-320 (see Chapter 3).
- One Micro USB-2 port when recovering firmware on the OpenCM9.04-C and its associated XL-320s (see Section 1.3.7 and Chapter 3) or when doing advanced C/C++ programming from a PC using the OpenCM or Arduino IDEs (see Section 1.3.9).

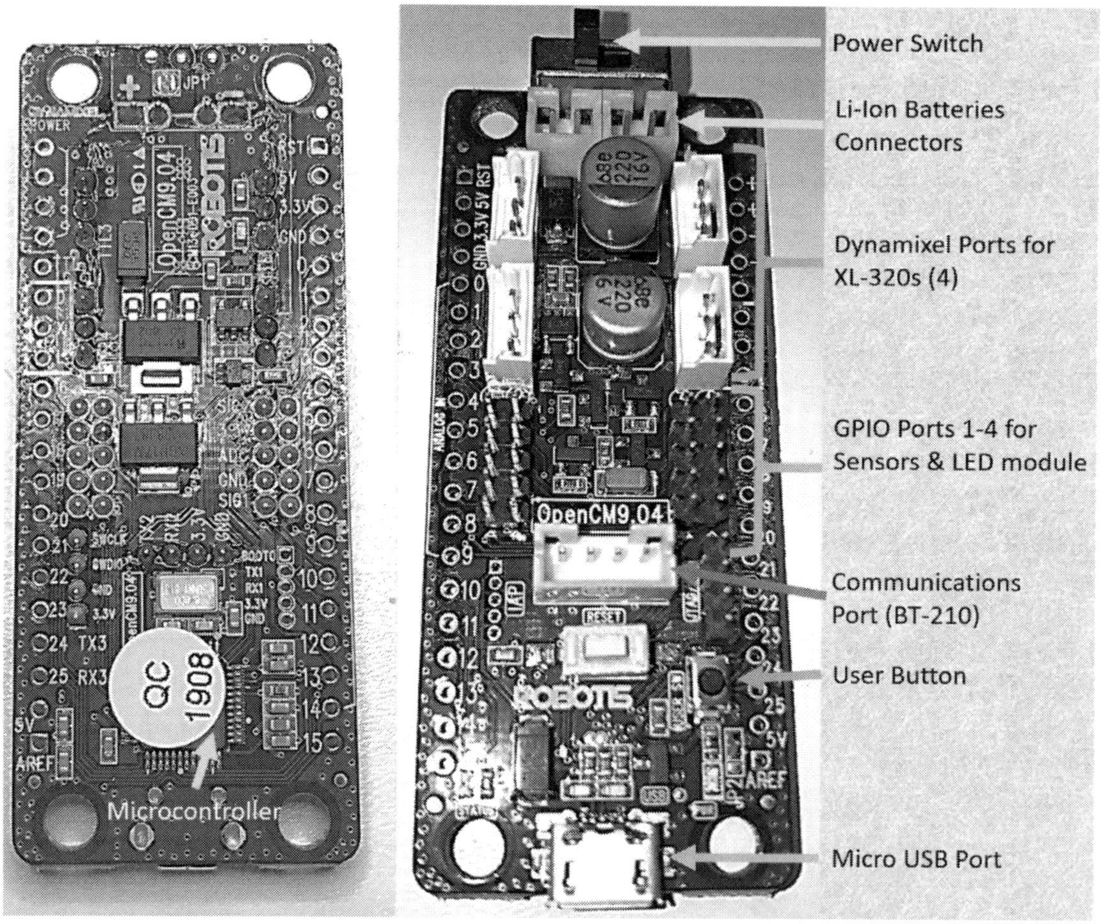

Fig. 1.1 Hardware Controller OpenCM9.04-C.

The reader is referred to the following ROBOTIS web page for more comprehensive details (http://emanual.robotis.com/docs/en/parts/controller/opencm904/) on many other aspects for the proper usage of the OpenCM9.04-C which will be discussed when appropriate in later sections of this book.

ROBOTIS CS also has a YouTube playlist of tutorials for the OpenCM9.04 at https://www.youtube.com/playlist?list=PLEf1s0tzVSnRgF2Cu9r91bU3bHuZU8Fxu

1.1.2 Servo Motor: XL-320 (Dynamixel Concept)

The MINI kit comes with 16 "real" XL-320s which are "smart" servo motors allowing it to perform human-like movements of its arms, legs and hip joints, and 1 "dummy" XL-320D serving as a component of its "unmovable" head assembly. The author would recommend readers to acquire a 17th XL-320 so that this head assembly can be motorized and controlled (see Section 2.2).

ROBOTIS trademarked most of their servo motors (including the XL-320) as "Dynamixel" or "Smart Servo" because each XL-320 has a dedicated microcontroller STM8S105K4T6C controlling the actual servo motor (see Fig. 1.2), and allowing each XL-320 to communicate with the Hardware Controller OpenCM9.04-C within a local wired (3-pin) communications network shared with other XL-320s (i.e. USART Ch.1, at a rate of 1 Mbps), all hooked up as a daisy chain (http://emanual.robotis.com/docs/en/dxl/x/xl320/).

Of course, in order to distinguish one XL-320 from another, each XL-320 in the MINI kit was pre-assigned a unique ID from 1 to 16 (see Section 2.2 for how to handle the situation when the reader decides to get that 17th XL-320). This arrangement allows the user to issue a simple assignment command such as *"ID[17]: Goal Position = 512"* to make the Head Servo (ID = 17) to go to Position 512 (i.e. commanding the MINI's head to face forward), without the need to program at the actual hardware level using Pulse-Width-Modulation (PWM).

Moreover, Hardware Controllers such as the OpenCM9.04-C are also considered as Dynamixels with a specially assigned **ID = 200** when using ROBOTIS provided software such as R+TASK, R+MOTION and OpenCM IDE, and consequently there can only be **one** Hardware Controller within each local wired Dynamixel network (i.e. for each ROBOTIS robot). In other words, the user should not use **ID = 200** for any of the XL-320s used.

Fig. 1.2 Printed Circuit Board (PCB) for XL-320.

The XL-320 is capable of 2 modes of operations: Continuous-Turn (i.e. as a wheel spinning CW or CCW) or Position-Control (i.e. as a true servo maintaining a given rotational position), but the Continuous-Turn mode is obviously not appropriate for the MINI kit which is a humanoid robot. However, the author will demonstrate the use of the 17th XL-320 (Head's servo) in both modes for the reader's benefit in Chapter 6.

Let's next look into how does the XL-320 "know its rotational position" at any one time:

- The left picture in Fig. 1.3 shows the inner details of an XL-320 that had been disassembled to show that when the Servo Motor turns, the Gear Train transmits this rotational motion into the Horn (i.e. to move the MINI's head for example) and also into the Position Encoder below it.
- The middle picture in Fig. 1.3 shows the PCB with the Position Encoder's cover removed (far right picture) so that its actual mechanism is exposed. This Position Encoder is essentially a rotational potentiometer using 2 graphite rings. The inner ring is continuous and serves as an analog voltage input signal for the microcontroller STM8S105K4T6C to digitize and send the resulting numerical value as "Present Position" to the OpenCM9.04-C. This "Present Position" value can then be used in the user's control program. The outer ring is "discontinuous" with one end connected to electrical ground and the other connected to V_{DD} = 7.4V, and because of this discontinuity the XL-320 can only provide positional information for only an arc of 300 degrees out of 360 degrees possible for a complete circle.

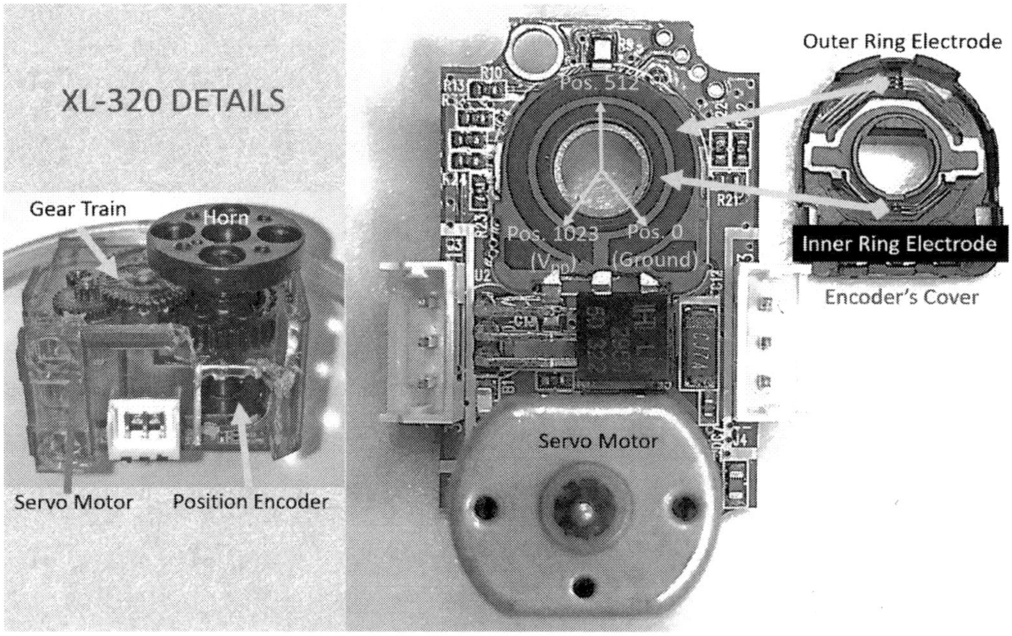

Fig. 1.3 Details of a Disassembled XL-320.

- The rightmost picture in Fig. 1.3 displays the special copper electrode assembly that electrically connects the two rings of the potentiometer together. This electrode assembly has built-in "brushes" to rub against the graphite rings when the cover is closed on top of the Position

Encoder in normal operation. This is how the position-dependent varying voltage on the outer graphite ring is electrically conveyed to the inner graphite ring and subsequently to the XL-320 microcontroller in order to digitize that voltage into a number between 0 and 1023.

- The "take-home" concept is that there is a "forbidden" zone of 30 degrees where the XL-320's horn cannot be sent into by the user program when the XL-320 is used in Position Control mode.

1.1.3 Batteries: LB-040/041

Currently, there is only one battery charger cradle available - LBB-041 (http://www.robotis.us/li-ion-battery-charger-set-lbb-041/) but two Li-Ion battery packs that the reader can purchase from ROBOTIS, LB-040 and LB-041 (http://www.robotis.us/li-ion-battery-3-7v-1300mah-lb-040/ and http://www.robotis.us/li-ion-battery-3-7v-1300mah-lb-041/). ROBOTIS recommends using the LB-040 for the MINI, because the MINI kit comes with the older cradle LBB-040 which is no longer available from ROBOTIS. **The user won't be able to fit an LB-041 battery into an LBB-040 cradle, but an LB-040 battery will fit OK into an LBB-041**.

The MINI's Quick Start manual on its pages 9-10 recommends removing the LBB-040 batteries from the MINI framework and to recharge them in separate LBB-040 or LBB-041 cradles, and not to charge them while they are still attached to the MINI in their normal configuration. This is to protect the OpenCM9.04-C PCB from possible power fluctuation coming from the power adaptor during the charging period.

1.1.4 Communication Hardware: BT-210, USB & Others

The MINI kit comes with a Bluetooth receiver called BT-210 supporting BT specification 2.1 (http://emanual.robotis.com/docs/en/parts/communication/bt-210/), meaning that it would work with most PCs directly using their built-in BT chips and also with Android mobile devices when the user wants to use the MINI Mobile App or to program the robot using ROBOTIS R+TASK and R+MOTION software (both desktop and mobile versions). The BT-210 plugs into the 4-pin white Molex connector shown in Fig. 1.1.

However, the BT-210 receiver will not work with iOS mobile devices like iPhone and iPad, and the user will need to purchase and use the BT-410 receiver instead (http://emanual.robotis.com/docs/en/parts/communication/bt-410/). To add to the "confusion", the BT-410 receiver will work fine with Android devices without additional hardware or software, and the BT-410 receiver can also work with PCs if the user is willing to additionally purchase the "BT-410 Dongle" (http://emanual.robotis.com/docs/en/parts/communication/bt-410-dongle/) which can be plugged to a USB port on the PC.

The reader also needs to be aware that the EDBOT software (Chapter 8) can only be used with the BT-210 receiver.

In the author's experience in using BT devices and Windows OS over the years, sometimes Windows updates can mess up the operation of BT devices, thus the author recommends users to also obtain the LN-101 to provide a secure wired communication line between a USB port and the 4-pin communications port on the MINI for those unpredictable situations (http://emanual.robotis.com/docs/en/parts/interface/ln-101/).

Alternatively, the reader can also use a "good" micro-USB cable to potentially perform these communications tasks with ROBOTIS R+TASK, R+MOTION and R+MANAGER software, but the author had noticed that the micro-USB cable that comes with the MINI kit sometimes is only good for charging the

Li-Ion battery packs and not good enough for data transfer. Furthermore, the author obtained "reliable" results when using the USB cable with the MANAGER tool only, and not with TASK and MOTION. Lastly, if the user plans to use the USB port on the OpenCM9.04-C, an access slot needs to be made into the MINI's torso frame part DMF-B01 (see Section 2.1.1). For advanced users who want to use the OpenCM or Arduino IDEs with the MINI, they would need to have this micro-USB access (see Section 1.3.9), if not the MINI's torso section would have to be disassembled and reassembled each time that the user wants to revise codes on the MINI.

1.1.5 Frame Parts & Mechanical Assembly

If the reader has access to CAD software and/or a 3D Printer, ROBOTIS has provided STL and STP files for all MINI frame parts at these web links:
http://www.robotis.com/download/darwinmini/darwin_mini_3dprinter.zip,
http://support.robotis.com/en/baggage_files/darwinmini/darwin_mini_stp.zip. Readers can use these STL/STP files to make their own spare parts or design custom parts.

The MINI kit includes a Quick Start manual that contains assembly instructions for this robot. ROBOTIS also provides a free 3D Viewer software called R+DESIGN which has usage instructions at this web link (http://emanual.robotis.com/docs/en/software/rplus2/design/). R+DESIGN works on Windows PC as well as iOS and Android/Chromebook devices, please use the appropriate links below to download this software:

- For Windows PC - http://en.robotis.com/service/download.php?no=11.
- For Android/Chromebook devices - https://play.google.com/store/apps/details?id=com.robotis.mdesign
- For iOS devices - https://apps.apple.com/us/app/r-m-design-robotis/id975779539

The R+DESIGN tool allows the user to change the user's 3D viewpoint on ROBOTIS parts and assemblies of existing ROBOTIS robots, which can be very helpful to users for visualizing and sequencing properly intricate assemblies for the hip and ankle joints of the MINI. However, R+DESIGN has one drawback because it cannot show the proper routing of the Dynamixel cables between XL-320s and frame parts, and at present only the printed Quick Start manual can do this job properly.

1.1.6 Compatible Actuators & Sensors (Port Concept)

The MINI kit does not include any sensor at all, but the OpenCM9.04-C can handle a variety of sensors and an LED module through four 5-pin GPIO connectors as shown in Fig. 1.1. For a list of compatible devices for each specific connector/port please check this web link (http://emanual.robotis.com/docs/en/parts/controller/opencm904/#robotis-5-pin-port). Please note that these devices are directly connected to the OpenCM9.04-C's microcontroller STM32F103CB through its I/O ports and not through the Dynamixel network like the XL-320s (see Section 1.1.2). For example, in this book the author uses 2 NIR sensors IRSS-10 (http://www.robotis.us/ir-sensor-irss-10/) mounted on either side of the MINI's head and a DMS-80 on the top of its head (http://www.robotis.us/distance-measuring-sensor-dms-80/) (see Fig. 1.4). Internally, the IRSS-10s are connected to Port 1 and Port 4, while the DMS-80 is connected to Port 2 and typical programming commands to access them are of this type *"Left_Head_Sensor = PORT[1]:IR Sensor"*, i.e. to read Port 1 for an IR Sensor input and to save this digitized value into Parameter "Left_Head_Sensor".

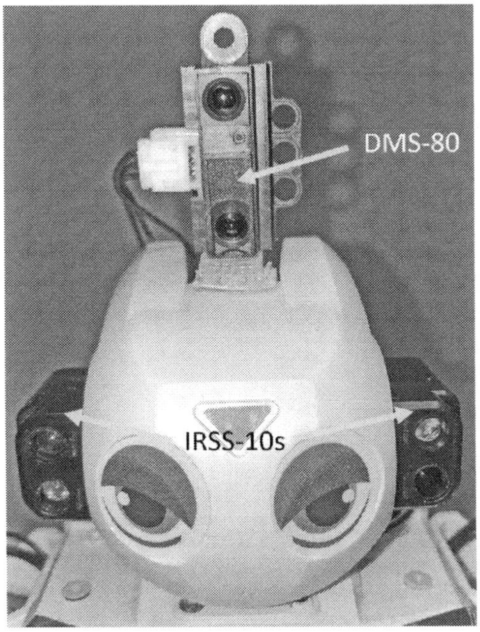

Fig. 1.4 MINI robot Zira with 2 IRSS-10s and 1 DMS-80 mounted on her head.

1.2 MINI Software Tools

There is a plethora of software tools that work with the MINI and they can come from ROBOTIS as free downloads or from other organizations or vendors which can be free or need to be purchased separately (see Section 1.2.2). ROBOTIS tools of course have extensive documentations at their e-manual web site http://emanual.robotis.com/ and users can also register at a public forum http://en.robotis.com/service/forum.php to get official answers to any technical issue that they may have.

1.2.1 ROBOTIS Provided Software

ROBOTIS offers many software tools that work with the MINI for free at their web site http://en.robotis.com/service/downloadpage.php?ca_id=10, notably:

- R+MANAGER V.2 which works only on Windows OS (or macOS with appropriate Windows Virtualization software).
- R+TASK and R+MOTION used to be 2 separate components for Versions 1 and 2 (Windows, Linux and macOS). In Summer 2019, ROBOTIS decides to upgrade and combine these two packages together and calls it R+TASK V.3 (see Fig. 1.5). R+TASK V.3 works with Windows, Android/Chromebook and macOS. **This book uses R+TASK V.3** and the Windows version can be directly downloaded from this link (http://www.robotis.com/service/download.php?no=1774). Mobile versions of R+TASK V.3 can be downloaded from these links (https://play.google.com/store/apps/developer?id=ROBOTIS or https://apps.apple.com/us/developer/robotis-co-ltd/id948481761).

- R+DESIGN as previously mentioned in Section 1.1.5 works on Windows, iOS and Android/Chromebook.
- DYNAMIXEL WIZARD V.2 which works with Windows, Linux and macOS (http://emanual.robotis.com/docs/en/software/dynamixel/dynamixel_wizard2/).
- For Mobile Apps (iOS and Android/Chromebook), the reader is already aware of the ROBOTIS MINI App, but the reader probably does not know that the R+m.PLAY700 App (iOS and Android/Chromebook), originally designed for the PLAY700 kit (i.e. Hardware Controller CM-50), does work fine with the OpenCM9.04-C using the TASK's SMART DEVICE features (see Chapter 7). The reader can download R+m.PLAY700 to his or hers compatible mobile device at these web links (https://apps.apple.com/us/app/play700/id1156037721, or https://play.google.com/store/apps/details?id=com.robotis.play700).
- For more advanced users using C/C++, they can use the package OpenCM IDE (http://emanual.robotis.com/docs/en/software/opencm_ide/getting_started/) which can be programmed to work with the R+m.PLAY700 App. OpenCM IDE works on Windows, Linux and macOS, unfortunately ROBOTIS is no longer supporting this package and is promoting the use of standard ARDUINO interface with their new tools called Dynamixel SDK and Dynamixel Workbench (http://emanual.robotis.com/docs/en/software/arduino_ide/) which work on Windows, Linux and MacOS.

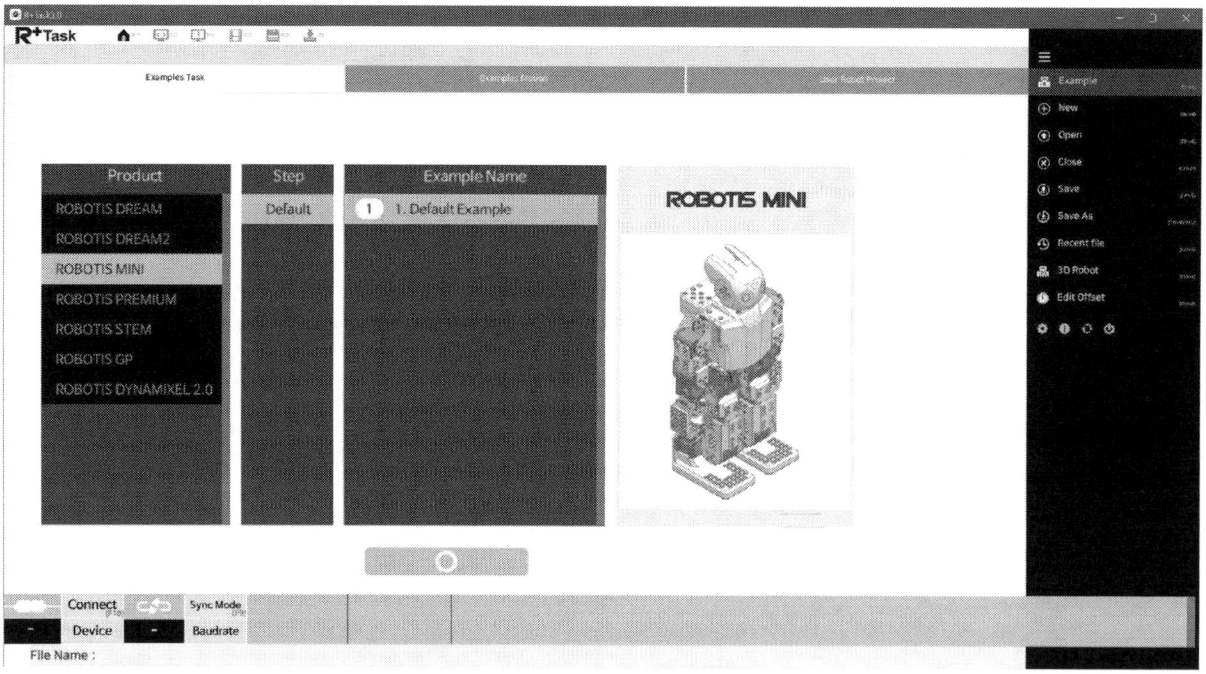

Fig. 1.5 Main Menu of the new R+TASK V.3.

1.2.2 ROBOTIS Partner Software

Currently, the author is aware of only 1 commercial software tool that works with the MINI and it is called EDBOT from Robots in Schools Ltd. (http://ed.bot/edbotsoftware). EDBOT runs on Windows, Mac, Linux (including Raspberry Pi) and Android/Chromebook. EDBOT allows the use of several development tools such as SCRATCH 2/3, Python, JavaScript and C# (http://support.ed.bot/) to work with ROBOTIS firmware and Motion files. A unique EDBOT feature is that it will allow the user to control several MINIs in the same user program in the languages listed above (https://www.youtube.com/watch?v=yUpx42eNR2I). Chapter 8 of this book will provide the reader with a firm foundation in using EDBOT V.5 with SCRATCH and Python.

The Synthiam organization supports the MINI with their free EZ-Builder environment (compatible with ARDUINO IDE tools from ROBOTIS) but it requires a different firmware than the standard one from ROBOTIS (https://synthiam.com/GettingStarted/Build-Robots/Robotis-OpenCM-9-04-17530). This is a good project for an advanced reader who wants to go beyond the standard MINI, with video camera control and speech recognition for example (https://synthiam.com/GettingStarted/Build-Robots/Robotis-Darwin-Mini-17529).

The European Poppy Project uses the XL-320s but with their own controller (Raspberry Pi), frame parts and free development software environments (https://forum.poppy-project.org/t/getting-started-with-poppy-project/362). This is also a good project for an advanced reader who wants to go beyond the standard MINI.

The PyPi organization hosts the free PYXL320 library to control XL-320 with Python (https://pypi.org/project/pyxl320/).

MATLAB has a toolbox for XL-320 (https://www.mathworks.com/matlabcentral/fileexchange/61541-xl_320-toolbox).

Kevin Walchko has a GITHUB page for Multiped Robotics using the XL-320 and AX-12 (https://github.com/MultipedRobotics).

Nyrio has a KickStarter campaign for a 6 axis robotic arm using the XL-320 (https://www.kickstarter.com/projects/niryo/niryo-one-an-open-source-6-axis-robotic-arm-just-f).

1.3 MINI Operating Configurations

Section 1.2 shows that there is quite a variety of software tools that can work with the MINI, and specific configurations of the firmware components on the OpenCM9.04-C will be needed for these software tools to work properly. **In this Section 1.3, the author will limit the discussion to the operational characteristics of ROBOTIS provided software tools and EDBOT V.5.**

The most important point of reference to understand the various MINI operating configurations is the Memory Map of its 128 KB of Flash Memory, so that is where we are going to start next.

1.3.1 Memory Map of OpenCM9.04-C

The Hardware Controller (HC) OpenCM9.04-C has 128 KB of Flash Memory partitioned out as shown in Fig. 1.6:

- The Boot Loader and EEPROM memory sections cannot be altered by the user.
- When the MANAGER tool is used to recover or update the OpenCM9.04-C's firmware, the memory sections from Address 0800-3000 and up to 0801-F800 are overwritten with new data (see Section 1.3.7).

Firmware Component	OpenCM9.04-C	
	Start Address	Size in KiloBytes
Boot Loader	0800 0000	12
ROBOTIS Firmware	0800 3000	42
ROBOTIS TASK	0800 D800	8
ROBOTIS MOTION	0800 F800	64
EEPROM	0801 F800	2

Fig. 1.6 Memory Map for the 128 KB Flash Memory of OpenCM9.04-C.

- When the TASK component of R+TASK V.3 is used to create/download behavior control programs for the MINI, that user code is loaded to the memory section starting at 0800-D800 and it is limited to 8 KB (see Section 1.3.4).
- When the MOTION component of R+TASK V.3 is used to create/download Motion data from the PC to the OpenCM9.04-C, these Motion data are downloaded into the 64 KB memory section starting at 0800-F800 (see Section 1.3.2).

Furthermore, the microcontroller STM32F103CB can be set into 4 Modes of operations: "Task Play", "Manage", "Motion Play" and "Boot Loader" (see Fig. 1.7), and these modes of operation activate/deactivate certain memory sections as shown in later sub-sections (*Note: OpenCM9.04-C does not use "Motion Play" or Mode 3*). By default, when the OpenCM9.04-C is turned on, it is set into "Task Play" mode (Mode 1) and if there is a valid TASK program already loaded into its 8 KB memory section, that TASK program will immediately execute.

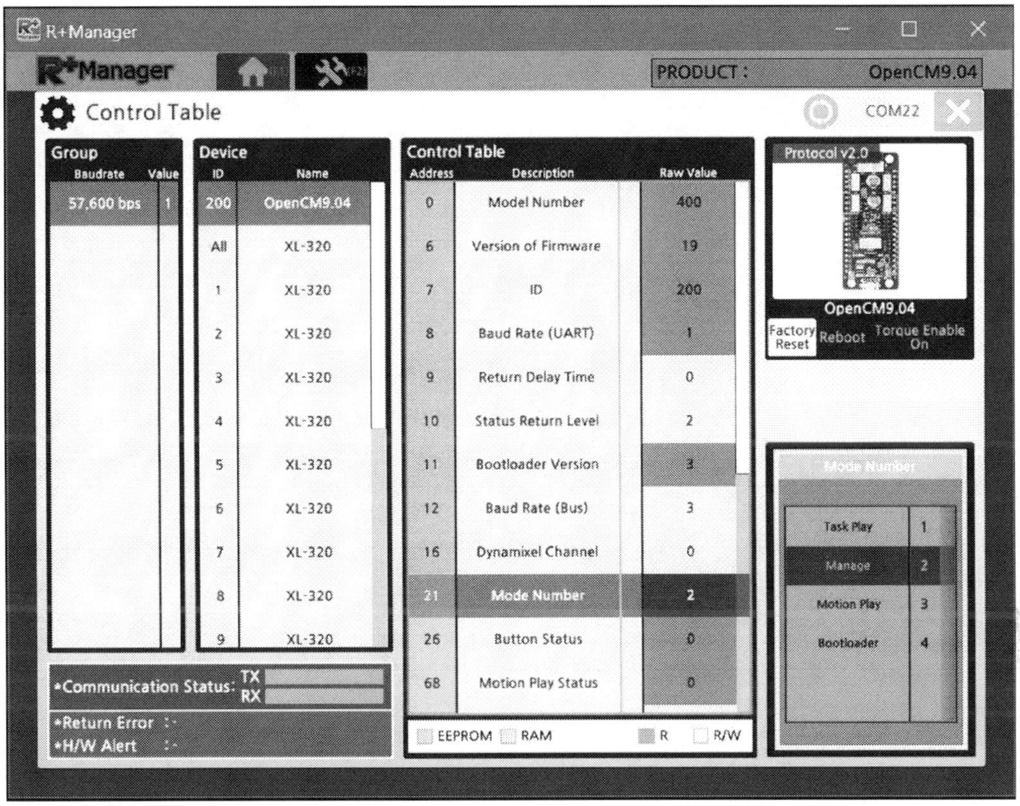

Fig. 1.7 MANAGER Tool showing "Mode Number" at Address 21.

1.3.2 Using MOTION 3 Component

The MOTION and TASK Sub-tools are designed to work together for a humanoid robot such as the MINI, but in the previous 2 versions they were provided as 2 separate applications using the same Serial Communication Port, so it was quite bothersome to switch back and forth to the proper COM Port when the user wants to develop a custom MINI project which usually requires both TASK and MOTION tools to be running. Then in Summer 2019, ROBOTIS releases R+TASK V.3 which combines TASK and MOTION into 1 application (see Figs. 1.8 and 1.11) and resolves that long-standing annoyance (**if the user does not use the "Synch" mode** inside MOTION 3 Component - for more details on the "Synch" mode see Chapter 4).

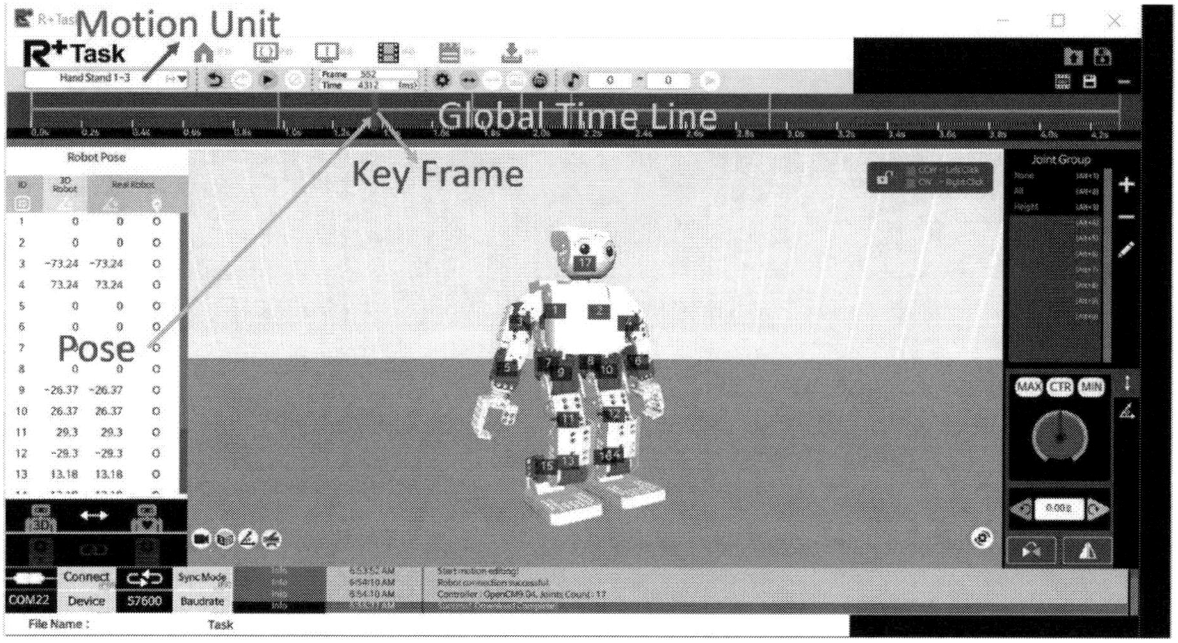

Fig. 1.8 MOTION 3 Sub-tool of R+TASK V.3.

By now, the reader is certain to have noticed the complex interface of the MOTION 3 Sub-tool as illustrated in Fig. 1.8. Chapter 4 will provide the reader with a firm foundation in using MOTION, so for now we'll just go over some overarching concepts:

- Creating human-like movements for the MINI is very similar to making cartoons which are essentially multiple "sequential" frames of "static images" that have minor changes done to the "object(s) in motion" in each frame. When this "multi-frame list" is played out at a proper speed, the human viewer then perceives "motion" for the "object(s)".
- Thus, when using MOTION, the user starts by creating POSEs of the MINI. A POSE is just a list of user-wanted servo positions for each of the 16/17 servos comprising the MINI at an instant in time, and the user can do this easily using a mouse/graphical interface for each servo or group of servos chosen by the user. When a POSE is inserted into a TIME-LINE, it is referred to as a KEY-FRAME.
- The TIME-STEP between each sequential KEY-FRAME is set by the user, however the smallest TIME-STEP allowable is 7.81 ms which corresponds to the refresh cycle time for the XL-320s attached to the OpenCM9.04-C. On a fast graphics card, MOTION will be able to synchronize the graphics of the simulated MINI's moves on the computer screen to the real-time execution of those moves by the physical robot.
- Several KEY-FRAMEs on a TIME-LINE form a MOTION-UNIT. Please note that each MOTION-UNIT should not be longer than 4.312 seconds. If the user-wanted robot movements would need more than 4.312 seconds to perform, additional MOTION-UNITs need to be created. Furthermore, several MOTION-UNITs can be organized into a MOTION-LIST or just MOTION in ROBOTIS nomenclature (see Fig. 1.9).

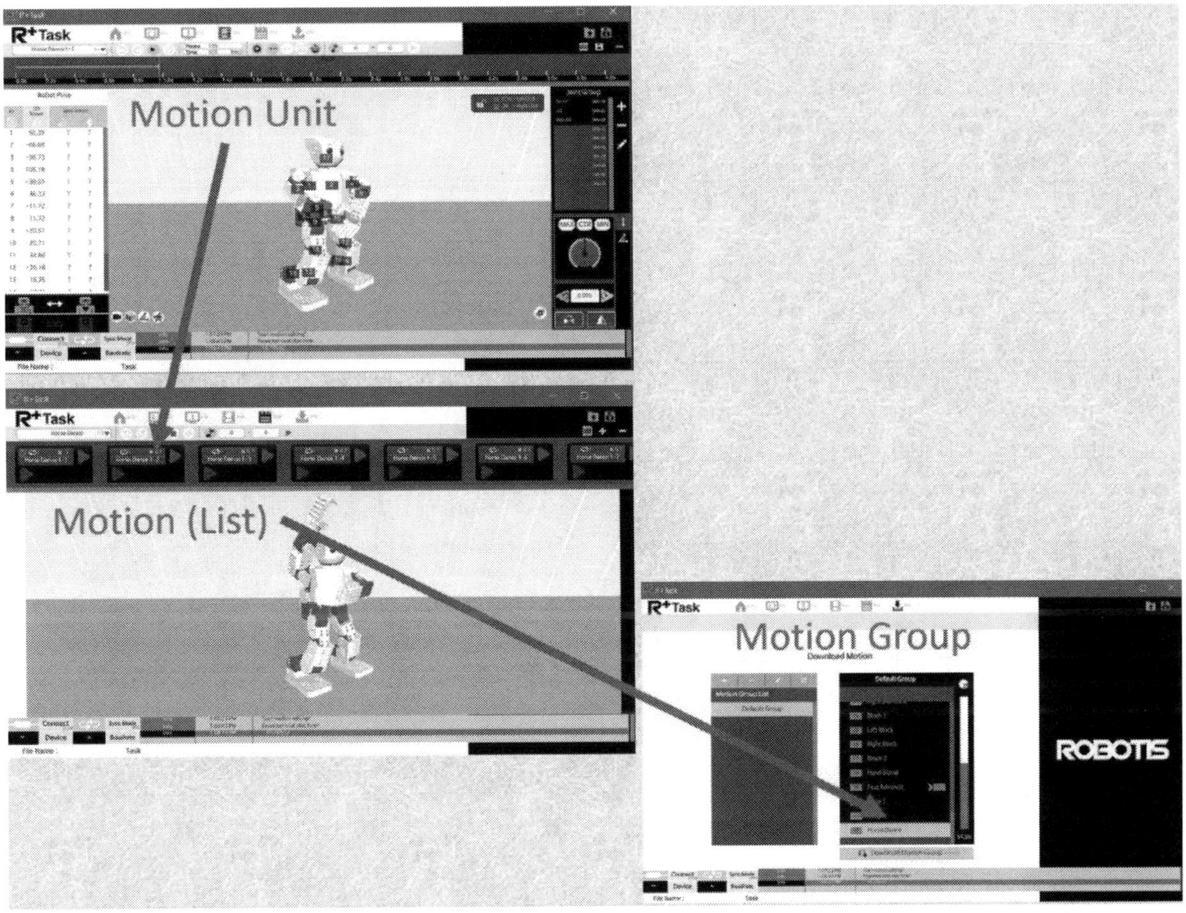

Fig. 1.9 MOTION UNIT >> MOTION LIST >> MOTION GROUP.

- Several MOTION-LISTs can then be merged into a MOTION-GROUP in the order determined by the user starting with MOTION-INDEX-NUMBER 1, then 2, etc... (by default). However, **starting with MOTION 3, the user can also redefine these MOTION-INDEX-NUMBERs as needed** (see Fig. 1.10 where "1" is being changed to "100" for "Initial Position"), before downloading the final MOTION GROUP to the MINI.

Fig. 1.10 Redefining MOTION INDEX NUMBER inside MOTION GROUP.

- The TASK 3 Sub-tool and the MINI App strictly refer to these MOTION-INDEX-NUMBERs when the user's program wants to activate specific MOTION-LISTs. **Starting with TASK V.3, these MOTION-INDEX-NUMBERs along with the MOTION-LIST-NAMES can also be reviewed once again in the new feature called MOTION LIST from inside the TASK Component** (see Chapter 6).
- Although several MOTION-GROUPs can be created within the MOTION 3 Sub-tool, only ONE MOTION-GROUP can be downloaded to the MINI at any one time (if its size is not greater than 64 KB).

The MOTION 3 Sub-tool has some new features such as:

- Insert a musical clip into a MOTION UNIT or a MOTION LIST/PAGE.
- Assign each servo motor with a nickname (instead of the default IDs 1 through 17).
- Reset the simulated robot to a default pose at any time.
- Set a motor position lock.
- Insert user's preferred landscape for the graphical background when working with user own robot creations.

Lastly, one important fact for the user to note is that as soon as MOTION 3 connects to the real MINI via its BT COM port, any TASK program that may be running up to that time will be disabled from then on (but the actual TASK code still remains in its allotted memory space). Essentially the OpenCM9.04-C is set into the "Manage" mode or Mode 2 (see Fig. 1.7). Afterwards, if the MINI's power is turned off and then back on, this TASK program will execute again as normal (i.e. "Task Play" or Mode 1).

The ROBOTIS e-Manual web site has instructional information for MOTION V.3 at (http://emanual.robotis.com/docs/en/software/rplustask3/motion_programming/) but it uses the ENGINEER Kit robot named Dr. R as the instructional model. Similar information for MOTION V.2 are still relevant and correct for the users to consult (http://emanual.robotis.com/docs/en/software/rplus2/motion/), if the reader prefers to see the MINI as the model robot.

ROBOTIS CS also publishes a series of seven YouTube videos showing how to use MOTION V.2 on the MINI (https://www.youtube.com/playlist?list=PLEf1s0tzVSnQSIj5MP0Uuss86BdjnlGAY) (still very useful for beginners).

1.3.3 Using MINI Mobile App

ROBOTIS has published instructional information for the MINI Mobile App at this link http://emanual.robotis.com/docs/en/software/mobile_app/mini_app/.

The user can use this App to do several basic tasks:

- Do an Assembly Check of the MINI and adjust each servo's OFFSET value.
- Associate a button on this App Interface to a specific MOTION LIST via its DEFAULT MOTION INDEX NUMBER. Thus, the user needs to make sure to use the MOTION Sub-tool to download the standard example MOTION GROUP to the MINI before using this App (especially if the user somehow had to recover the firmware on the OpenCM9.04-C, as this action would erase all existing MOTION data and TASK code).
- Associate a user gesture to a specific MOTION LIST.
- To use Voice Recognition to activate a specific MOTION LIST.
- To use Messenger's notifications to activate specific MOTION LISTs.
- This App also offers a Virtual Controller interface for the user to access the Soccer or Fighter modes.
- The user can also set up a chosen mobile device as a "server" for other "client" mobile devices to access the same MINI robot (via WiFi). The "server" still connects to the MINI via Bluetooth.
- The user can also add a sound clip to a certain MOTION LIST to be played when that MOTION LIST is performed.

Lastly, an important note for the user is that as soon as the MINI App connects to the real MINI via its BT COM port, any TASK program that may be running up to that time will be disabled from then on (but the actual TASK code still remains in its allotted memory space). Essentially the OpenCM9.04-C is set into the "Manage" mode or Mode 2 (see Fig. 1.7), similarly to as when the MOTION Sub-Tool connects to the robot (see Section 1.3.2).

1.3.4 Using TASK 3 Sub-tool

As previously mentioned, the new R+TASK V.3 application combines into one interface the two Sub-tools TASK and MOTION that used to be separate software packages. In this sub-section, the TASK 3 Sub-tool is described in general concepts and features (see Fig. 1.11) while Chapter 6 will take the reader into more detailed usage procedures.

TASK V.3 kept the functionalities of TASK V.2 which are documented at this link (http://emanual.robotis.com/docs/en/software/rplus2/task/) and added a new command for the OpenCM9.04-C controller:

- POLYNOMIAL, i.e. " **? = ? + (? - ?) * ?** "

Please note that TASK Sub-tool is the "true" programming tool to use for controlling the algorithmic behavior of the robot, while the MOTION Sub-tool is more of a generator of data as MOTION UNITs and MOTION LIST/PAGEs:

- The TASK 3 Sub-tool provides all programming flow control structures such as Sequence, Condition, Loop and Function, but no explicit arrays are supported. It provides one CALLBACK function which is activated by the hardware timer every 7.81 ms (which interestingly is the same value for the smallest TIME STEP allowed between consecutive MOTION KEY FRAMES – see Section 1.3.2).

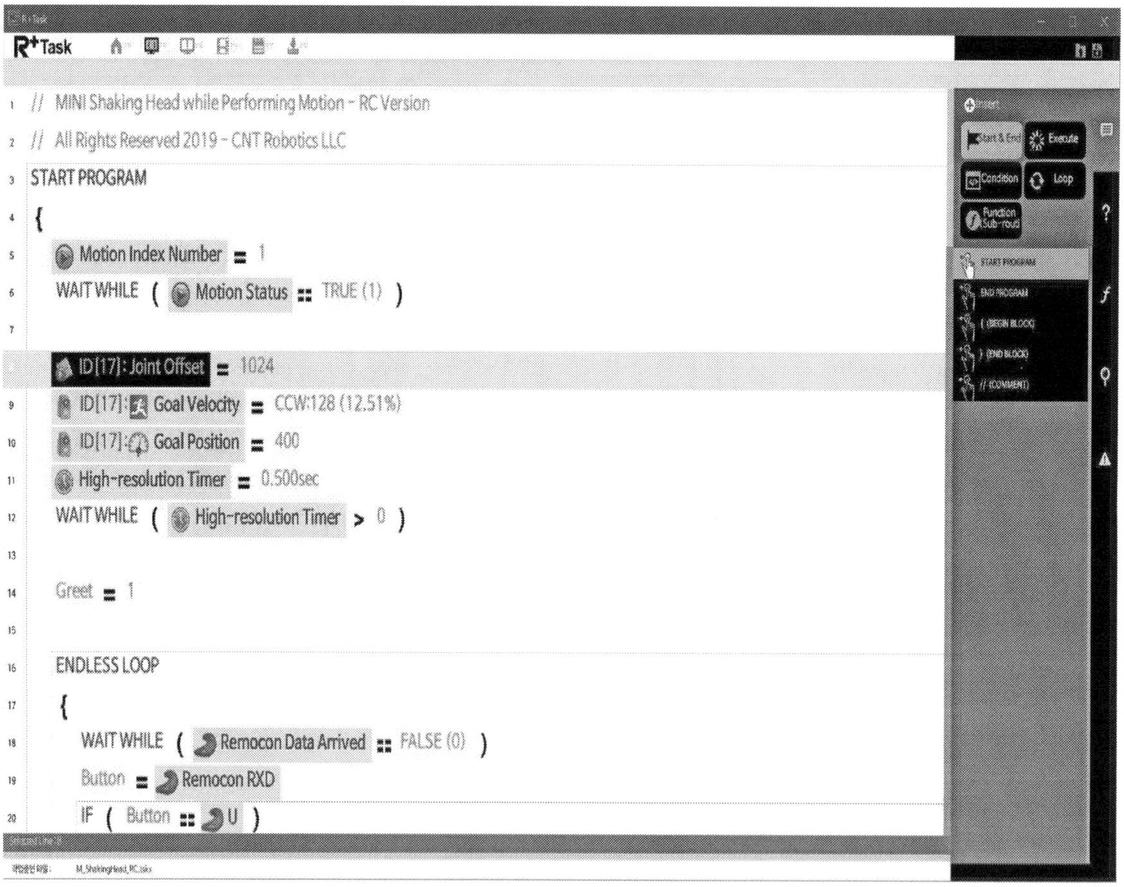

Fig. 1.11 TASK Sub-tool of R+TASK V.3.

- A TASK program can activate MOTION LIST/PAGEs by referring to the corresponding MOTION INDEX NUMBER (see Statement 5 in Fig. 1.11). When it sets the JOINT OFFSET of a given servo to "1024" (Statement 8 in Fig. 1.11), then it can set the GOAL VELOCITY and

GOAL POSITION of that servo directly and independent of any MOTION-LIST/PAGE being played, i.e. independent of other servos' pre-programmed movements (see Statements 9-10 in Fig. 1.10 and more in Chapters 4 and 6).

- Using SMART DEVICE commands, the TASK Sub-tool can leverage graphics, video, multimedia and communication services of a connected mobile device via Bluetooth (see Fig. 1.12 and section 1.3.5). Please note that the Mobile Device is assigned a Dynamixel **ID = 100** in Fig. 1.12.

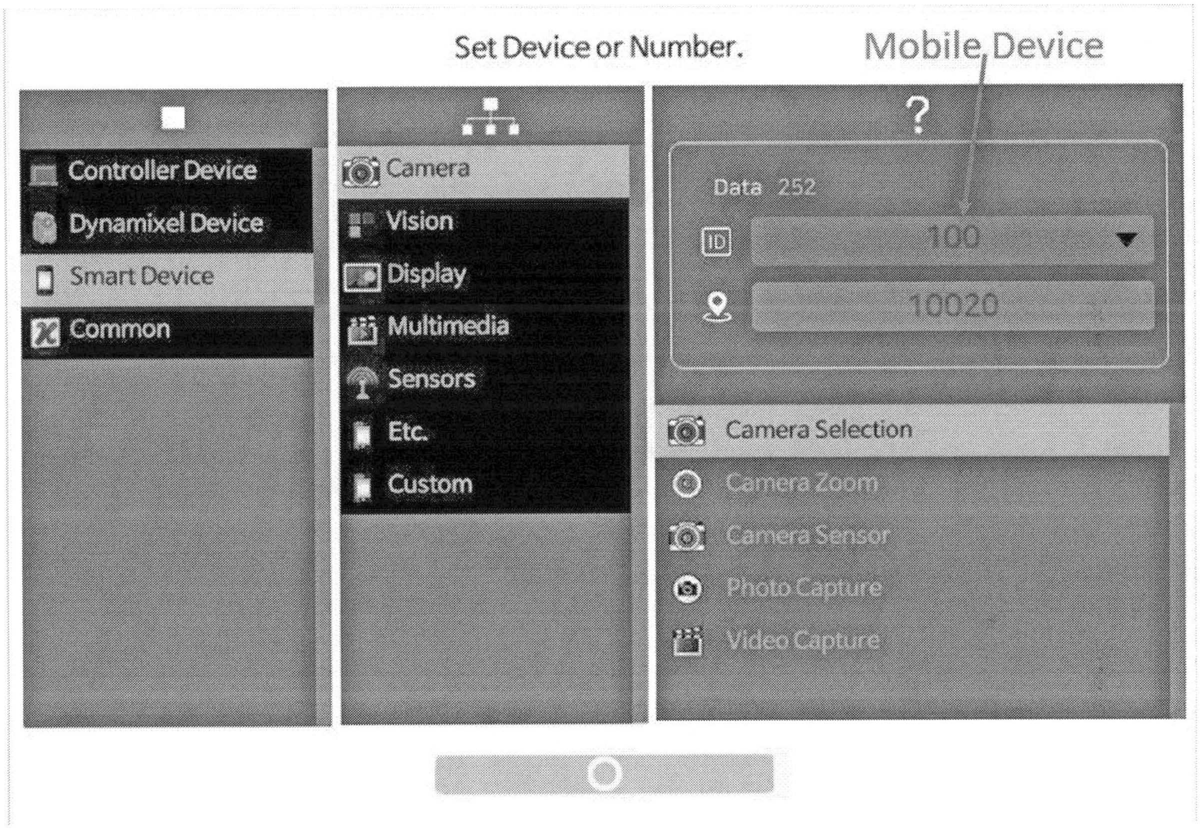

Fig. 1.12 SMART DEVICE commands available when using TASK Component of R+TASK V.3.

In short, the TASK Sub-tool can access all firmware components of the OpenCM9.04-C (see Fig. 1.6) plus the services of a mobile device. A loaded TASK program will execute as soon as power is applied to the OpenCM9.04-C, i.e. the Hardware Controller's default operating mode is "Task Play" or Mode 1 (see Fig. 1.7).

The TASK V.3 user manual can be viewed at (http://emanual.robotis.com/docs/en/software/rplustask3/).

1.3.5 Using PLAY700 Mobile App from TASK programs

The R+m.PLAY700 Mobile App was released in 2016 and it was originally designed for the CM-50 which is the Hardware Controller for the PLAY700 kit (http://www.robotis.us/robotis-play-700-ollobot/, http://emanual.robotis.com/docs/en/edu/play/play-700/), but the OpenCM9.04-C can use this Mobile App as well via the SMART DEVICE commands as part of a TASK program (see Fig. 1.13 and YouTube video https://www.youtube.com/watch?v=Z0JeShxiolc). Chapter 7 will take the reader into the details of using these SMART DEVICE functions.

Fig. 1.13 Example of an OpenCM9.04-C carbot using TASK V.2 and R+m.PLAY700.

R+m.PLAY700 is available for Android and iOS devices and the user can download this App from the following appropriate links: (https://play.google.com/store/apps/details?id=com.robotis.play700, https://apps.apple.com/us/app/play700/id1156037721).

R+m.PLAY700 was itself derived from a R+SMART application designed for the SMART series of robots which are only available in S. Korea at present (http://www.robotis.com/shop/list.php?ca_id=103020), thus the current usage instructions for the SMART DEVICE commands are written in Korean, but the reader can use Google Chrome to get this Korean website translated into "useable" English or perhaps into the reader's own language (http://emanual.robotis.com/docs/kr/software/mobile_app/rplussmart/).

From the Dynamixel network viewpoint (see Section 1.1.2 and Fig. 1.12), the Mobile Device running R+m.PLAY700 is assigned an ID = 100, as compared to the OpenCM9.04-C which has an ID = 200. For the MINI, the Mobile Device can be considered as an advanced Sensor/Actuator module when using TASK and R+m.PLAY700. However when using a more advanced tool such as the OLLOBOT SDK (https://github.com/ROBOTIS-GIT/OLLOBOT), the Mobile Device takes on a supervisory role in this hierarchical Dynamixel network: Mobile Device >> OpenCM9.04-C >> Servo Motors, whereas TASK/MOTION functionalities are completely disabled and the Remote Device commands "flow" only through the ROBOTIS Firmware memory section in Fig. 1.6, i.e. in "Manage" Mode (see YouTube video at https://www.youtube.com/watch?v=Co1sMo8j-Tk and more discussions in Section 1.3.10).

1.3.6 Using EDBOT 5

EDBOT (http://ed.bot/edbot) is a software tool designed to use ROBOTIS MOTION data and to issue direct commands to the OpenCM9.04-C Firmware with non-ROBOTIS software such as SCRATCH,

Python, JavaScript or C# (i.e. **EDBOT puts the OpenCM9.04-C into "Manage" Mode**). It can be purchased at https://shop.ed.bot/collections/products separately if the user already possessed a MINI kit (https://shop.ed.bot/collections/products/products/edbot-software), or together with a completely assembled MINI robot equipped with one NIR sensor (https://shop.ed.bot/collections/products/products/edbot).

EDBOT is currently at Version 5 and it can be freely downloaded from http://support.ed.bot/edbot-software-download.html but it won't work properly until the user obtains an install key from the parent company Robots in Schools Ltd. EDBOT works on Windows, Linux, macOS and Chromebook. It is a "server" tool so it can be used to control several MINI robots connected within the same Bluetooth network (see Fig. 1.14 and YouTube video at https://www.youtube.com/watch?v=yUpx42eNR2I). Chapter 8 will show the reader how to use the EDBOT tool with SCRATCH and PYTHON.

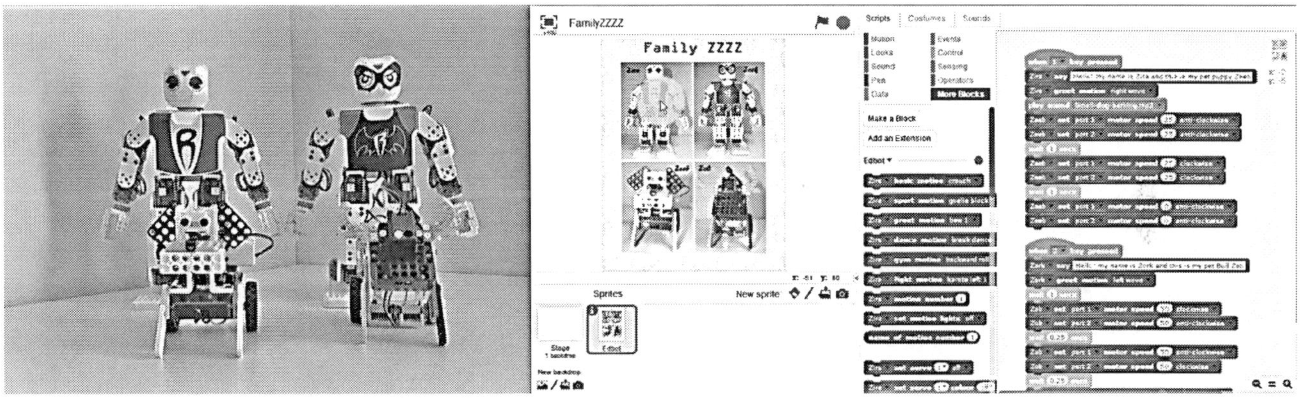

Fig. 1.14 Example of a SCRATCH application with EDBOT Tool.

1.3.7 Using MANAGER 2 (Windows OS)

The ROBOTIS MANAGER Tool works well on Windows OS (less well on macOS with Windows Virtualization software) (see Fig. 1.7). **When it connects to the MINI, it sets the OpenCM9.04-C into "Manage" mode, meaning that all TASK and MOTION functionalities cease.** It can be used to test and/or update the standard ROBOTIS firmware functionalities of the OpenCM9.04-C and of the XL-320s connected (at least those servos that are detected to have a hardware connection with the microcontroller STM8S105K4T6C). It can also perform a very critical operation called "Firmware Recovery" with on-screen instructions when the firmware is somehow degraded wholly or partially (see YouTube video at https://www.youtube.com/watch?v=PyQCzc73uh8). After this firmware is recovered, all ROBOTIS provided tools should be working well together.

1.3.8 Using DYNAMIXEL WIZARD 2

The DYNAMIXEL WIZARD 2 is a very sophisticated real-time diagnostic tool for all ROBOTIS servo motors (see Fig. 1.15). Unfortunately, the author had found that it can only keep a secure connection with the OpenCM9.04-C using a wired connection like one with the LN-101. When using the BT-210, Fig. 1.15 shows that it could not find all 17 XL-320s, and for those that it found, their connections were tenuous. In short, for the MINI the author recommends the use of the MANAGER Tool via Bluetooth for beginners, and the DYNAMIXEL WIZARD 2 only for advanced users who have special needs to tweak the XL-320s at a low level and using a wired connection.

The DYNAMIXEL WIZARD 2 can be downloaded at this web link http://en.robotis.com/service/downloadpage.php?ca_id=10. It works with Windows, Linux and macOS. It also sets the OpenCM9.04-C into the "Manage" mode.

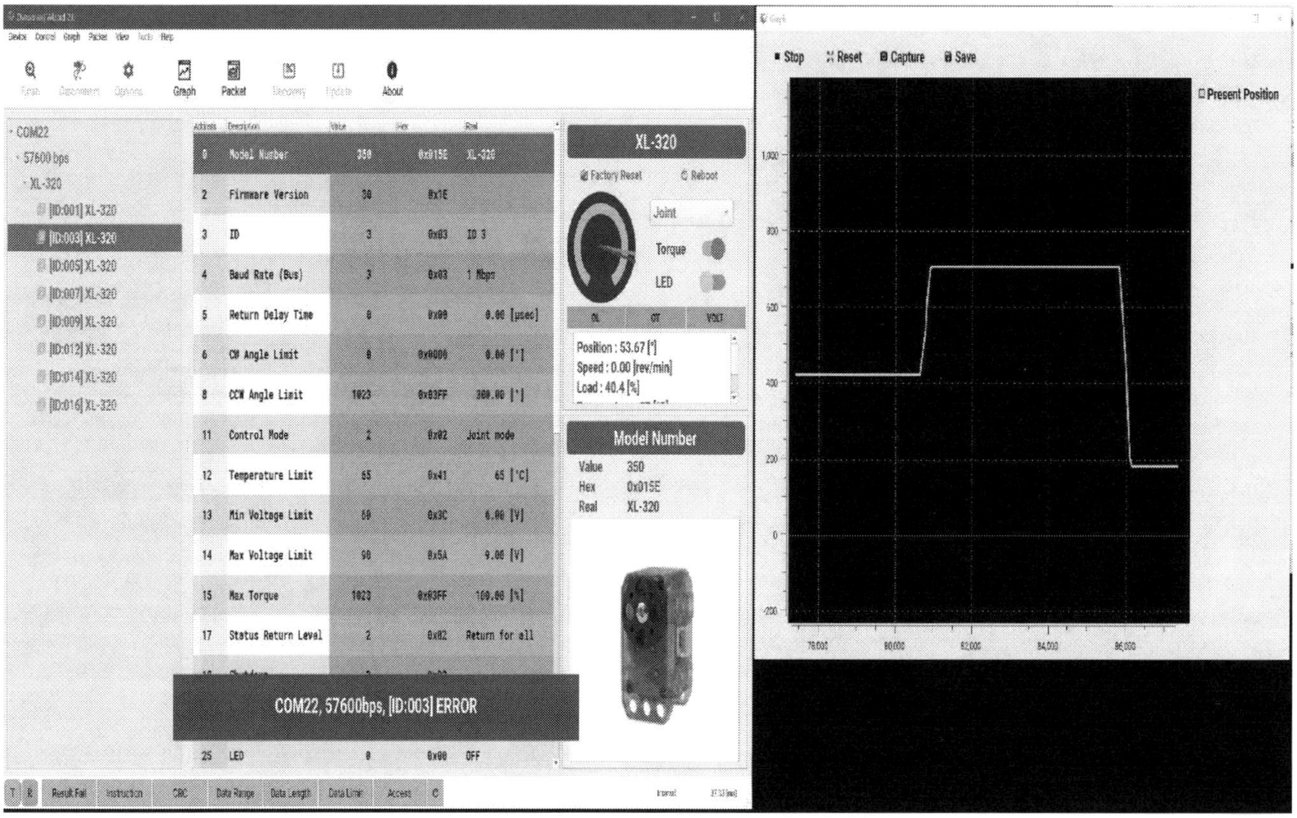

Fig. 1.15 Using the DYNAMIXEL WIZARD 2 on a MINI robot.

1.3.9 Using OpenCM IDE & Arduino IDE

If an advanced user is interested in using Embedded C/C++ with the OpenCM9.04-C then he or she needs to look at either the OpenCM IDE or the standard ARDUINO IDE. Both IDEs work on Windows, Linux and macOS.

The OpenCM IDE was derived from an ARDUINO IDE with some special functionalities for interfacing with mobile devices via SMART DEVICE functions like for the TASK Sub-tool. It can be downloaded at this link http://en.robotis.com/service/downloadpage.php?ca_id=10, however ROBOTIS has decided not to support this tool anymore and recommends users to use the standard ARDUINO interface. ROBOTIS still carries OpenCM IDE usage instructions at this link http://emanual.robotis.com/docs/en/software/opencm_ide/getting_started/. This web page also provides a link for downloading the ARDUINO IDE. Instructions for using the ARDUINO IDE are available at http://emanual.robotis.com/docs/en/software/arduino_ide/.

It is important to note that these 2 IDEs will set the OpenCM9.04 in "Boot Loader" mode and erase all ROBOTIS Firmware components. Thus, once the user starts programming with these IDEs, and if the user wants to get back to the ROBOTIS Software Sub-tools TASK and MOTION, the user would need to perform the "Firmware Recovery" process using the R+MANAGER Tool (see Section 1.3.7). The reader can watch this YouTube video for a side by side comparison of programming techniques between R+TASK and OpenCM IDE (https://www.youtube.com/watch?v=0C2TCVnhqgA).

The author plans to showcase the applications of these IDEs to the MINI in an "advanced" MINI book due in late 2020.

1.3.10 Using OLLOBOT SDK

The OLLOBOT SDK is another advanced tool for creating Android mobile apps for the OpenCM9.04-C controller and it can be downloaded from GITHUB (https://github.com/ROBOTIS-GIT/OLLOBOT). The reader can watch this YouTube video for an application of the OLLOBOT SDK to an OpenCM9.04-C based CarBot (https://www.youtube.com/watch?v=Co1sMo8j-Tk). The OLLOBOT SDK requires the use of Android Studio and advanced knowledge of JAVA.

The author also plans to showcase the applications of the OLLOBOT SDK to the MINI in an "advanced" MINI book due in late 2020.

1.4 Dynamixel System Design Paradigm

By now, the reader is probably overwhelmed by the various options to operate the MINI, thus the goal of this section is to provide an integrative view of all ROBOTIS Educational Robotics kits - a sort of a zoom out to see the "forest" so that the reader won't feel lost among the "trees"!

To be clear, the following materials are the author's personal view and understanding of ROBOTIS systems from his many years of using them. Officially, ROBOTIS so far had not published any details about their robotics system design approach, although they did patent the concept of "Dynamixel".

The author believes that the MINI kit shares into a common ROBOTIS Dynamixel System Design Paradigm which considers a typical ROBOTIS robot to be a Computer Network with four major components:

1) A Hardware Controller, e.g. OpenCM9.04-C for the MINI, which is the "Main Brain" for the robot (Dynamixel ID = 200). It contains a firmware to allow low-level functionalities and additionally stores the user-programmed instructions for the robot to execute at run time.

2) The Sensors component which helps the Hardware Controller get information about its environment, from a simple sensor such as the Color Sensor RCS-10 (connected to local I/O ports of Hardware Controller) to a more complex sensor such as the IR Sensor Array (from Bioloid STEM kit) which requires its own embedded controller with its unique Dynamixel ID.

3) The Actuators component which allows the Hardware Controller to perform the appropriate robot actions upon the real world as programmed by the user, from a simple geared motor such as the GM-10A to the more sophisticated dual servo module such as the 2XL430-W240-T with its own embedded controller and unique Dynamixel IDs.

4) A Remote Device (Dynamixel ID = 100) which can be a mobile device or even a desktop PC. The Remote Device has a more flexible role depending on the specific device used and on its current role in this

network of robotic components. For example, if a smartphone (running the MINI App) is used with the OpenCM9.04-C, it can act as a Bluetooth Remote Controller (i.e. a "glorified" Sensor) telling the Hardware Controller about which U-D-L-R Touch Area on the phone screen had been pressed by the user. However, if the same smartphone (running the R+m.PLAY700 App) is used with the MINI, it can additionally act as an Actuator displaying appropriate graphics or videos as commanded by the OpenCM9.04-C according to its TASK programmed logic (see Chapter 7). When a desktop PC is used as a Remote Device, the interactions become more complex. For example, if the TASK's Virtual Remote Controller is used (see Fig. 1.16), then the PC acts as a glorified Sensor informing the Hardware Controller about which "U-D-L-R-1-2-3-4-5-6" buttons had been pushed by the user. However, if the SCRATCH-PYTHON/EDBOT tool chain is instead used on the PC (see Chapter 8), then the PC takes on the role of a Primary Hardware Controller as SCRATCH-PYTHON codes run on the PC, and not on the OpenCM9.04-C which now has a secondary role as it is just letting SCRATCH-PYTHON commands flow from the PC (via EDBOT) to pass through to its attached Actuators and Sensors (via the OpenCM9.04-C's Firmware).

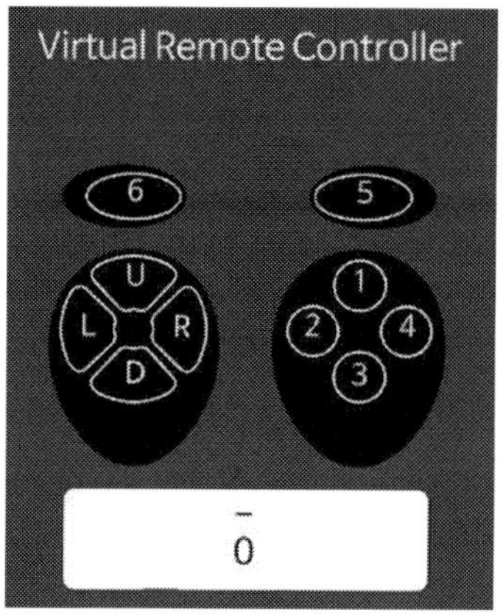

Fig. 1.16 Using the Virtual Controller of R+TASK V.3.

To achieve hierarchical control in this evolving network of robotic components, ROBOTIS software uses two types of communication packets: Instruction and Status. Instruction packets can only be issued by the Top-Level Controller which can be either the Hardware Controller (e.g. OpenCM9.04-C) or the Remote Device, and there can be only ONE Top-Level Controller at any one time (if ones use the proprietary firmware provided by ROBOTIS for its controllers). Once a network component receives an Instruction packet, it will execute the prescribed instruction and will optionally send back Status packets to the originator of that Instruction packet.

Currently, there are 3 possible configurations of this Dynamixel network:

1. If the reader so far has only used the MINI robot with its MINI M obile App, then the reader is familiar with the first type of network configuration (see Fig. 1.17). In this first network configuration, the Top-Level Controller is the Remote Device which can be a PC using the SCRATCH-PYTHON/EDBOT tool chain or a mobile device using the MINI App (see Fig. 1.17 for additional situations). In this configuration, the Hardware Controller (CM-XXX) is put into "**Manage**" mode so the downloaded TASK program is suspended. Furthermore, all Instruction packets flow from the SCRATCH-PYTHON/EDBOT tool chain or the MINI App, via a Bluetooth utility, first to the Firmware component of the Hardware Controller before reaching

the Actuators and Sensors, meaning that communication delays would likely occur within this network configuration due to the extra layers of communication software (especially when Bluetooth serial communication is restricted to a rate of 57.6 Kbps via BT-210/410). Furthermore, the EDBOT tool allows for multiple robots control from a single PC (see Chapter 8) and thus makes the issue of communications delays even more critical.

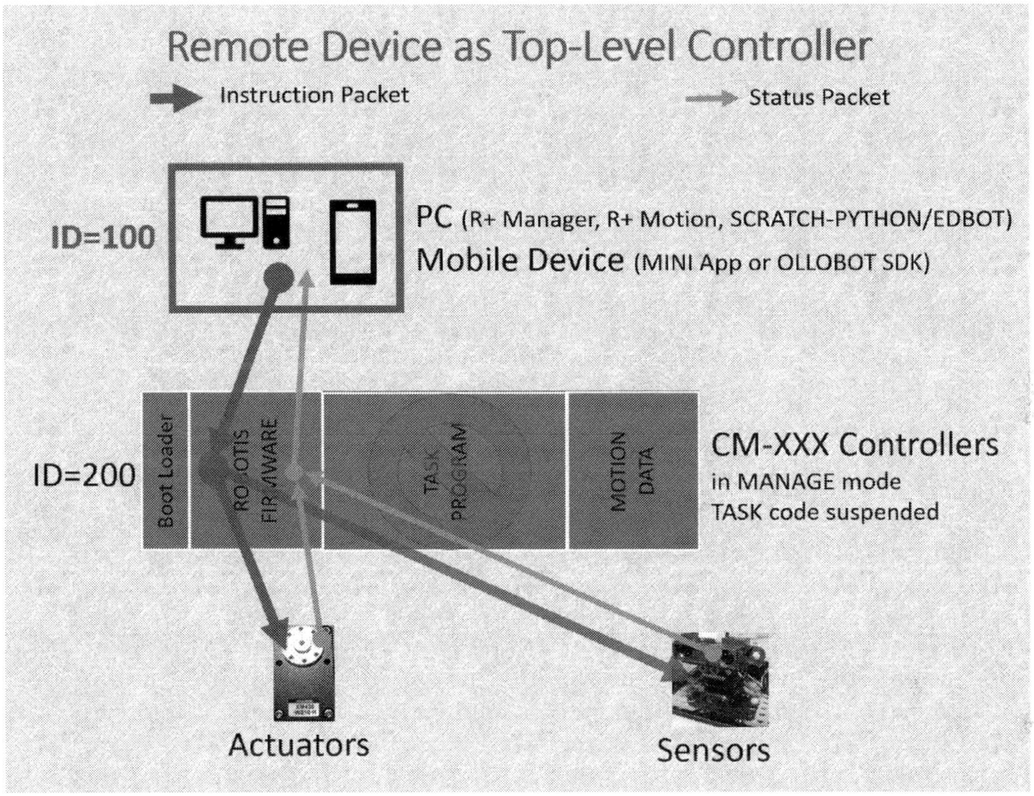

Fig. 1.17 Dynamixel Network Configuration with Remote Device as Top-Level Controller.

2. If the reader had used the PLAY700, DREAM or BIOLOID kits before, then the reader is familiar with the second type of network configuration (see Fig. 1.18) where the Top-Level Controller is the Hardware Controller (CM-XXX) running in **"Task Play"** Mode, while the Remote Device can be any of the following devices: Physical RC-100B or Virtual RC-100 on the PC, or Mobile Device running the R+m.PLAY700 App. In this configuration, the Hardware Controller can issue Instruction packets to Local Actuators and Sensors as well as to the Mobile Device which can be considered a Remote Actuator/Sensor module.

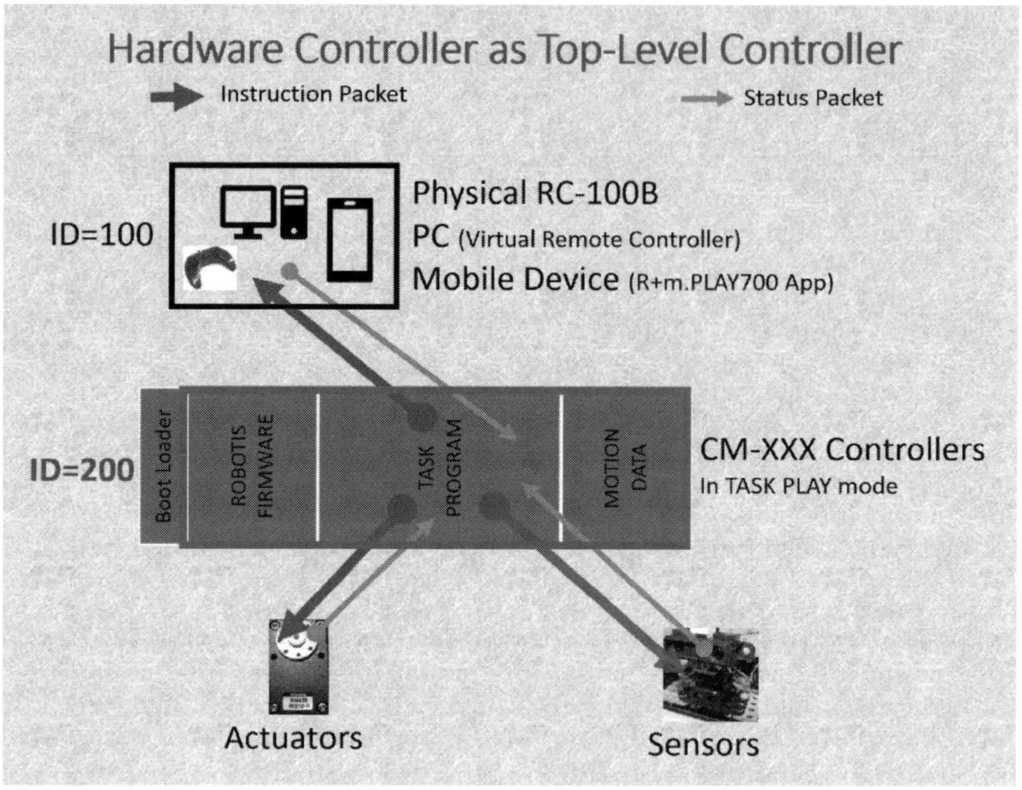

Fig. 1.18 Dynamixel Network Configuration with Hardware Controller as Top-Level Controller.

3. There is a third configuration for a more advanced and flexible use of ROBOTIS robotic systems with Embedded C/C++ (see Fig. 1.19), which is outside the scope of this book, but interested readers can refer to this web link for more details (https://www.cntrobotics.com/robotis-evolving-dynamixel-concept). Advanced readers can learn more about Instruction and Status packets at (http://support.robotis.com/en/product/actuator/dynamixel_pro/communication/instruction_status_packet.htm). In this book, we'll be using the software tools such as R+TASK V.3 and SCRATCH-PYTHON/EDBOT tool chain wherein the tasks of handling Instruction and Status packets are performed "behind the scene", thus readers won't have to deal with them directly.

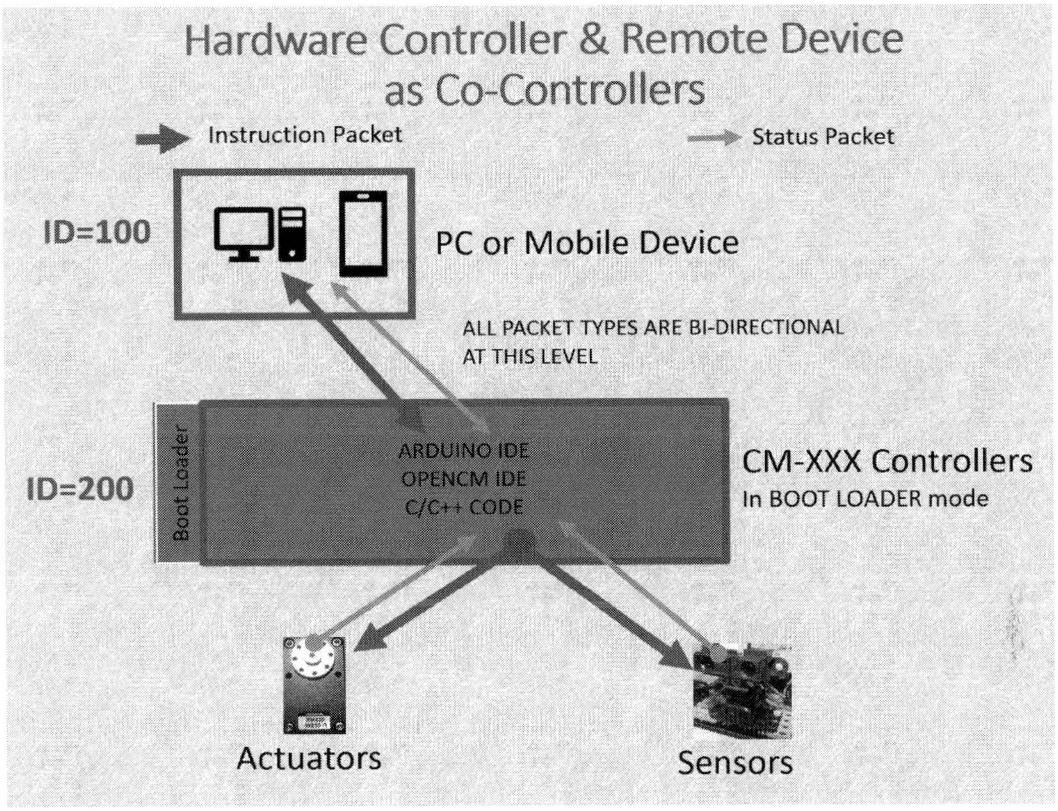

Fig. 1.19 Dynamixel Network Configuration with Hardware Controller and Remote Device as Co-Controller.

Please note that there is a third type of communications packet, called **"RemoCon"**, used in Remote Control schemes whereas the user will have the opportunity to have a more hands-on practice in Chapter 6 (http://emanual.robotis.com/docs/en/software/rplustask3/task_parameters/#remote-controller and http://emanual.robotis.com/docs/en/parts/communication/rc-100/#communication-packet).

1.5 Review Questions

1. What is the name of the Hardware Controller for the MINI kit?
2. How many modes can this Hardware Controller be set to?
3. What is the name of the microcontroller used for the MINI Hardware Controller?
4. What is the name of the actuators used for the MINI?
5. List 3 sensors that can be used with the MINI.
6. Can SCRATCH 2 be used with the MINI? If yes, how?
7. How much RAM memory does the MINI have?
8. Can the phone app R+m.PLAY700 work with the MINI system?
9. What is the name of the tool that displays the mechanical assembly of the MINI robot?
10. Can the MINI implement C-based computer programs from the user? If yes, how?
11. List the 4 software tools that come with the RoboPlus Software Environment.
12. List 4 software packages that can work with the MINI.

13. List 3 possible communication/interface protocols that are available to the user for use with the MINI.
14. List 3 communication ports available on the MINI and describe their purposes and functions.
15. Can the MINI support user-based firmware and boot loader?
16. Can the MINI support an Arduino-based IDE?
17. Which controller(s) support an expansion board?
18. Describe/Explain the Dynamixel concept from ROBOTIS.
19. What are the reserved Dynamixel IDs in robotics kits from ROBOTIS? What hardware components are they assigned to?
20. Describe how the XL-320 knows where its Current Servo Position is at.
21. What is the allowable range of degrees turn allowed for the XL-320?
22. How many types of battery can the MINI used for its own power?
23. What are the possible Operating Conditions for the MINI?
24. At first boot, in which mode is the MINI put in?
25. What is the relationship between TASK and MOTION sub-tools?
26. List 4 features offered in the MOTION sub-tool.
27. List the Mobile Apps that work with the MINI.
28. What functions can the Dynamixel Wizard perform for the MINI robot?
29. How many types of communication packets does the MINI support?

Chapter 2: Kit Assembly Tips

The MINI kit comes with a Quick Start manual to guide the user in the assembly of the MINI robot. The author recommends the user to also use the R+DESIGN software for certain steps that may be unclear in this Quick Start manual as it can only show **one view angle** of the 3D sub-assemblies, while this view angle can be controlled by the user inside R+DESIGN (see Fig. 2.1 and Video 2.1). However only the Quick Start manual would show the proper routing of the 3-pin cables connecting all the XL-320s into a daisy chain.

The goal of Chapter 2 is to offer some practical information to make the robot construction process easier for the first-time builder. The reader is recommended to read Section 2.1 before starting on building the MINI robot for the first time.

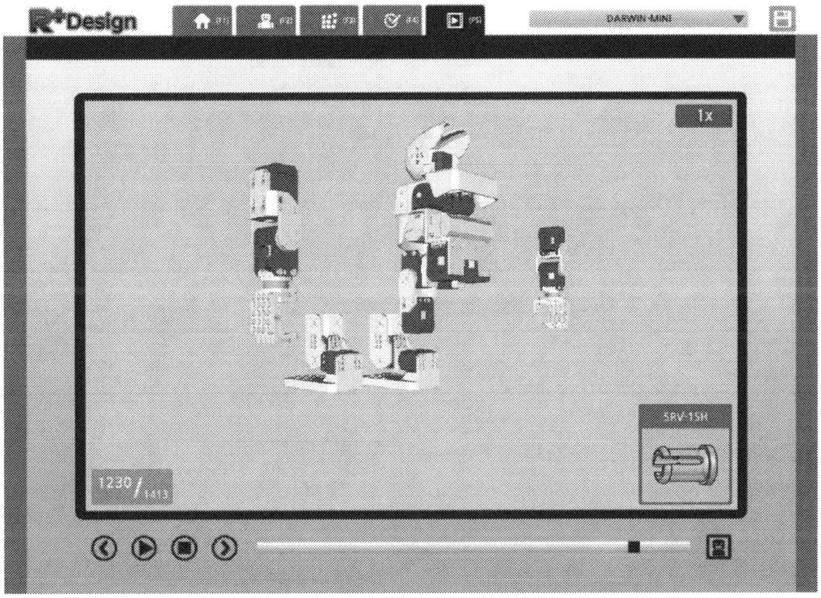

Fig. 2.1 Using R+DESIGN to ease the assembly of the MINI robot.

2.1 Preparations Before Assembly

Depending on the user's needs such as planning to use the Micro-USB port or to motorize the head piece and/or use extra ROBOTIS sensors, certain steps have to be performed in advance of the construction phase of the MINI robot as shown in the Quick Start manual.

2.1.1 Creating USB Access Slot

If in addition to the regular MANAGER/TASK/MOTION tools, an advanced user plans to use the OpenCM IDE, then this user would need to access the micro USB port on the OpenCM9.04-C at all times and this will require the user to create an access slot on the MINI torso frame part DMF-B01 (see Fig. 2.2). This slot can be created using a drill bit or any other rotating carving tool.

Fig. 2.2 Micro USB slot carved into MINI frame part DMF-B01.

2.1.2 Firmware Update and Recovery

For all users, the author recommends to check/update via R+MANAGER the firmware for the OpenCM9.04-C Controller and also for the XL-320s, including checking their pre-assigned Dynamixel IDs before assembling them together to create the MINI robot, as it would be difficult to manipulate the XL-320 cabling system once the physical robot is constructed (see Fig. 2.2 and the video at https://www.youtube.com/watch?v=0EPdpZHudOs for an illustration of the **Firmware Update** process using MANAGER V.1 and V.2). ROBOTIS CS also publishes a video tutorial on the **Firmware Recovery** process at https://www.youtube.com/watch?v=PyQCzc73uh8 using MANAGER V.2).

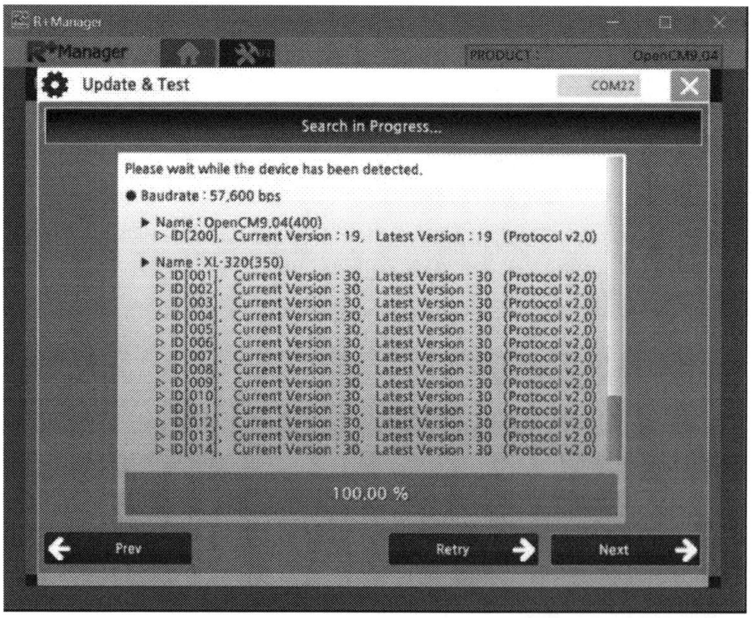

Fig. 2.3 Updating Firmware on OpenCM9.04-C and attached XL-320s via R+MANAGER.

It is also recommended to perform this check with only groups of 4 to 5 XL-320s at a time, for easier troubleshooting in case some errors pop up.

2.1.3 Dynamixel Check

R+MANAGER can also help the user check each single XL-320 for its proper functions (see Fig. 2.4) before assembling them into the framework for the MINI, especially to make sure that all XL-320s are set to Position 512 (i.e. 0 degree) at first. The MINI Mobile App can also perform this check (see Chapter 5).

Please note that the Quick Start manual also shows special configuration for Servos 15 (Step 24), 16 (Step 32), 7 and 8 (Step 45) for the actual subassembly tasks.

Fig. 2.4 Checking each XL-320 for proper functionality before assembly into the MINI robot.

2.1.4 Optional Sensors & Cables

The MINI kit does not come with any sensor at all, but the author wanted to use two NIR sensors (IRSS-10) mounted on each side of the Head frame part DMF-F01 (see Fig. 2.5) and they are to be connected to GPIO Ports 1 and 4 on the OpenCM9.04-C PCB. These NIR sensors would need to be attached to the DMF-F01 part before attaching Servo 17 to serve as a motorized head (see Section 2.2). A DMS-80 sensor is also mounted on top of the robot's head using Dual-Lock tape (see Fig. 2.5). This sensor is connected to Port 2 of the OpenCM9.04-C PCB. Fig. 2.5 also shows how to route the servo cables and the DMS-80 cable via the center slot of the frame part DMF-F01 to provide freedom of rotation to Servo 17 and to minimize cable entanglements.

This web link http://emanual.robotis.com/docs/en/parts/controller/opencm904/#robotis-5-pin-port) has important information about the proper connection of the 5-pin cable onto the OpenCM9.04-C controller and also the compatibility of specific sensors to each of the four 5-pin GPIO Ports.

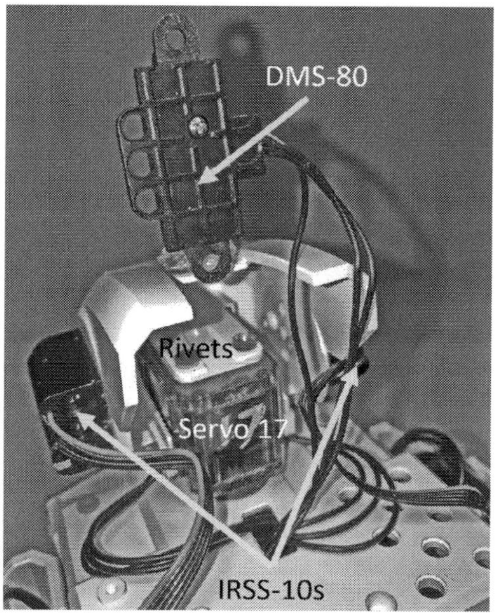

Fig. 2.5 Attaching IRSS-10 and DMS-80 Sensors to MINI Head Part.

This ROBOTIS link (http://emanual.robotis.com/docs/en/edu/mini/#customize) displays a list of other sensors that can be used with the MINI. Please note that the DMS-80 NIR sensor and the Gyro Sensor GS-12 come with special 3-pin to 5-pin cables (http://en.robotis.com/shop_en/item.php?it_id=903-0186-000).

The reader may also need to purchase extra 5-pin cables for those customizing projects, ROBOTIS offers the standard 150 mm long cables (http://en.robotis.com/shop_en/item.php?it_id=903-0086-000) and an extra-long set at 400 mm (http://en.robotis.com/shop_en/item.php?it_id=903-0187-000).

2.2 Assembly of Head Servo

The MINI kit comes with a "dummy" XL-320 for its Head subassembly, but for this book the author needs it to be motorized so it can be swung right and left, and along with the NIR Sensors on both sides of its head (Section 2.1.4), an "object scanning" feature can then be developed (see Chapter 6).

The author had created a video to help interested readers with the addition of this Head Servo to the standard MINI robot at https://www.youtube.com/watch?v=rVBDah3DjDQ.

2.3 Mobile Device Attachment

To use the R+m.PLAY700 with the MINI (via TASK programming – see Chapter 7), ones would need to use a smart phone (either Android or iPhone). The author used some Dual-Lock tape to attach an older Samsung S2 to the robot (see Fig. 2.6) and the reader can see that the extra weight made the robot lean further forward. This effect would need a readjustment of the various **offset** values of the **servos serving as ankle, knee and hip points of the MINI before the default MOTION PAGEs, as provided by the ROBOTIS MOTION tool, can be used (see Section 4.4 and Chapter 7 for more details)**, especially the ones used for walking. Thus, the user is recommended to use the smallest and lightest smartphone possible.

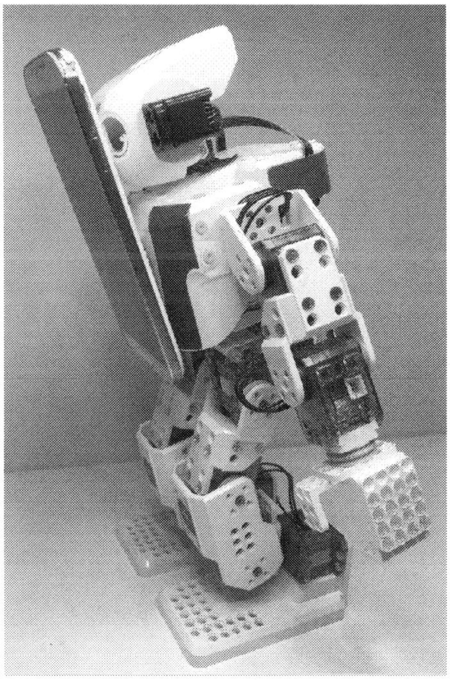

Fig. 2.6 Attaching a mobile device to the MINI robot will make it lean more forward.

The following appropriate links https://play.google.com/store/apps/details?id=com.robotis.play700 and https://apps.apple.com/us/app/play700/id1156037721 provide the minimum Operation System versions for the user-chosen smartphone, i.e. Android 2.3.3 and later, or iOS 8.0 and later.

A USB webcam is also used in Chapter 8, and its attachment scheme will be illustrated in that chapter.

2.4 Post-Assembly Troubleshooting

ROBOTIS hosts a Web Forum to help users troubleshoot their products at http://en.robotis.com/service/forum.php#, where the user has to register to ask for technical help.

The author through his own personal experiences have found two post-assembly main issues:

1. Some servos "did not go where they were supposed to go" when playing Motions: this was most likely due to the misorientation of the horn during assembly. Most often the only proper solution

is to dismantle the servos and frame parts involved and reassemble them correctly this time around. But sometimes, a simpler procedure may suffice, such as the case of the shoulder connections (see Fig. 2.7):

a. Unscrew completely the four M2 screws to remove the physical connection between the servo's horn plate and the shoulder frame part, but make sure to keep "in place" the long rivet on the opposite side of the horn. Now the servo can be commanded to turn on its own axis without taking the frame parts with it.
b. Next use the MANAGER Tool (see Chapter 3) to set this servo to its proper Goal Position for the wanted Pose *(the reader can check with the MOTION Tool to determine what is the proper Goal Position value to use in this case – the Initial Position Motion Unit is always a good choice)*, next set up the shoulder frame part involved in the proper orientation for this wanted Pose then screw back in the four M2 screws to connect these frame parts to the horn of this servo and "Mission Accomplished!".

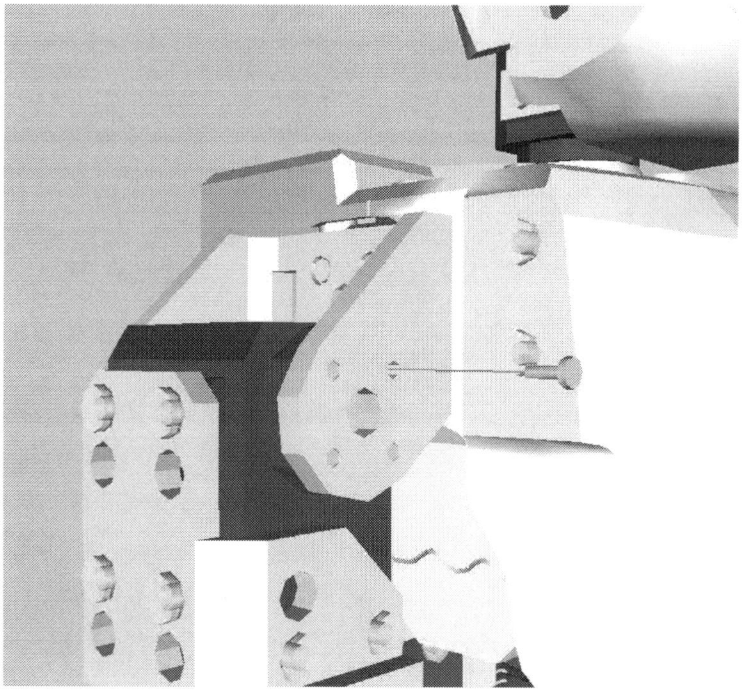

Fig. 2.7 MINI Right Shoulder connections.

2. The robot started a Motion Page properly and then some servos stopped working: this was most likely due to some 3-Pin cables being pulled or flexed so electrical contacts were lost temporarily. For this case, the user needs to check on the two cables coming out of the affected servo and unkink/free them up as much as possible.

Chapter 3: Using MANAGER (Windows OS only)

The MANAGER V.2 tool (http://emanual.robotis.com/docs/en/software/rplus2/manager/) is designed to allow the user to perform a comprehensive hardware/firmware check on the entire MINI robot. This capability had been useful to the author to troubleshoot actuators and sensors when they did not seem to work properly, and to determine that it was due to actual hardware malfunctions and not to some programming errors that the author may have done inside TASK or other programming IDEs.

The goal of this Chapter is to get the reader familiarized with the main features of the tool MANAGER V.2 and to show how they are related to the complete Control Tables of the OpenCM9.04-C and individual XL-320 actuators as well as other ROBOTIS sensors. The reader is also shown a preview of how to access some features of R+TASK V.3 before using them fully later in Chapters 4 and 6.

3.1 Connection Options between PC and MINI robot

The MINI kit comes with the BT-210 Bluetooth Receiver and a micro USB Cable which can be used to connect to the MINI's Hardware Controller OpenCM9.04-C, however there are other connection options that provide other advantages (and disadvantages) such as the BT-410 Bluetooth set and LN-101 USB Dongle.

3.1.1 Using BT-210

The BT-210 is provided to the user with the purchased MINI kit and it is based on Bluetooth 2.0 which uses more electrical power from the Hardware Controller that it is connected to. Furthermore, all users of the BT-210 could not use this receiver with ROBOTIS provided software for most of the year 2018 due to a Windows 10 update which was not compatible with the BT-210. Since Spring 2019, ROBOTIS upgrades the BT-210's firmware and this BT-210's problem no longer exists with ROBOTIS controllers such as the MINI, but who knows, this issue may pop up again in the future! Pairing the PC's Bluetooth server to the BT-210 receiver on the bot is done via the normal Bluetooth setup tool from Windows OS.

Fig. 3.1 shows that the BT-210 would show up as a "Bluetooth Serial Port" inside MANAGER which only uses the 57.6 kbps rate although quite a few connection speeds are available for the BT-210 (http://emanual.robotis.com/docs/en/parts/communication/bt-210/#specifications).

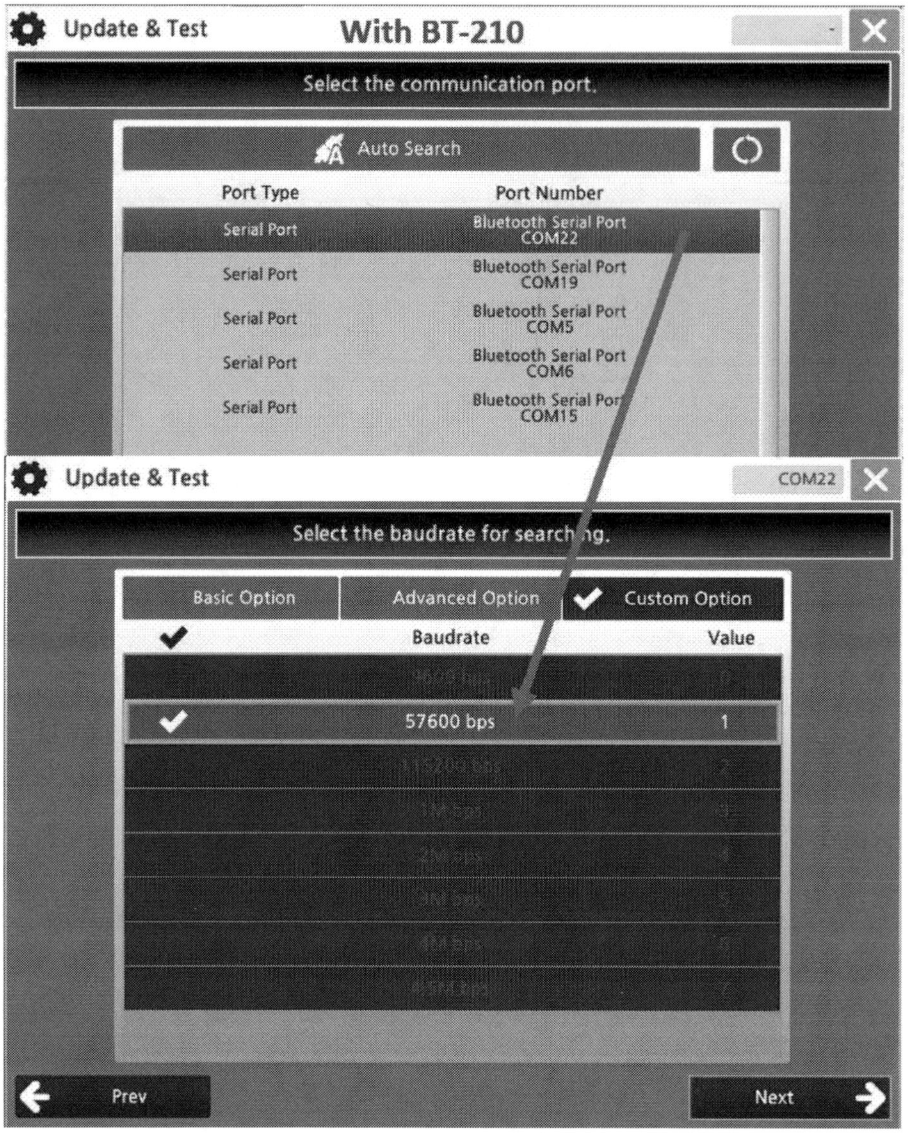

Fig. 3.1 Only 57.6 Kbps rate is allowed with BT-210 via MANAGER.

The BT-210 is not compatible with iOS devices, so if the reader intends to use iOS devices and the R+m mobile versions of TASK and MOTION or the MINI App, then the reader needs to invest in getting the BT-410 receiver and the BT-410 USB dongle (see Section 3.1.2).

3.1.2 Using BT-410 and BT-410 Dongle

The user may already have the BT-410 (http://emanual.robotis.com/docs/en/parts/communication/bt-410/) because he or she wanted to use the MINI mobile App on iOS devices such as iPhones and iPad. Then in order to use MANAGER on a Windows PC, the user would need to also purchase the BT-410 USB Dongle (http://emanual.robotis.com/docs/en/parts/communication/bt-410-dongle/). The BT-410 Dongle would show up as a "USB Serial Port (1.0)" inside MANAGER. Only one serial connection rate is allowed with the BT-410 Receiver/USB-Dongle combination, i.e. 57.6 Kbps (see Fig. 3.2), although the BT-410 Dongle is capable of other rates. The user also needs to make sure that the check mark is set for using the Dongle (see top picture inside Fig. 3.2).

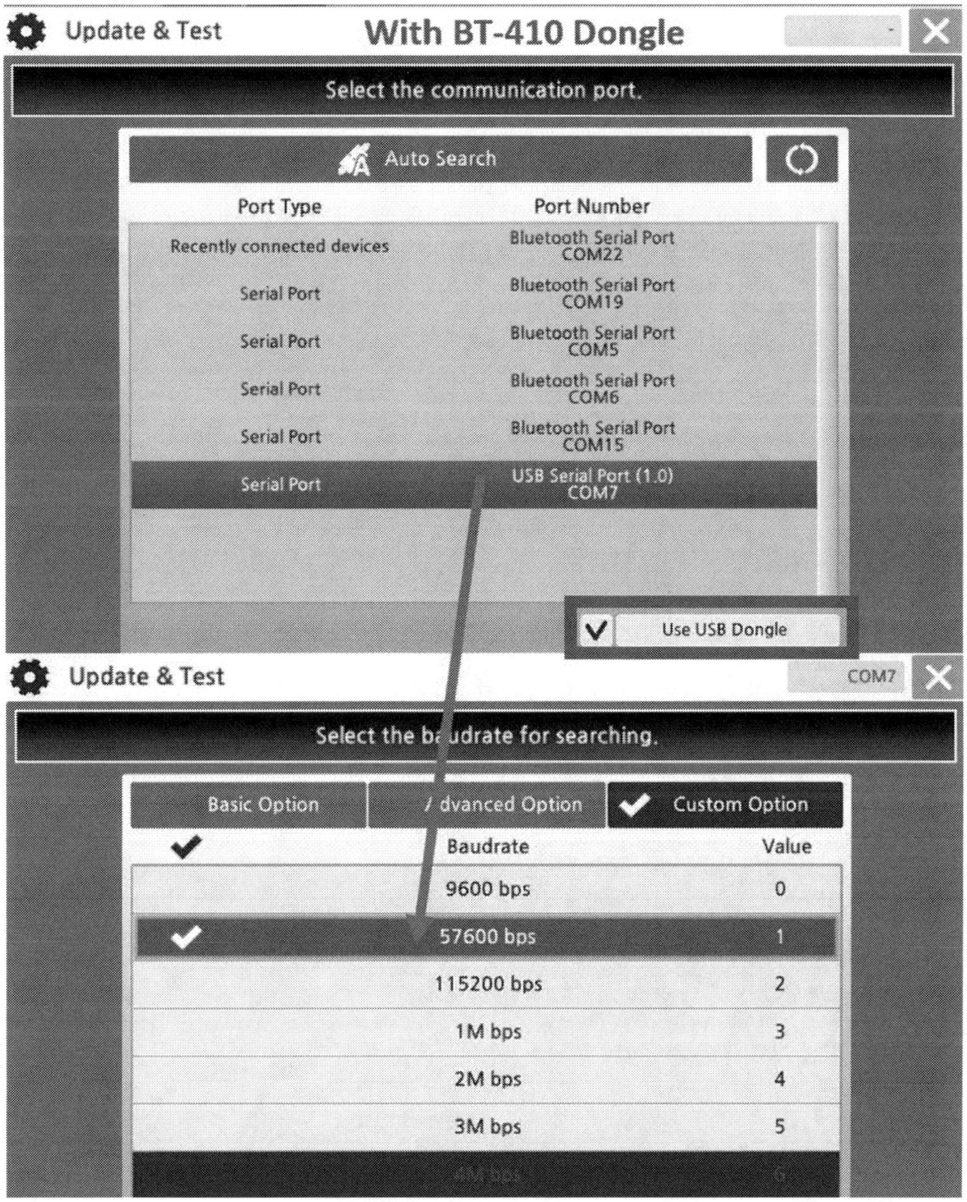

Fig. 3.2 Only 57.6 Kbps is allowed with BT-410 USB dongle via MANAGER.

Please note that for the first time use of this BT-410 set, the BT-410 Receiver needs to be very close to the BT-410 USB Dongle (about 1-2 inches away) for the automatic first pairing to occur (when blue LEDs on both devices stay continuously lit). After this first pairing, both devices will pair up automatically from then on at normal use distances (but no more than 10 meters between the two devices). If later, the user wants to use a different BT-410 Receiver, the close-up first pairing process needs to be redone for the new BT-410 Receiver so that the USB BT-410 Dongle can "forget" the previous BT-410 Receiver (in other words, only one specific BT-410 Receiver per BT-410 Dongle at any one time).

3.1.3 Using Micro-USB Cable

A **good** Micro-USB cable, i.e. one capable of doing data transfer reliably in addition of a power charging capability, can allow MANAGER to connect to the OpenCM9.04-C Controller from a PC. However, the Micro-USB cable is incompatible with the BT-210 and BT-410 receivers, so the author uses the micro-USB cable with MANAGER only in the situation when all Bluetooth connections are not possible with the OpenCM9.04-C. Furthermore, using the micro-USB cable requires the practitioner to create an access slot to the torso frame part of the MINI robot (see Section 2.1.1). The OpenCM9.04-C would show up as "ROBOTIS Virtual COM Port" when the Micro USB connects the PC to the OpenCM-9.04-C, and only the 57.6 Kbps rate is allowed with the micro-USB cable via MANAGER (see Fig. 3.3).

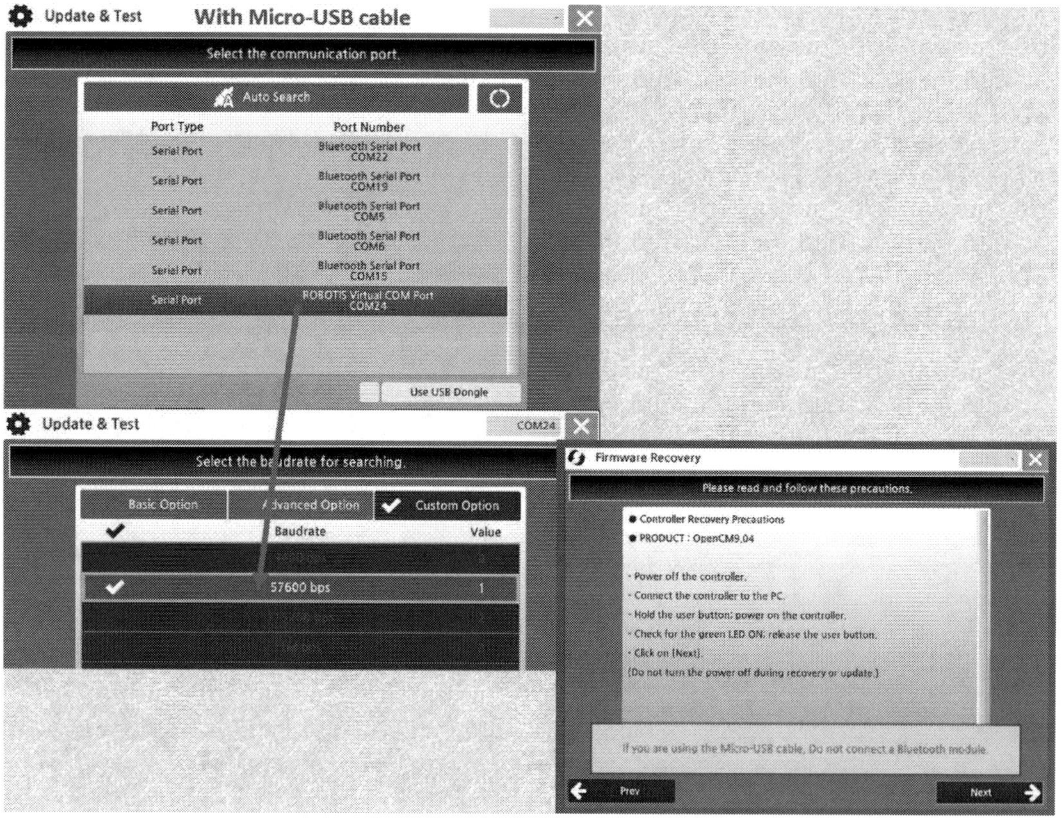

Fig. 3.3 Only 57.6 Kbps is allowed with the micro-USB cable via MANAGER.

The author also needs to caution the readers that he had connection success with the Micro-USB cable in "Update & Test" mode, only when it was "**just**" used for "Firmware Recovery" of the OpenCM9.04-C (see Section 3.2.2), otherwise MANAGER would usually respond with "Cannot Find Device" after it timed out after a few minutes search via the Micro USB Cable.

3.1.4 Using LN-101

The LN-101 (http://emanual.robotis.com/docs/en/parts/interface/ln-101/) is a USB device and the second wired option that the user can use when the Bluetooth options are not possible. The BT-210/410 Receivers also need to be removed from the OpenCM9.04-C's 4-pin Communication Connector (see Fig. 1.1), because the 4-pin cable of the LN-101 will need to go to that same connector. From the author's personal experiences, he had found that the micro-USB cable connection gets flimsy with use, while the LN-101 keeps good pin connections always.

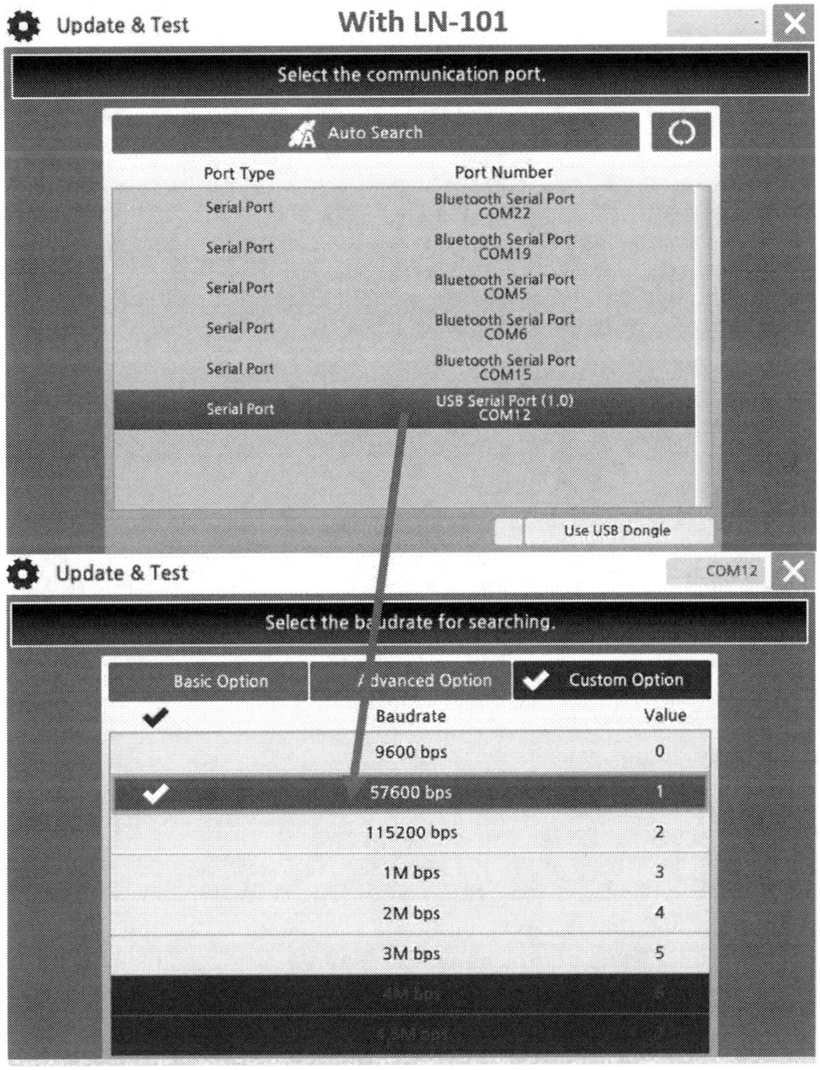

Fig. 3.4 Possible serial connection rates with the LN-101 USB Dongle.

The LN-101 would show up as a "USB Serial Port (1.0)" to MANAGER and quite a range of serial connection rates is possible with the LN-101 (see Fig. 3.4) but once again MANAGER only uses the 57.6 Kbps rate. The author likes this wired connection option the best, but the LN-101 is not a practical option once the MINI robot is fully assembled due to its long 4-pin ribbon cable dangling behind the robot.

3.2 Main Features of MANAGER V.2

Three main features are shown for MANAGER in Fig. 3.5, although only "Update & Test" and "Firmware Recovery" are currently operational. The "Self-Checklist" option just takes the user to a temporary web site.

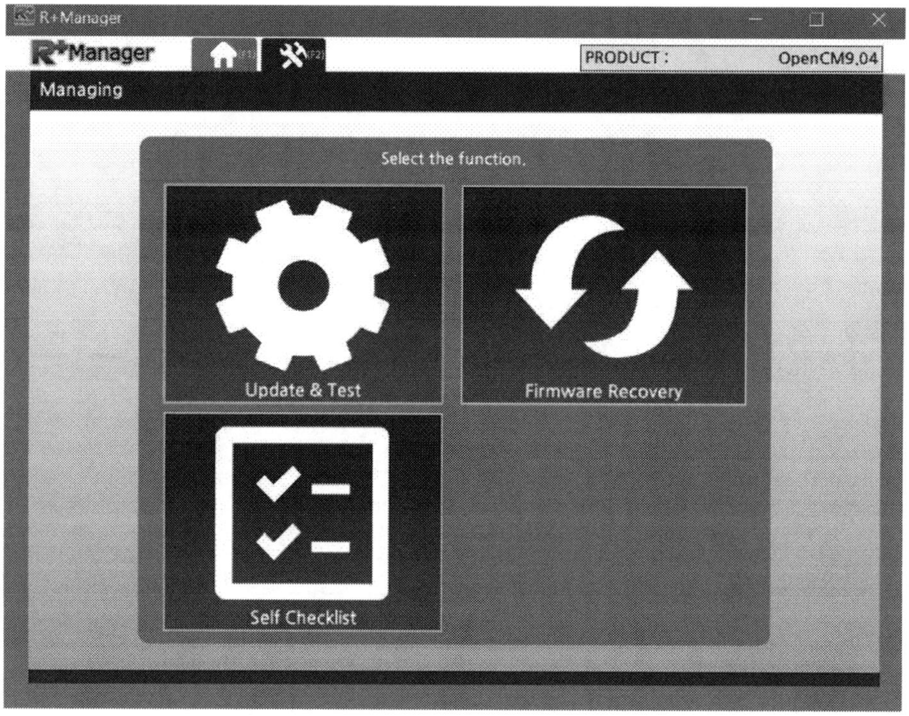

Fig. 3.5 Current Operational Features for MANAGER: "Update & Test" and "Firmware Recovery".

3.2.1 Using "Update & Test"

Once the user chooses the "Update & Test" option, MANAGER would go through several screens to check that the user is using the correct communications port and if the OpenCM9.04-C has the most current firmware version and if it does not, that will trigger another set of on-screen prompts to help the user update to the most current version of the firmware for the **OpenCM9.04-C**. Once this process is done, the user needs to choose the "Update & Test" option again so that MANAGER can now check on the current firmware of the **individual XL-320s** and then updates them one by one while keeping the original IDs of the XL-320s intact.

If all installed firmware are current, the final operational screen for MANAGER is shown in Fig. 3.6 and in its "Control Table" sub-window (third one from the left), the reader can see that some items are greyed out (such as "Mode Number" or "Motion Play Status"), this means that these parameters are "Read-Only" to the user when using MANAGER. Other items however are not greyed out (such as "Baud Rate" and "Dynamixel Channel") meaning that they can be modified by the user while inside the MANAGER tool (or by software commands inside TASK programs - see Chapter 6).

Fig. 3.6 MANAGER's main window when connected to the OpenCM9.04-C.

The author's MINI robot happens to have two IRSS-10 NIR Sensors (see Fig. 2.5) mounted on Ports 1 and 4 (Read-Only), thus Fig. 3.7 shows that ones can see live data coming through those ports at Addresses 360 and 366. Likewise, if the user had mounted actual DMS-80 Sensors onto Ports 3 and 4 which then would show live data coming in from those sensors (Addresses 370 and 372), otherwise the numerical data shown at those ports would have no meaning (i.e. an open circuit). If the user had mounted other 5-pin compatible sensors onto Ports 1 through 4 (see Section 2.1.4 and http://emanual.robotis.com/docs/en/parts/controller/opencm904/#robotis-5-pin-port), then the user can click on the right-pointing triangle corresponding to a specific port to see a pull-down menu where the user can choose the appropriate sensor used (see Fig. 3.8).

Fig. 3.7 MANAGER's main window showing live data from Ports 1 through 4 (IRSS-10 and DMS-80 by default).

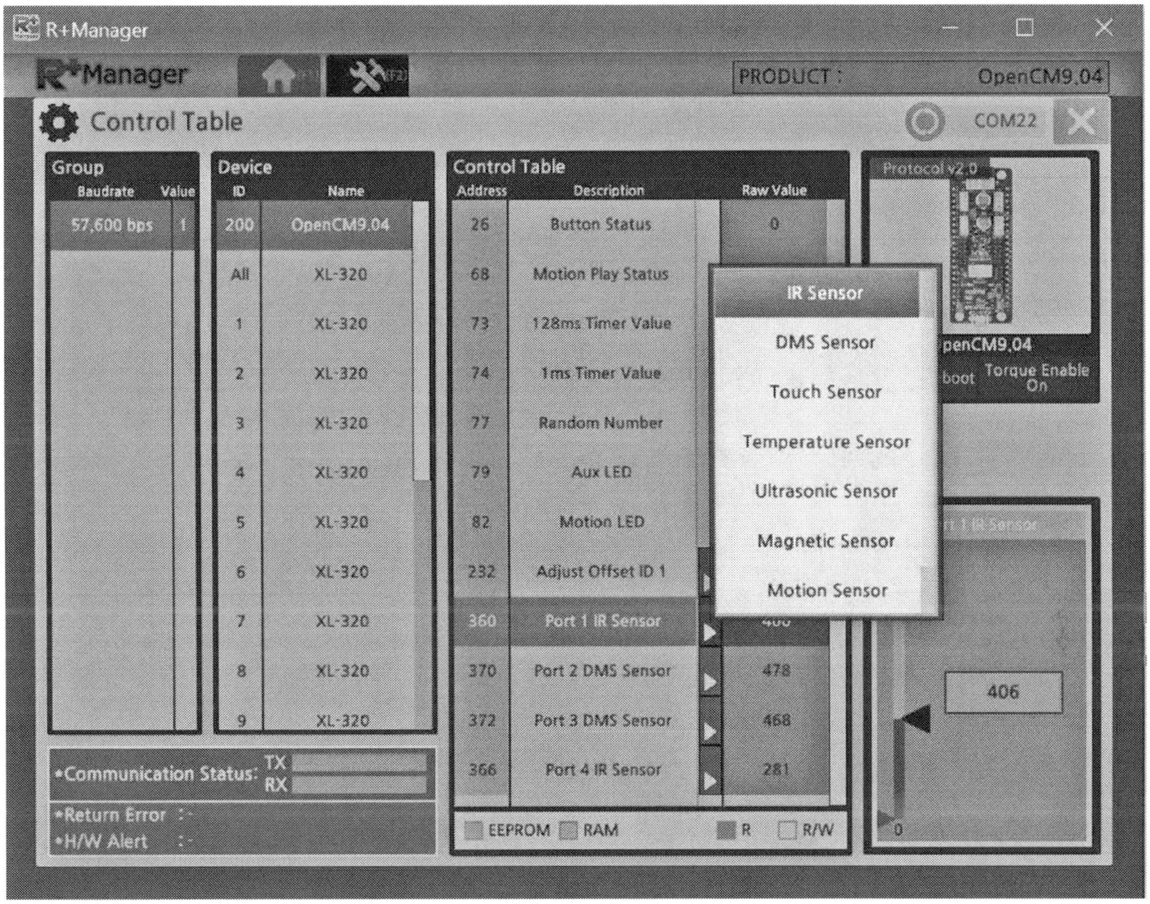

Fig. 3.8 Pull-down menu to choose appropriate 5-pin sensors used inside MANAGER tool.

In Figs. 3.7 and 3.8, a sharp-eyed reader may have noticed the items "Adjust Offset ID 0" (Address 230) and "Adjust Offset ID 1" (Address 232) which correspond to the individual Offset Values of the individual XL-320s used with the MINI. Furthermore, the user can change those values inside MANAGER, but the author recommends to only change these values inside the MOTION Sub-tool instead (see Chapter 4), as this task would be easier within the MOTION's graphical interface illustrating various poses of the MINI.

The user can access more details regarding this "Control Table" for the OpenCM9.04-C at this web link (http://emanual.robotis.com/docs/en/parts/controller/opencm904/#control-table) which shows that there are many more parameters that are not shown inside the MANAGER tool. **This web version of the Control Table is the definitive and complete guide to programming the OpenCM9.04-C with the TASK Sub-tool and other more advanced IDEs (OpenCM and Arduino)**, however beginner users need to be aware of **inconsistencies in the nomenclature used**, for example:

- Address 66 or "Motion Play Page" did not show up in the MANAGER tool (see Fig. 3.8), but it was displayed in the web-based Control Table, and as a matter of fact, it can be accessed inside a TASK program (see Statement 5 in Fig. 1.10), however it is labeled as "Motion Index Number" inside TASK Sub-tool, thus creating unfortunate confusion for beginner users.
- Similarly, Addresses 404 to 407 correspond to the use of a "Motion Sensor" via Ports 1 through 4 respectively in the web information, but inside a TASK program this type of sensor is

referred to as “Passive Infrared Sensor” (see Fig 3.9), which is also a correct technical characterization of this sensor, but this results in another unfortunate confusion for beginner users.

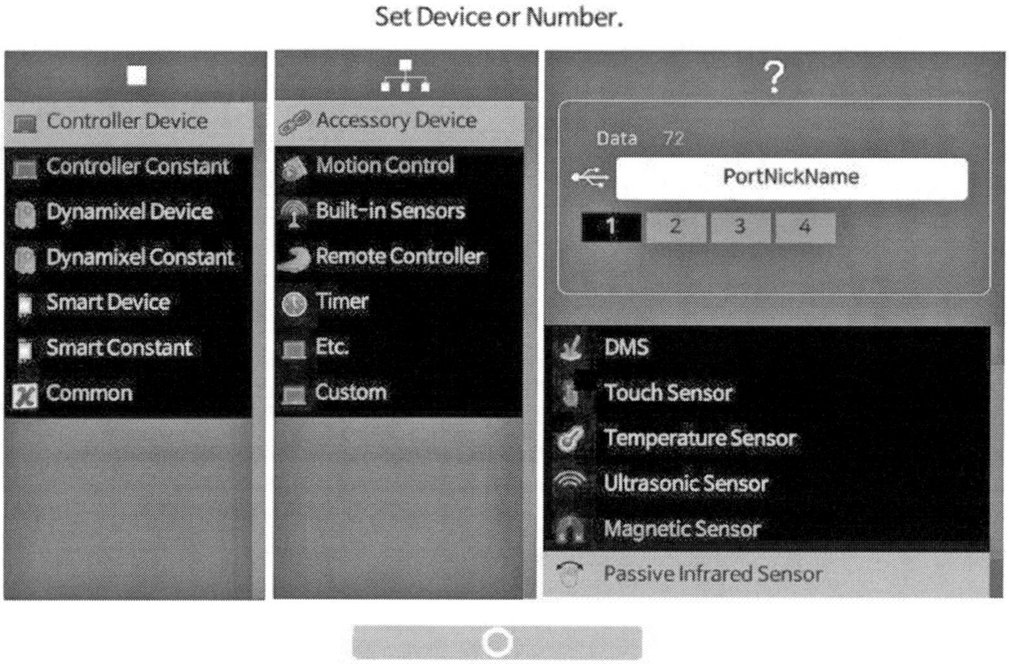

Fig. 3.9 “Motion Sensor” in Web Control Table is referred to as “Passive Infrared Sensor” in TASK.

The user can also “drill down” to individual XL-320 to check their settings and test their actual operations (see Figs. 3.10 and 3.11), for example:

- “Operating Mode” at Address 11 is greyed out in Fig. 3.10 meaning that it cannot be changed inside MANAGER, but in actuality this parameter can be changed inside a TASK program (see Chapter 6), meaning that an XL-320 can be programmed to change from Wheel Mode or Joint Mode (and vice-versa) as needed inside a programming tool such as TASK or OpenCM IDE.

Fig. 3.10 Read-Only settings for individual XL-320 from MANAGER Tool.

- In Fig. 3.11, addresses 27-29 indicate that the XL-320 can be controlled using Proportional-Integral-Derivative (PID) control protocols (https://en.wikipedia.org/wiki/PID_controller), but we will use them in **Proportional Control mode only in this book**, thus we'll be programming via the P parameter only which is defaulted to 64.
- The author recommends readers to play with the parameters "Goal Position" (Address 30), "Goal Velocity" (Address 32) and "Goal Torque" (Address 35) just to see how the chosen XL-320 behave physically. However, if the tested XL-320 is already mounted onto the MINI framework, the reader needs to be aware that its "Goal Position" is then "mechanically" restricted by the frame parts that it is attached to, so the tested servo will be able to comply only to a more restricted range of Servo Position than its full range [0-1023] (**please note that forcing the servo to go beyond its physical restraints is damaging to the servo**).
- **This web link (http://emanual.robotis.com/docs/en/dxl/x/xl320/#control-table) has the definite and complete Control Table for the XL-320 for advanced users when programming via OpenCM and Arduino IDEs. Once again, the reader needs to be aware that inconsistencies in nomenclature for certain items exist between the web-based information and the same item's description inside the MANAGER Tool, when in doubt it is best to refer to the item's Memory Address.**

Fig. 3.11 Some user-modifiable settings for individual XL-320 from MANAGER Tool.

Please see Video 3.1 for an "Update & Test" demonstration session with "older" OpenCM9.04-C and XL-320s.

3.2.2 Using "Firmware Recovery"

The user would need to use the "Firmware Recovery" option if he/she notices that the OpenCM-9.04-C and/or some of the XL-320 actuators behave abnormally.

For a tutorial on the Firmware Recovery of an OpenCM9.04-C via a Micro-USB cable please see (https://www.youtube.com/watch?v=PyQCzc73uh8) – **please make sure that all XL-320s are disconnected from the OpenCM9.04-C before performing this process**. Personally, the author had more reliable performances using the LN-101. Once the OpenCM9.04-C's firmware is recovered successfully, the user should next reconnect the XL-320s back to the OpenCM9.04-C and then choose the "Update & Test" option to update/test the firmware of the individual XL-320s and fortunately this process will keep the original IDs of the XL-320s being updated.

If the above procedure was not enough and some XL-320s were still misbehaving, then each of those faulty XL-320s would need to have their firmware recovered **one by one**, i.e. to connect only one XL-320 at a time to the OpenCM9.04-C and to perform the following procedure with the **LN-101** (*the author had no success when using the Micro-USB cable for this task*):

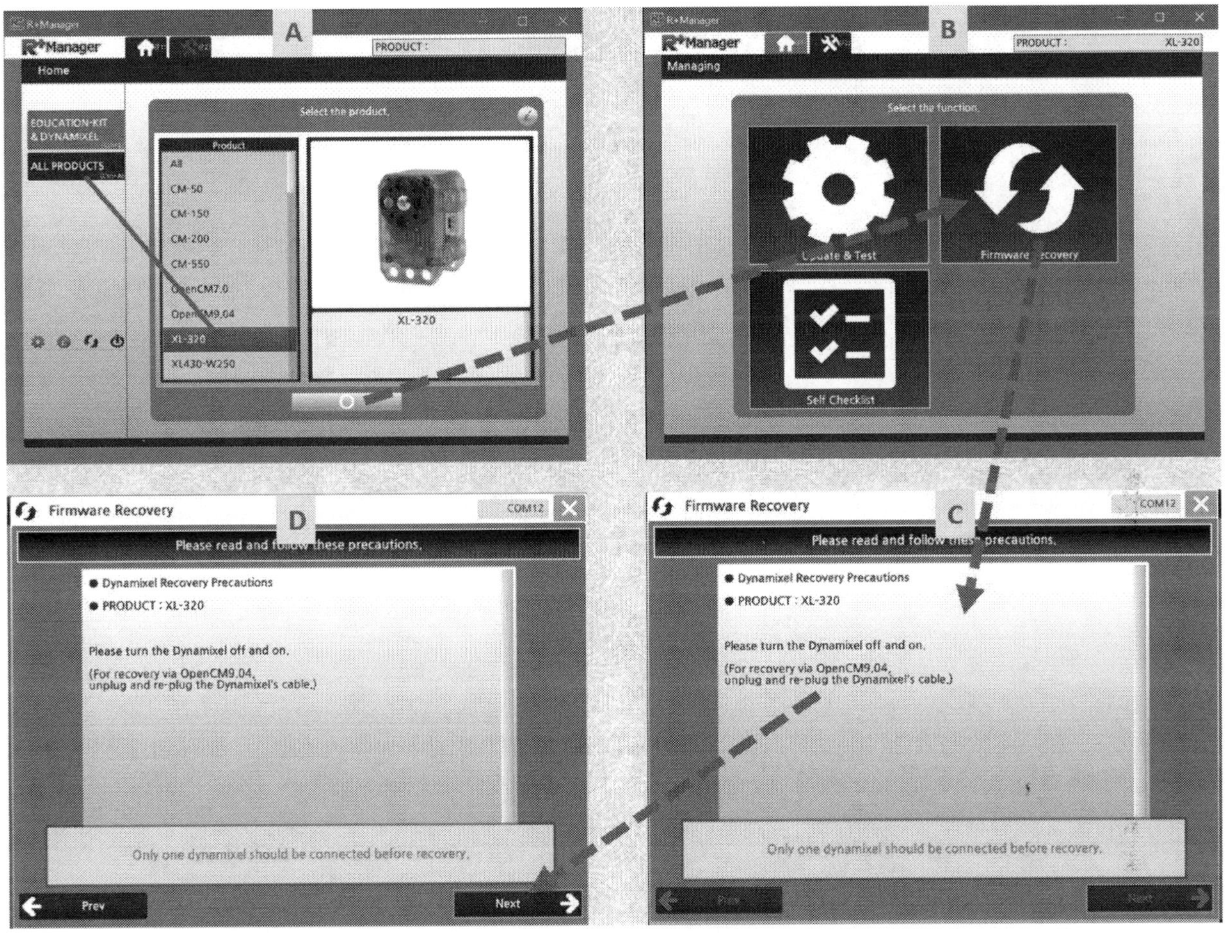

Fig. 3.12 Firmware Recovery for XL-320 using MANAGER V.2 (Part 1).

1. At the opening window of MANAGER, please go to "ALL PRODUCTS" (left panel) and choose the item "XL-320", and then click "O" (see Fig. 3.12-A).
2. Next choose "Firmware Recovery" (see Fig. 3.12-B) and follow on-screen instructions until the user can see Fig. 3.12-C.
3. At this point just unplug the XL-320's ribbon cable from the 3-pin connector on the OpenCM9.04-C's PCB, and then re-plug this cable back in. If all goes well, the user should see Fig. 3-12-D with the "NEXT" box highlighted. Then the user should click on "NEXT" to continue. If the user had to wait for a while and nothing much happened, then the user should restart this whole procedure (maybe there were some bad electrical contact issues).
4. Lastly, the user should keep on following the next on-screen instructions (see Fig. 3.13) and wait on the recovery process to finish.

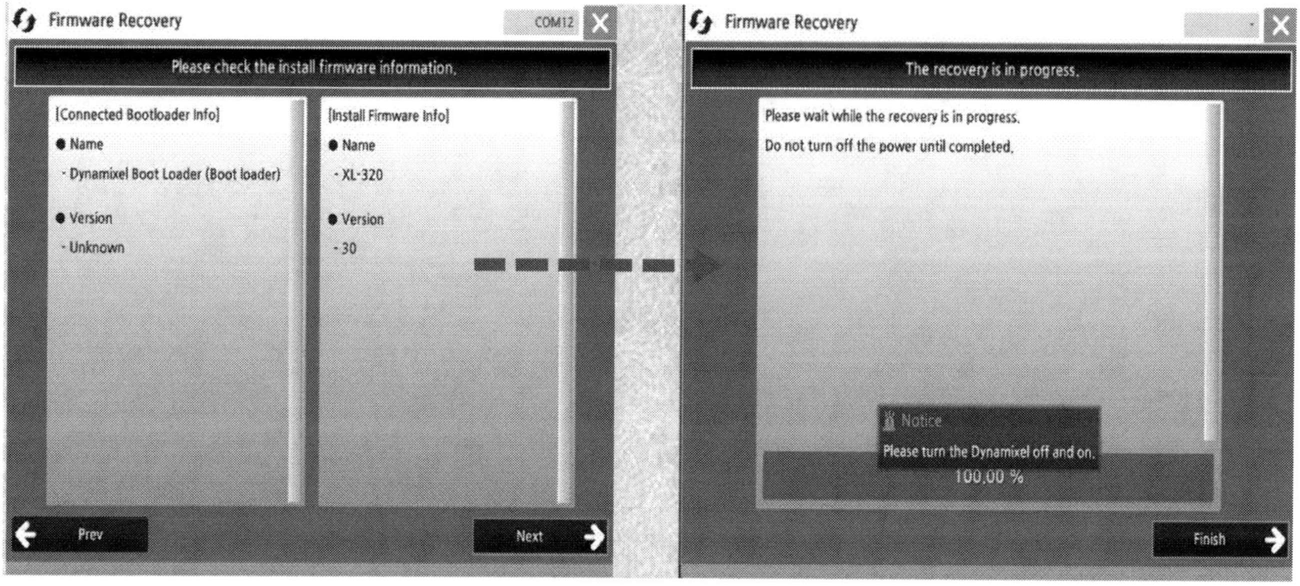

Fig. 3.13 Firmware Recovery for XL-320 using MANAGER V.2 (Part 2).

One important fact to remember is that if any servo has gone through the firmware recovery process, its Dynamixel ID is reset to "1", so the **next must-do** step is to use MANAGER in "Update & Test" mode to relabel its ID with the proper ID number. Then onto the next XL-320 to be recovered. On the practical side, the author had seen XL-320s beyond recovery or repairs, when he just had to purchase new ones! Please see Video 3.2 for a "Firmware Recovery" demonstration session with OpenCM9.04-C and XL-320.

Lastly, if an advanced user had tried out the OpenCM or Arduino IDEs, then the "Firmware Recovery" process for the OpenCM9.04-C **is a must if this user wants to get back to using the RoboPlus Suite of tools**: MANAGER, TASK and MOTION (see Section 1.3.9).

3.3 Using MANAGER on completed MINI

In this situation, it is best to lay the MINI flat on its back or its front side before turning it on and connect MANAGER to it to prevent any unforeseen servo malfunction from making the MINI collapse and damage itself. Once MANAGER connects successfully to the OpenCM9.04-C, the user should see a screen such as the one shown in Fig. 3.7 and the user needs to remember to scroll down to see all servos listed. If some servos are missing, this may mean that some cable connections may be loose or that some servos have become non-operational, and sometimes a firmware recovery would solve this later problem but remember the "ID = 1" issue mentioned in Section 3.2.2. Once again, the user needs to remember that the full range of servo positions [0 – 1023] are not available due the mechanical constraints of the MINI's framework, i.e. the user should try to vary the servos' positions only a little bit at a time while watching for possible jam situations.

3.4 Review Questions

1. On which Operating System(s) can the MANAGER Tool operate?
2. Which communications port(s) can the MANAGER Tool be used with?
3. What kind of communications devices are compatible with the MANAGER tool?
4. What are the main functions/features of the MANAGER Tool?
5. What is the allowed communication rate that can be used with the MANAGER Tool?
6. Can the MANAGER Tool be used to update the firmware on individual actuators?
7. What is the "Dynamixel Channel" setting? What can it be used for?
8. What the memory address that corresponds to the JOINT-OFFSET for Servo #17?
9. When ones perform "Firmware Recovery" on individual actuators via MANAGER, their Dynamixel IDs are not modified (True or False).

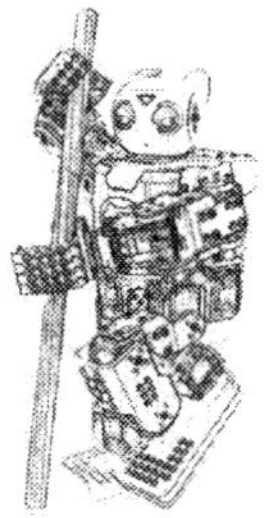

Chapter 4: Using MOTION Tool

As previously mentioned in Chapter 1, the current V.3 of the MOTION Sub-tool is combined with the TASK Sub-tool into a single application called R+TASK V.3, available in Windows, Android and Mac OSes (http://www.robotis.com/service/downloadpage.php?ca_id=10). This integration allows a more efficient way to create, test and program MOTION data sets. The technical manual for R+TASK V.3 is available at (http://emanual.robotis.com/docs/en/software/rplustask3/). This e-manual contains the step-by-step instructions for all basic features of the TASK and MOTION Sub-tools for R+TASK V.3, thus this book will not try to re-write these How-To's but only to refer to or expand on specific ones as needed.

The goal of this Chapter is to get the reader familiarized with the concept of "MOTION" as implemented by ROBOTIS and get in some practices in creating KEY-FRAME, MOTION-UNIT, MOTION-LIST and MOTION-GROUP.

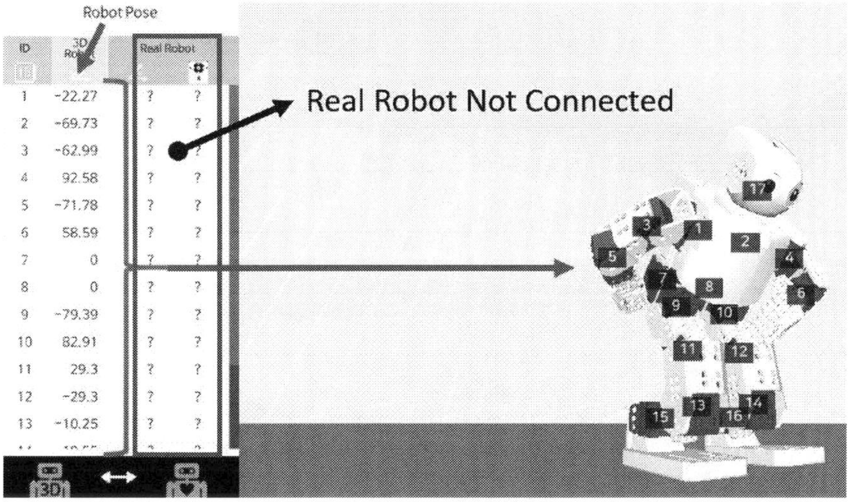

Fig. 4.1 A predefined POSE/KEY-FRAME from the "Greet 2" MOTION-UNIT.

First, the reader is recommended to view Video 4.1 for a quick tour of the Motion Editing features of the MOTION V.3 tool before reading on the rest of this chapter.

The ROBOTIS MOTION data structure is quite complex and requires several layers of understanding to be proficient at it. There are two fundamental concepts that need to be understood clearly: a POSE/KEY-FRAME and a TIME-LINE (i.e. timing) for these KEY-FRAMEs:

- Creating head/arm/leg movements for the MINI is very similar to making cartoons which are essentially multiple "sequential" frames of "static images" that have minor changes done to the "object(s) in motion" in each frame. Each of these static images of the MINI is called a POSE which can be defined as a 1-dimensional array of positional values (i.e. angle values from -300 degrees to +300 degrees) of the various servos used on the MINI. This array uses the servos' IDs (1 to 17 in the case of the MINI with head joint) as its array index (see left side of Fig. 4.1).

- When the physical MINI robot performs these predesigned movements in real time, the OpenCM9.04-C needs to have the poses for each of its servos updated every 7.81 ms (a hardware requirement) (*Note: human muscle signals get updated every 4 to 50 ms*). Fortunately, the MOTION tool does not require the user to predefine robot poses every 7.81 ms, so the user needs to only define Key Poses, also referred to as KEY-FRAMEs, on a TIME-LINE every few seconds apart or so (see Fig. 4.3). Then, at run-time, the OpenCM9.04-C firmware will generate as needed INTERPOLATED-FRAMEs between adjacent KEY-FRAMEs for every 7.81 ms time interval. This procedure "guarantees" that going from 1 KEY FRAME to the next will keep up with the time schedule of KEY-FRAMEs as defined by the user on the TIME-LINE.

4.1 Motion Data Structure and Motion Design

For this section, we'll just use the simulated 3D graphics MINI robot and we won't need the real MINI until Section 4.4. Let's first explore various components provided by the MOTION tool to make this Motion Programming work possible using the Default Example Motion File for the MINI, Thus would the reader please run the R+TASK V.3 application and open up the MINI's Example Motion File with Head Joint (see Fig. 4.2). This would take the user to the MOTION-UNIT Window (partially shown in Fig. 4.3) – which can also be reached from anywhere inside TASK 3 by pushing F4 on the keyboard.

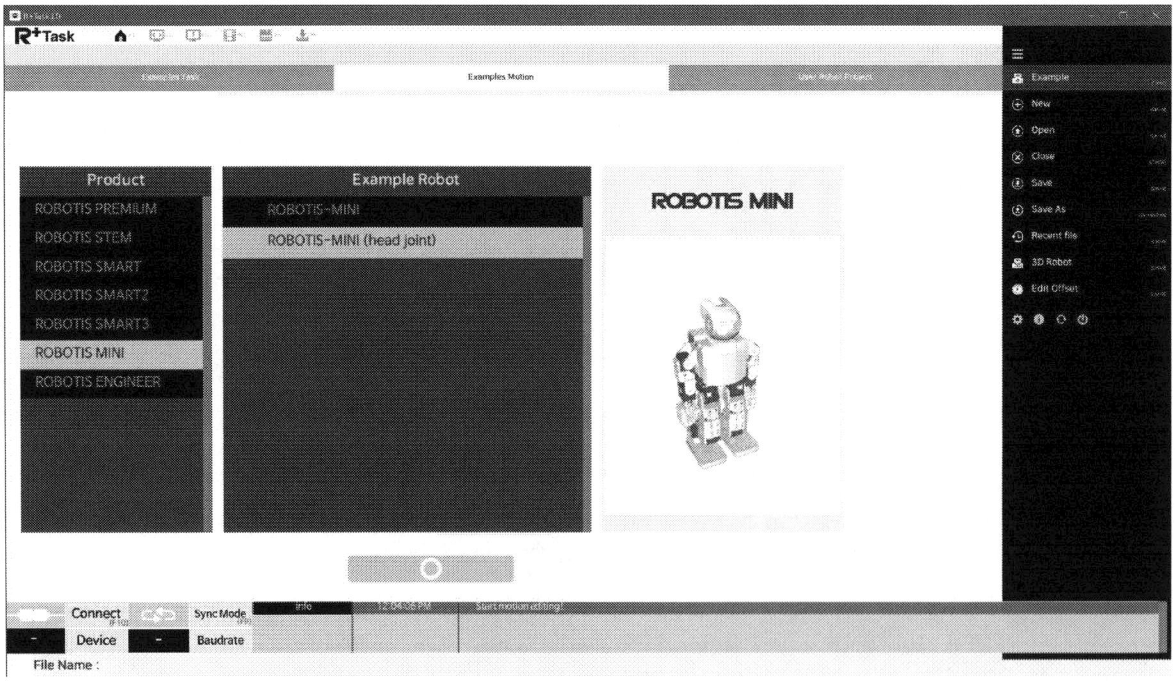

Fig. 4.2 Opening the MINI's Example Motion File with Head Joint.

4.1.1 Key-Frame, Time-Line & Motion-Unit

Next, would the reader please left click with the mouse on the Pull-Down Menu for the Motion Unit List and choose the Motion Unit named "Advance 1-1" (see Fig. 4.3) which was designed to make the graphical (3D) robot walk forward with the left leg, then make another step forward with the right leg.

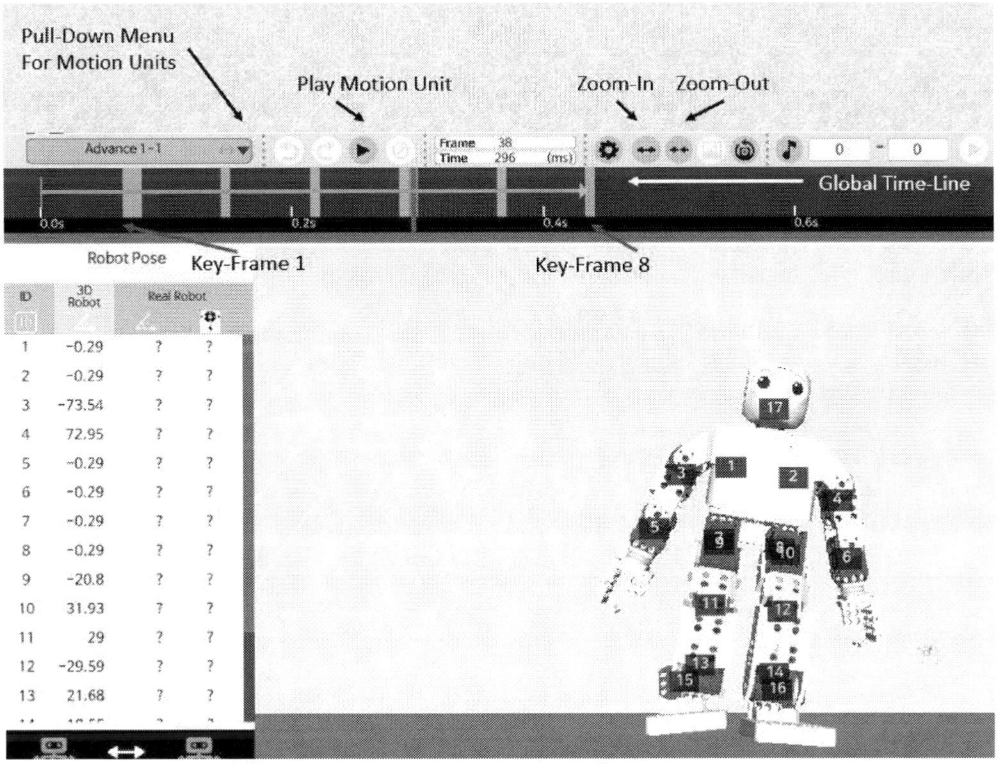

Fig. 4.3 Displaying the Motion Unit named "Advance 1-1".

The reader can experiment with the "Play Motion Unit" button to verify what this MOTION-UNIT really does, and the "Zoom-In" and "Zoom-Out" buttons to enlarge/reduce the Time Scale on the TIME-LINE.

In the MOTION Sub-tool, putting various KEY-FRAMEs on a TIME-LINE creates a MOTION-UNIT. Of course, the user needs to observe human movements and learn to decompose them into sequential KEY-FRAMEs that make sense kinematics wise and dynamics wise. The author had found that watching video replays at slow speed is a great help.

In Fig. 4.3, the reader needs to notice that:

- There are 8 KEY-FRAMEs defined for this MOTION-UNIT "Advance 1-1".
- KEY-FRAMEs 1 and 2 are next to each other (i.e. only 7.81 ms apart).
- KEY-FRAMEs 5 and 6 are also next to each other.
- There are no KEY-FRAME defined for TIME = 0.

The specific reason for why there is no KEY-FRAME at TIME = 0 is because MOTION-UNIT 1 (i.e. Initial Position) is assumed to be already played. As a matter of fact, when we get into programming motions for the MINI with the TASK Component later in Chapter 6, one of the first program statements would always be to play the Initial Position MOTION-UNIT to put the robot into a Ready pose:

Motion Index Number = Motion[1]: Initial Position

Furthermore, the general reason for MOTION-UNITs not to have a KEY-FRAME defined at TIME = 0 is to facilitate the combination of MOTION-UNITs together to form a more complex choreography called a MOTION-LIST (see Section 4.2) such as the one for MOTION-LIST "Horse Dance" **whereas the "last" KEY-FRAME of the preceding MOTION-UNIT effectively becomes the "initial" KEY-FRAME for the following MOTION-UNIT. This is a key concept that the user needs to apply when the user decides to create his/her own MOTION-UNITs.**

4.1.2 Basic Motion Design Concepts

Let's look at some basic Motion Design concepts using MOTION-UNIT "Advance 1-1" as an example:

- Fig. 4.4 shows more details for KEY-FRAME 1 and KEY-FRAME 2. KEY-FRAME 1 occurred at TIME = 70 ms and KEY-FRAME 2 occurred at TIME = 78 ms.

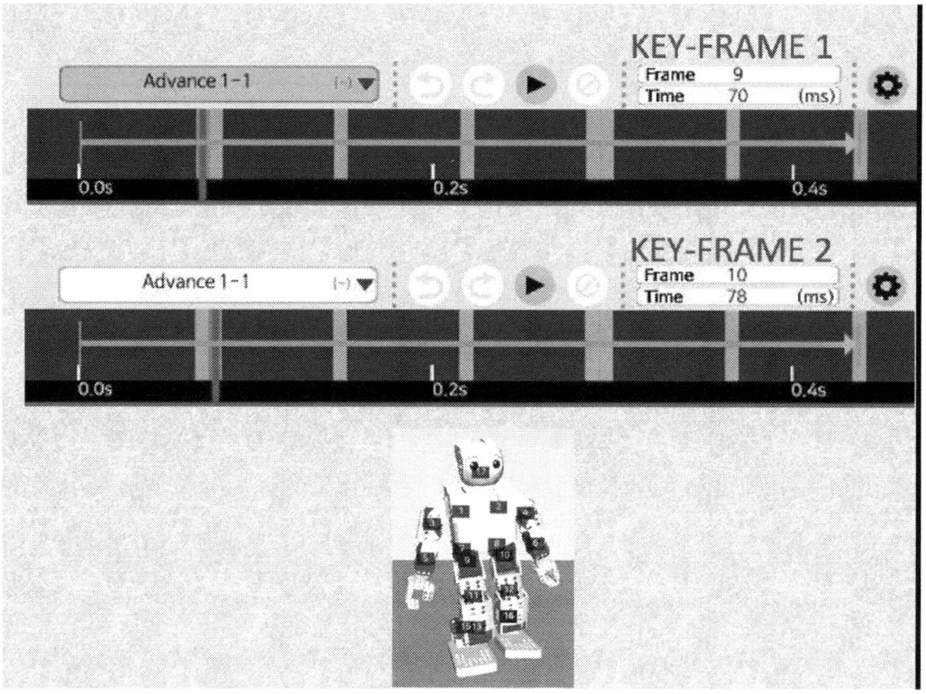

Fig. 4.4 Details for KEY-FRAME 1 and KEY-FRAME 2 for MOTION-UNIT "Advance 1-1".

- The most interesting fact to notice is that both KEY-FRAMEs have the same positional angle values for all 17 servos, this means that **for 8 ms the robot was programmed to stay still**. Here the author surmises that the ROBOTIS folks wanted to stabilize the robot in a mechanical sense. The reader should remember that the robot was in the "Initial Position" with both feet flat on the ground and its CG between its feet (see left picture in Fig. 4.5) and most importantly it was at rest. Then from this rest position, the goal of KEY-FRAME 1 was to **make the robot raise its left foot and shift its CG to the inside of its right foot** (so that it won't fall to its left side), i.e. to accelerate Servos 9-10 and 13 through 16 from the Initial Position and then to decelerate them when they were close to KEY-FRAME 1's goal positions (see right picture in Fig. 4.5).

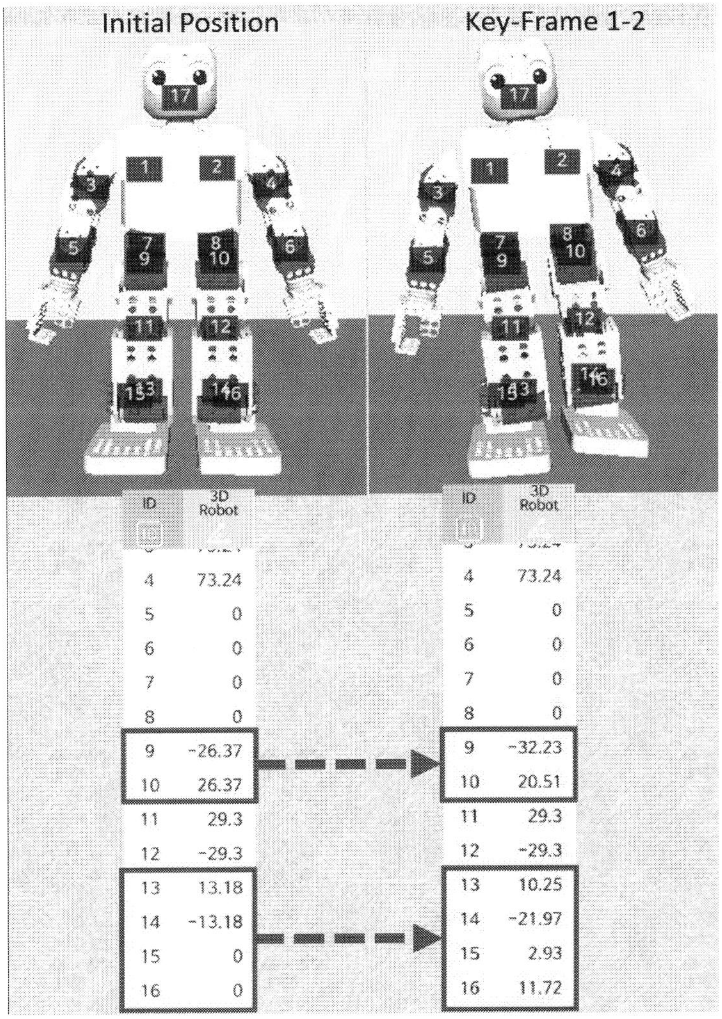

Fig. 4.5 Details for INITIAL POSITION and KEY-FRAMEs 1 and 2 of MOTION-UNIT "Advance 1-1".

- These servo dynamics would generate vibrations throughout the mechanical framework of the robot, and now that it stood only on 1 leg, it needed a little bit of rest time (7.81ms) to let these vibrations dissipate before making another dynamic move as programmed by KEY-FRAME 3 (which lifted the left leg higher and bent the left knee some more). KEY-FRAME 4 then put the left leg down and forward of the right leg, but no motion rest was used for KEY-FRAMEs 3 and 4 (see Fig. 4.6) because we wanted to keep that forward momentum to help us raise the right leg next with KEY-FRAME 5 (and also because the robot was more stable with both legs on the ground when at KEY-FRAME 4 – in a way it was OK to let the robot "fall into a stable position"). Then as before, KEY-FRAME 6 was used to help stabilize the right-leg-lift move triggered by KEY-FRAME 5 (i.e. the robot stayed still for 7.81 ms).

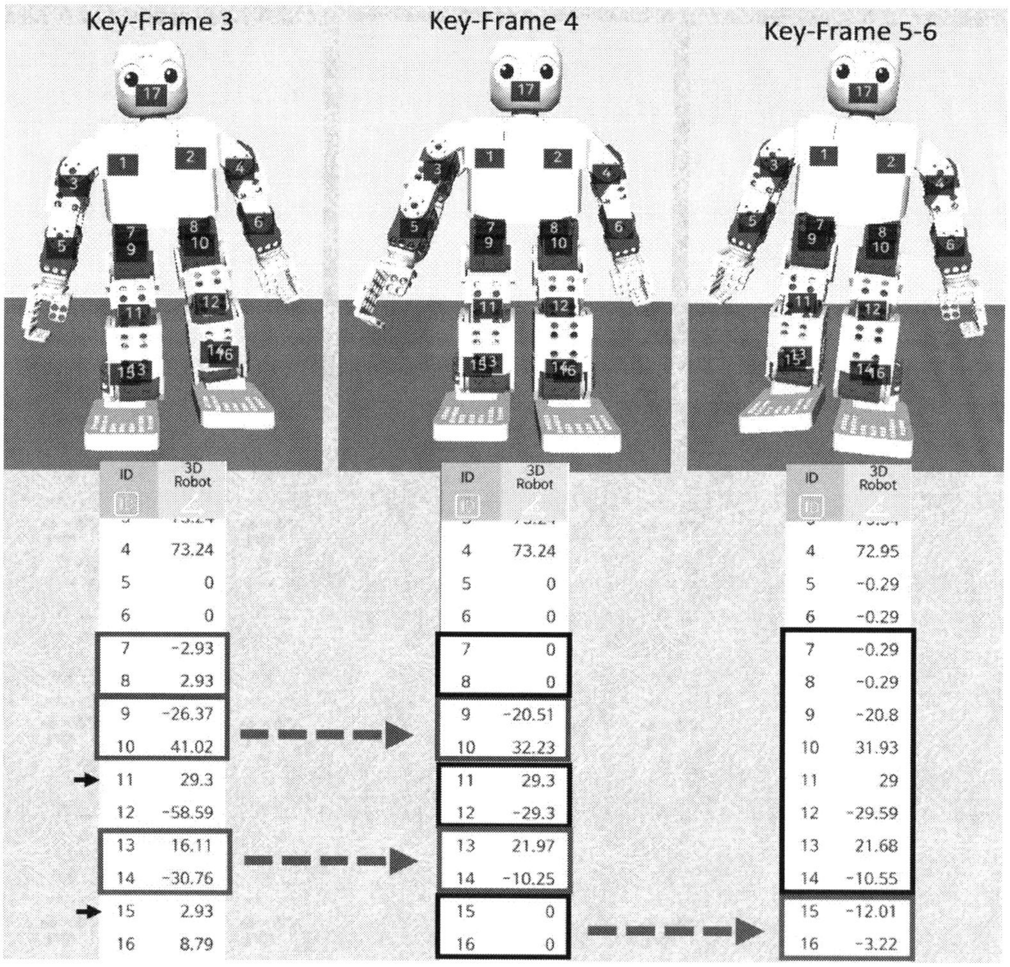

Fig. 4.6 More details for KEY-FRAMEs 3 through 6 of MOTION-UNIT "Advance 1-1".

- In Fig. 4.6, Black Arrows denoted servos' values that did not change from the previous KEY-FRAME, while Blue and Red Frames denoted the servos that got changed appreciably, and finally Black Frames denoted servos' values that are reset to the values used for the Initial Position (see Fig. 4.5) or did not change appreciably from the previous KEY-FRAME.
- In KEY-FRAME 3 of Fig. 4.6, the values taken by Servos 7 and 8 effectively "spread" the legs apart at the hip joint, while Servos 10 and 12 lifted the left thigh and knee higher and more

forward. Servos 14 and 16 worked together to make the left foot more level with the ground in preparation for the "landing" of the left leg by KEY-FRAME 4.

- In KEY-FRAME 4 of Fig. 4.6, please note that the values for Servos 7-8 (Hip Sideways), 11-12 (Knee), 15-16 (Ankle Sideways) are back to the values used by the Initial Position, while the values for Servos 9-10, and 13-14 corresponded to a "left-leg forward" pose instead of the pose when both legs are aligned for the Initial Position.
- In KEY-FRAME 5-6 of Fig. 4.6, only Servos 15 and 16 changed their values appreciably from KEY-FRAME 4 as the robot was starting to lean on the left ankle and to sweep in its right foot.

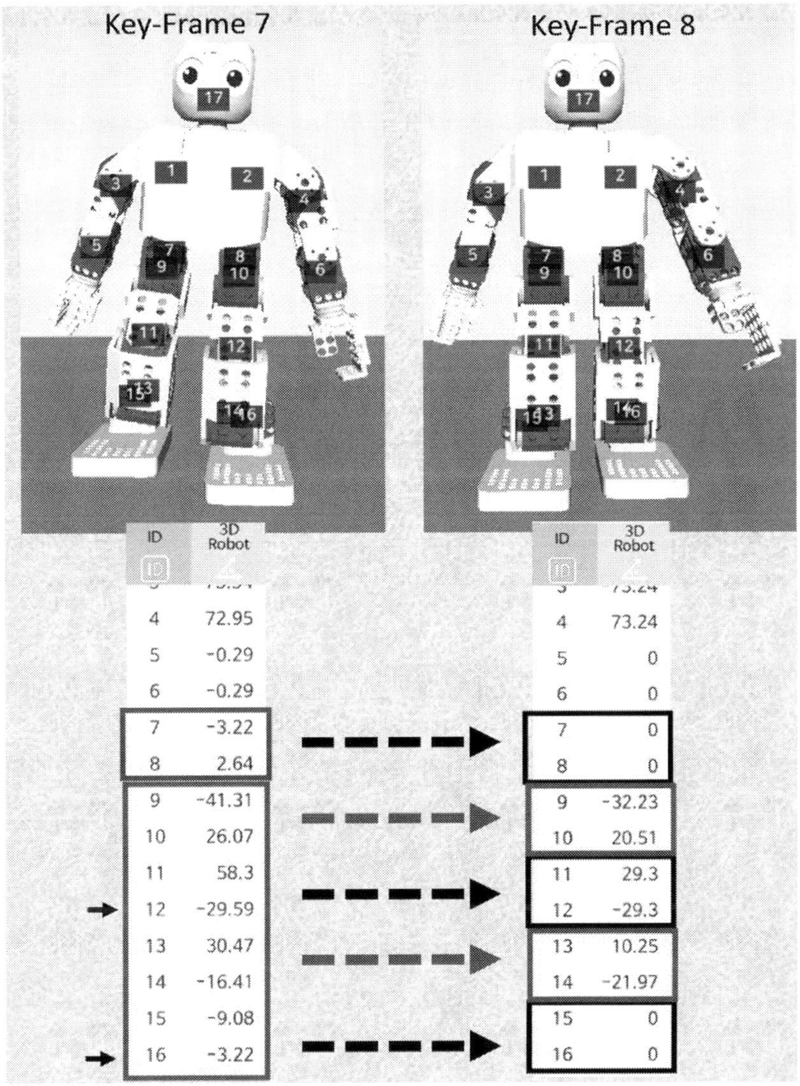

Fig. 4.7 More details for KEY-FRAMEs 7 and 8 of MOTION-UNIT "Advance 1-1".

- Fig 4.7 shows more details for the KEY-FRAMEs 7 and 8, and similarly, Black Arrows denoted servos' values that did not change "much" from the previous KEY-FRAME, while Red Frames denoted the servos that got changed "appreciably", and finally Black Frames denoted servos' values that are back to the values used for the Initial Position (see Fig. 4.5) or did not change appreciably from the previous KEY-FRAME.

- KEY-FRAME 7's actions (Fig. 4.7) were quite similar to KEY-FRAME 3, so Servos 7-8 spread the legs out, while Servos 9-14 moved the right thigh and knee higher and forward while maintaining overall balance on the left leg, at the same time Servo 15 made the right foot more level with the ground to get it ready for the "landing" move by KEY-FRAME 8. The reader should also note that KEY-FRAME 7 was not the perfect mirror solution to KEY-FRAME 3 in terms of the actual numerical values of the XL-320s, thus reminding the reader that there can be many possible solutions for the same robot movements that the reader wants to design.
- In KEY-FRAME 8 (Fig. 4.7), please note its similarity with KEY-FRAME 4 whereas the values for Servos 7-8 (Hip Sideways), 11-12 (Knee), 15-16 (Ankle Sideways) are back to the values used by the Initial Position, while the values for Servos 9-10, and 13-14 corresponded to a "left-leg forward" pose instead of the "right-leg forward" pose for KEY-FRAME 4.

So that was Motion Design in the ideal world of computer graphics, now how about in the real world where the physical robot was going through these same KEY-FRAMEs? In an actual test, the robot did make the left and right steps as commanded to, but it was more "shuffling" than "walking" in the real world:

- Fig. 4.8 shows pictures of a real MINI robot holding poses defined by the KEY-FRAMEs 1 through 4. In KEY-FRAMEs 1-2-3, this particular robot's right leg was "too weak" (due to the real weight of the robot and slacks in the servos' gear trains and various mechanical fasteners) to hold the left foot **above** the walking surface as previously shown for the graphical robot. The robot still used the outer edge of its left foot as a contact line, but this pose reduced the surface friction enough that the robot could shuffle its left leg forward without tripping on itself to get into Key-Frame 4.

Fig. 4.8 Poses made by a real MINI robot at KEY-FRAMEs 1 through 4 of MOTION-UNIT "Advance 1-1".

- Fig. 4.9 shows a slightly different situation for moving the right leg (i.e. KEY-FRAMEs 5 through 8). In KEY-FRAME 7, this robot's left leg was "strong" enough to hold the right foot's toes **above** the walking surface, but the left heel was still resting on the walking surface. But this reduced friction enough for the robot to shuffle the right leg forward as needed to get into KEY-FRAME 8.

Fig. 4.9 Poses made by a real MINI robot at KEY-FRAMEs 5 through 8 of MOTION-UNIT "Advance 1-1".

From the previous exercise, the author has learned a few pointers:

- The MINI robot can "walk" (i.e. shuffle) on a hard and smooth surface but not likely on carpets.
- Users should not expect a "symmetrical" performance for any real robot, but with some tweaking in the Graphical Motion Design phase, this robot could have more similar left and right performances. **It will always be more challenging to make a humanoid robot walk "straight" than to make a "wheeled" robot drive straight.**
- During the writing of this book (Fall 2019), ROBOTIS released a new robotics kit named ENGINEER 1 which has a humanoid robot named MAX-E1 with a new mechanical design (needing only 12 servos instead of 16) and an alternate gait solution allowing this robot to lift its walking leg much higher than the MINI (see Fig. 4.10). The reader can also watch this YouTube web site for a video of an earlier prototype of this type of humanoid robot (https://www.youtube.com/watch?v=9YtieresealI).

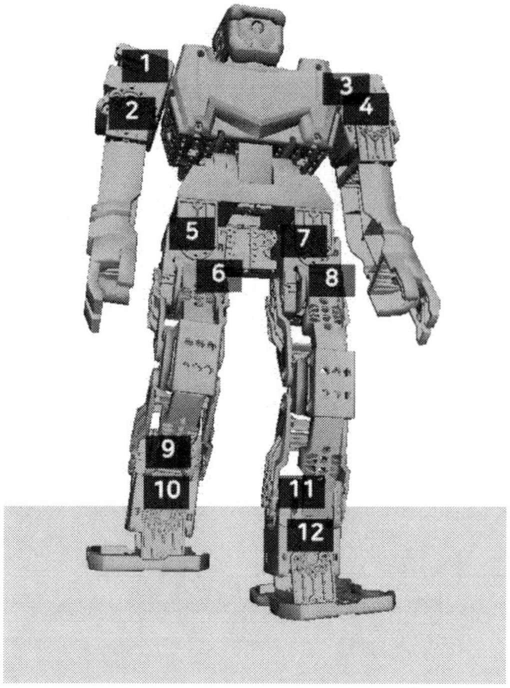

Fig. 4.10 New MAX-E1 Humanoid Robot can lift its walking leg much higher than the MINI.

The author further noticed that the following MOTION-UNITs also had "resting" poses designed into them for the MINI:

- "Sidestep Right" and "Sidestep Left".
- "Advance 1-1" and "Advance 1-2".
- "Reverse 1-2" (but surprisingly not "Reverse 1-1").

The author was then curious about what would happen without these "resting" poses or that the "resting" poses lasted a longer time period? Hence, he copied and modified the MOTION-UNIT "Advance 1-1" into 2 variants: Version A where there was no "resting" periods and Version B where the resting period was 3 times longer than 7.81 ms, i.e. 23.43 ms (see Fig. 4.11). The reader can open the enclosed sample file "MINI_Custom.mtn3" file to access and try out these modified MOTION-UNITs "Advance 1-1-a" and "Advance 1-1-b". The reader can also watch Video 4.2 for a tutorial of these steps and for a performance comparison between these 3 versions of the MOTION-UNIT "Advance 1-1". The author could not honestly discern any performance differences among the 3 versions.

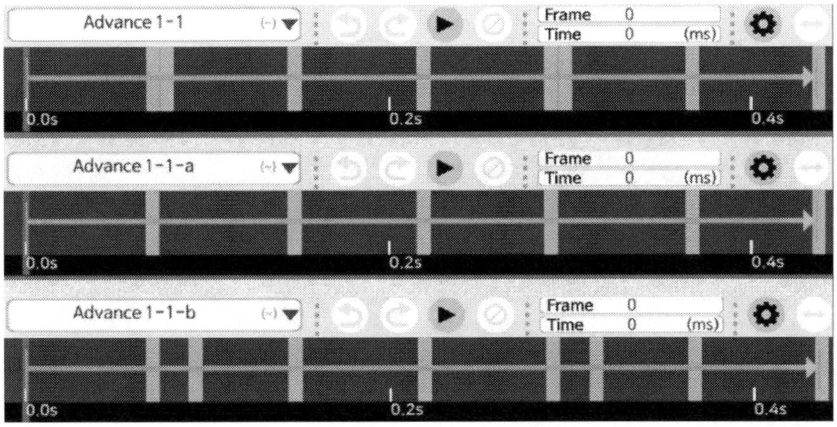

Fig. 4.11 Original and Versions A & B of the MOTION-UNIT "Advance 1-1".

Let's next look at the "counter-balancing" motion of the arms when the MINI is walking. Figs. 4.12 and 4.13 show more details for the "counter-balancing" actions for the arms between KEY-FRAME 2 and KEY-FRAME 5:

- In KEY-FRAME 2 of Fig. 4.12, the MINI has partially lifted its left leg up, so the arms were still aligned with the body. During the transition to KEY-FRAME 3 when the left knee would be at its highest position, the right arm (Servo 1) swung forward while the left arm (Servo 2) swung backward to keep overall balance, both by the same amount of 8.79 degrees.

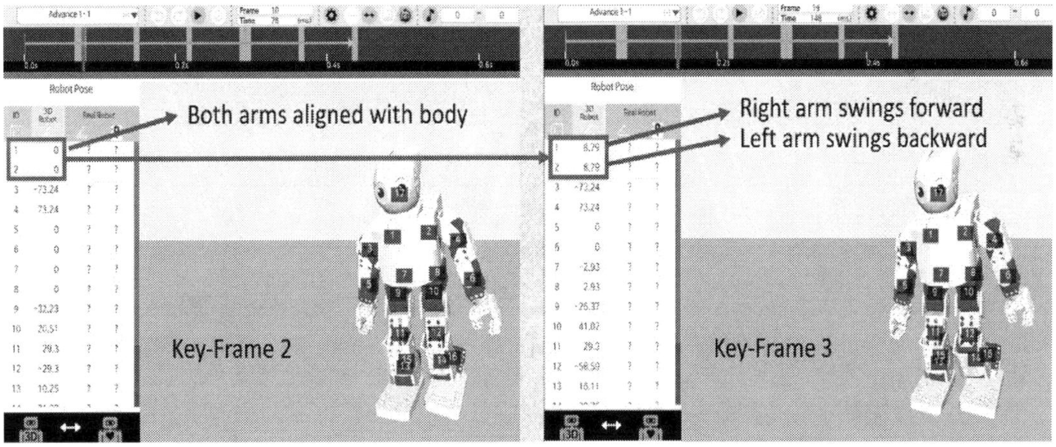

Fig. 4.12 Arms balancing when lifting Left Leg in MOTION-UNIT "Advance 1-1".

- Next during the transition to KEY-FRAME 4, when the left foot would land forward of the right foot, the right arm swung to its most forward position of 29.3 degrees, while the left arm swung also to its most backward position, but only to 14.65 degrees (see Fig. 4.13). Lastly, when shifting weight to the left leg and starting to lift the right leg (i.e. transitioning to KEY-FRAME 5), the arms were withdrawn back to their body-aligned positions with both servos almost back to their 0 degree value (actually overshooting by 0.29 degrees, but no human can really discern such a small angular difference in the real world).

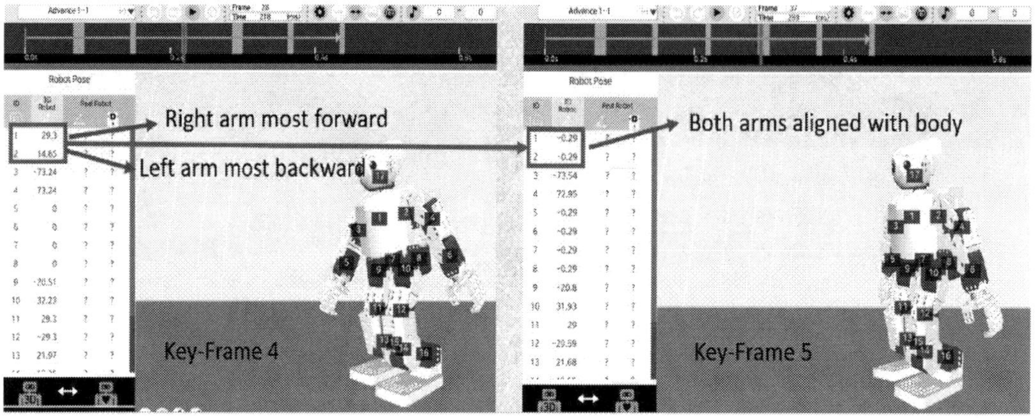

Fig. 4.13 Arms balancing when landing Left Leg and starting to lift Right Leg in MOTION-UNIT "Advance 1-1".

- For the remaining KEY-FRAMEs 6-7-8, the same procedure was repeated to balance the Right Leg's "Lift-and-Forward" motion with the "Forward-Swinging" of the Left Arm.

The author would recommend that the reader does his/her own analysis of the MOTION-UNIT "Advance 1-2" which makes the MINI make a half step forward with its left leg so that both feet are aligned at the end of this MOTION-UNIT (i.e. in Initial Position again).

4.2 Motion-Units' Flow Control with Motion-List Tool

For this section, we'll also just use the 3D graphics simulated MINI robot. In a way, we can consider the MOTION-UNITs to be the different "MOTION LETTERS" in a user-defined "MOTION ALPHABET", and the MOTION-LIST is the tool to create "MOTION WORDS" from these "MOTION LETTERS". The MOTION-LIST tool can be accessed from anywhere inside TASK 3 by pushing the F5 key (see Fig. 4.14). Officially ROBOTIS named this tool "MOTION", but the author added "LIST" to create "MOTION-LIST" to distinguish it from "MOTION-UNIT" and "MOTION-GROUP". In older ROBOTIS documents (e.g. for TASK V.1), "MOTION-PAGE" was used to refer to the same data construct as "MOTION/MOTION-LIST".

4.2.1 Basic Use of Motion (List) Tool

The MOTION-LIST tool by default opens on the first item which happens to have only 1 MOTION-UNIT i.e. "Initial Position", of course other MOTION-LIST items can have a combination of MOTION-UNITS as shown in later paragraphs. The most used menu buttons are shown in Fig. 4.14:

- PLAY/PAUSE (F7) toggles between "PLAY" and "PAUSE" for the current MOTION-LIST item.

- Once "PLAY" mode is activated, the two menu buttons on its right get highlighted: STOP (F8) and EMERGENCY STOP (Esc). The difference between these two options are subtle and will be explained in Section 4.2.2. For now, we'll just use F7 and F8 keys for this section.
- The user can also insert music clips to play along with one's MOTION-LIST ("wav" or "mp3" files).

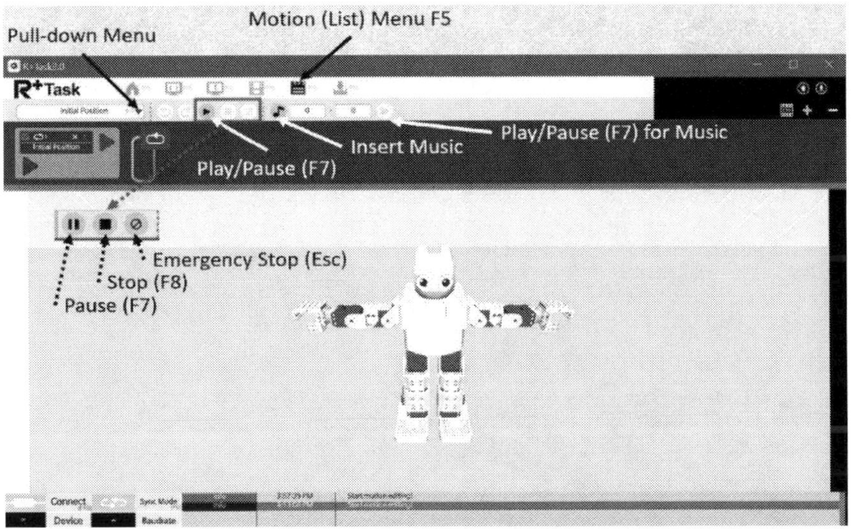

Fig. 4.14 MOTION-LIST tool of TASK 3 can be accessed via F5 on the keyboard.

Next, the reader needs to use the Pull-down Menu and choose the item named "Advance", then the reader would see Fig. 4.15 displayed:

- MOTION-LIST "Advance" is shown to start with MOTION-UNIT "Advance 1-1" followed by MOTION-UNIT "Advance 1-2".
- There is an "Endless Loop" object which is always positioned as the last item in a MOTION-LIST item. Its use will be shown in Section 4.2.2.

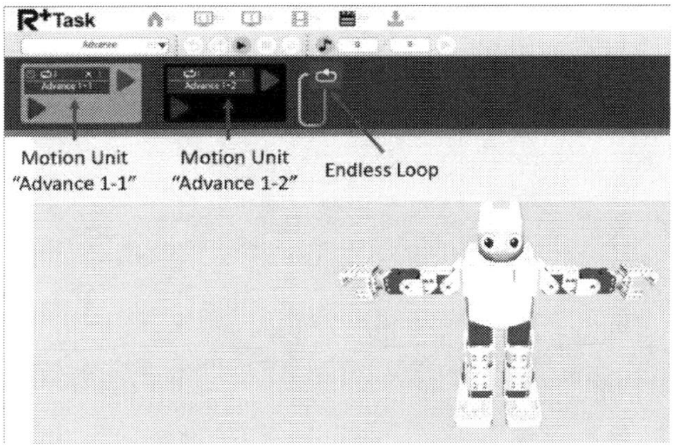

Fig. 4.15 MOTION-LIST Item named "Advance".

Fig. 4.16 describes the details of the graphical block representing MOTION-UNIT "Advance 1-1":

- If the user double-left-clicks the mouse in the red-bordered rectangular area labeled "Advance 1-1", the corresponding "Edit Linked Motion Unit" window pops up. This "Edit" window has two tabs: the one with the "Film Clip" icon shows the Motion Unit List (the reader may have to scroll down to see the item "Advance 1-1" already highlighted), and the other with the triangular "Play" icon shows the Play Parameters associated with the chosen Motion Unit item (i.e. "Advance 1-1").
- The Play Parameters for "Advance 1-1" were:
    - Playing Speed set at "1" (i.e. use the "as-is" timeline for "Advance 1-1").
    - Setting the number of Loops to "3" (i.e. execute "Advance 1-1" three times when the "PLAY/PAUSE" (F7) button is pushed by the user – see Fig. 4.14).
    - The "Insert Call Index" field comes from R+MOTION V.2 and this feature is not used in this book.

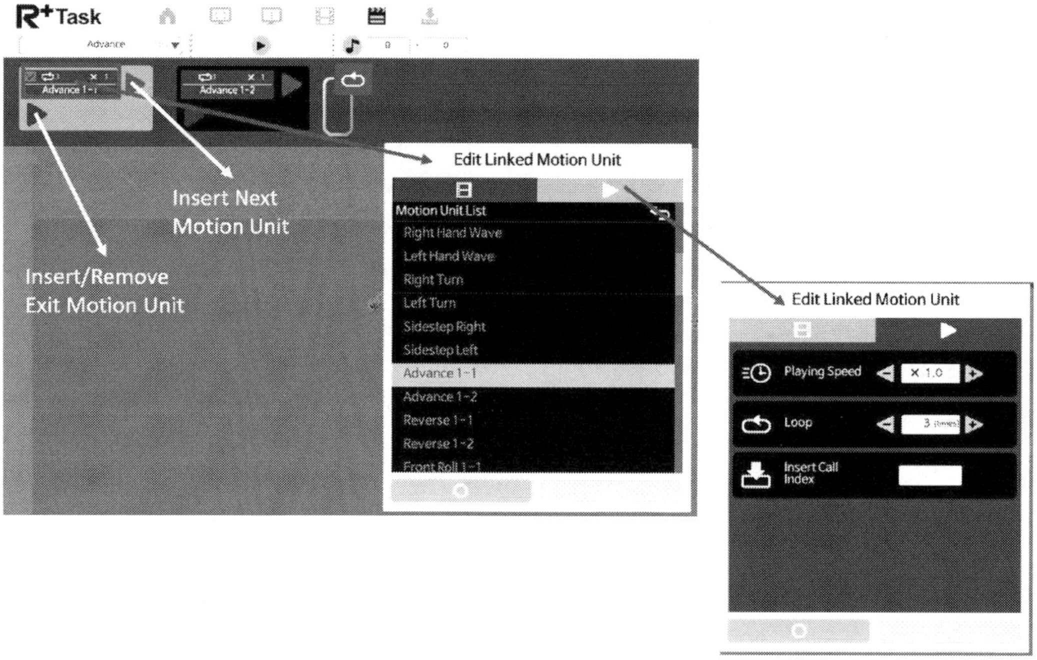

Fig. 4.16 "Edit" windows for MOTION-UNIT "Advance 1-1".

- There are also two "PLAY" icons (which are not used for "Advance 1-1"):
    - The one on the right of block "Advance 1-1" can be used to insert another MOTION-UNIT right after "Advance 1-1" with a mouse's double-left-click.
    - The one below and left of block "Advance 1-1" can be used to INSERT/REMOVE an "EXIT MOTION UNIT" (topic to be further elaborated in Section 4.2.2).

Fig. 4.17 describes details of the graphical block representing MOTION-UNIT "Advance 1-2":

- If the user double-left-clicks the mouse in the black rectangular area labeled "Advance 1-2", the corresponding "Edit Linked Motion Unit" window pops up. Same as before, this "Edit" window has two tabs: the one with the "Film Clip" icon shows the Motion Unit List and the other with the triangular "Play" icon shows the Play Parameters associated with MOTION UNIT "Advance 1-2".
- The Play Parameters for "Advance 1-2" were:
    - Playing Speed set at "1" (i.e. use the "as-is" timeline for "Advance 1-2").
    - Number of Loops set to "1" (i.e. play "Advance 1-2" once when user pushed the "PLAY/PAUSE" (F7) button – see Fig. 4.14).
    - The "Insert Call Index" field is also not used for "Advance 1-2".

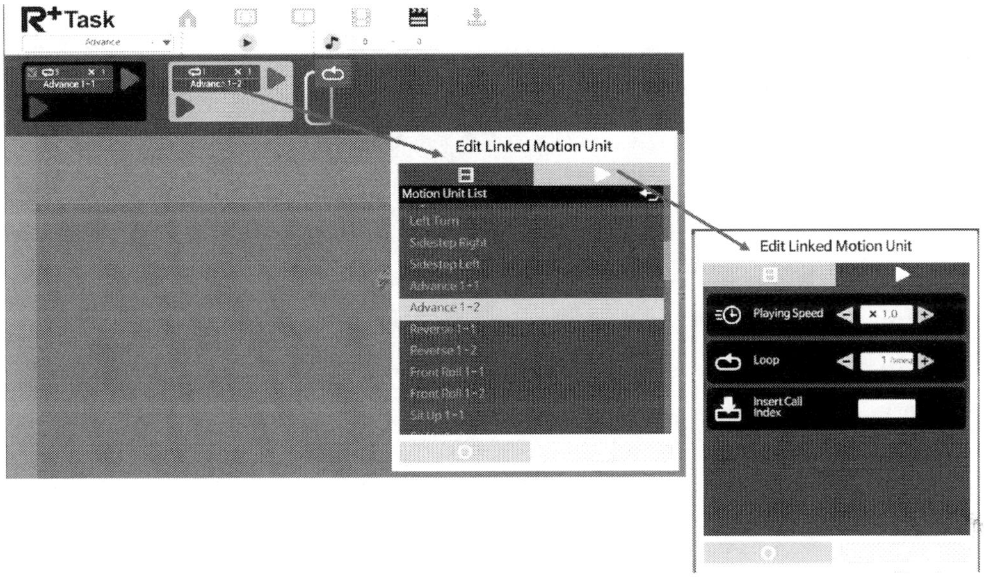

Fig. 4.17 "Edit" windows for MOTION-UNIT "Advance 1-2".

Going back to Fig. 4.15, most readers likely have already figured out that when the F7 button is pushed, "Advance 1-1" will execute 3 times (i.e. the MINI will make 6 full steps forward starting with the left leg), then "Advance 1-2" will execute once (i.e. the MINI will make a half-step forward with the left leg so that both feet would be aligned at the end of the "Advance" MOTION-LIST/PAGE Item).

4.2.2 Advanced Use of Motion (List) Tool

First, the reader should go back to the MOTION-UNIT tool (F4) and check out the MOTION-UNITs "Fast Advance 1" and "Fast Advance 2" where Motion Rests were not used, and where at first glance they seemed to be similar poses with "Advance 1-1" and "Advance 1-2" using similar time frames but their activations would use different kind of concepts such as "Double Speed" and "Exit Motion" which can only be used inside the MOTION-LIST tool.

Next, let's get back to MOTION-LIST tool by pushing F5 and choose the MOTION-LIST Item "Fast Advance". Fig. 4.18 shows that the MOTION-UNIT "Fast Advance 1" would be played at Double Speed and only once, but "Fast Advance 1" was enclosed within the Endless Loop object, meaning that once "PLAY" (F7) is pushed, the MINI will keep on walking forward at double speed **forever**. At this point, if the user pushes F7 to pause this process, the MINI would stop at whatever "INTERPOLATED" KEY-FRAME it happened to be at the time when the user pushed F7 (see Fig. 4.19), and the Emergency Stop (ESC) button would also stay highlighted. Next, if the user also pushed the Emergency Stop (ESC) button, the current "PLAY" process would stop with the MINI holding this latest pose. Here, we still are in the graphical domain so the graphical MINI can hold any pose, but later in Section 4.6 when we use the "SYNCH" mode while playing this MOTION-LIST Item, the little exercise won't be allowed by MOTION 3 at all as it can put the real MINI in an unstable pose that would make the real MINI fall and damage itself!

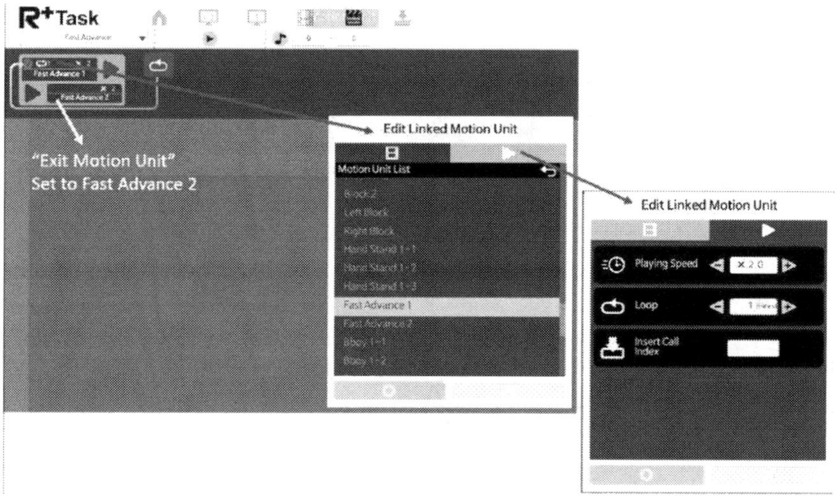

Fig. 4.18 "Edit" windows for MOTION-LIST/PAGE Advance 1-2".

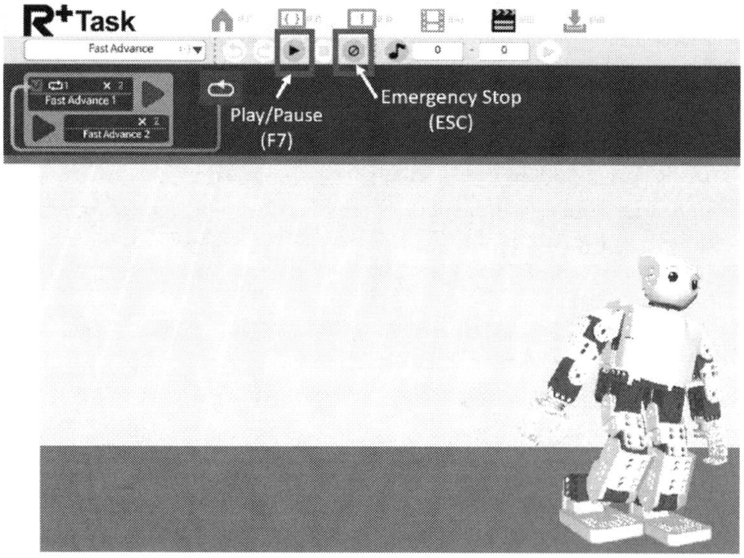

Fig. 4.19 Pausing the MOTION-LIST/PAGE Item "Fast Advance" at a random time.

To avoid the previous situation, ROBOTIS provided a more systematic way to stop the MINI (graphical and real robots) while they are "in motion" with the use of the STOP (F8) button and the EMERGENCY STOP (ESC) button. Fig. 4.20 shows that:

- Pushing on the STOP (F8) button would stop the MINI (graphical and real robots) in the **pose defined by the last KEY-FRAME** of MOTION-UNIT "**Fast Advance 2**" (i.e. the **EXIT MOTION-UNIT** as defined in Fig. 4.18). Later in Chapter 6, the user can achieve this effect using the special MOTION-LIST/PAGE "**-1**" inside a TASK program.
- Pushing on the EMERGENCY STOP (ESC) button would stop the MINI (graphical and real robots) in the **pose defined by the last KEY-FRAME** of the **current** MOTION-UNIT "**Fast Advance 1**". Later in Chapter 6, the user can achieve the same result using the special MOTION-LIST/PAGE "**0**" inside a TASK program.

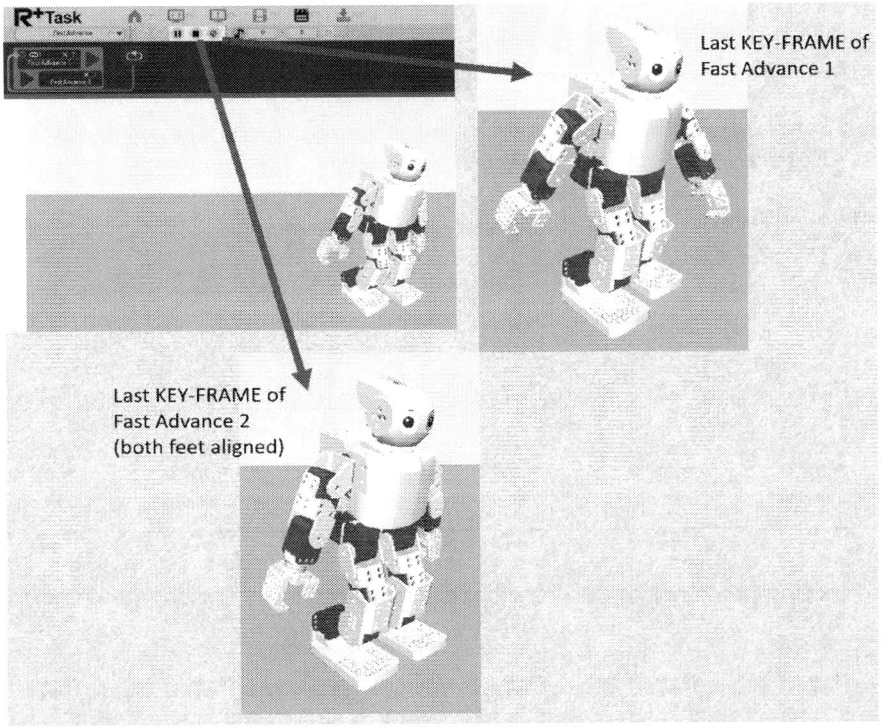

Fig. 4.20 Using STOP (F8) and EMERGENCY STOP (ESC) while MINI is "in motion".

- In the situation where NO EXIT MOTION-UNIT is defined then STOP (F8) has the same effect as EMERGENCY STOP (ESC).

Later in Chapter 6, the author will show how to achieve these effects when programming the MINI with the TASK Sub-tool.

Although MOTION 3 does not have a Physics Engine, it can do a decent job illustrating such motions as "Push Up" or "Horse Dance". However, it cannot properly illustrate motions that are highly dynamic such as "Front Roll" or "Bboy2" where the MINI would do full body roll forward or backward which can only be performed by the real MINI. These highly dynamic motion examples also show the exemplary skills of the ROBOTIS folks who had created these MOTION-UNITs and MOTION-LIST/PAGEs.

The last feature to be discussed in this Section 4.2.2 is the "Endless Loop" stretchable icon (see Fig. 4.21):

- If the reader left-clicks with the mouse on the arrowhead at the top left corner of the MOTION-UNIT "Fast Advance 1", a "right-pointing" white triangular icon would appear. Next without releasing the left-click, the reader can drag the mouse to the right so that the "Endless Loop" icon is not encompassing any MOTION-LIST items at all. Then, from this point forward, the MOTION-LIST Item "Fast Advance 1/Fast Advance 2" will perform only once when F7 is pushed.

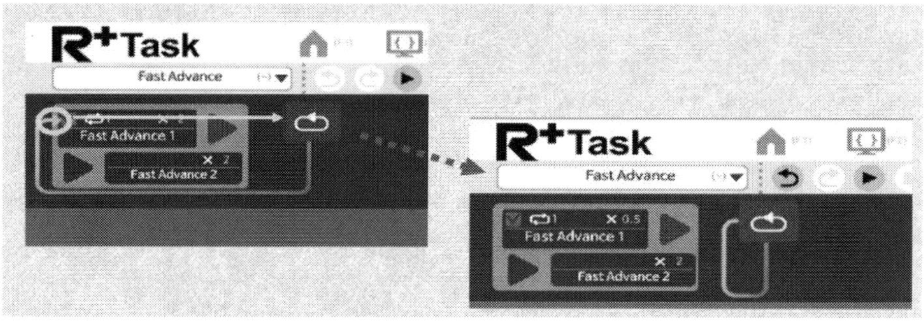

Fig. 4.21 Using the Endless Loop stretchable icon.

4.3 Motion Lists Organization with Motion Group

The next level of organization is with the MOTION-GROUP tool which can be accessed by pushing F6. The MOTION-GROUP tool can be used to gather user-chosen MOTION-LIST/PAGE Items into different MOTION-GROUPs, although only one MOTION-GROUP can be downloaded to the MINI at any one time (see Fig. 4.22).

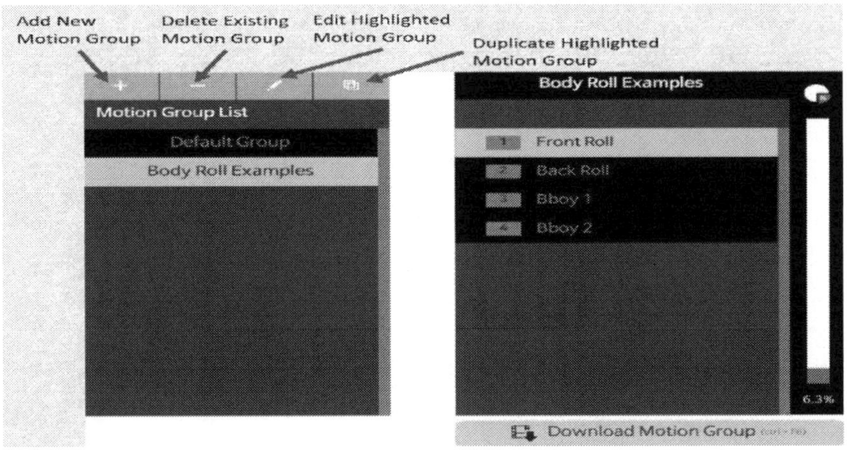

Fig. 4.22 Using MOTION-GROUP Tool to organize MOTION-LIST Items.

When editing a given MOTION-GROUP, the user can use the right and left pointing arrowheads to transfer chosen MOTION-LIST/PAGE Items back and forth between the Motion Candidate List panel on the left and the current Motion Group panel on the right (e.g. "Body Roll Examples") (see Fig. 4.23). The user can also re-edit/re-order the Motion Index Number of any MOTION-GROUP item and the current memory usage is also shown (see Fig. 4.23).

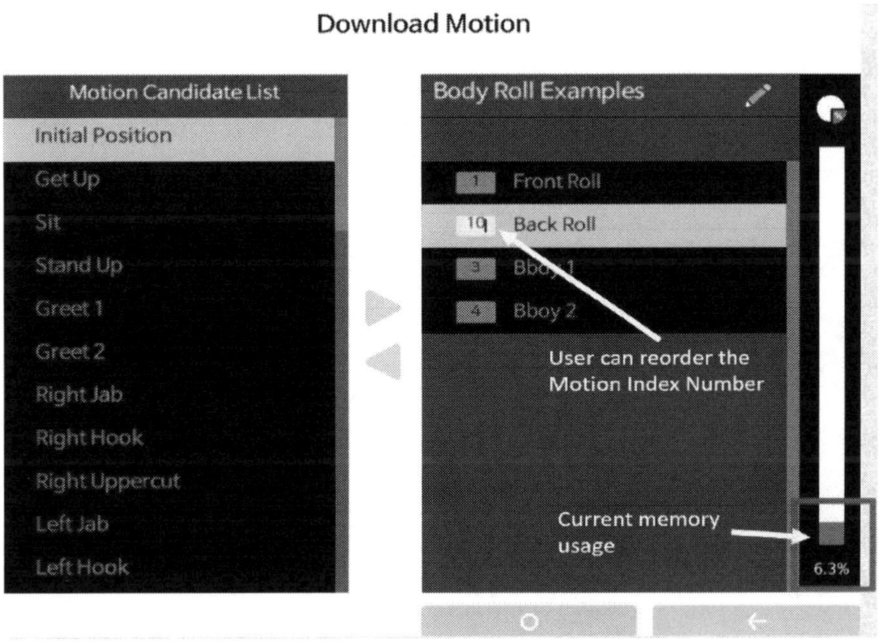

Fig. 4.23 Editing current MOTION-GROUP with existing MOTION-LIST/PAGE Items and reordering their Motion Index Numbers.

Only available with R+TASK V.3 is a new feature when using its TASK Sub-tool: the TASK program can now access these MOTION-GROUPs details using the MOTION-LIST item from the standard editing pop-up window (see Fig. 4.24).

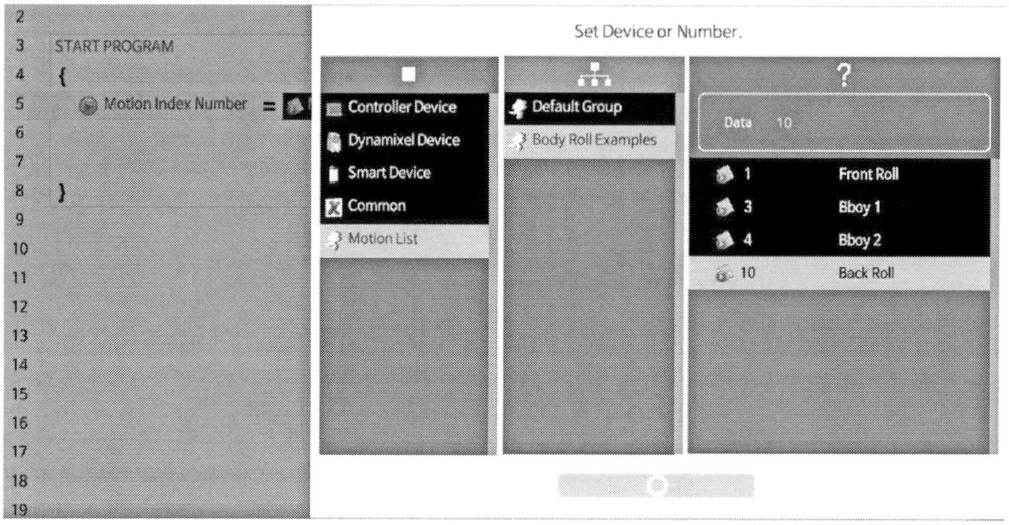

Fig. 4.24 Accessing MOTION-GROUP details from TASK Sub-tool.

4.4 Servo Offset Calibration

In this section, we will need to use the real MINI robot in synchronized mode with the MOTION Tool which can be achieved with the following procedure described in Fig. 4.25:

- Push F10 to connect MOTION Sub-tool to the MINI with the proper Bluetooth COM Port.
- Next, push F9 to synchronize the 3D graphical robot to the real robot.
- Next, PLAY the MOTION-UNIT named "Calibration" provided by the author in his file "MINI_Custom.mtn3" (it is located towards the bottom of the Motion Units list).
- Then, if the MINI robot had been well constructed, the real robot would match its pose with the 3D graphical robot as shown in Fig. 4.25. This means that for this given MINI robot, the existing OFFSET POSE is properly set, but where can we find this so-called OFFSET POSE?

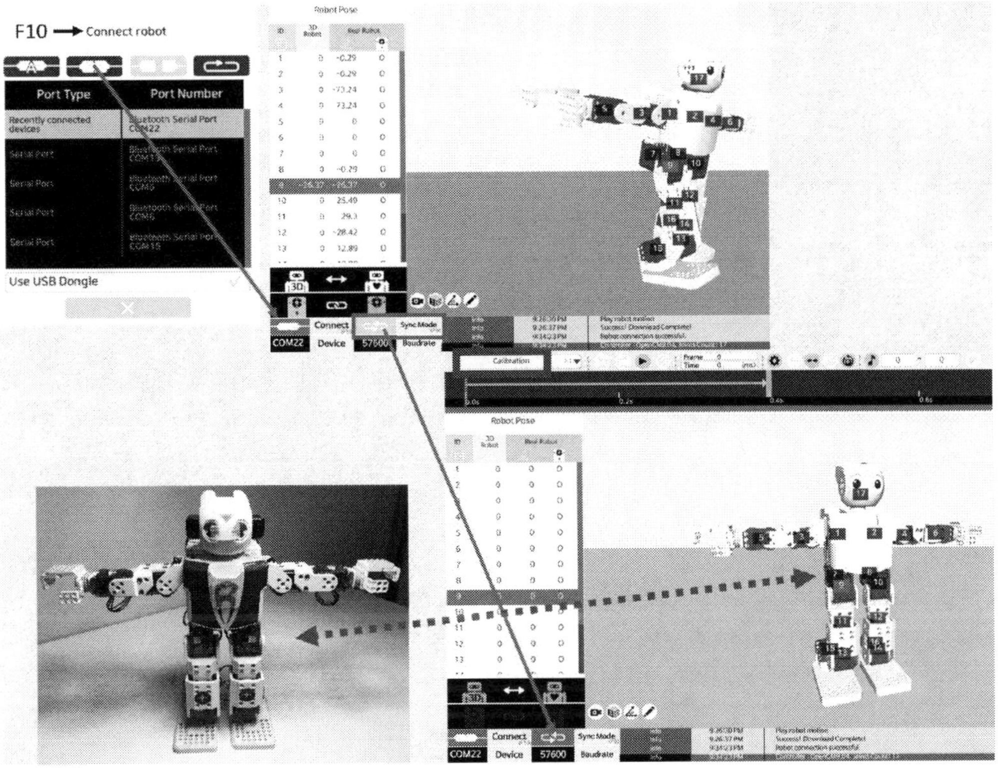

Fig. 4.25 Getting MOTION Tool and MINI robot ready for Calibration Pose.

Thus, we next check the OFFSET POSE for the MINI which can be accessed by pushing F1 to get to the HOME menu and then by choosing "Edit Offset" in the bottom right area of the HOME menu (see Fig. 4.26).

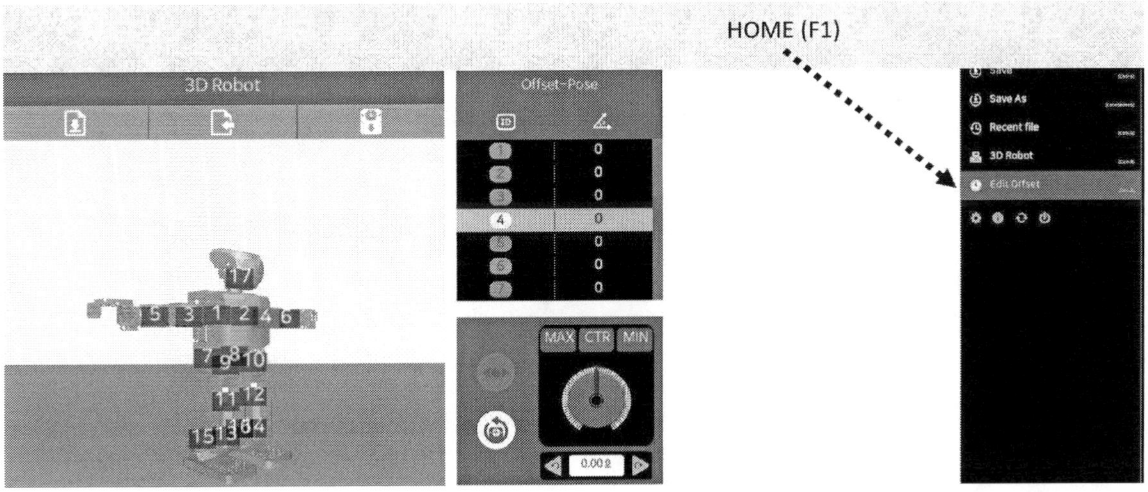

Fig. 4.26 Checking on the OFFSET POSE of the MINI via the HOME Menu (F1).

In the panel named "Offset-Pose" in the middle of Fig. 4.26, the reader can see that all servos have default zero values in this OFFSET POSE. If the previous step of "PLAYING" the MOTION-UNIT "Calibration" did not result in matching poses between the real and graphical robots, then the impacted servos will have to be adjusted one by one for their OFFSET POSE values until there is a pose-match between the real and graphical robots. These adjusted values can be positive or negative depending on the situation.

For example, the reader may recall from Section 2.3 where the author mentioned that he would like to attach a smart phone in front of his MINI robot so that he would be able to use its camera via SMART DEVICE programming with the TASK Sub-tool. Well, when the author attached the lightest phone available to his MINI and then played the "Calibration" MOTION-UNIT, the real robot was leaning forward some as expected (see Fig. 4.27), and from previous experiences, the author knew that this configuration would lead to an easy fall when trying to make this MINI walk.

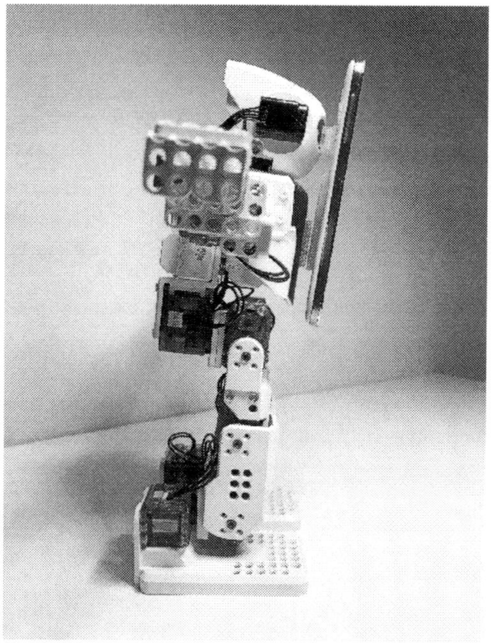

Fig. 4.27 MINI with attached smart phone was leaning forward in its "Calibration" Pose.

Thus, the author had to adjust the OFFSET POSE for this current configuration, and it turned out that (for the author's robot) only 3 servos had to be adjusted: ID = 12, 13 and 14 (see Fig. 4.28). After downloading these new offset values to the MINI, the MOTION-UNITs "Calibration" and "Initial Position" would yield the respective physical poses on the real robot as shown in Fig. 4.28.

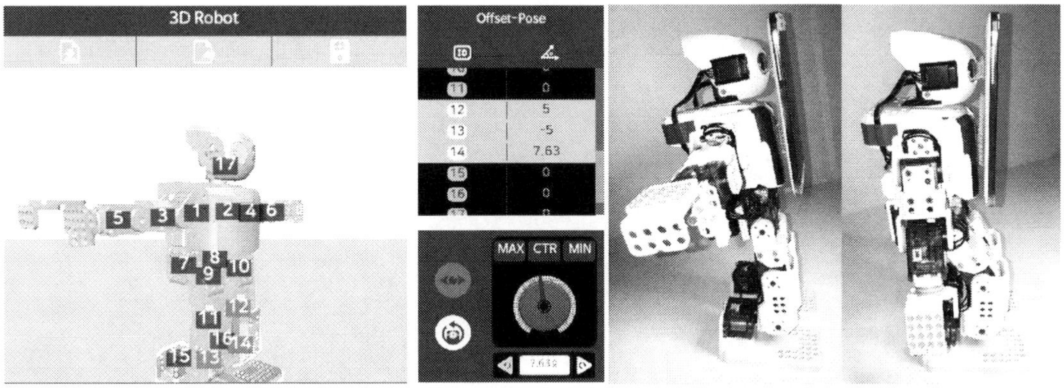

"Calibration" Pose "Initial" Pose

Fig. 4.28 MINI with revised offset values for Servos 12, 13 and 14 and resulting physical "Calibration" and "Initial" poses on the real MINI robot.

Of course, the real test was for the MINI to walk by executing the MOTION-LIST?PAGE Item named "Advance" previously analyzed in Section 4.2.1, and the disappointing result is shown in Video 4.3 where the MINI could make only 2 steps before falling forward.

The next idea was to stiffen up the leg servos to see if that could help the robot hold its body more upright when it walked. So the "Joint Parameters" feature was next used to set the "Flexibility Level" parameters for the leg servos (ID from 9 to 14) – for more information please visit this web link http://emanual.robotis.com/docs/en/software/rplus1/motion/#joint-softness (in MOTION V.1 the "Flexibility Level" was called "Joint Softness"). Fig. 4.29 shows how to access this feature and the author used values of 2 for the Flexibility Level of Servos 9 to 14 (their default values were "4").

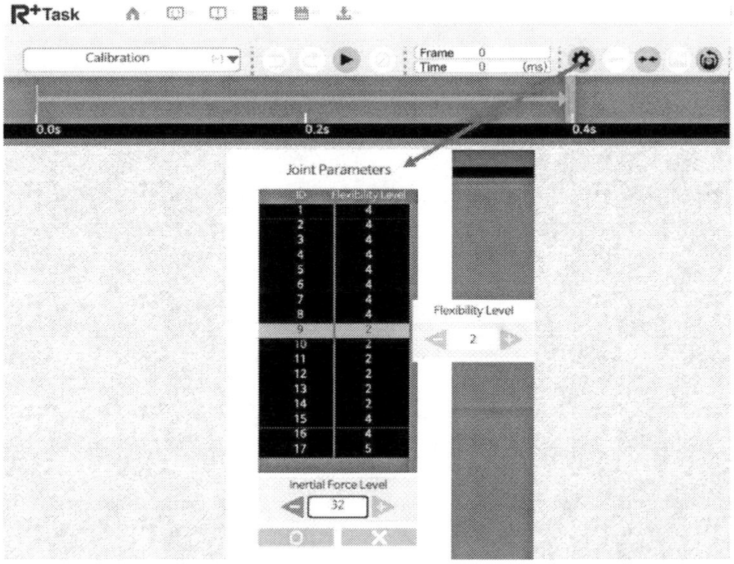

Fig. 4.29 Adjusting Flexibility Level parameters to give more "strength" to the MINI leg servos.

Video 4.4 shows that the MINI performed better, even when using the MOTION-LIST/PAGE Item named "Fast Advance", but still the MINI fell forward after a few seconds of walking. These results indicated that the MINI would not be able to handle the extra weight of a smart phone when walking, thus the author had to revise applications that use this SMART DEVICE feature in Chapter 7.

Later in Chapter 8, where a wired USB webcam is used in conjunction with the PYTHON/EDBOT tool chain, a different hardware attachment and appropriate OFFSET configuration will be illustrated.

4.5 Motion Creation Practicum

When the author first saw the Motion Unit "Greet 2" (see Fig. 4.30), it reminded him of a "Crouching Bow" move made by practitioners of one of the Escrima schools in the Phillippines whereas:

1. The arms usage is reversed, i.e. left arm in the back and right arm in front.
2. The right first is also raised to the forehead level.
3. The left leg is shifted to the back to yield a slight crouch of the body with the right knee bent.

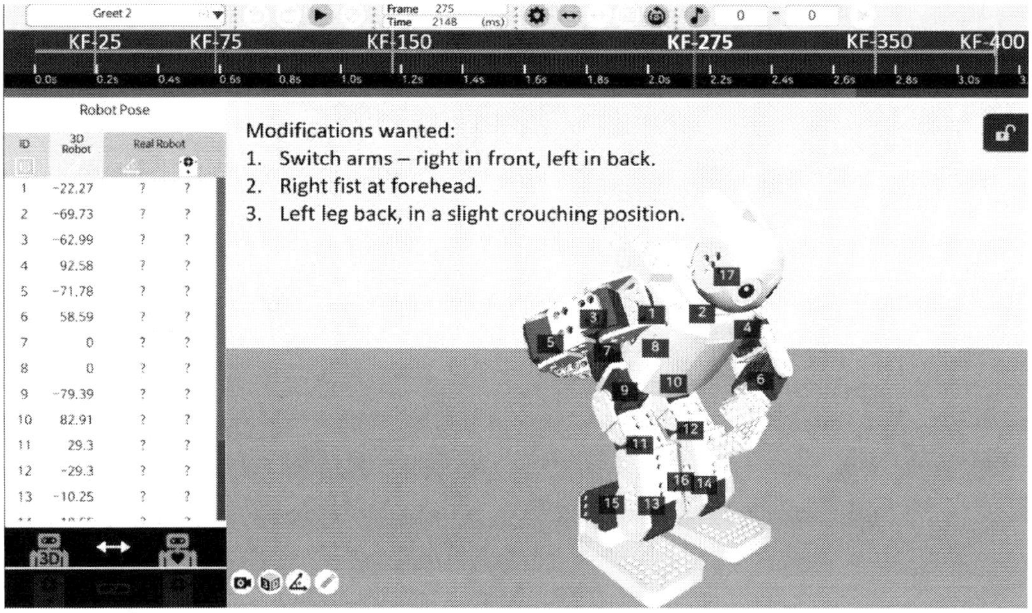

Fig. 4.30 Modifications to be made to Motion Unit "Greet2" to create a "Crouching Bow".

Let's first review how a human would perform this "Crouching Bow":

1. From a feet-together standing position, a human shifts his/her weight to the right leg and curl the toes of the left leg to reduce its contact area with the ground and to get it ready to be slid back to the rear.
2. As the left leg slides to the back, the right knee also bends to lower the whole body. At the same time the arms move appropriately (left arm to the back and right arm to the front).
3. The moves timing would have to be such when the body gets to the final crouching position, the right first also reaches the forehead level (i.e. key frame KF-150 in Fig. 4.30).

The author immediately realized that the MINI robot had "rigid" feet, i.e. "no toes curling" was possible, so he had to figure out a different way to reduce the contact area of the MINI left leg in order to slide it backwards. In theory, the rest of the maneuvers should be feasible via the MOTION Sub-tool from what we have learned so far. The reader may also review various editing techniques by consulting with the materials presented at this web link http://emanual.robotis.com/docs/en/software/rplus2/motion/#edit-motion-unit3d-robot before continuing with the rest of this Chapter.

The author divided this project in three parts:

1. Modify "Greet 2" into "Crouching Bow 1" whereas the bowing moves stayed the same except that the arms were switched so that the left arm went to the back while the right arm went to the front (with no changes to the feet servos).
2. Modify "Crouching Bow 1" into "Crouching Bow 2" whereas KEY-FRAMEs KF-150 and KF-275 would now make the MINI robot raise its right first to its forehead (also with no changes to the feet servos).
3. Modify "Crouching Bow 2" into "Crouching Bow 3" by adding the feet maneuvers to get the MINI robot into a slight crouching position by KF-150 and then to stand back up by KF-400. This would be the most complicated phase of this project as the real MINI robot could not be allowed to fall during these leg moves.

4.5.1 From "Greet 2" to "Crouching Bow 1"

First let's review "Greet 2" using Fig. 4.30 which shows that it has six KEY-FRAMEs defined at KF-25, KF-75, KF-150, KF-275, KF-350 and KF-400.

Next, let's duplicate the Motion Unit "Greet 2" and name this duplicate as "Crouching Bow 1" (see Fig. 4.31-A).

Then, let's edit "Crouching Bow 1" so that the KEY-FRAME at time = 0 (KF-0) would be the same as the ending KEY-FRAME KF-400 (which was a "ready" pose anyway) (see Fig. 4.31-B). The alert reader would remember from Section 4.1.1 which mentioned that in general MOTION-UNITs have no KEY-FRAME defined at TIME = 0 to facilitate the combination of MOTION-UNITs together to form a more complex choreography called a MOTION-LIST. But here the author had to violate this rule, as later in Section 4.5.2 he would need to add two additional KEY-FRAMEs between KF-0 and KF-25 using the "POSE-INTERPOLATING" feature of the MOTION tool (see Introduction section to this Chapter 4). Without defining KF-0, the author was not able to perform this task which had 3 steps (see Fig. 4.31-B):

1. Step 1 was to "left-click" with the mouse on KF-400, the graphical MINI would take the pose as shown in Fig. 4.31-B.
2. Step 2 was to move the mouse's cursor to "Time = 0" and "left-click" there. This would establish the linkage between the current graphical MINI pose (i.e. "Ready Pose") and KEY-FRAME KF-0.
3. Step 3 was to left-click on the "+" icon found on the right side of the main menu bar to **actually save** this new KEY-FRAME KF-0 into the Motion Unit "Crouching Bow 1" (the author himself had forgotten to do this step several times!).

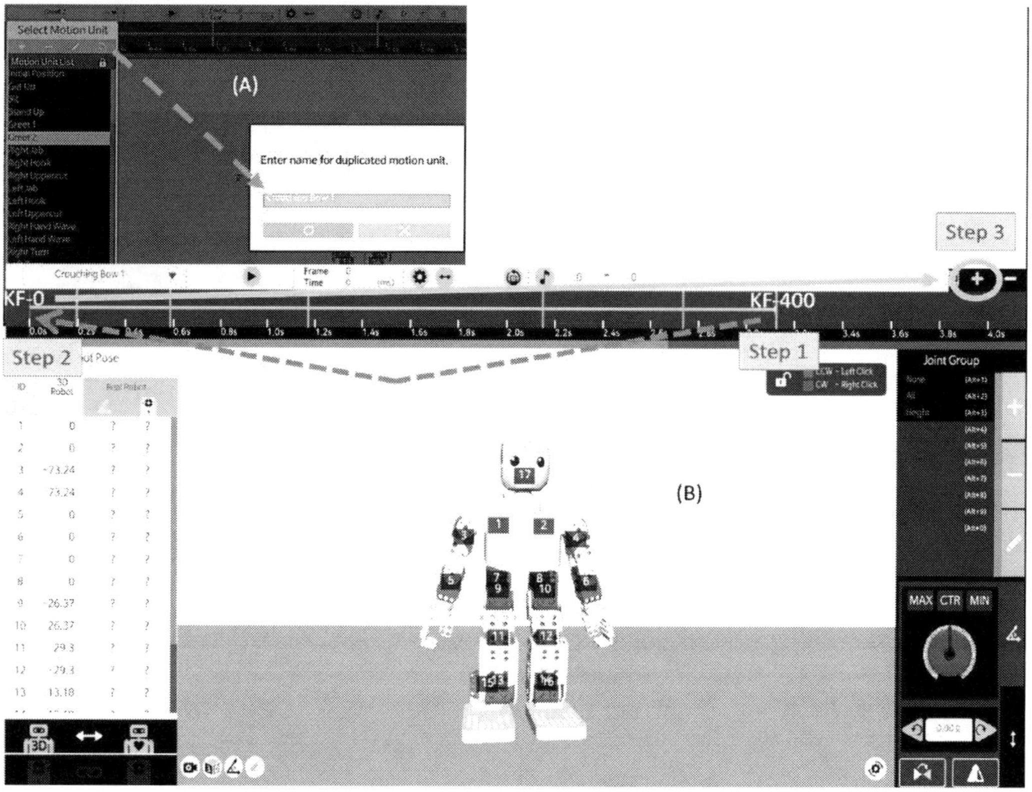

Fig. 4.31 A) Duplication of "Greet 2" into "Crouching Bow 1",

B) Copying of KF-400 into KF-0 (i.e. when Time = 0).

The last procedure for Motion Unit "Crouching Bow 1" was to switch the usage of the arms for KEY-FRAMEs KF-25 through KF-350. As an example, Fig. 4.32 showed the required 4 steps for KF-25 which were also repeated for the other KEY-FRAMEs KF-75, KF-150, KF-275 and KF-350:

1. Step 1 was to choose KF-25 on the timeline with the mouse's cursor.
2. Step 2 was to "Control-left-click" servos 1 through 6 (i.e. servos for both arms) in the "Robot Pose" panel.
3. Step 3 was to "left-click" on the MIRROR Tool located near the bottom right area of the MOTION tool, and ones can immediately see the switching of the arms (a real cool MOTION feature). In actuality, for KF-25 (and KF-350), only the servos 1 and 2 got "mirrored" (see Fig. 4.32 again). However, for KF-75, KF-150 and KF-275, all six servos got "mirrored".
4. Step 4 was to "left-click" on the "floppy" icon found in the top right area of the MOTION tool to save this "mirrored" pose into Motion Unit "Crouching Bow 1" permanently.

Fig. 4.32 Typical steps needed to mirror the arms poses for KF-25 in "Crouching Bow 1".

The author recommends that the reader repeat these steps for KF-75, KF-150, KF-275 and KF-350 for the sake of practice (see Fig. 4.33).

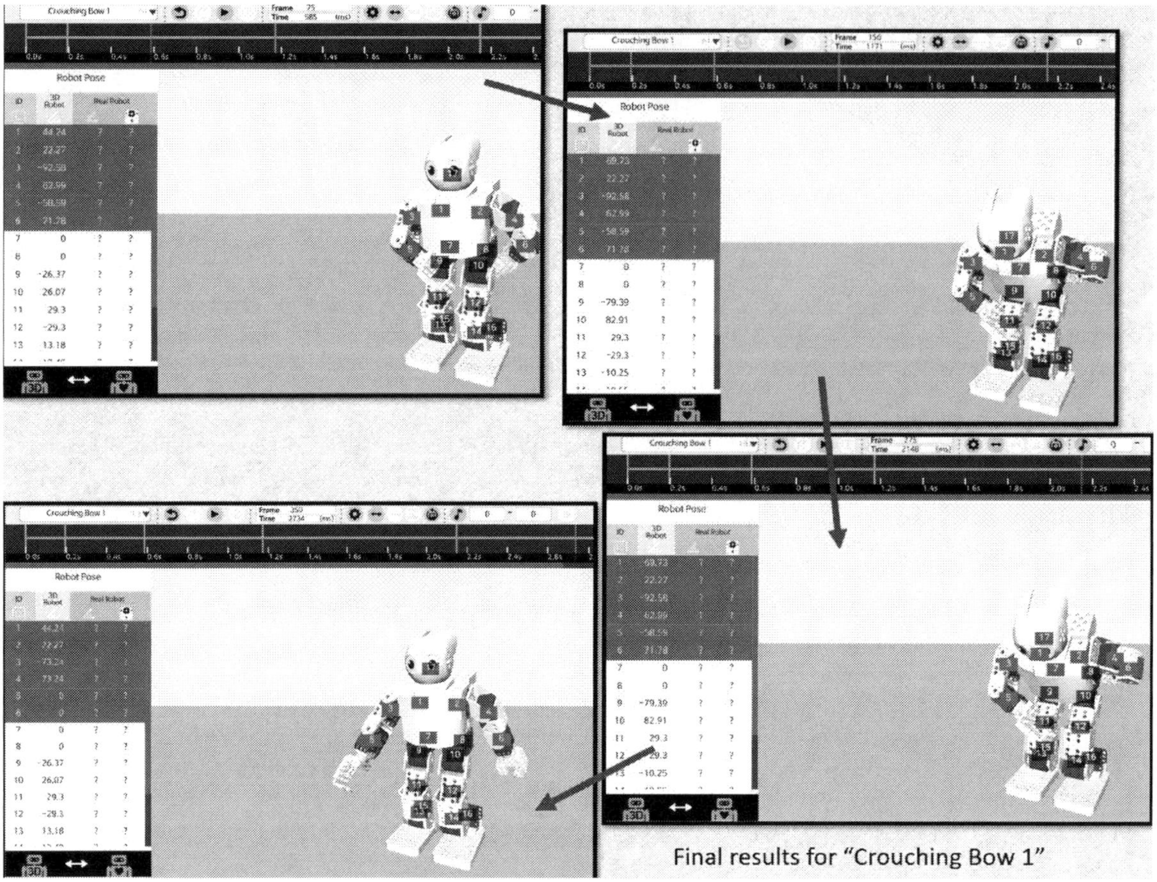

Fig. 4.33 Final results for KF-75, KF 150, KF-275 and KF-350 for Motion Unit "Crouching Bow 1".

The results of all previous procedures were saved as Motion Unit "Crouching Bow 1" in the author-provided "MINI_Custom.mtn3" file and it was located towards the bottom of the Motion Units list. Please see Video 4.5 for a demo of "Crouching Bow 1" on the real MINI robot.

4.5.2 From "Crouching Bow 1" to "Crouching Bow 2"

For this section, let's duplicate the previously obtained Motion Unit "Crouching Bow 1" and name this duplicate as "Crouching Bow 2".

"Crouching Bow 2" differed from "Crouching Bow 1" in terms of KF-150 and KF-275 only (see Fig. 4.34), and only in the values of Servos 1, 3 and 5 in order to bring the right fist to the forehead level. For the practicing reader, please remember to save the modified poses that you made in "Crouching Bow 2".

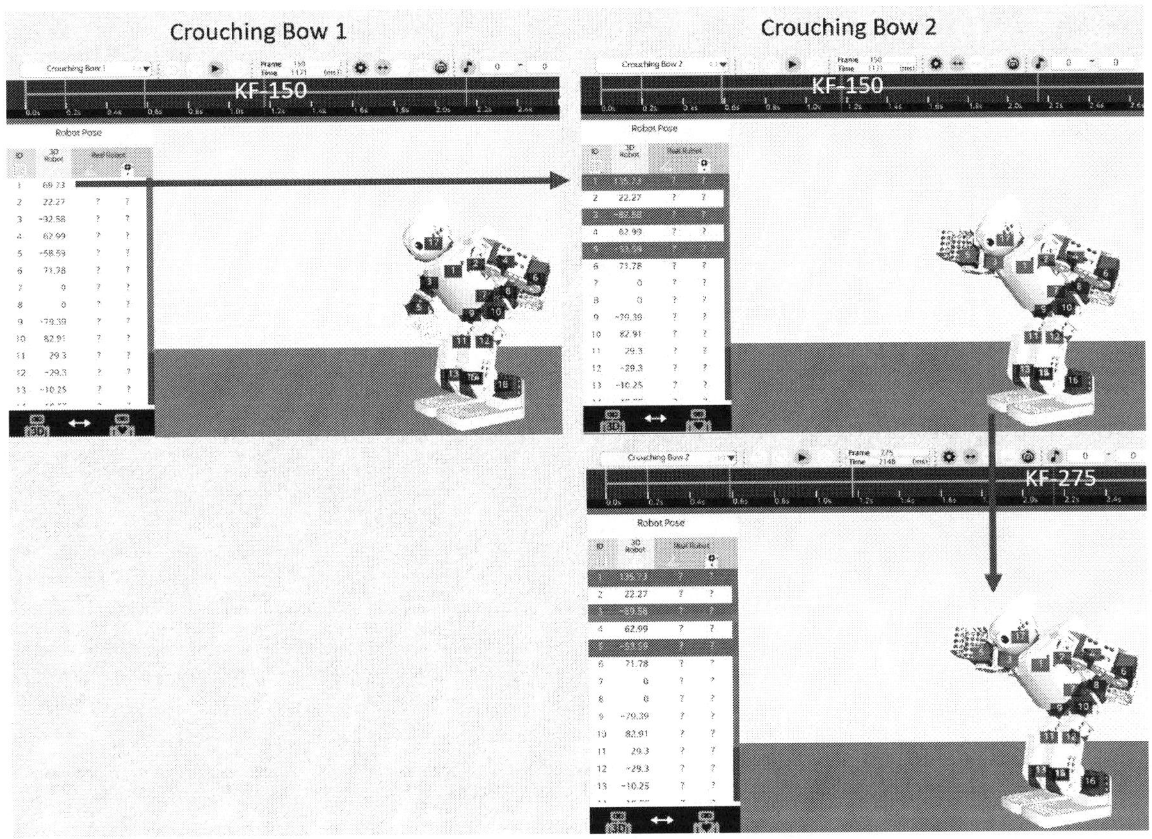

Fig. 4.34 Modifications for Servos 1, 3 and 5 for KF-150 and KF-275 in Motion Unit "Crouching Bow 2".

The results of the previous procedure were saved as Motion Unit "Crouching Bow 2" in the author-provided "MINI_Custom.mtn3" file and it was located towards the bottom of the Motion Units list. Please see Video 4.6 for a demo of "Crouching Bow 2" on the real MINI robot.

Before going to Section 4.5.3, the author needs to point out that all work shown in Sections 4.5.1 and 4.5.2 were done "off-line" on the "graphical" MINI only, before verifying the obtained Motion Units "Crouching Bow 1" and "Crouching Bow 2" on the "real" MINI robot because the legs servos were not involved (see Videos 4.5 and 4.6). This approach would be applicable only to motions that involved the arms and head servos only. In the coming Section 4.5.3 where the legs servos were to be manipulated, the "real" MINI robot had to be used in "**Sync Mode**".

4.5.3 From "Crouching Bow 2" to "Crouching Bow 3"

For this section, let's duplicate the previously obtained Motion Unit "Crouching Bow 2" and name this duplicate as "Crouching Bow 3".

Fig. 4.35 illustrated the main goal which was to modify the KF-150 found in "Crouching Bow 2" on the left side into a new KF-150 for "Crouching Bow 3" on the right side. To maintain stability at all times for the "real" MINI robot, the author had found the need to create "intermediary" key-frames KF-11, KF-18, KF-100 and KF-128, and of course the "real" MINI robot was set in "Sync Mode" with the MOTION tool during these modifications to the Motion Unit "Crouching Bow 3". If the reader wants to retrace these steps along with the presented materials, please make sure that spare and charged batteries are available, as the procedures are lengthy!

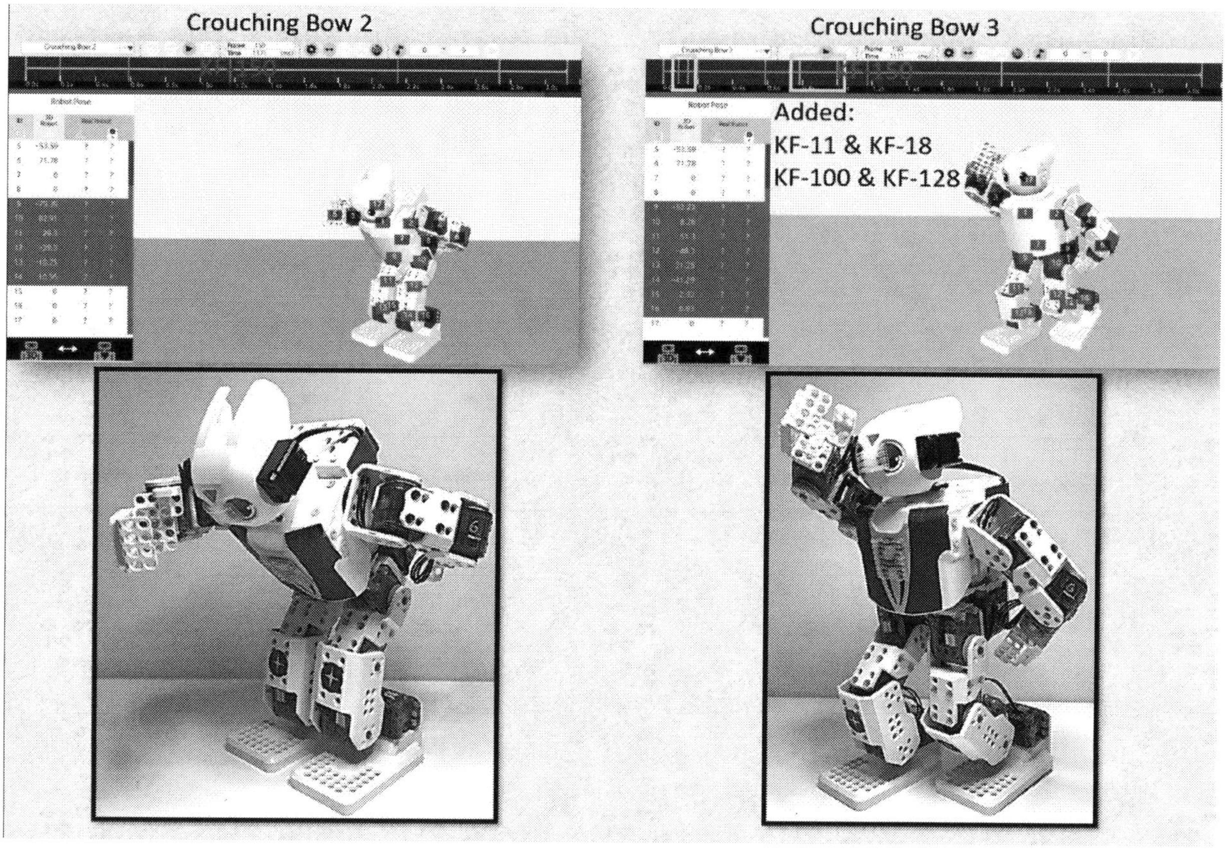

Fig. 4.35 Main goal in modifying "Crouching Bow 2" into "Crouching Bow 3".

A. KF-11:
   KF-11 was inserted at Time = 85 ms, whereas the arms poses were already in "correct" positions by the "automatic interpolating of poses" feature inside the MOTION tool (see previous materials described in Sections 4.5.1 and 4.5.2, but the legs were still aligned as before in Motion Unit "Greet 2" (e.g. "start" KF-11, see left side of Fig. 4.36). The goals for the "final" KF-11 (see right side of Fig. 4.36) were:
   - to shift most of the robot weight to the right leg (Servos 9, 11, 13 and 15). Here the author used the solution described in Motion Unit "Advance 1-1" (see Fig. 4.5).
   - to start shifting the left leg backwards using Servos 10, 12 and 14.
   - to reduce the ground contact area for the "rigid" left foot by rotating it inward using Servo 16 so that the left leg was supported by the outside edge of the left foot.

   A photo of the real robot showed that this "final" KF-11 pose was stable (bottom right of Fig. 4.36).

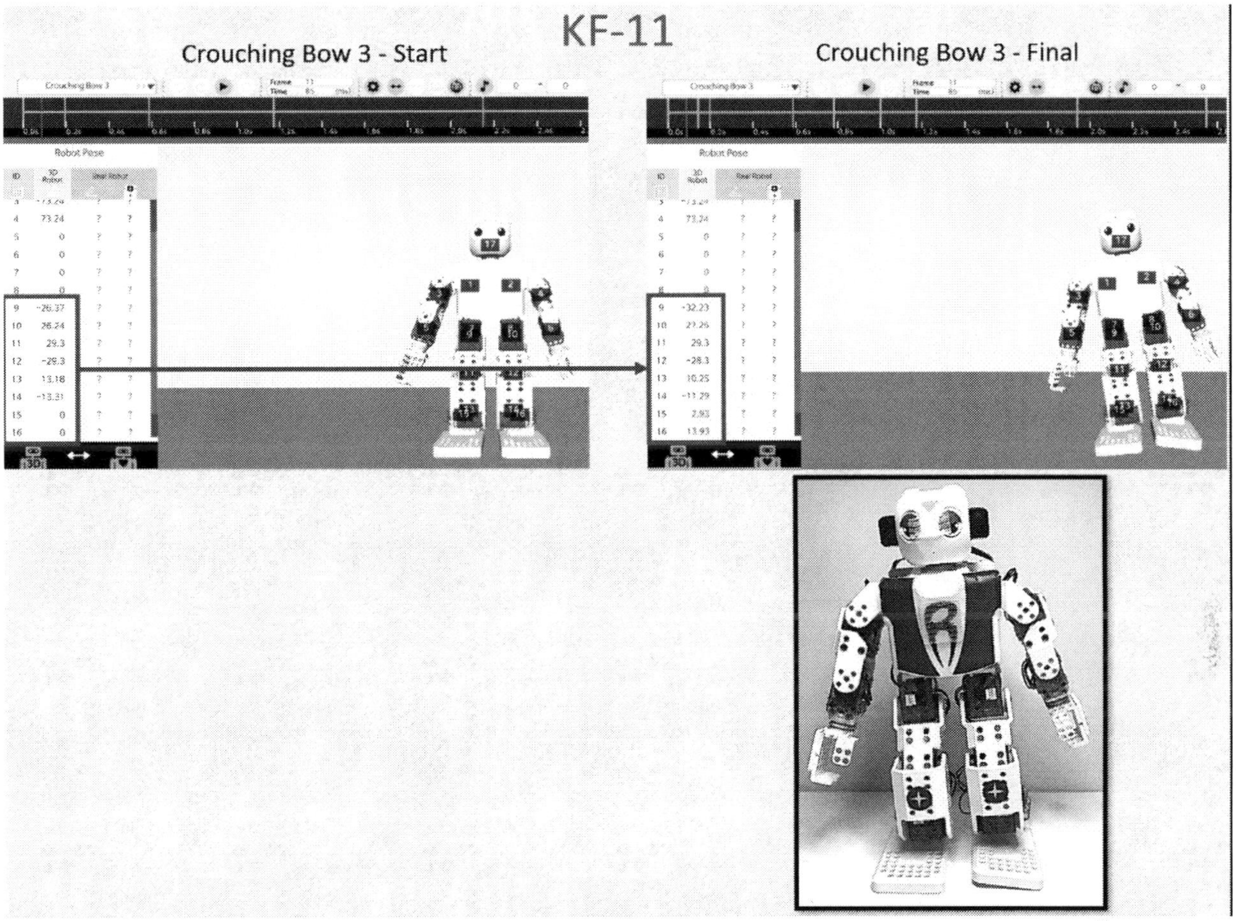

Fig. 4.36 Details for Legs Servos in KF-11 of Motion Unit "Crouching Bow 3".

B. KF-18:
KF-18 was inserted at Time = 140 ms and, as discussed before, the arms poses were already set appropriately but the legs were still "incorrectly" aligned (see left side of Fig. 4.37). For lack of space, the author did not show in Fig. 4.37 an intermediary step that the author had to do which was to copy the servo position values of Servos 9 through 16 from KF-11 (Fig. 4.36) to the corresponding ones on the "start" KF-18 (left side of Fig. 4.37). Essentially, the author wanted to start modifying the legs servos from their "last-known" stable solution KF-11 into the "proper" pose needed for the "final" KF-18. The goals for the "final" KF-18 (see right side of Fig. 4.37) were:

- to bend the right knee to get it into a crouching position.
- to slide the left leg as far back as possible without loss of robot stability.

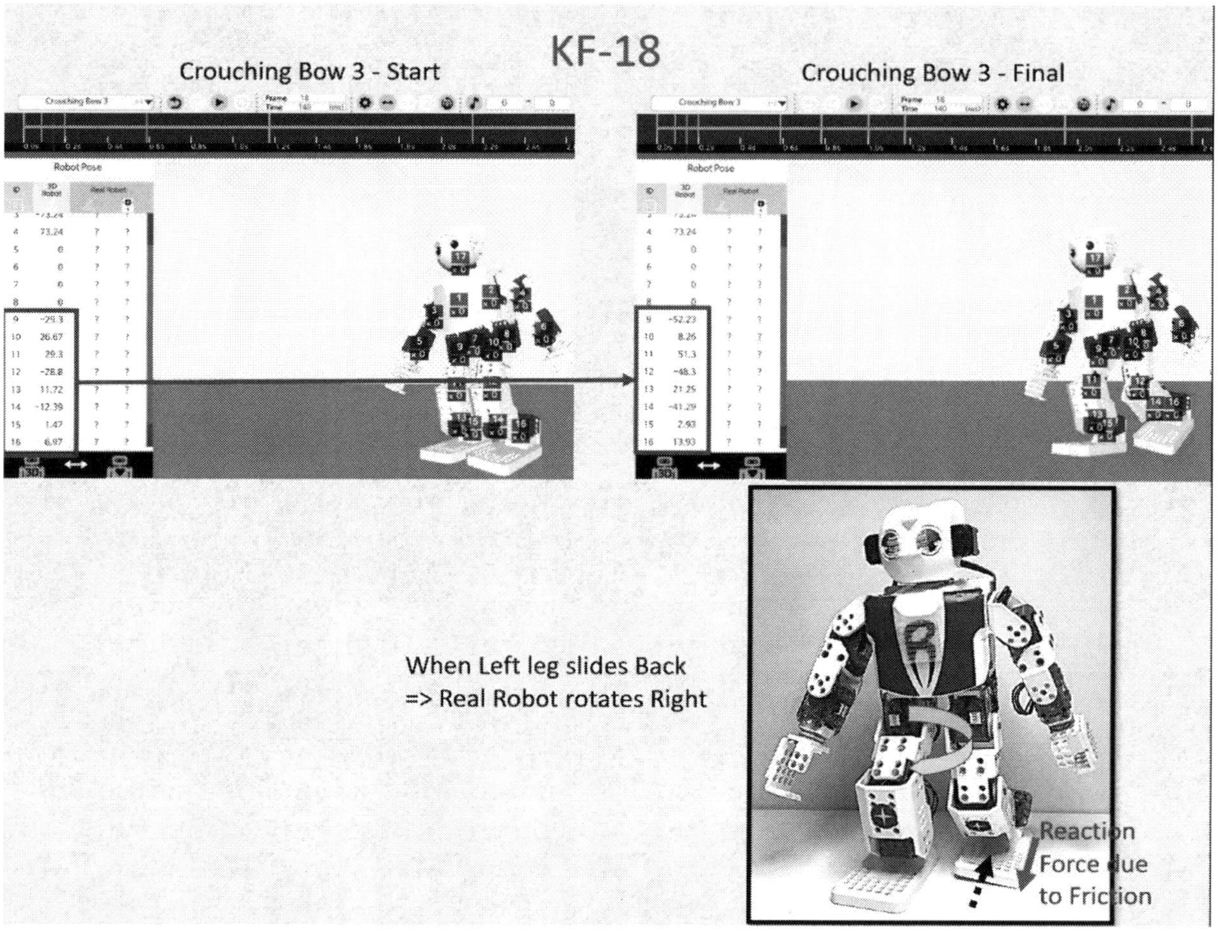

Fig. 4.37 Details for Legs Servos in KF-18 of Motion Unit "Crouching Bow 3".

To bend the MINI's right knee, the author used the "Joint Group" and "Differential Positioning" features provided in the MOTION tool (see Fig. 4.38):

- First, the author took the real MINI robot off the "Sync Mode" (i.e. used only the graphical 3D robot) before performing the following procedures which would put the real robot into some temporary unstable situations which are described further below.
- The author used the Joint Group named "Height – Right Only" (middle panel in Fig. 4.38) which was modified from the Joint Group named "Height" already provided by ROBOTIS (right panel in Fig. 4.38). In general, the Joint Group feature is used when the user wants to move a certain group of servos together within a user-defined constraint which is implemented by using **a set of multipliers** with a specific **non-zero** value for each servo of a given Joint Group. For the "Height – Right Only" Joint Group, the author used a multiplier value of "1" for Servo 9, then "-2" for Servo 11, and "-1" for Servo 13 (see middle panel of Fig. 4.38). This means that if Servo 9 position gets changed by 1degree CW, Servo 11 will change by 2 degrees CCW and Servo 13 will change by 1degree CCW. To implement these servo changes, the user needs to put the MOTION tool into a **Differential Positioning** editing mode by pushing F12 (which is set automatically when the user chooses any Joint Group, except "None").

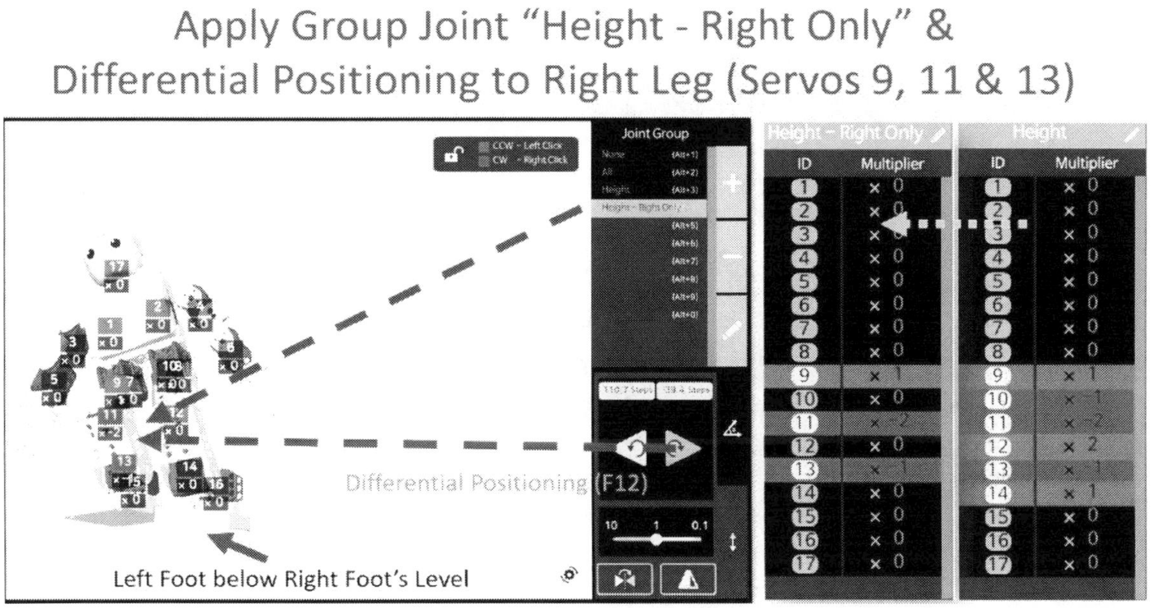

Fig. 4.38 Using "Joint Group" and "Differential Positioning" to bend the MINI's Right Knee.

- Next, the author left-clicked on the right-pointing triangle shown in Fig. 4.38 several times until the graphical MINI showed that the right knee has bent down sufficiently. However, the reader can see from Fig. 4.38 that these actions would also put the left foot below the right foot's horizontal plane. If the real MINI robot was in "Sync Mode", it would have tipped over to the right, as in the physical world the ground surface would not be yielding to this "downward" motion of the left foot. Thus, this "unstable" situation had to be fixed on the graphical 3D MINI before the author could save this new pose for the robot. The author's solution is described in the next paragraph.

To slide the left leg back, the author manually adjusted the Servos 10, 12 and 14 until the left foot was positioned "mostly" at the same horizontal level as the right foot (see Fig. 4.39):

- Here, the author relied extensively on the adjustable 3D view angle on the graphical robot using the mouse's right button to monitor that the left foot reached the same horizontal level as the right foot. The author did adjust in order Servos 10, 12 then 14, but this order may not be necessary. In practice, the reader may also find a different solution for his/her robot, in any case the reader needs to do the checking procedure outlined below.
- Next, the MINI was put back in "Sync Mode" to verify that the real robot was still stable in this new pose. The author did need a couple of trials to get to the final KF-18 shown in Fig. 4.37, before saving this pose into Motion Unit "Crouching Bow 3" (**very important step!**). The author was keen on keeping the real robot resting its left leg only on the outside edge of its left foot to minimize its contact surface area (i.e. the same as for KF-11 – see Fig. 4.36).

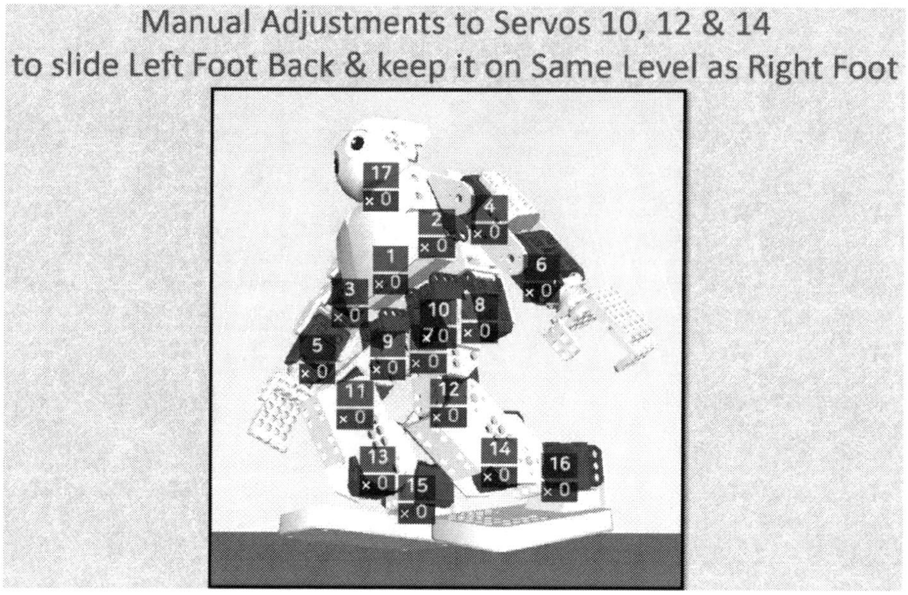

Fig. 4.39 Manual adjustments used for Servos 10, 12 and 14 for posing the MINI's left leg.

- As a final check and with the real MINI in "Sync Mode", the author brought the MINI back to KF-11 on the Time-Line of the MOTION Tool, and then went back to KF-18 to make sure that the real MINI could make the transition back and forth between KF-11 and KF-18 in a stable fashion. **This was to ensure that the entire Motion Unit "Crouching Bow 3" would be a stable solution for the real robot when it was completely defined with later steps.**

Fig. 4.37 also has a photo of the real robot illustrating that this KF-18 solution was workable, but with a "twist": the real robot rotated slightly to the right when ones compare Fig. 4.37 to Fig. 4.36. Of course, the "graphical" 3D robot showed no such behavior when the author was editing the legs servos' positions!

The reader may have used the ROBOTIS provided Motion Unit named "Right Turn" whereas the graphical robot's "Right Turn" is enacted by first putting the left leg in front of the right leg, and then pulling the left leg back to align it with the right leg. On the real robot, the actual "right turn" happens during this "pulling back" of the left leg. With the Motion Unit "Crouching Bow 3", KF-11 and KF-18 were doing the same job of moving the left leg backwards but from a different start position for the legs, that was all! The "Physics" explanation for this phenomenon was about friction between the robot's foot and the walk surface: as the real robot dragged its left leg back, the reaction forces due to friction acted in the other direction on the robot's contact surface (i.e. the outer edge of its left foot), and Fig. 4.37 showed that this reaction force would create a right moment with respect to its right foot, thus making it turn right as seen. Interested readers may want to watch these videos illustrating the same turning technique applied to the BIOLOID PREMIUM robots called BiPed (https://www.youtube.com/watch?v=W-GEsjRaDYs) and GerWalk (https://www.youtube.com/watch?v=7990W425nQU). Interested readers may also want to check out a different turning technique used for the Walking Droid (https://www.youtube.com/watch?v=BW4GpdSRVrU). The difference in techniques used was simply due to the different configurations of servos used in constructing the different robot legs.

C. KF-25:
KF-25 was inserted at Time = 195 ms, whereas the arms poses were already in "correct" positions by automatic pose interpolations. The author needed only to copy the legs servos positions from KF-18 to KF-25 to achieve the motion goal for KF-25 (see Fig. 4.40).

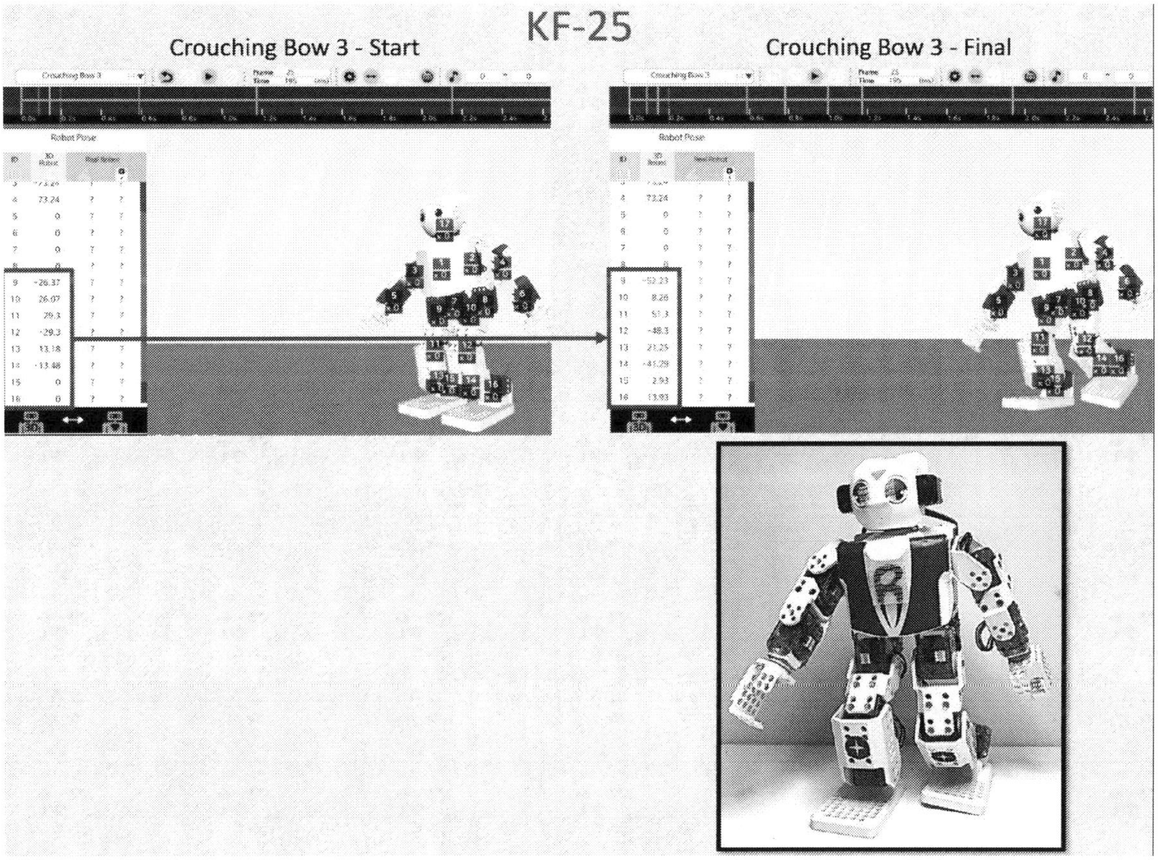

Fig. 4.40 Details for Legs Servos in KF-25 of Motion Unit "Crouching Bow 3".

The author chose not to slide the left leg further back as compared to KF-18, because the real robot started to show some signs of unstable behaviors when the author checked the playing of this Motion Unit from KF-0 to KF-25. Interested readers may want to try to slide the left leg further back just for your own practice.

D. KF-75:
KF-75 was inserted at Time = 585 ms, whereas, once again, the arms poses were already in "correct" positions by automatic pose interpolations. Similarly, the author only needed to copy the legs servos positions from KF-25 to KF-75 to achieve the motion goal for KF-75 (see Fig.

4.41). Interested readers may also want to try to slide the left leg further back just for your own practice.

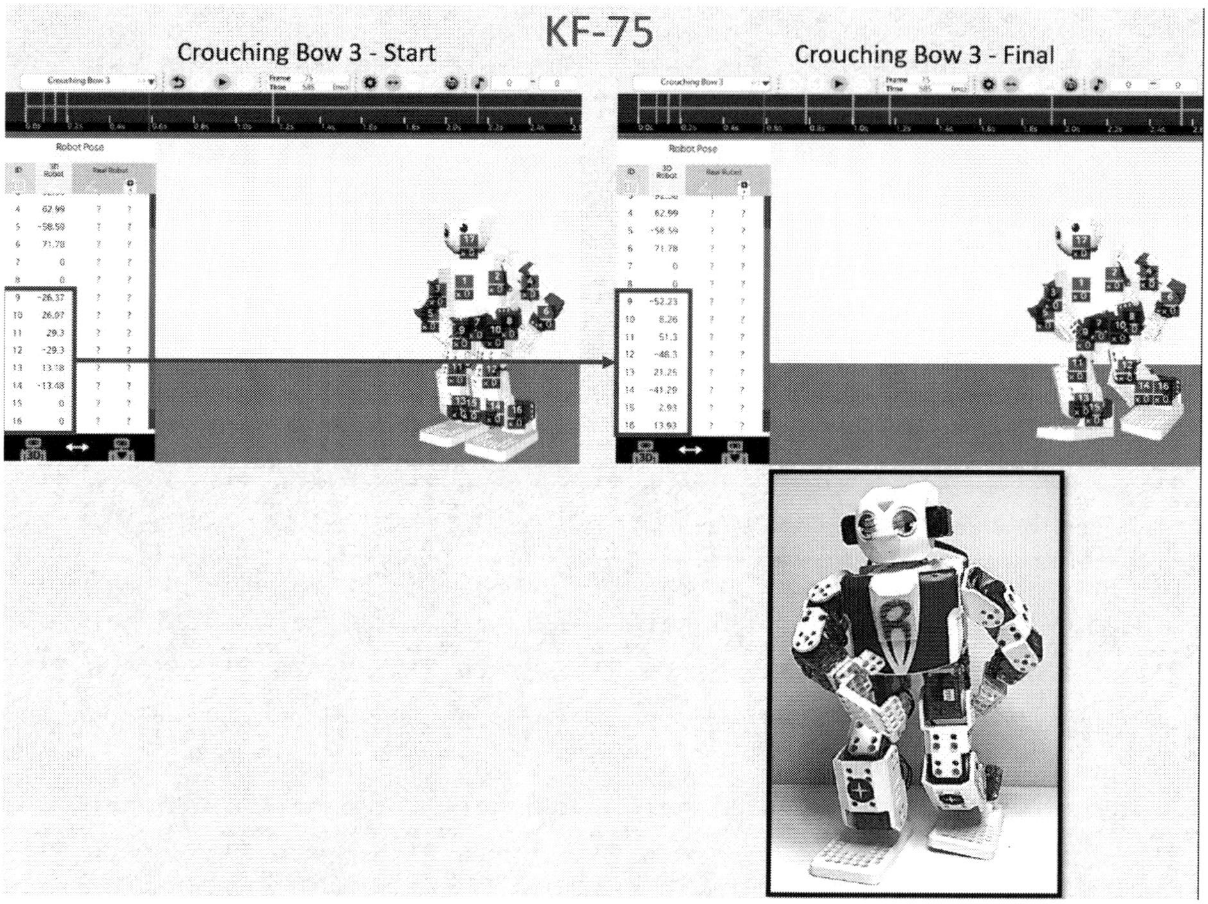

Fig. 4.41 Details for Legs Servos in KF-75 of Motion Unit "Crouching Bow 3".

E. KF-100:
KF-100 was inserted at Time = 1171ms, whereas, once again, the arms poses were already in "correct" positions by automatic pose interpolations. Similarly, the author only needed to copy the legs servos positions from KF-75 to KF-100 to achieve the motion goal for KF-100 (see Fig. 4.42). Interested readers may also want to try to slide the left leg further back just for your own practice.

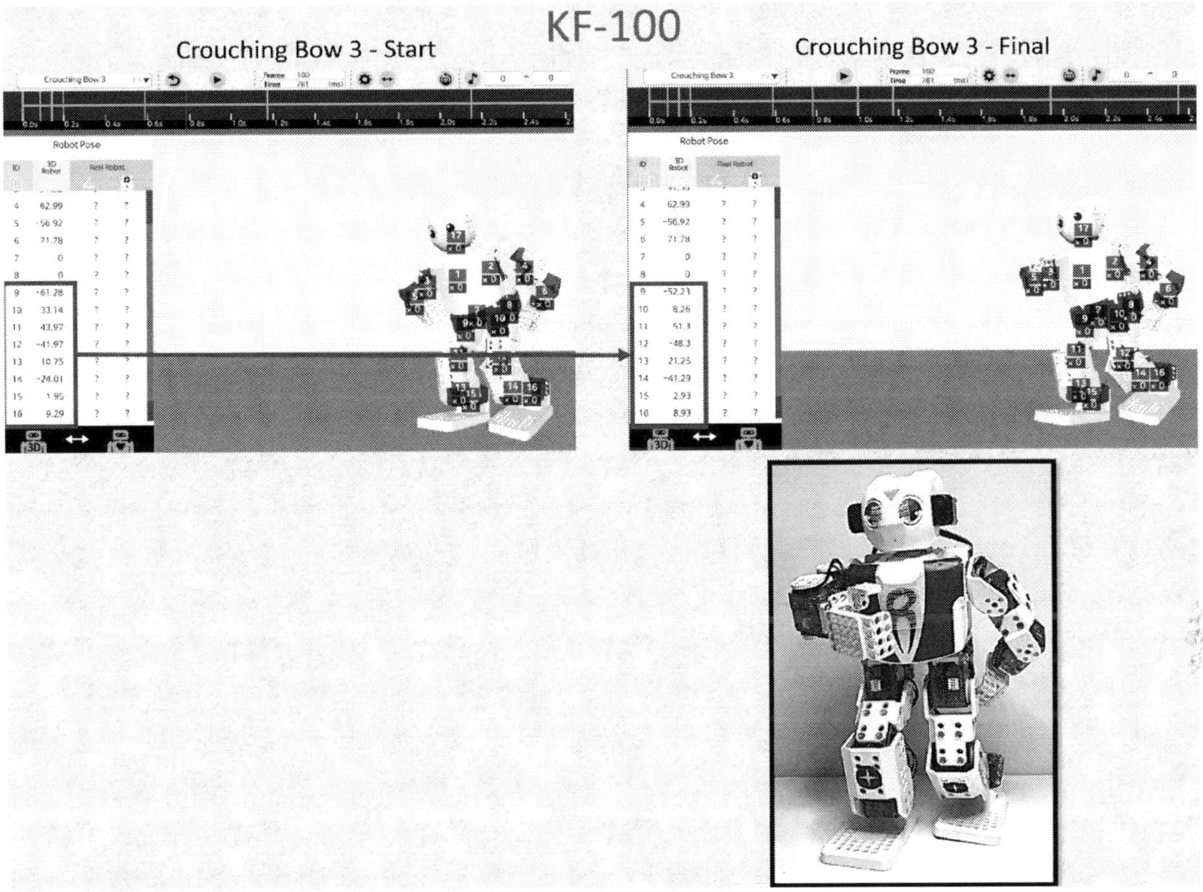

Fig. 4.42 Details for Legs Servos in KF-100 of Motion Unit "Crouching Bow 3".

F. KF-128:
The same procedure was used for KF-128 as for KF-75 and KF-100 to achieve the pose illustrated in Fig. 4.43.

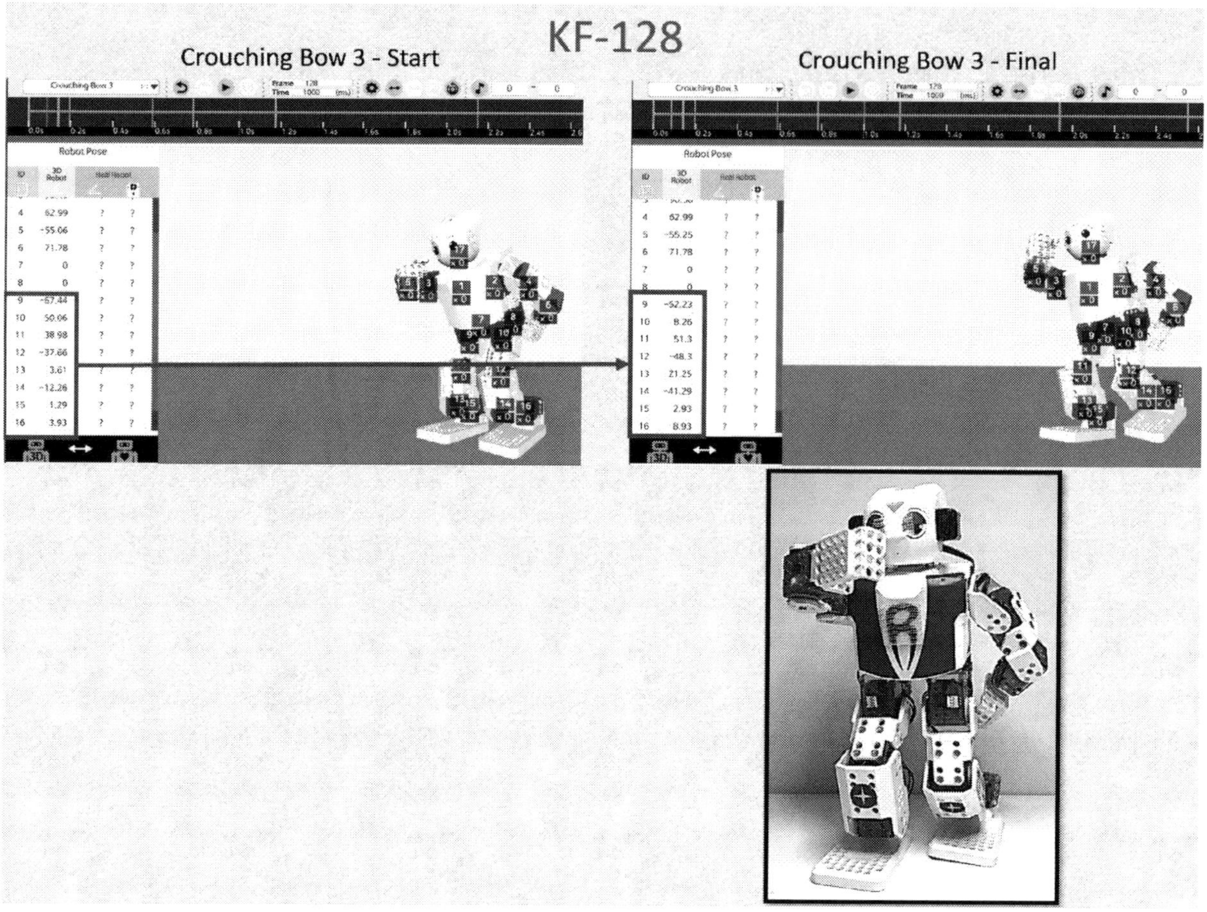

Fig. 4.43 Details for Legs Servos in KF-128 of Motion Unit "Crouching Bow 3".

G. KF-150:
Again, almost the same procedure was used for KF-150 as for KF-75, KF-100 and KF-128 to achieve the pose illustrated in Fig. 4.44 for KF-150. But now with the right arm extended in front of the robot, the real robot started to exhibit some occasional "forward falling" results when the playing of Motion Unit "Crouching Bow 3" reached this pose during the author checking procedure (i.e. during "real dynamic" motion). Thus, the author had to modify Servo 16 to make the left foot "flatter" with the ground surface, so it went from 8.93 degrees at KF-128 to 0.93 degree at KF-150.

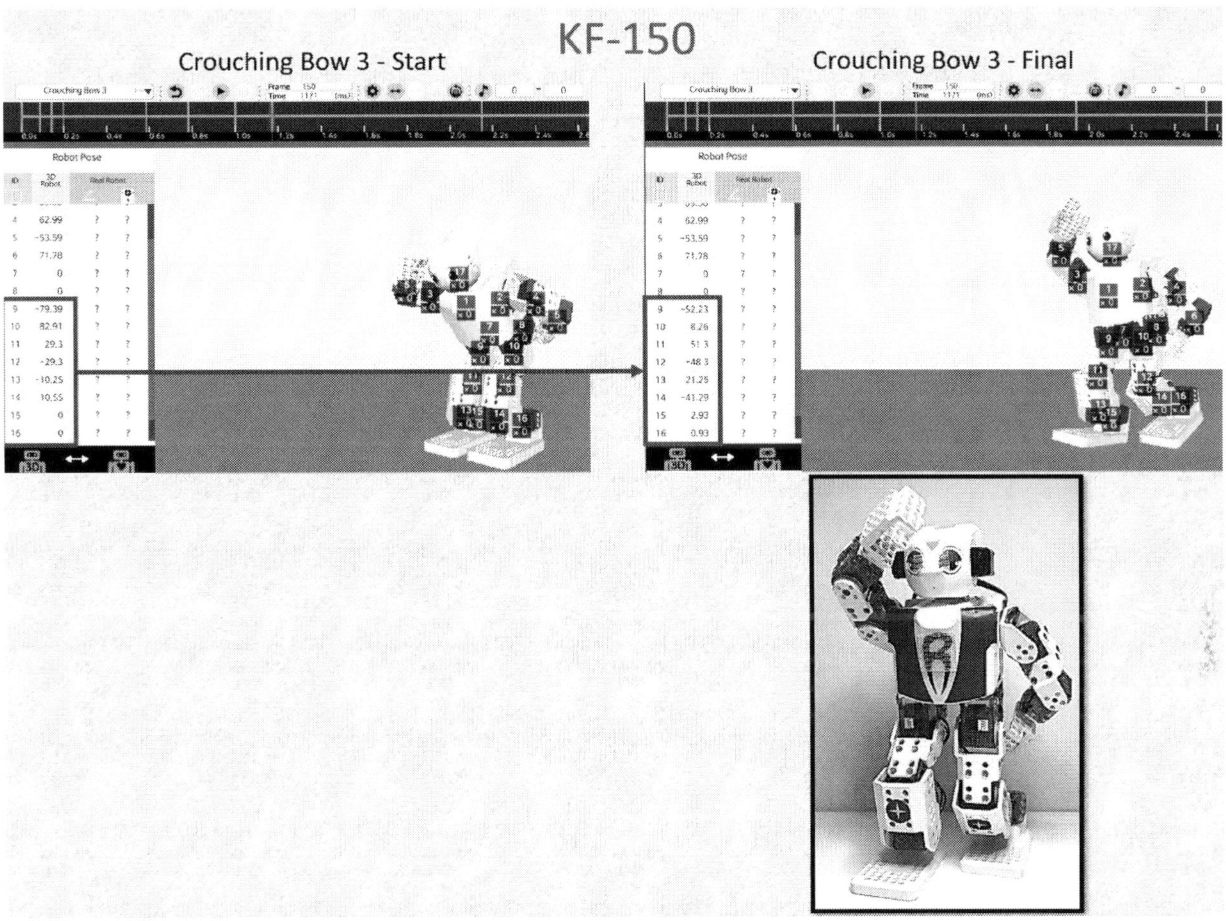

Fig. 4.44 Details for Legs Servos in KF-150 of Motion Unit "Crouching Bow 3".

H. KF-275 >> KF-248:
In "Greet 2" and "Crouching Bow 2", KF-150 and KF-275 were exactly the same robot pose, and so at first the author copied KF-150's servo values into KF-275, and when the author "statically" checked on KF-150 and KF-275 by clicking on their respective positions on the timeline, there were no problem at all. However, if this current "Crouching Bow 3" were "played" from the beginning KF-0 to KF-275, i.e. under "real dynamic" motion, occasionally the real robot was swaying slightly forward and fell forward. Thus, the author had to modify Servos 9 and 10 slightly: Servo 9 changed from -52.23 to -56.23 degrees and Servo 10 changed from 8.26 to 21.26 degrees (see Fig. 4.45). The author's goal was to straighten up the robot's torso and to shift the robot's Center of Gravity slightly backwards, but this movement also moved the left leg slightly forward. Here, friction forces again exerted their effects by making the real robot starting to rotate left. Furthermore, the author decided to shorten this "crouching bow" pose from KF-275 to KF-248 (i.e. by 211 ms).

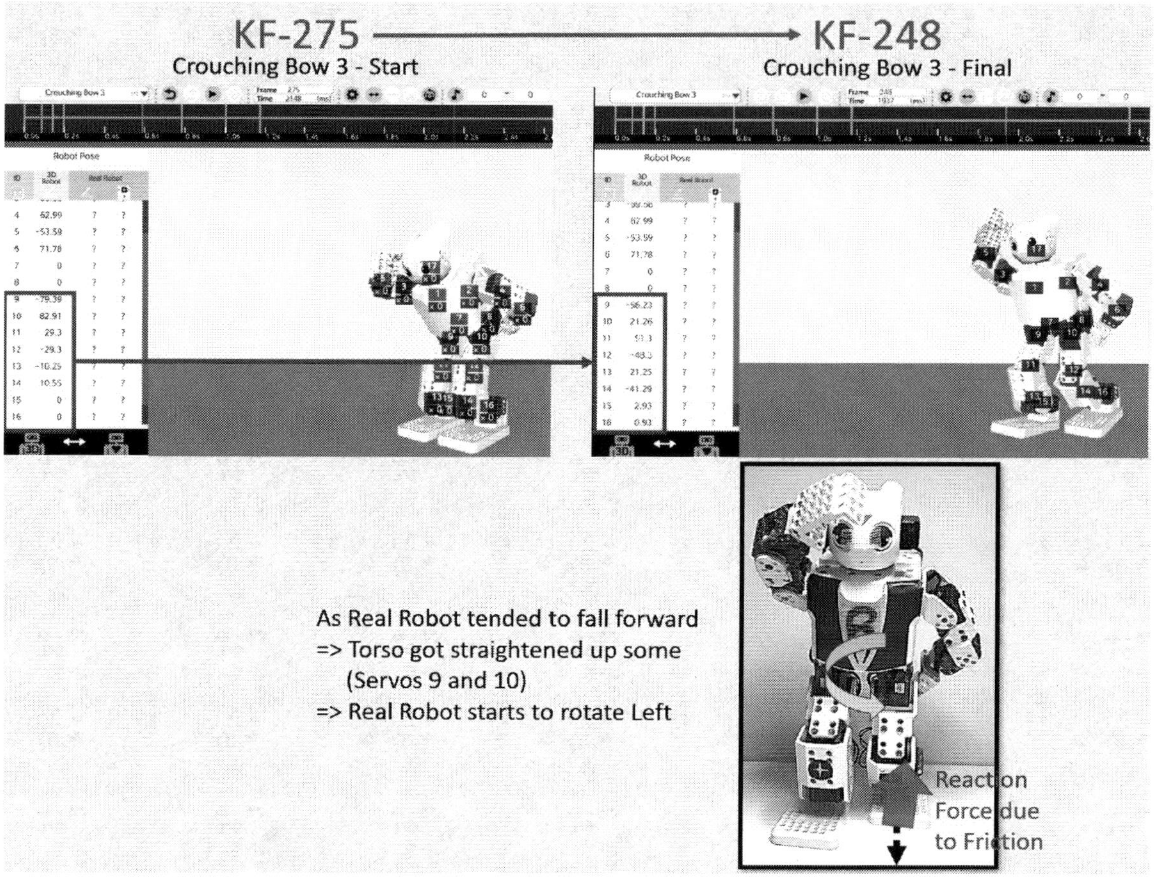

Fig. 4.45 Details for Legs Servos in KF-248 of Motion Unit "Crouching Bow 3".

I. KF-350 >> KF-326:
In "Crouching Bow 2", KF-350 was the pose when the robot straightened itself up from the bowing. So correspondingly for "Crouching Bow 3", both legs should be brought back together by moving the left foot forward to meet the right foot and both feet should also be flat with the ground surface when KF-350 was reached. Consequently, please note that for KF-350 the legs servos values were the same between "Crouching Bow 3 Start" (i.e. "Crouching Bow 2") and "Crouching Bow 3 Final" in Fig. 4.46. Once again, under "real dynamic motion", frictional reaction forces rotated the real robot to the left, but this time a lot more left when compared to KF-18 (see Fig. 4.37): this was because a larger frictional contact area was involved as the feet were flatter to the ground surface from Fig. 4.44 to Fig. 4.46, as compared to Fig. 4.37. The author also decided to shorten this recovery maneuver from KF-350 to KF-326.

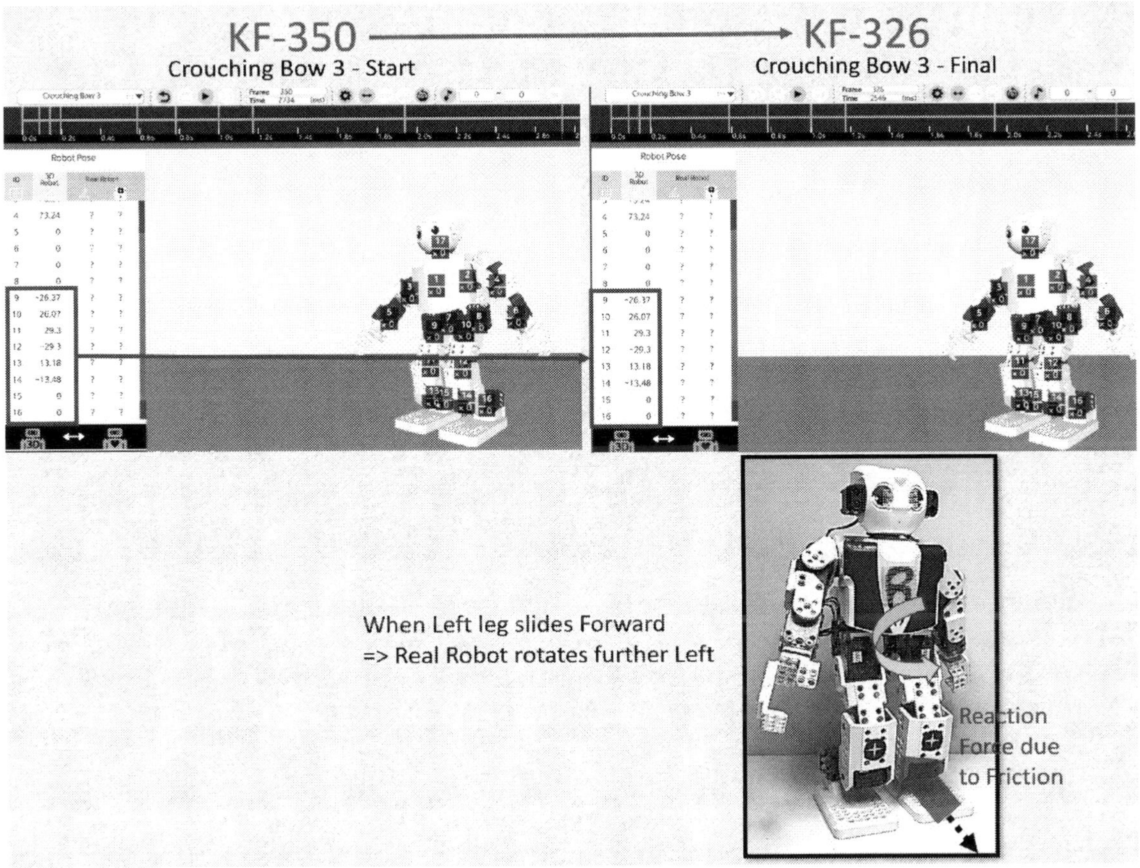

Fig. 4.46 Details for Legs Servos in KF-326 of Motion Unit "Crouching Bow 3".

J. KF-400 >> KF-395:
In "Crouching Bow 2", KF-400 was merely the "Ready" pose for the MINI robot, and so Servos 13 and 14 were brought back to 13.18 and -13.18 degrees respectively in "Crouching Bow 3 Start" of Fig. 4.47. However, under "real dynamic" motion and under a shortened time schedule (by 40 ms – from KF-400 to KF-395), the author found that the real robot would sway and then fall backwards for these values for Servos 13 and 14. Thus, they were readjusted to 16 and -16 instead (see Fig. 4.47), and these final values prevented the real robot from falling backwards.

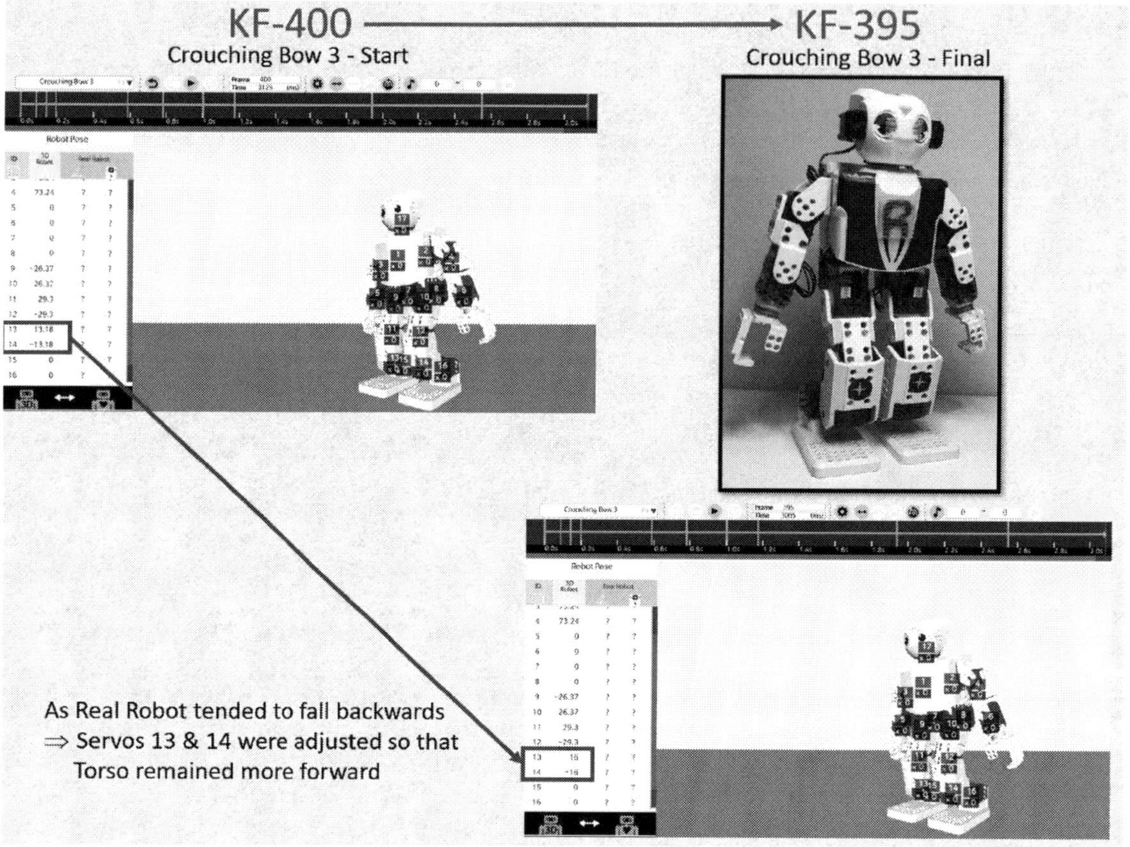

Fig. 4.47 Details for Legs Servos in KF-395 of Motion Unit "Crouching Bow 3".

Video 4.7 shows the performance of the final Motion Unit "Crouching Bow 3" as provided in the author's Motion file named "MINI_Custom.mtn3" and it was located towards the bottom of the Motion Units list. The author also created a new Motion Page/List named "Crouching Bows" listed after the "Horse Dance" (see Fig. 4.48-A) and a new Motion Group named "CT Custom Group" which was the Default Group with the addition of "Crouching Bows" at Motion Index = 45 (see Fig. 4.48-B).

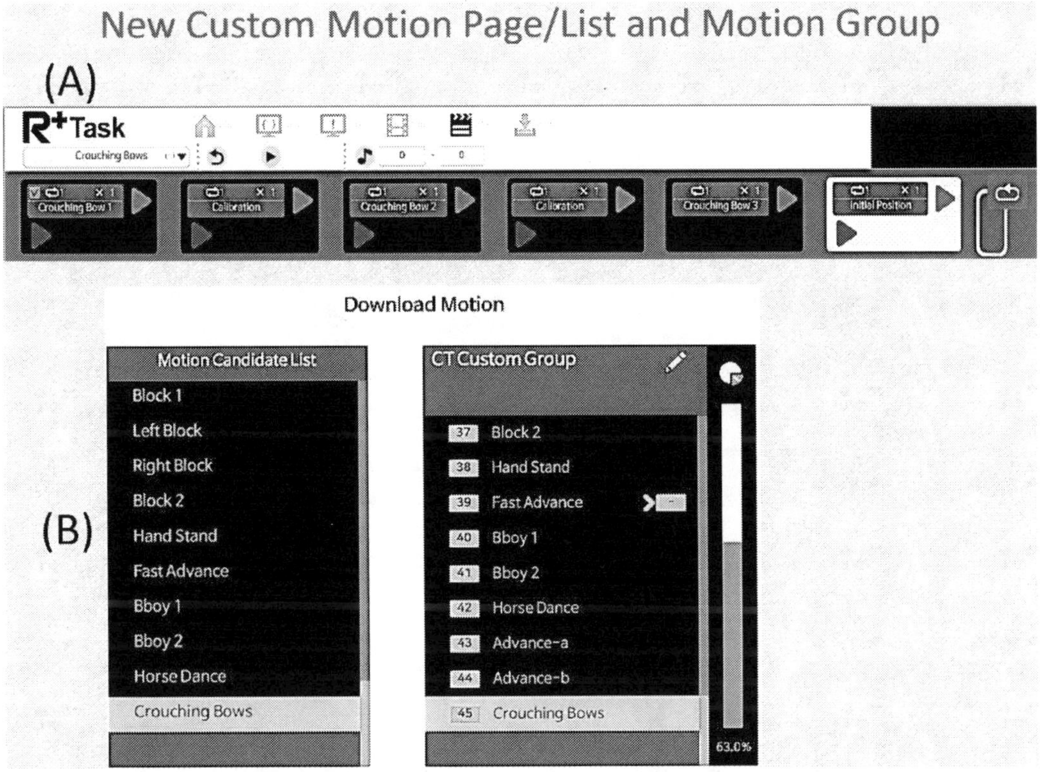

Fig. 4.48 New Motion Page/List and Motion Group created by author in "MINI_Custom.mtn3".

4.5.4 Additional Motion-Group Examples

The author's file "MINI_Custom.mtn3" also contains additional MOTION-GROUP examples:

1. "RC Walk 1" (see Fig. 4.49) where the 6 MOTION-LIST/PAGEs used were the unchanged original examples from ROBOTIS.

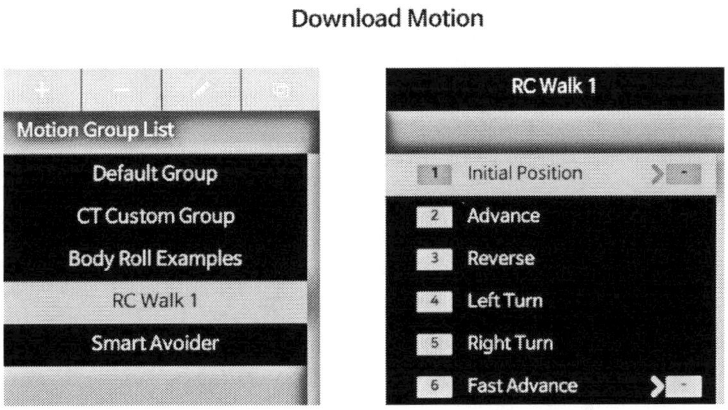

Fig. 4.49 New Motion Group "RC Walk 1" in "MINI_Custom.mtn3".

2. "Smart Avoider" where the MOTION-LIST/PAGEs used were the original ROBOTIS examples, however they are further modified with matching EXIT MOTION-UNITs to be used for Emergency Motion Stoppage Programming in Chapter 6 (see Fig. 4.50).

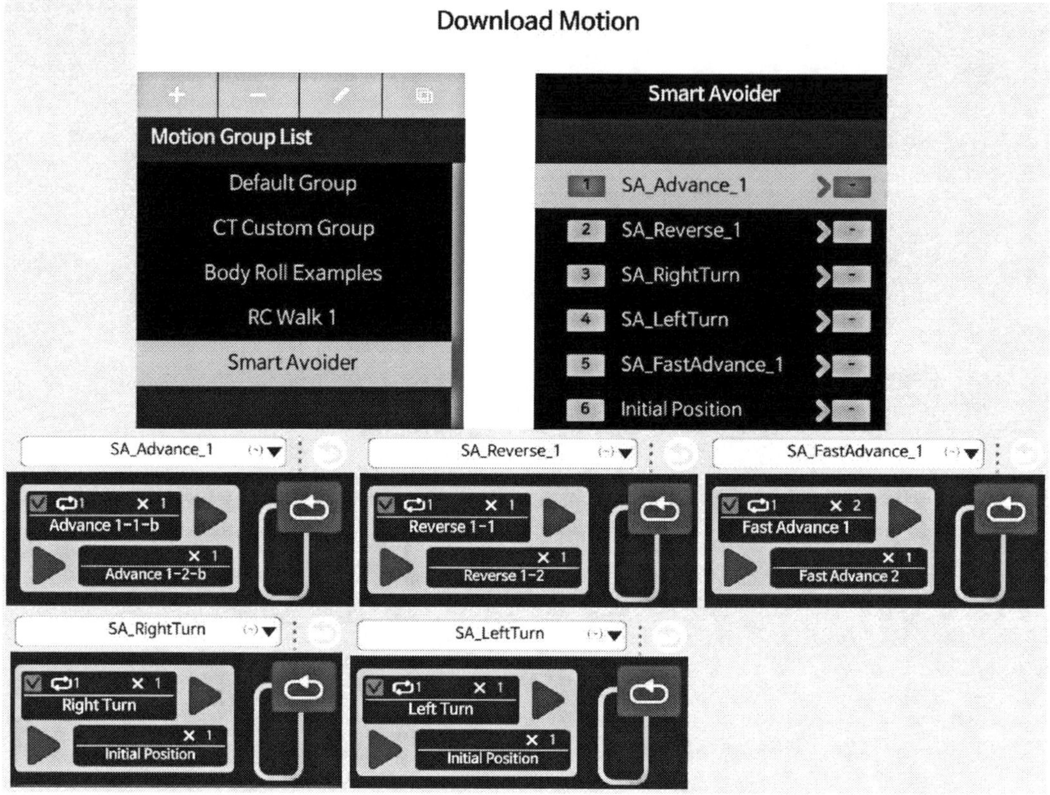

Fig. 4.50 New Motion Group "Smart Avoider" in "MINI_Custom.mtn3" using EXIT MOTION-UNITs.

With this "Practicum" section, the author hopes that the main Motion Design techniques had been showcased so that the readers can be comfortable in using them when the needs arise in their applications of the Motion Sub-tool, especially when the legs need to be in motion (beware of the dynamics effects of the ground surface!).

A final note for interested readers: currently the MOTION's simulation engine does not implement a physics engine as some other more advanced robot simulator like Webots from Cyberbotics (https://www.cyberbotics.com and https://github.com/cyberbotics) – unfortunately no one has created a MINI model in Webots yet! Perhaps some readers of this book want to pick up the challenge?

4.6 Motion Editing vs. Motion Programming

ROBOTIS literature has always been careful to use the words "MOTION EDITING" when referring to the usage of the MOTION Sub-tool, and the MOTION FILE is considered as DATA. Only a TASK file (see Chapter 6) is considered as a "real" Robot Control PROGRAM. But in the author's opinion, the

MOTION tool does support the 3 fundamental control structures used in all computer programming languages, albeit in a limited way:

1. SEQUENCE CONTROL – When working with MOTION-UNITs, the user learned how to sequence the robot's poses in a choreography sense instead of a mathematical logic sense as in standard computer programming (see Fig. 4.3). The user also learned how to account for the laws of Physics which is usually not included in Computer Science lessons for beginners.
2. REPETITION CONTROL – When working with MOTION-LIST/PAGEs, the user learned how to use the COUNTING LOOP (see Fig 4.18) and the ENDLESS LOOP (see Fig. 4.21). Only the CONDITIONAL LOOP (WHILE loop) is not implemented in the MOTION Sub-tool.
3. SELECTION CONTROL – This concept is introduced to the reader in a "sneaky" way via the use of the EXIT MOTION UNIT by way of the buttons STOP (F8) and EMERGENCY STOP (F9) (see Fig. 4.20).

Thus, the author thinks that it is reasonable to associate the concept of MOTION PROGRAMMING with the MOTION Sub-tool. If the reader happens to be an instructor or teacher, he or she may be interested to know that the author had noticed among his robotics students that their MOTION PROGRAMMING skills are independent from their COMPUTER PROGRAMMING skills, and to be a well-rounded "roboteer", students would need competency in both sets of skills.

As a final note for this chapter, back in 2009-2010, ROBOTIS published the User Manual for the BIOLOID PREMIUM (http://www.robotis.us/robotis-premium-programming-guide-en/) and it is still available at present. This manual was written using the TASK and MOTION tools in their Version 1, but it does have interesting and useful information regarding humanoid robot gait control, however the reader would have to do the extra work of adapting the contained information to TASK Version 3 and the MINI. For more advanced treatments of this subject area, the reader is referred to Kajita et al. (2014) and Goswami and Vadakkepat (2019).

4.7 Review Questions

1. In the context of the MOTION Sub-tool:
    a. What is a Joint-Offset?
    b. For what purpose can they be used for?
    c. How can they be changed?
    d. Where are the Joint-Offset Parameter located/saved?
2. What is a Key-Frame? What is an Interpolated-Frame?
3. Why is the smallest time-step is equaled to 7.81 ms?
4. What is a Motion-Unit?
5. What is the maximum time period that can be used for the Time-Line of a Motion-Unit?
6. How does one create a Motion-List/Page item?
7. What are the two methods of stopping a Motion-List/Page during its playing?
8. What is a Joint-Group? How are they defined? What is its purpose?
9. What is an Exit-Motion?
10. How can one modify the speed of execution of a Motion-List/Page?
11. Can a given Motion-Unit be repeated for a fixed number of times or forever? In the case of "forever", how can one stop this Motion-Unit?
12. How many Motion-Groups can be created in a Motion File?
13. How many Motion-Groups can be downloaded to the MINI at any point in time?
14. What is "Differential Positioning"?

15. What is “Pose Mirroring”?

4.8 References

Goswami A., Vadakkepat P. (Editors) (2019) Humanoid Robotics: A Reference. Springer Nature B.V., Dordrecht.

Kajita et al. (2014) Introduction to Humanoid Robotics. Springer, Heidelberg.

Chapter 5: Using MINI Mobile App

With the MINI Mobile App, a first-time user can access the ROBOTIS provided "Default" Motion Group (e.g. the "default.mtnx" file) and put the MINI robot through its paces for "edutainment" purposes. It has many utility functions built-in: Actuator Test, Motion Offset, Motion File Setting, Server/Client Settings, just to list a few popular ones. The robotics programming capabilities for this App are limited to:

1. Assigning a "Button", i.e. a touch area on the mobile device, to a user-assigned Motion Page/List via its Motion Index Number.
2. Assigning a particular "Gesture" with the mobile device to a given Motion Page/List via its Motion Index Number.
3. Assigning a Voice Command to the mobile device to activate a given Motion Page/List via its Motion Index Number.
4. Using a Virtual Remote Controller (U-D-L-R-1-2-3-4-5-6 buttons) to access two modes of control: "Soccer" (Defense or Offense) and "Fight" modes.

As of the writing of this chapter, the current MINI Mobile App is V.0.9.2.6 for the Android version (https://play.google.com/store/apps/details?id=com.robotis.darwinmini) and 0.9.15 for the iOS version (https://apps.apple.com/us/app/robotis-mini/id948481762), and **both are still using the MTNX files**, instead of the MTN3 files as created by the new MOTION 3 Sub-tool of the R+TASK V.3 application (see Chapter 4). Fortunately, there was NO DIFFERENCE between the formats of an MTNX file and an MTN3 file, thus the author just changed the extension MTN3 from his file "MINI_Custom.mtn3" (from Chapter 4) to MTNX to obtain the file "MINI_Custom.mtnx" which can then be used by the current MINI App directly. The interested reader can use Windows NOTEPAD to open and read inside these MTNX/MTN3 files to see the details in ASCII text (they are XML files in actuality). Perhaps in a future update of the MINI App, it can read MTN3 files directly.

The MINI App is well documented at these web links: http://emanual.robotis.com/docs/en/software/mobile_app/mini_app/ and http://emanual.robotis.com/docs/en/edu/mini/, thus the author did not see the point of rewriting these informational details back into this Chapter 5. **Thus, the goal of this Chapter is to help the reader go through a practicum in using the MINI App to add a new button with the author's provided file "MINI_Custom.mtnx".**

5.1 Transferring "MINI_Custom.mtnx" to Mobile Devices

For the purpose of writing this Chapter, the author used either an Android tablet (Samsung Tab A) or an iPad as the mobile platforms.

If the reader uses an Android device and the file "MINI_Custom.mtnx" file is on a Windows PC, then Windows Explorer can be used to drag and drop this file from the PC into the appropriate "ROBOTIS MINI" folder on the Android tablet (see Fig. 5.1).

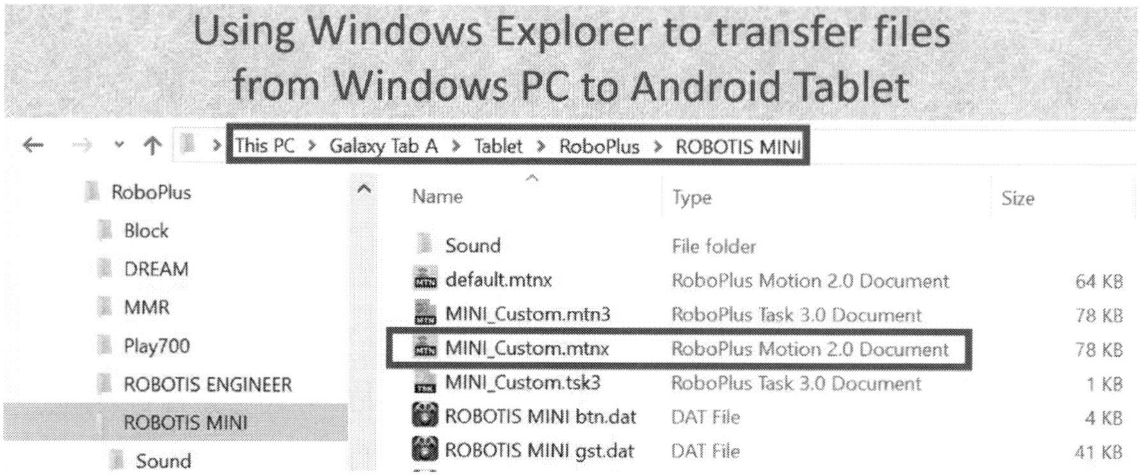

Fig. 5.1 Transferring "MINI_Custom.mtnx" from Windows PC to Android Tablet.

If the reader uses an iPad tablet (or iPhone), then "iTunes" has to be used to transfer the file "MINI_Custom.mtnx" from the Windows PC to the iPad (see Fig. 5.2). The procedure is a little more involved:

1. After connecting the iPad to the Windows PC via iTunes, the user needs to choose the "File Sharing" tool.
2. Next, in the middle panel, scroll to the "Robotis MINI" App, at that time the right panel may be empty or it may show the "Default.mtnx" file.
3. Then just drag and drop the "MINI_Custom.mtnx" file from Windows Explorer into the right panel, and this file will be copied to the user's iPad (or iPhone) at the proper location for the MINI App to access it as needed.

One important point to remind the reader here is that the **BT-410 needs to be used with an iOS device**, not the original BT-210 that came with the MINI kit (see Section 1.1.4).

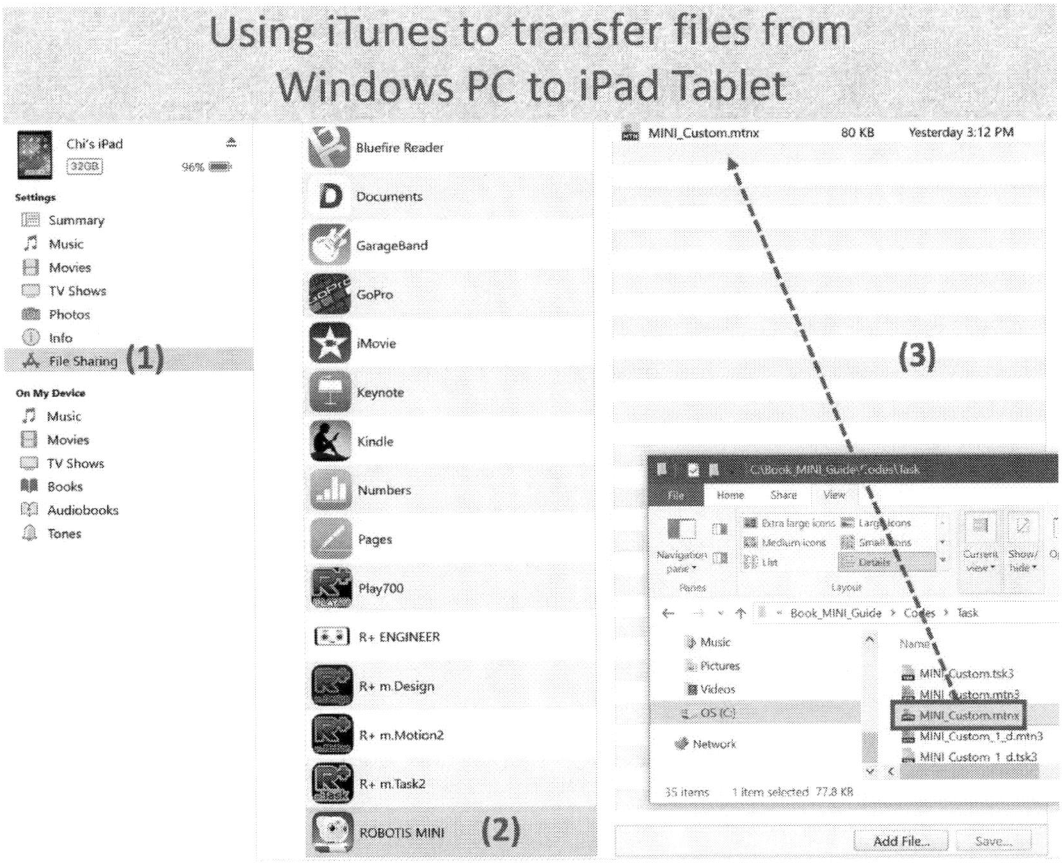

Fig. 5.2 Transferring "MINI_Custom.mtnx" from Windows PC to iPad Tablet via iTunes.

5.2 Connecting to Bluetooth & Performing Actuator Test

This web link shows how to connect the MINI robot to the MINI App via Bluetooth (http://emanual.robotis.com/docs/en/edu/mini/#connecting-the-robot-with-the-app-using-bluetooth).

Next, it would be a good idea to use the "Actuator Test" function to check out the overall assembly of the servo motors to see if they were at the right locations on the robot's frame and if they were controllable from the reader's mobile device (http://emanual.robotis.com/docs/en/edu/mini/#assembly-check-using-the-app). A minor note to remember that the current **MINI App can only test up to 16 XL-320s** on the MINI robot, as the author's robot had 17 servos, he had to test the 17th servo using the MANAGER tool (see Section 2.1.2).

5.3 Associating "MINI_Custom.mtnx" to MINI App

The MINI App by default uses the Motion File "default.mntx", and as we want to use the "MINI_Custom.mtnx" Motion File, we have to modify the "Motion File Settings" which can be found under the Setting icon located near the top right corner of the App's screen when the MINI App is activated (see Fig. 5.3).

Changing Motion File Settings

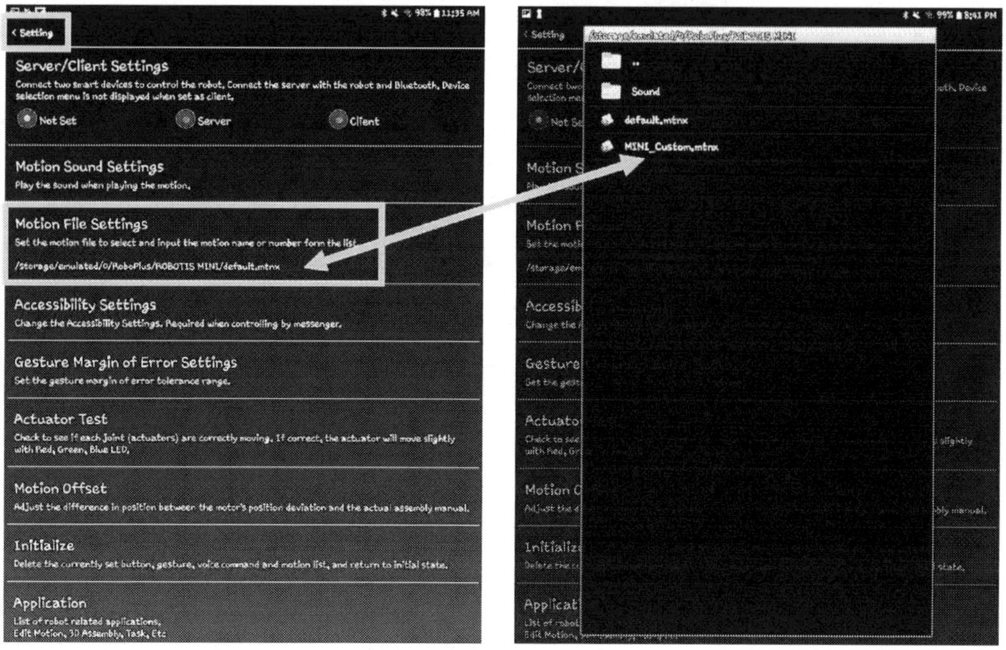

Fig. 5.3 Changing the default setting for “Motion File Settings” in SETTING Menu.

Unfortunately, the reader will have to do this step every time that the reader wants to use any Motion File other than the “default.mtnx” file, as this file change cannot be made permanently within the MINI App. One way to work around this issue is to rename “default.mtnx” to “default_old.mtnx” and then to rename “MINI_Custom.mtnx” as the new “default.mtnx” (the author leaves that choice to the reader).

Another important point to mention is that the Motion File “MINI_Custom.mtnx” has several Motion Groups defined within it such as “Default Group”, “CT Custom Group” or “Body Roll Examples”, thus the reader needs to be sure to choose the right Motion Group as indicated in the rest of this Chapter (see Fig. 5.4), and to download it to the MINI robot from inside R+TASK V.3, **JUST BEFORE using the MINI App with it**. **The reason for this cautionary step is because the author had found that working with the MOTION tool in SYNC Mode will replace any Motion Group previously downloaded to the MINI robot with the one that the author was experimenting with in SYNC Mode!**

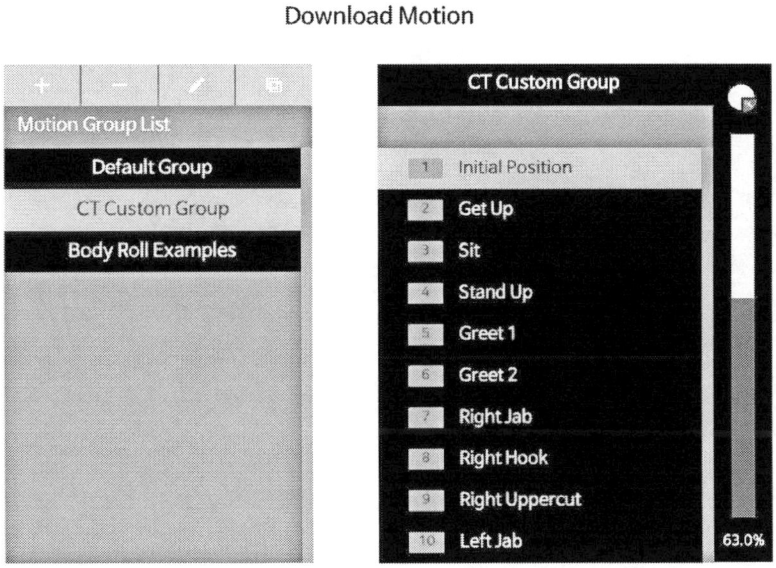

Fig. 5.4 Motion Groups defined in "MINI_Custom.mtnx" Motion File.

5.4 Assigning a New Button for Motion Page/List #45 "Crouching Bows"

In this section, we'll practice how to assign a new "Button" (i.e. a Touch Area on the mobile device) to the Motion Page/List named "Crouching Bows" in the "**CT Custom Motion Group**".

This web link (http://emanual.robotis.com/docs/en/edu/mini/#control-with-buttons) gives good information about how to Add/Edit/Delete/Arrange Buttons inside the MINI App – the reader is recommended to review these steps in this web link first before reading on.

Once connected to the real robot, the MINI App main screen looks like what is shown in Fig. 5.5, whereas the user can either choose to "Run" or to "Edit", but please note that the bottom 3 buttons of the "Run" Screen, i.e. "Stop Motion", "Init Pose" and "Stand Up", are not editable (for further information please review this web link http://emanual.robotis.com/docs/en/edu/mini/#operation). **Furthermore, there is a typo in the right-most label: it should have been "Get Up" (i.e. Motion Unit #2) instead of "Stand Up" which corresponds to Motion Unit #4.**

Fig. 5.5 Main Screens for the MINI App.

The procedure to assign a new button to the Motion Page/List "Crouching Bows" is as follows:

1. Let's first choose the "Edit" tab (Step 1 in Fig. 5.6).
2. Once in the Edit Screen, beneath the large "Add Button" button, there are 3 smaller buttons with the "+" sign in them, signifying that they are available to be assigned to a user-defined Motion Page/List. And let's say that the reader wants to use the middle button for the current task (Step 2 in Fig. 5.6).
3. Once this middle button is pushed, the "Edit Button" window pops up. Next, if the user taps on the "List" area (Step 3 in Fig. 5.6), a scrollable menu comes up displaying a list of all 3 Motion Groups associated with the file "MINI_Custom.mntx": e.g. "Default Group", "CT Custom Group" and "Body Roll Examples". If the user only sees the "Default Group", then the user needs to go back to Section 5.2 to set the Motion File Setting to "MINI_Custom.mtnx".
4. Then, the user can scroll up or down this menu to find the Motion Page/List named "Crouching Bows" belonging to the "CT Custom Group" as shown in Step 4 of Fig. 5.6.
5. Next, if the user taps on "Crouching Bows", the user is brought back to the "Edit Button" window as shown on the bottom left picture of Fig. 5.6. If the user wants to use a different text than "Crouching Bows", that field can be re-edited as needed.
6. The final step is to tap the "OK" button, and the new "Edit" and "Run" Screens would then look like those shown in Fig. 5.7, and the user can now test out "Crouching Bows" on the real robot.

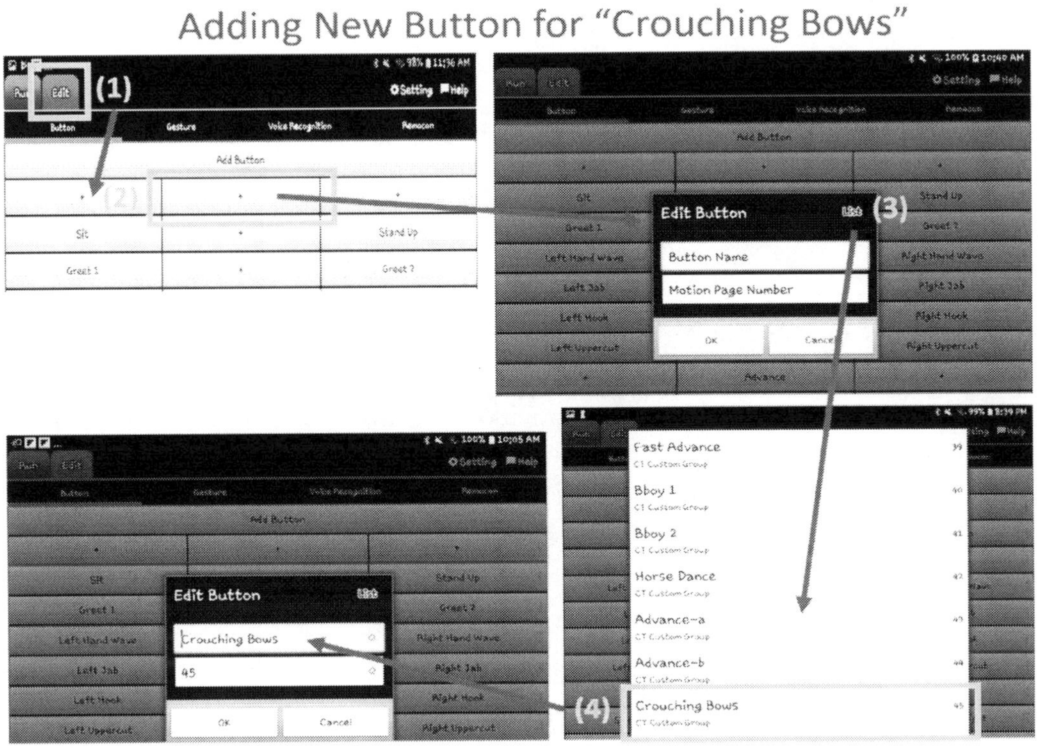

Fig. 5.6 Assigning New Button to Motion Page/List "Crouching Bows".

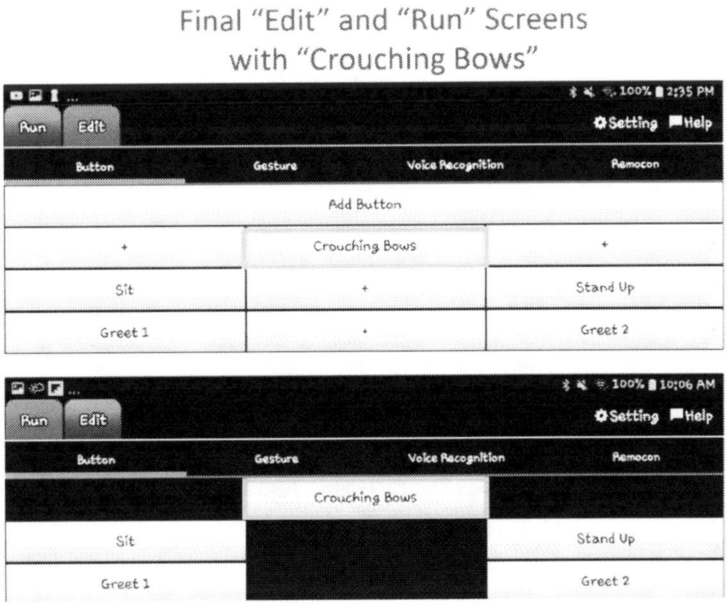

Fig. 5.7 Final "Edit" and "Run" Screens obtained.

One important note for the reader is that when the robot is playing a Motion Page/List, all other Motion buttons will be ignored, **except for the "Stop Motion" button** which can be found at the bottom left corner of the "Run" screen back in Fig. 5.5. When the "Stop Motion" button is used, the MINI robot would stop at the end of the current Motion Unit being played, and sometimes at the current Key-Frame being played: this is the so-called **Motion Page/List 0** (for more details please review Section 4.2.2 regarding the STOP (F8) button provided in the MOTION tool). This feature will be revisited in Chapter 6 when programming with TASK codes (see http://emanual.robotis.com/docs/en/software/rplustask3/task_parameters/#motion-index-number and http://emanual.robotis.com/docs/en/software/rplus1/task/programming_02/#motion-parameters).

5.5 Using Gestures, Voice Recognition, Messenger and Remote Controller

This web link (http://emanual.robotis.com/docs/en/software/mobile_app/mini_app/#app-usage) has good information about how to control the MINI using the other features: Gestures, Voice Recognition, Messenger and Remote Controller, and the author does not have any more useful information to add.

Overall, the MINI App is a good tool to use for beginning robot users but it is too restrictive as a robot programming tool, thus in Chapters 6, 7 and 8, the reader will be shown in-depth and flexible ways to control the MINI robot using the ROBOTIS TASK and PLAY700 tools along with the integration of selected ROBOTIS sensors to the OpenCM9.04-C, as well as other third-party and open-source software such as EDBOT and OPENCV.

Chapter 6: Using TASK

To reach the broadest range of readers, the author will assume that the reader is a complete beginner in the areas of computer programming and robot control, thus Chapter 6 would start with broad robot control concepts before getting into the finer details of programming the MINI robot with the TASK Sub-tool of the R+TASK V.3 Tool.

The R+TASK V.3 Tool offers most of the standard programming structures found in other languages such as C/C++. Below is a short description of the most useful features for a beginning programmer (more details on specific items will be provided at appropriate sections in this chapter):

1) ASSIGNMENTS – TASK provides three types of assignment statements: LOAD, COMPUTE and POLYNOMIAL.

   The syntax for LOAD is “A = B” with the usual meaning of “Operand A is assigned the Value of Operand B”.

   The syntax for COMPUTE is “A = B *Op* C” where the RHS represents the Value obtained when performing the “*Op*” operation between Operands B and C. The “*Op*” operations supported are the 5 basic arithmetic operators “+, - , * , / , and % (remainder operation)”, and the two bit-level operators such as “&” (AND) and “|” (OR). Only INTEGER arithmetic is supported. VARIABLE parameters are supported but arrays are not.

   POLYNOMIAL is “format-free”. It is the usual “mathematical expression” in standard computer languages but it can only include the arithmetic and logical operators listed in the previous paragraph and of course the parentheses “(“ and “)” are used to group sub-expressions as needed.

2) LABEL and JUMP statements are supported.

3) CONDITIONS – For Conditional statements, TASK provides the usual logical operators: “&&” and “||” respectively for logical AND and logical OR operations, and also “==”, “!=” , “<”, “<=”, “>”, “>=” for equality and inequality tests. The standard IF, IF-ELSE-IF, ELSE structures are supported, but nested conditional statements using parentheses are not supported (so the user will have to apply DeMorgan’s theorems to expand complex logical statements as needed).

4) LOOPS – for Repetition statements, TASK provides “WAIT WHILE”, “LOOP WHILE”, “LOOP FOR”, “ENDLESS LOOP” and “BREAK LOOP”. Standard computer languages normally offer a DO-WHILE loop structure which is not supported in TASK.

5) FUNCTION definition and calling are provided in TASK. A special CALLBACK function can also be used: it is triggered by a hardware timer every 7.81 ms independently of the main TASK program. CALLBACK can only handle up to a maximum of two external device calls.

6.1 Learning TASK Programming Basics

The user is recommended to first view Video 6.1 for a quick working tour of the TASK V.3 application and to learn about the editing basics for TASK V.3 component by following the author's steps in creating the example TASK program called "M_LEDControl_Servo6.tsk3".

The following sub-sections, 6.1.1 through 6.1.5, were written for a "complete beginner" user in mind, but they could be used as "review materials" for readers with more experiences in computer programming from elsewhere.

6.1.1 LED Control with Timer

Video 6.1 just showed the mechanics of creating a TASK program such as "M_LEDControl_Servo6.tsk3" without much explanations for why certain statements are needed and why a certain order of these statements was used. In this sub-section 6.1.1 each statement used will be explained in more details (see Fig. 6.1).

```
4   START PROGRAM
5   {
6       // Overall robot initialization
7       ID[All]: Torque ON/OFF = FALSE (0)
8       ID[All]: LED = FALSE (0)
9       // Set LED to Red for 2 seconds
10      ID[6]: LED = 1
11      High-resolution Timer = 2.000sec
12      WAIT WHILE ( High-resolution Timer > 0.000sec )
13      // Set LED to Blue for 5 seconds
14      ID[6]: LED = 4
15      High-resolution Timer = 5.000sec
16      WAIT WHILE ( High-resolution Timer > 0.000sec )
17      // Set LED to White and quit the program
18      ID[6]: LED = 7
19  }
```

Fig. 6.1 TASK program "M_LEDControl_Servo6.tsk3".

The overall goal of "M_LEDControl_Servo6.tsk3" was to turn on the LED of the Servo with ID=6 in Red color for 2 seconds, then switch it to Blue for 5 seconds, and finally switch it to White before quitting the program:

1. The Main program should be enclosed within the two outermost curly brackets (Statements 5 and 19) – and there should be only ONE Main program. At run time, the OpenCM9.04-C will execute the actionable statements enclosed within these curly brackets in sequence and in the order that they were written (i.e. from Statement 6 to Statement 18).

2. Statements 6, 9, 13 and 17 are comments inserted by the user to help other readers to follow his or her logic flow. These statements are ignored by the compiler and will not be executed by the MINI Controller at run time.
3. Statements 7, 8, 10, 11, 14, 15 and 18 are LOAD (i.e. assignment type) Statements belonging to the EXECUTE group as used in TASK's nomenclature.
4. Statement 7 turns off power to ALL 17 Servos, i.e. it will make the real MINI go "limp", thus the user needs to lay the MINI flat on his frontal or back side before running this program to avoid any mishap to the robot.
5. Statement 8 turns off the embedded LEDs to ALL Servos.
6. Statements 10-12 work together to produce the effect of lighting up the Servo-6's LED in Red for 2 seconds. Statement 10 turns on Servo-6's LED in Red color by using the special value of "1" (for other color settings please consult this web link http://emanual.robotis.com/docs/en/dxl/x/xl320/#led25). Statements 11 and 12 need to be used together and in that order. Statement 11 sets a value of 2 seconds, to be exact 2000 milliseconds, into the High-Resolution Timer which immediately starts counting down, per millisecond, from 2000 towards 0. Statement 12 is a WAIT-WHILE loop and it does its part by checking this Hi-Res Timer's counter against Zero. If this counter value is greater than 0 (i.e. the check's result is TRUE), this WAIT-WHILE loop (i.e. Statement 12) is executed once again – physically, this means that the 2000 ms duration previously set by Statement 11 has not elapsed yet. But eventually this counter would reach 0 and stay at 0, at that time the check's result is FALSE which makes the controller quit the WAIT-WHILE loop, and then the controller goes on to execute the next actionable statement which is Statement 14.
7. In a similar manner as described in the previous paragraph, Statements 14-16 work together to turn on Servo-6's LED in Blue for 5 seconds.
8. The last actionable statement is Statement 18 which turns on Servo-6's LED in White, and then the MAIN program terminates right after it. So, what will happen to this LED then? The answer is that Servo-6's LED will stay lit in White as long as power is still applied to the MINI robot.

Please see Video 6.2 for a demonstration of "M_LEDControl_Servo6.tsk3" at run time.

The program "M_LEDControl_Servo6.tsk3" illustrates the first two Robot Control Concepts:

A. **By default, a TASK program executes actionable statements in a sequential manner from beginning to end (i.e. SEQUENCE CONTROL). Consequently, a robot will keep on executing the same programmed action until a new actionable statement changes its last action. For example, the LED will stay Red (Statement 10) until it is commanded to turn Blue (Statement 14). Some of my younger beginner students thought that the Hi-Res Timer turned off the LED (Statements 12 and 16).**
B. **A robot's action is not terminated, just because the TASK program has stopped execution. In the current version of "M_LEDControl_Servo6.tsk3", the author had specified two INITIALIZATION actions (Statements 7 and 8) to put the robot in a "known" condition before writing out the "real program" (Statements 9 to 18). He should have also specified the FINALIZATION actions by issuing/inserting a command such as Statement 8 to turn off all LEDs just before the closing curly bracket "}" (currently Statement 19). This will put the robot in a "known" condition before its program finishes. In this particular case, a lit LED does not carry much significance, but let's say that we happen to be dealing with a program that has the robot moving around, then we all can see that it will be very**

important for the robot to be at a standstill (or in another safe position) before the program finishes - whether intentionally or not!

6.1.2 Sequential Control of Servo Positions

The goal of "M_SequentialControl_Servo6.tsk3" is of course to learn the effect of SEQUENCE CONTROL on the motions performed by Servo-6 (see Fig. 6.2):

```
4   START PROGRAM
5   {
6       // Overall robot initialization
7       ID[All]: Torque ON/OFF = FALSE (0)
8       ID[All]: LED = FALSE (0)
9       // Special initialization for Servo 6
10      ID[6]: Torque ON/OFF = TRUE (1)
11      ID[6]: Operating Mode = 2
12      ID[6]: Torque Limit = 1023
13      ID[6]: Goal Velocity = 1023
14      // Command Servo 6 to move to 0 degree position
15      ID[6]: Goal Position = 512
16      WAIT WHILE ( ID[6]: Is Moving == TRUE (1) )
17      // Command Servo 6 to move to -90 degree position
18      ID[6]: Goal Position = 205
19      WAIT WHILE ( ID[6]: Is Moving == TRUE (1) )
20      // Command Servo 6 to move to +90 degree position
21      ID[6]: Goal Position = 819
22      WAIT WHILE ( ID[6]: Is Moving == TRUE (1) )
23  }
```

Fig. 6.2 TASK program "M_SequentialControl_Servo6.tsk3".

1. The use of Statements 7 and 8 are already explained in the previous Section 6.1.1.
2. Statement 10 turns on electrical power to Servo 6 (while electrical power to all other servos is still disabled by Statement 7).
3. Statement 11 puts Servo 6 into "Operating Mode 2" which is its Position Control mode. "Operating Mode 1" makes Servo 6 behave like a wheel (i.e. continuously turning in a CW or CCW direction or stopping at its current and arbitrary position). This Mode 1 is obviously not appropriate for a humanoid robot which is constrained by its body frame parts.
4. Statement 12 sets the "Torque Limit" for Servo 6, i.e. the "Maximum" electrical current allowed into the motor used for Servo 6 (review Fig. 1.3). The allowable value for "Torque Limit" varies from 0 to 1023. Statement 12 uses 1023 for maximum holding power at any position that Servo 6 is programmed to GET TO and HOLD. Obviously, we all want our robot to be

"strong" when moving around or standing still, so the 1023 value is often used. But the reader may be surprised to know that under certain circumstances a "Torque Limit" value of 0 is also useful, if the user needs to rotate any XL-320 by hand. This special application will be showcased later in Section 6.7 for the "Mimicking MINIs" project.

5. Statement 13 sets the "Goal Velocity" of Servo 6 to its maximum value of "1023" (its minimum value being of course "0"). This "Goal Velocity" controls how fast Servo 6 will go from one programmed servo position to the next. For now, we'll just use its maximum value of "1023". Later in Section 6.7.3, an appropriate use of a "Goal Velocity" value of "0" will be illustrated.
6. Statement 15 sets the "Goal Position" for Servo 6, i.e. the servo position where Servo 6 should be, if it is not there yet. The allowable values for "Goal Position" span from 0 to 1023 and please review Fig. 1.3 to locate where physically these possible Goal Positions correspond to on the real servo. Statement 15 sends Servo 6 to Position 512 (i.e. 0 degree when setting it in the appropriate TASK sub-menu – see Fig. 6.3).

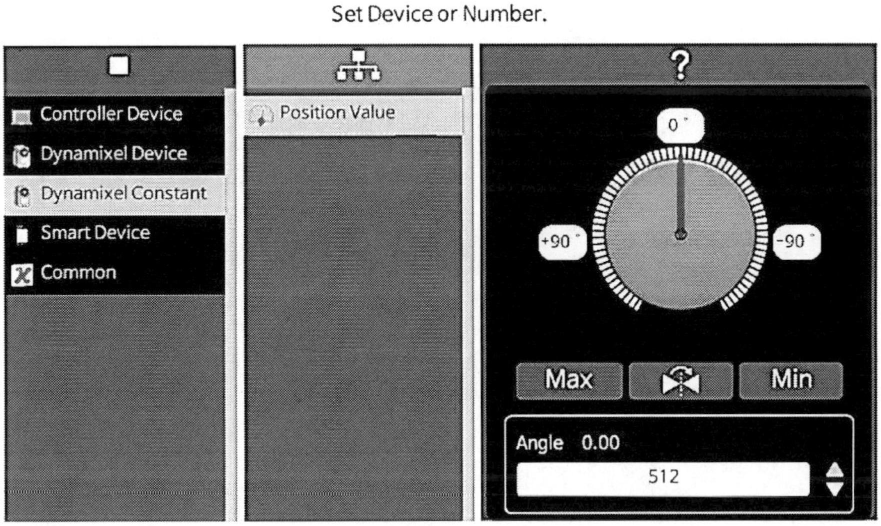

Fig. 6.3 Setting a Position Value for a Servo's Goal Position inside TASK sub-menu.

7. Statement 16 illustrates an operational concept that most beginner "roboteers" need to get used to when programming ROBOTIS Dynamixel Servos in their Position Control mode and it is simply that the controlling program needs to wait until the targeted servo reaches the Goal Position as previously commanded (by Statement 15). This wait time cannot be determined in advance as the Servo's Starting Position is not known in advance and can be anywhere between 0 and 1023. But this "indeterminate" wait time can be accommodated by using a WAIT-WHILE loop that monitors for the TRUE/FALSE logical value of the "Is Moving" flag (see Fig. 6.4). When Servo 6 is in motion towards its Goal Position, this flag is set to TRUE. When it reaches this Goal Position, this flag will be set to FALSE and thereby will terminate this WAIT-WHILE loop and the robot controller goes on to execute Statement 18 next. A sharp reader may have already noticed that this program did not have to specify Servo 6 to turn CW or CCW at all to get to the commanded Goal Position, because the embedded controller inside each XL-320

servo "knows" where its starting and ending positions are and it would switch the power polarities on its electrical motor as needed to turn CW or CCW.

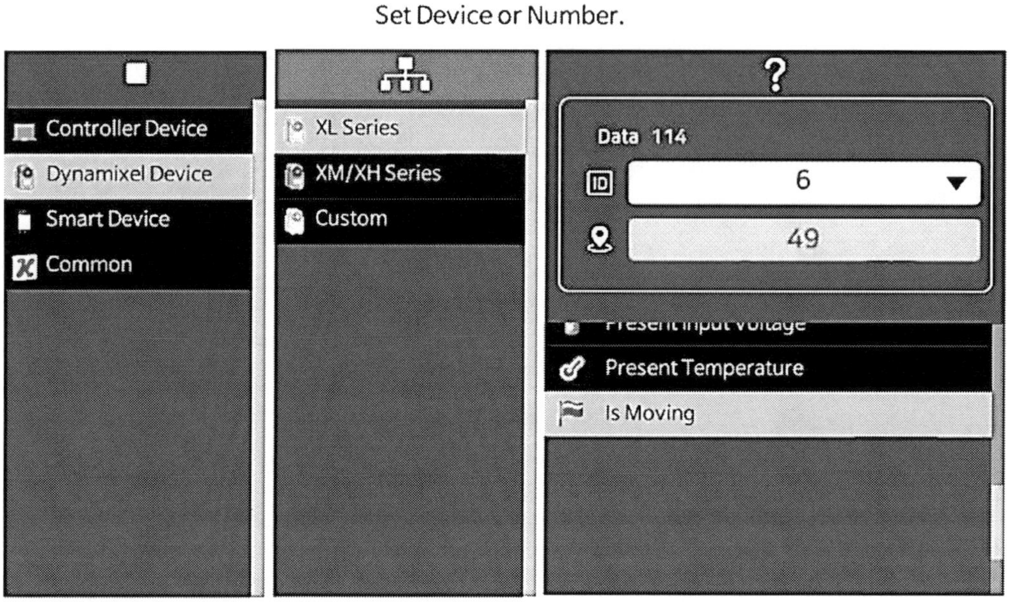

Fig. 6.4 Accessing Servo-6's "Is Moving" Flag inside TASK sub-menu.

8. If there are some other tasks (unrelated to Servo-6's Goal Position) that the robot needs to do right away after Statement 15, the programmer can insert them between the current Statements 15 and 16, so that the XL-320 controller does not have to "waste" time "doing nothing" while waiting for Servo 6 to reach its Goal Position.
9. Next, Statement 18 sends Servo 6 to a new Goal Position of 205, and Statement 19 does the job of waiting for it to get there.
10. Finally, Statement 21 sends Servo 6 to a new Goal Position of 819, and Statement 22 does the job of waiting for it to get there, and the Main program terminates. The interested reader can try to comment out (i.e. Control-E) or delete Statement 22 to see if the final position for Servo 6 remains Position 819! Why yes? Or why not? Please refer to Computing Concept B in Section 6.1.1.

Please see Video 6.3 for a demonstration of "M_SequentialControl_Servo6.tsk3" at run time.

The program "M_SequentialControl_Servo6.tsk3" illustrates two new Robot Control Concepts:

C. The proper use of an XL-320 servo requires quite a few preparation steps:
   1) Enable its Torque function (i.e. power it up) – Statement 10.
   2) Set its Operating Mode to 2 (i.e. Position Control mode) – Statement 11.
   3) Set its Torque Limit to a value between 0 and 1023 (i.e. maximum current allowed into its electrical motor) – Statement 12.
   4) Set its Goal Velocity to a value between 0 and 1023 (i.e. how fast for the servo to move from its Present Position to its Goal Position) – Statement 13.

D. Once the 4 preparation steps are done, the programmer can move the targeted servo to any allowable Goal Position [0-1023] by issuing commands such as Statements 15, 18 or

21, but the programmer needs to include appropriate wait times for the servo to get to its destination using the "Is Moving" flag.

6.1.3 Endless Loop Control of Servo Positions

The "M_EndlessLoopControl_Servo6.tsk3" program showcases the use of an Endless Loop with Position Control commands within it (see Fig. 6.5). It uses the previous Servo Control concepts described in Section 6.1.2.

```
4  START PROGRAM
5  {
6     // Overall robot initialization
7     ID[All]: Torque ON/OFF = FALSE (0)
8     ID[All]: LED = FALSE (0)
9     // Special initialization for Servo 6
10    ID[6]: Torque ON/OFF = TRUE (1)
11    ID[6]: Operating Mode = 2
12    ID[6]: Torque Limit = 1023
13    ID[6]: Goal Velocity = 1023
14    ENDLESS LOOP
15    {
16       // Command Servo 6 to move to 0 degree position
17       ID[6]: Goal Position = 512
18       WAIT WHILE ( ID[6]: Is Moving == TRUE (1) )
19       // Command Servo 6 to move to -90 degree position
20       ID[6]: Goal Position = 205
21       WAIT WHILE ( ID[6]: Is Moving == TRUE (1) )
22       // Command Servo 6 to move to +90 degree position
23       ID[6]: Goal Position = 819
24       WAIT WHILE ( ID[6]: Is Moving == TRUE (1) )
25    }
26 }
```

Fig. 6.5 TASK program "M_EndlessLoopControl_Servo6.tsk3".

The ENDLESS-LOOP is one among the 4 loop structures that are supported in a TASK program in its LOOP sub-menu (see Fig. 6.6), the other 3 being LOOP-WHILE, LOOP-FOR and WAIT-WHILE. This ENDLESS-LOOP begins at Statement 14 which effectively is just a label. Statements 17 through 24 are set inside the curly brackets (Statements 15 and 25) marking the beginning and end of this ENDLESS-LOOP and they are the controller's commands that are going to be executed in the sequence that they are defined during each cycle of the LOOP and also for ever or until the user turns off the power to the robot.

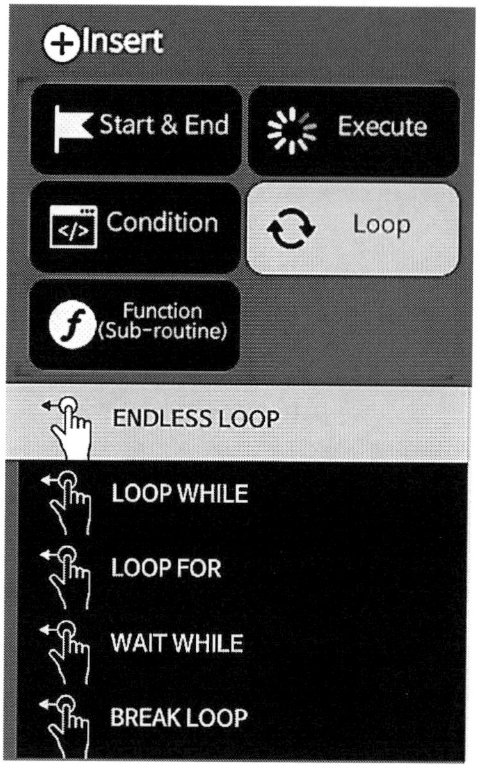

Fig. 6.6 LOOP sub-menu in a TASK program.

Video 6.4 shows the run-time performance of the program "M_EndlessLoopControl_Servo6.tsk3" to illustrate that an Endless Loop will run forever as true to its namesake and the only way to stop this program is to turn off the power to the whole robot. Video 6.4 also shows that when using an Endless Loop, the WAIT-WHILE loops are even more crucial to the proper sequencing of servo motions, because without them the controller would try to switch so fast between the three commanded Goal Positions that physically the user can only see a "trembling" of the MINI's hand (i.e. Servo 6). Later, Section 6.4 will illustrate the proper use of the Endless Loop which is UBIQUITOUS in ALL robot control programs.

The program "M_EndlessLoopControl_Servo6.tsk3" illustrates one new Robot Control Concept:

E. **The ENDLESS-LOOP is difficult to use as the programmer needs to think/plan about a proper situation/event to break out of the loop, as turning off the power to the whole robot to stop the execution of its controlling program is not acceptable for any self-respecting roboteer! Later sections in Chapter 6 will demonstrate several ways to create "smooth" exits from endless loops.**

6.1.4 Counting Loop Control of Servo Positions

The "M_CountingLoopControl_Servo6.tsk3" program illustrates the use of a Counting Loop (labeled as LOOP-FOR in TASK program nomenclature). This type of loop is used when the programmer knows exactly how many times that a group of statements needs to be executed by the robot's controller to accomplish a given task (see Fig. 6.7 and Statements 14 through 27).

```
4  START PROGRAM
5  {
6      // Overall robot initialization
7      ID[All]: Torque ON/OFF = FALSE (0)
8      ID[All]: LED = FALSE (0)
9      // Special initialization for Servo 6
10     ID[6]: Torque ON/OFF = TRUE (1)
11     ID[6]: Operating Mode = 2
12     ID[6]: Torque Limit = 1023
13     ID[6]: Goal Velocity = 150
14     LOOP FOR ( Count = 1 ~ 3 )
15     {
16         // Set a new Goal Velocity for Servo 6
17         ID[6]: Goal Velocity = ID[6]: Goal Velocity * Count
18         // Command Servo 6 to move to 0 degree position
19         ID[6]: Goal Position = 512
20         WAIT WHILE ( ID[6]: Is Moving == TRUE (1) )
21         // Command Servo 6 to move to -90 degree position
22         ID[6]: Goal Position = 205
23         WAIT WHILE ( ID[6]: Is Moving == TRUE (1) )
24         // Command Servo 6 to move to +90 degree position
25         ID[6]: Goal Position = 819
26         WAIT WHILE ( ID[6]: Is Moving == TRUE (1) )
27     }
28 }
```

Fig. 6.7 TASK program "M_CountingLoopControl_Servo6.tsk3".

Same as for the ENDLESS-LOOP, the LOOP-FOR structure displays its looping mechanism in a single statement (Statement 14). The LOOP-FOR structure employs a user defined Variable (named "Count" in Statement 14), and the programmer needs to also specify the **Start Value** (i.e. 1) and the **Last Value** (i.e. 3) for this Variable "Count" to take on during each iteration of this loop structure. Statement 14 is written so that the group of statements from Statement 16 to 26 will be executed exactly 3 times. This LOOP-FOR then terminates at Statement 27 which leads to Statement 28 which is the end of this TASK program.

The reader needs to note that the Servo-6's Goal Velocity is set to (150 * 1 = 150) for the first loop when Count is set to 1 (see Statements 13 and 17). Next, this Goal Velocity is set to (150 * 2 = 300) for the second loop as Count is set to 2, and then for the third loop this Goal Velocity is set to (300 * 3 = 900) as Count is set to 3.

Here, an attentive reader would have taken a pause and wondered that shouldn't the Goal Velocity take on the value of (150 * 3 = 450) instead for the third loop? And this reader would be incorrect, as he/she had forgotten the Computing Concept that a Parameter will retain its "old" value until it gets changed to a "new" one. Thus, when the first loop is entered, the old/current value for Goal Velocity is "150", as Statement 13 is the last assignment statement affecting Goal Velocity, and the current value for Count is "1". Therefore Statement 17 instantiates to "150 * 1 = 150" on its Right-Hand-Side (RHS), so Goal Velocity is reassigned to the same value of "150". Next, when the second loop is entered, the current value for Goal Velocity is "150" and the current value for Count is "2", therefore the new value for Goal Velocity is "150 * 2 = 300". Lastly, when the third loop is entered, the current value for Goal Velocity is "300" and the current value for Count is "3", thus the effective Goal Velocity for Servo 6 is now "300 * 3 = 900" during the execution of the third loop.

Please see Video 6.5 for a run-time demonstration of "M_CountingLoopControl_Servo6.tsk3" with an initial value of 150 for Goal Velocity for the first Part and the reader can see that Servo 6 is running at higher and higher speed for each of the three loops. For Part 2 of Video 6.5, the starting value for Goal Velocity is changed slightly to 200, and the reader can see that the second loop is running at a higher Goal Velocity than the first loop (same result as Part 1), but the third loop was running at a very slow Goal Velocity! So, what is going again with the MINI robot?

During Part 2 of Video 6.5 and mathematically, the Goal Velocity values should have been "200" (= 200 * 1) for the 1st loop, then "400" (= 200 * 2) for the 2nd loop, then "1200" (= 400 * 3) for the 3rd loop – meaning that Servo 6 should have run at the highest speed at the 3rd loop as in Part 1! But it sure did not!

Some readers may have figured out the reason by now, if they recall that the allowable values for the Goal Velocity are from 0 to 1023, which means that Goal Velocity uses only the first 10 bits of its actual binary storage space inside the OpenCM9.04-C's memory space:

- "1023" in decimal translates to "11 1111 1111" in binary.
- For the third loop, "1200" in decimal translates to "100 1011 0000" in binary requiring 11 bits, but because Goal Velocity can only use the first 10 bits, a truncated value of "00 1011 0000" is set to Goal Velocity instead. As "00 1011 0000" translates into "176" in decimal, this results in a much slower speed for the third loop than for the first two loops.

The program "M_CountingLoop_Servo6.tsk3" illustrates two new Robot Control Concepts:

F. **When using loops (of any type), the programmer needs to check on the Variables used inside the loops and to make sure that they take on the intended numerical values during each iteration of the loops. In TASK, this check can be done using PRINT or PRINTLINE statements as needed.**
G. **The programmer needs to be aware that internally TASK programs use 16 bits to store user-defined parameters, however many robot control parameters only use the contents in the first 10 bits [0-1023].**

6.1.5 Conditional Loop Control of Servo Positions with Remote Controller

The TASK program "M_ConditionalLoopControl_Servo6.tsk3" has two goals: the first one is to illustrate the proper use of the LOOP-WHILE structure, and the second one is to introduce the usage of the Virtual Remote Controller RC-100 (see Fig. 6.8).

The usage of a LOOP-WHILE structure involves more "coordination" than for the other two loops: ENDLESS-LOOP and LOOP-FOR. But it is also the most flexible and powerful one to use. From Section

6.1.4, a reminder that a LOOP-FOR structure defines its Starting/Entering Condition (Count = 1), Continuing Condition (Count <= 3) and Ending/Exiting Condition (Count > 3) in a single statement (see Statement 17 in Fig. 6.7).

```
4   START PROGRAM
5   {
6       // Overall robot initialization
7       ID[All]: Torque ON/OFF = FALSE (0)
8       ID[All]: LED = FALSE (0)
9       // Special initialization for Servo 6
10      ID[6]: Torque ON/OFF = TRUE (1)
11      ID[6]: Operating Mode = 2
12      ID[6]: Torque Limit = 1023
13      ID[6]: Goal Velocity = 1023
14      // Read in Remocon Input
15      DataIn = Remocon RXD
16      LOOP WHILE ( DataIn == -- )
17      {
18          // Command Servo 6 to move to 0 degree position
19          ID[6]: Goal Position = 512
20          WAIT WHILE ( ID[6]: Is Moving == TRUE (1) )
21          // Command Servo 6 to move to -90 degree position
22          ID[6]: Goal Position = 205
23          WAIT WHILE ( ID[6]: Is Moving == TRUE (1) )
24          // Command Servo 6 to move to +90 degree position
25          ID[6]: Goal Position = 819
26          WAIT WHILE ( ID[6]: Is Moving == TRUE (1) )
27          DataIn = Remocon RXD
28          Print Screen with Line = DataIn
29      }
30  }
```

Fig. 6.8 TASK program "M_ConditionalLoopControl_Servo6.tsk3".

In Fig. 6.8, it looks like that Statement 16 defines the LOOP-WHILE structure which has a logical expression defined as DataIn == --) (i.e. DataIn == 0), and it is used as the Starting/Entering, Continuation and Ending/Exiting Conditions. This logical expression, when evaluated to TRUE, translates into every-day language as "**no button had been pushed on the Remote Controller**", and when it is evaluated to FALSE, it translates into "**some button(s) had been pushed on the Remote Controller**".

If this logical expression evaluates to FALSE when Statement 16 is **first executed**, then the loop's contents defined between Statements 17 and 29 are skipped completely and Statement 30 is next executed, meaning that the program terminates. However, if this expression evaluates to TRUE when Statement 16 is **first executed**, then the loop's contents defined between Statements 17 and 29 are executed in the shown sequence, and whenever Statement 29 is reached, the OpenCM9.04-C returns to Statement 16 and re-evaluates this expression. If this expression remains TRUE then the looping process continues, but if this expression evaluates to FALSE then the loop is immediately exited to Statement 30, and the Main program terminates.

Let's take a closer look at how this LOOP-WHILE's Starting Condition is controlled by Statement 15 which essentially reads in the latest input from the Remote Controller RC-100 and saves it into Parameter "Dataln". If the user downloads and then starts this program, he/she should then avoid touching any of the buttons of the Remote Controller RC-100 (whether Virtual or Physical), so that Parameter "Dataln" is assigned a value of "0" which would make the logical expression (Dataln == --) evaluate to TRUE, ensuring that the LOOP-WHILE structure at Statement 16 will be entered and therefore will make Servo 6 move to its three programmed positions as before. When Statement 27 is reached, a new value is read in from the Remote Controller and saved into "Dataln". If the user still does not touch any button on the RC-100, then "Dataln" retains its old/current value of "0", and another iteration of this loop structure happens. But if the user had touched some button(s) on the RC-100, then "Dataln" would be set to some number different from "0". Thus, this logical expression (Dataln == --) would evaluate to "FALSE" and would make this loop exit, and subsequently the Main program would also stop.

Statement 15 is important for it controls whether this LOOP WHILE is skipped or entered for the first time. Please note that this Initialization step for the Loop Control Variable "Dataln" is always defined "before" the LOOP WHILE statement (Statement 16). In later programming examples, this Initialization step may be performed elsewhere in the MAIN program or even in other FUNCTIONs.

Once this LOOP WHILE is entered, then Statement 27 is important for it controls whether this LOOP WHILE is continued once more or that it is time to exit this LOOP WHILE and go on to Statement 30. Statement 27 is used to update the Loop Control Variable "Dataln" with the latest data input. Beginner programmers may forget to program in statements such as Statement 27, which means that "Dataln" retains its old data yielding an "equivalent" ENDLESS LOOP situation as the Loop Logical Expression would always evaluate to TRUE. This Update step is usually programmed in just before the closing curly bracket defining the LOOP WHILE. However, in later programming examples, this Initialization step may be performed elsewhere in the MAIN program or in other FUNCTIONs also.

Please see Video 6.6 for a runtime performance of the "M_ConditionalLoopControl_Servo6.tsk3" program.

The program "M_ConditionalLoop_Servo6.tsk3" illustrates two new Robot Control Concepts:

- **H. To read in the latest input from the Remote Controller RC-100, just access the Remocon RXD device as shown in Fig. 6.9 (Lines 15 or 27) and save the result in a user-defined parameter such as "Dataln".**
- **I. The logical expression used in defining the LOOP WHILE Condition is always tested by the OpenCM9.04-C controller <u>before</u> it decides whether to <u>skip</u> (when evaluated to FALSE) or <u>execute</u> (when evaluated to TRUE) the statements contained within the curly brackets defining the LOOP WHILE. The programmer also needs to ascertain that the VARIABLE(s) used in this logical expression are properly "initialized" and "updated" to ensure that the LOOP WHILE behave as intended by the programmer.**

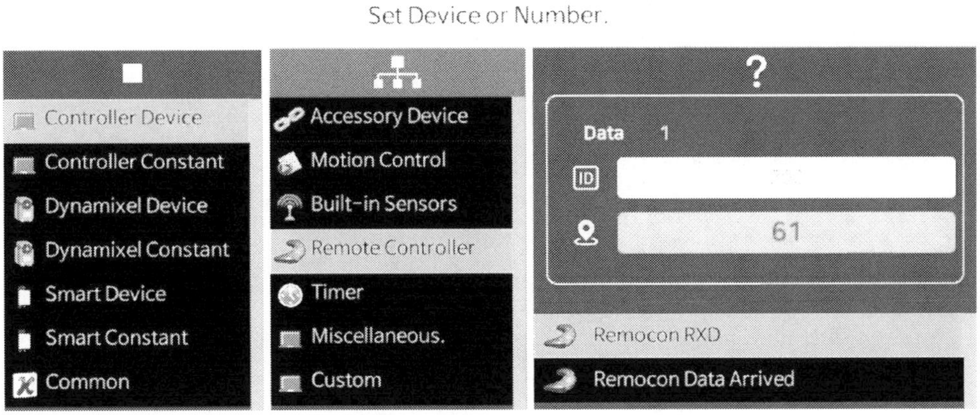

Fig. 6.9 Accessing "Remocon RXD".

6.2 Structure of a complete TASK program

In this Section 6.2, the TASK program "M_RC_Walker_Basic_a.tsk3" is used just to illustrate the various components of a typical TASK code. A detailed procedure will be provided in Section 6.3 to illustrate a general approach to solve a typical robotics project resulting in such a TASK program.

Please note that the program "M_RC_Walker_Basic_a.tsk3" uses the "RC_Walk_1" MOTION-GROUP from the "MINI_Custom.mtn3" file (see Fig. 6.10). **The reader needs to make sure to download this MOTION-GROUP before downloading and running the "M_RC_Walker_Basic_a.tsk3" code.**

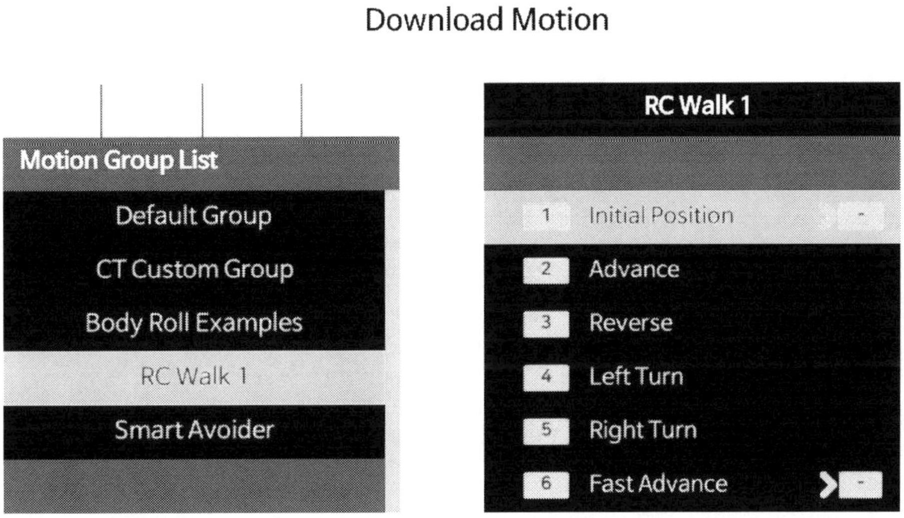

Fig. 6.10 Accessing "RC_Walk_1" MOTION-GROUP.

Fig. 6.11 shows Part 1 of the program "M_RC_Walker_Basic_a.tsk3".

```
// MINI RC Walk V. 1
// Walk in only 1 direction U-D-L-R at any one time using IF-ELSE-IF
// All Rights Reserved - CNT Robotics LLC 2019

START PROGRAM
{
    // Overall robot initialization
    CALL Init
```

Fig. 6.11 Part 1 of "M_RC_Walker_Basic_a.tsk3".

Fig. 6.12 shows Part 2 of the program "M_RC_Walker_Basic_a.tsk3".

```
    ENDLESS LOOP
    {
        // Remote Control allowed only when robot is NOT IN MOTION
        IF ( Motion Status == FALSE (0) )
        {
            // Wait if no new Remocon Data Received
            WAIT WHILE ( Remocon Data Arrived == FALSE (0) )
            // Read in Remocon Input
            DataIn = Remocon RXD
            IF ( DataIn == U )
            {
                // Go forward
                Motion Index Number = Motion [2]: Advance
            }
            ELSE IF ( DataIn == D )
            {
                // Go backward
                Motion Index Number = Motion [3]: Reverse
            }
            ELSE IF ( DataIn == L )
            {
                // Go left
                Motion Index Number = Motion [4]: Left Turn
            }
            ELSE IF ( DataIn == R )
            {
                // Go right
                Motion Index Number = Motion [5]: Right Turn
            }
        }
    }
}
```

Fig. 6.12 Part 2 of "M_RC_Walker_Basic_a.tsk3".

Fig. 6.13 shows Part 3 of the program "M_RC_Walker_Basic_a.tsk3.tsk3".

Fig. 6.13 Part 3 of "M_RC_Walker_Basic_a.tsk3".

6.2.1 Comment & Command Statements

"Comment" statements are Statements 1-3, 7, 12, 15, 17, 21, 26, 31 and 36 which are used as minimal annotations to help the reader follow the logic of the program. They will be ignored by the TASK tool during the compiling step of the coding process.

The true "Command" statements are the remaining Statements found in Figs. 6.11, 6.12 and 6.13. The TASK tool will check these statements for proper syntax and if no compile error is found, it will convert these statements into proper machine code to download to the OpenCM9.04-C controller which in turn would execute this machine code at runtime, when power is reset to the OpenCM9.04-C controller.

6.2.2 Start & End Commands

Matching opened "{" and closed "}" curly brackets are used to mark various blocks of code to convey their logical sequencing and grouping, as well as their hierarchical levels which are denoted by subsequent indentations towards the right side of the program listing.

The (one and only) MAIN program begins with the opened curly bracket at Line 6 (Fig. 6.11) and ends with its matching closed curly bracket at Line 41 of Fig. 6.12.

Within the MAIN block, the next-level sub-block is the ENDLESS LOOP defined at Line 10 (Fig. 6.12) and its contents begin with the opened curly bracket at Line 11 (Fig. 6.12) and ends with its matching closed curly bracket at Line 40 of Fig. 6.12.

Within the above ENDLESS LOOP, the IF "Motion Status" sub-sub-block is defined at Line 13 (Fig. 6.12) and its contents begin with the opened curly bracket at Line 14 (Fig. 6.12) and ends with its matching closed curly bracket at Line 39 of Fig. 6.12.

Within the above IF "Motion Status" sub-sub-block, there is an IF-ELSE-IF structure which begins at Line 19 and ends at Line 38 in Fig. 6.12. This IF-ELSE-IF structure has 4 independent components (depending on the actual value for "DataIn" at runtime) that can be executed as **one and only one instance** at any given time during the runtime of this TASK program. Each of those specific cases for "DataIn" is defined within its matching curly brackets such as Lines 30 and 33 for the case when (DataIn == L) .

Fig. 6.13 illustrates a user-defined FUNCTION named "Init" which can act as an independent sub-program that can be "called into execution" from the MAIN program (Line 8 of Fig. 6.11) or from other user-defined FUNCTIONS. It is defined at Line 43 in Fig. 6.13 and its contents are defined within the matching curly brackets at Lines 44 and 52.

In later example programs, the use of a special function named CALLBACK will be illustrated. The CALLBACK function is executed every 7.81 ms by the OpenCM9.04-C independently of the MAIN program and all other user-defined FUNCTIONs (i.e. it is under hardware control).

6.2.3 Execute Statements

The most common type of EXECUTE Statements is the LOAD type and it is used throughout this example program at Lines 18, 22, 27, 32, 37, 45 to 50 in Figs. 6.12 and 6.13. EXECUTE Statements affect the informational data flow/status from sensors/parameters and the subsequent robot's actions. Without them, the robot would be a "no-fun" static object.

6.2.4 Loop Statements

Only two types of LOOP Statements are used in this example TASK code and they are of the type ENDLESS-LOOP (Lines 10 in Fig. 6.12) and WAIT-WHILE (Line 16 in Fig. 6.12 and Line 51 in Fig. 6.13). Section 6.3 will show that the ENDLESS-LOOP is a direct consequence of the use of the "Sense-Think-Act" paradigm. LOOP Statements allow the creation of compact code blocks that can be conditionally repeated instead of having to lay them out sequentially in the controller's memory space and thus make more efficient use of this limited resource.

6.2.5 Condition Statements

Only two types of CONDITION Statements are used in this example TASK code and they are of the type IF (Lines 13 and 19 in Fig. 6.12) and ELSE-IF (Lines 24, 29 and 34 in Fig. 6.12. Section 6.3 will show that the CONDITION Statements flow from the use of the EVENT-PROGRAMMING approach. In a way, CONDITION Statements allow the mapping of all possible event/action pairs that are currently considered by the programmer for the robot to deal with. CONDITION Statements allow the robot to respond to a changing environment, in the real world, that it must work within.

6.2.6 Function Statements

The only FUNCTION Statements used are the FUNCTION (definition) used at Line 43 of Fig. 6.13) and the (function) CALL at Line 8 of Fig. 6.11. FUNCTION Statements allow the programmer to use a top-down and modular approach to develop an efficient solution to a given robotics project with defined goals and tasks (see Section 6.3).

6.3 Beginner's Approach for Robotics Projects

In this section, the author strives to provide beginning "roboteers" with a general approach to thinking about robotics projects, as well as concrete sequential steps that ultimately result in an efficient TASK program that can perform the tasks/behaviors that are required of the MINI robot for a given project.

6.3.1 "Sense-Think-Act" Paradigm

Most if not all my beginning robotics students were surprised when I started my robotics short courses by mentioning that they had been doing "Robotics" in their everyday activities and doing it very well too, all by themselves. I used Fig. 6.14 to explain how humans use our Senses to let us know about the external World (i.e. Perception) and apply those Sensations into our Cognitive process (i.e. Thinking) to devise proper Actions/Reactions to the situation at hand. These Actions/Reactions in turn would change some aspects of the external World resulting in new Sensations triggering the next World-Human-Interactions cycle and so forth.

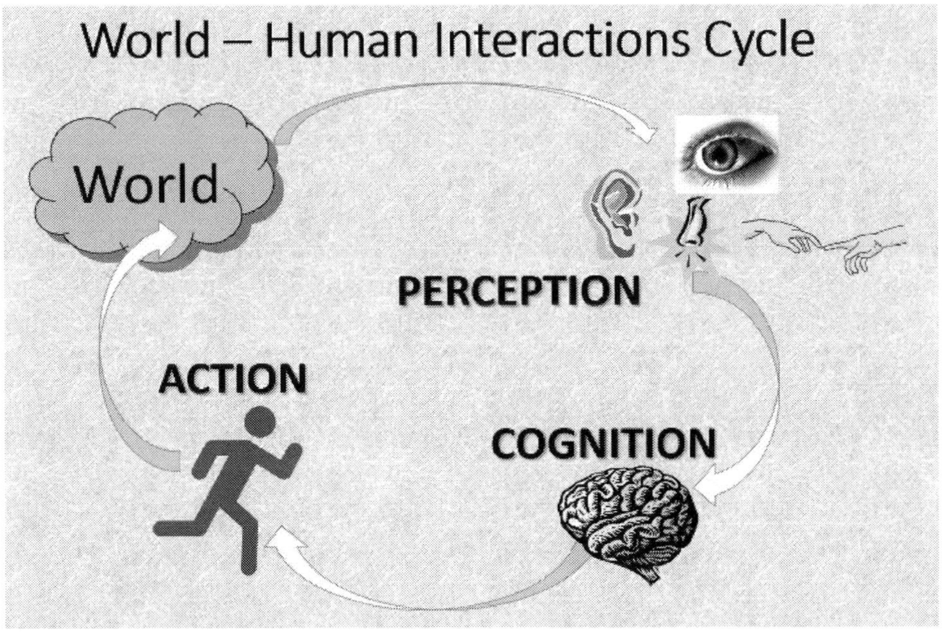

Fig. 6.14 World-Human Interactions Cycle.

Thus, in a way, doing Robotics is just trying to reproduce this interactions cycle onto a robot as shown in Fig. 6.15. For the "Autonomous" option (inner loop), mechanical or electronic Sensors would convert their interactions with the World into Input Data to be sent to the Hardware Controller which was previously Programmed by the user to deal with Events deemed important to the operation of the Robot. Using this user-programmed Logic, the Hardware Controller Outputs appropriate commands to its Actuators which then could change the Sensors' perspective onto the World and therefore generate new Input Data for the Hardware Controller to process anew through another interactions cycle.

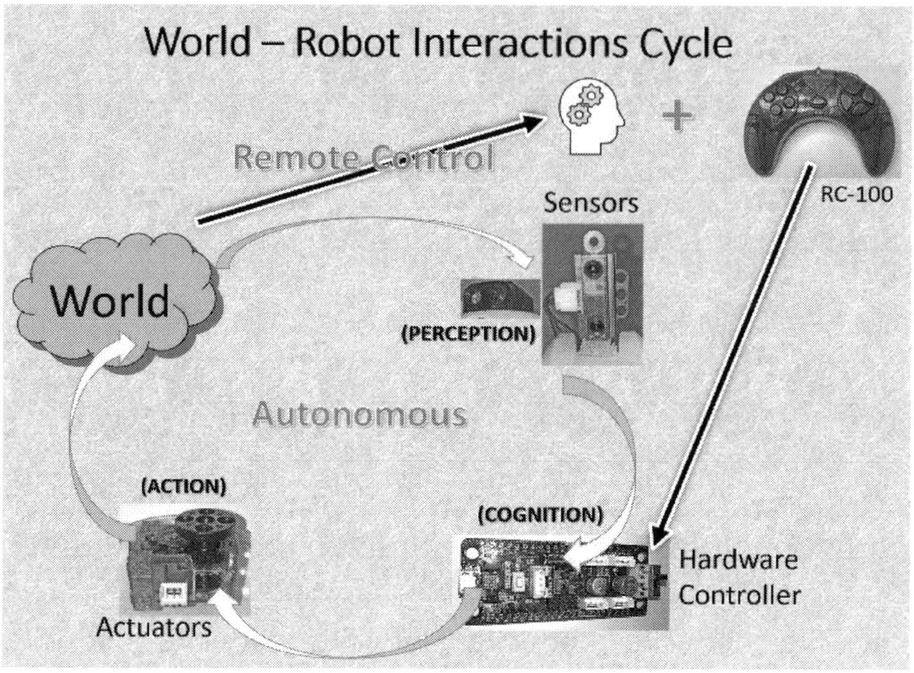

Fig. 6.15 World-Robot Interactions Cycle.

In the case, when the robot does not have enough sophistication in its sensing and processing capacities, the human user then handles all the sensing and decision-making processes and through a remote-control device he/she just sends direct commands to the robot telling it what to do (i.e. "Remote Control" outer loop in Fig. 6.15). Of course, depending on the robotic application's needs, ones can have a gradual variation between "100% autonomous" and "100% remote-control" (for example, consider the current situation for "smart cars" whereas various control approaches are being developed).

Another approach to increase the sensing and processing capacities of the robot is to just build those extra capacities into the robot/hardware-controller itself at great costs. But for consumer robotics, a cheaper "Cooperating-Controllers" approach is usually followed and that is to leverage other computing platforms that already have the features that the robot designer would need such as machine vision, multimedia services, speech synthesis or recognition, etc. (see Fig. 6.16). ROBOTIS follows this approach, for example:

- For the PLAY700 system, ROBOTIS created the PLAY700 Mobile App to leverage many features of the ubiquitous smartphone for the CM-50. Fortunately, the MINI system can also interface with the PLAY700 App (https://play.google.com/store/apps/details?id=com.robotis.play700&hl=en_US) (see more details in Chapter 7).
- Although outside the scope of this book, a more advanced ROBOTIS system that uses this approach is the TurtleBot3 Waffle Pi that combines the use of a Local Controller (OpenCR) and a Single Board Computer (Raspberry Pi 3). The interested reader can read more details at this web link (http://emanual.robotis.com/docs/en/platform/turtlebot3/specifications/#hardware-specifications).

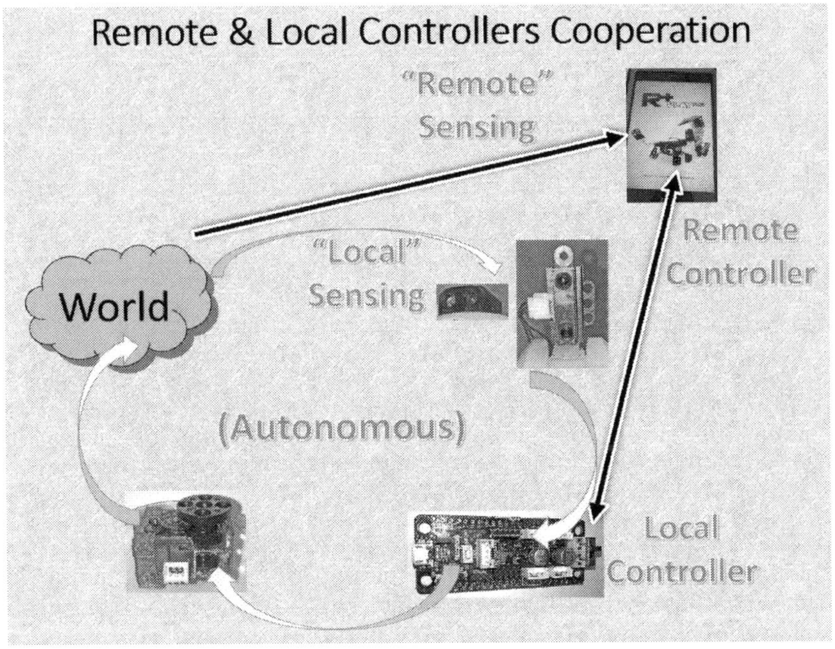

Fig. 6.16 "Cooperating-Controllers" Approach for ROBOTIS robots.

The previous concepts are usually encapsulated within the "Sense-Think-Act" Paradigm for Robotics which is very useful for us "roboteers" as it helps us keep the "big picture" of any robotics project in mind, as we start our "drilling down" journey into this paradigm, in order to get to the actual robot-controlling code that works on the OpenCM9.04-C.

6.3.2 Reactive-Control Approach

Next, let's spell out the **goals** for the current project "RC Walker Basic" that can lead to the example code "M_RC_Walker_Basic_a.tsk3" which used MOTION-LISTs/PAGEs defined in the MOTION-GROUP "RC_Walk_1" (see Fig. 6.10):

1. **We want to use the Virtual Remote Controller (VRC) feature of the TASK 3 Windows Application to control the movement of a MINI robot (see Fig. 1.16).**
2. **If no button is pushed on the VRC, the MINI should stand still in a Ready Pose.**
3. **If the "U" button is pushed, the MINI should make 3 steps forward and then stop with both feet in-line.**
4. **If the "D" button is pushed, the MINI should make 3 steps backward and then stop with both feet in-line.**
5. **If the "L" button is pushed, the MINI should make 3 steps to the left and then stop with both feet in-line.**
6. **If the "R" button is pushed, the MINI should make 3 steps to the right and then stop with both feet in-line.**

Let's see how we can use the "Sense-Think-Act" (STA) Paradigm to help us program the MINI robot to perform the previous set of tasks. Fig. 6.17 is just another illustration of STA, but we'll be concentrating on the dashed red boundary in its lower right corner (i.e. zooming in on the Computer). STA still applies at this level, but different robotics practitioners would start using different labels such as "Events", "Actions" and

"Reactive Control" as they are more useful in the next steps that we need to take in order to create the actual computer code to control the MINI robot. Interested readers should read Matarić (2007) for a more complete description of other robotics control approaches, or Ben-Ari and Mondada (2018) for more details about "Reactive Control" and "Finite State Machines".

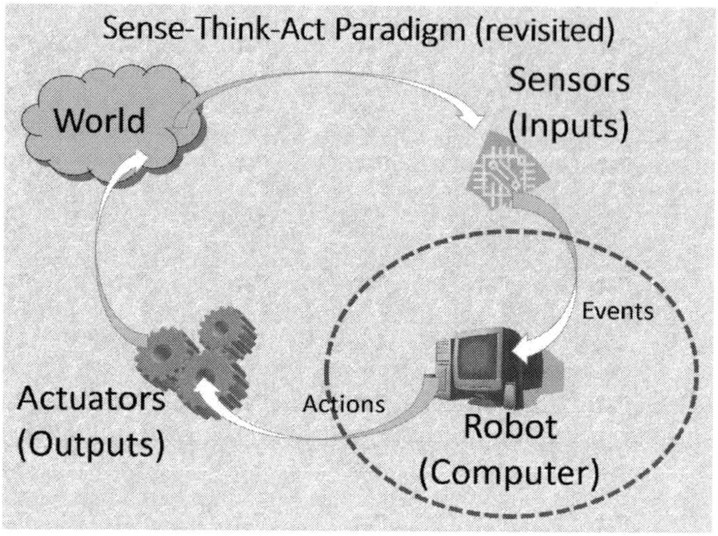

Fig. 6.17 "Sense-Think-Act" Paradigm revisited.

6.3.3 Event-Action Pairs

With the "Reactive Control" approach, the **first step RC1** is to translate each project goal from the above list into a specific Condition/Event matched with an appropriate Action for the robot to do (see Fig. 6.18).

Reactive Control Approach

Given Condition (Event) >> Appropriate Robot Action

Condition/Event	Action
No button pushed	MINI in Ready pose
"U" button pushed	MINI walks Forward
"D" button pushed	MINI walks Backward
"L" button pushed	MINI turns Left
"R" button pushed	MINI turns Right

Fig. 6.18 Event-Action Pairs.

At this stage of the process, the reader needs not consider any hardware or software systems in mind at all, perhaps just needing a "thought experiment" or "visualization exercise" approach whereas the reader would imagine to be "one" with the robot and would listen to specific verbal commands from another person (acting as Remote Controller), and then would perform the commanded movements.

Before we go on to the next step, we need to have a closer look at the last four Event-Action pairs listed in Fig. 6.18. Although it may be possible for the user to push several buttons on the RC-100 at the same time, from Chapters 4 and 5 we know that the MINI can only execute ONE MOTION-LIST/PAGE at any one time: for example, it will be impossible for the MINI to walk 3 steps forward and to the left at the same time. Therefore, we can constraint these five Event/Action pairs to be **mutually exclusive** in this project, meaning that the user can only push 1 button on the VRC when controlling the robot at runtime.

6.3.4 Basic Remote-Control Concepts (revisited)

Although some Remote-Control concepts were introduced in Section 6.1.4, let's now look more formally at how the OpenCM9.04-C can receive data from the VRC (i.e. the PC) via the BT-210 receiver:

1. Via its firmware, the Hardware Controller OpenCM9.04-C is "always" listening via its BT-210 for any "new" data coming in from the PC. When this event happens, it sets a flag named "**Remocon Data Arrived**" to TRUE, otherwise if there is no "new" data coming in, "Remocon Data Arrived" is set to FALSE.
2. When "Remocon Data Arrived" is set to TRUE, the Hardware Controller also saves the data received into a memory block named "**Remocon RXD**" accessible to the programmer who should then save this data into a user-defined **TASK Variable** (e.g. "DataIn").
3. The last processing step is to **compare "DataIn" to specific VRC constants** to determine which button U-D-L-R on the VRC had been pushed by the user. Fig. 6.19 illustrates where to find these Remote-Control parameters in the TASK menu system.

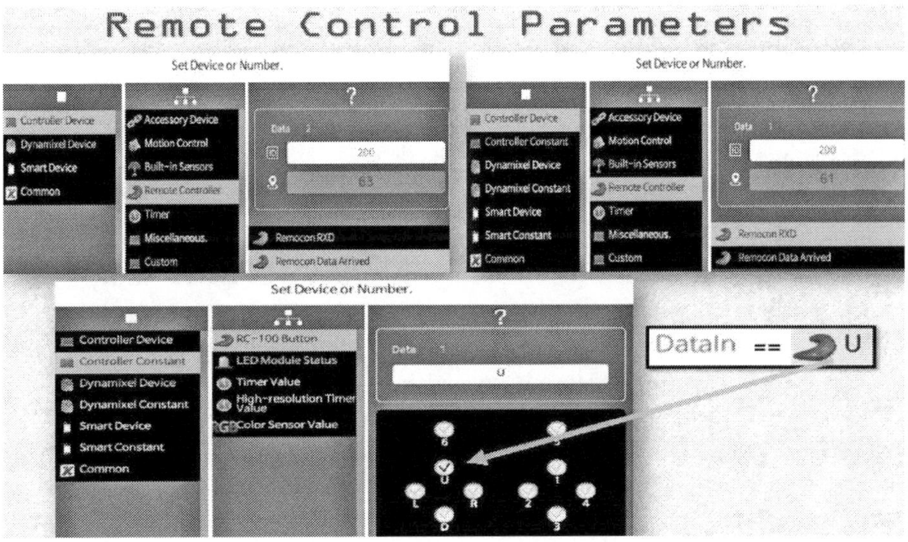

Fig. 6.19 Using Remote Control parameters.

1. Back in Section 6.1.4, this logical expression DataIn == -- was meant to be "no button had been pushed on the Remote Controller", but for this "Basic RC Walker" project, it can also stand for "all buttons had been released on the Remote Controller".

6.3.5 Sensor-Actuator Pairs

The **second step RC2** is to build upon the results of the previous step RC1 by mirroring its Event-Action pair into a matching Sensor-Actuator pair as shown in Fig. 6.20. **The goal is to specify the kind of Sensors/Actuators (in a general sense) to be used and to prescribe how they are to be used.**

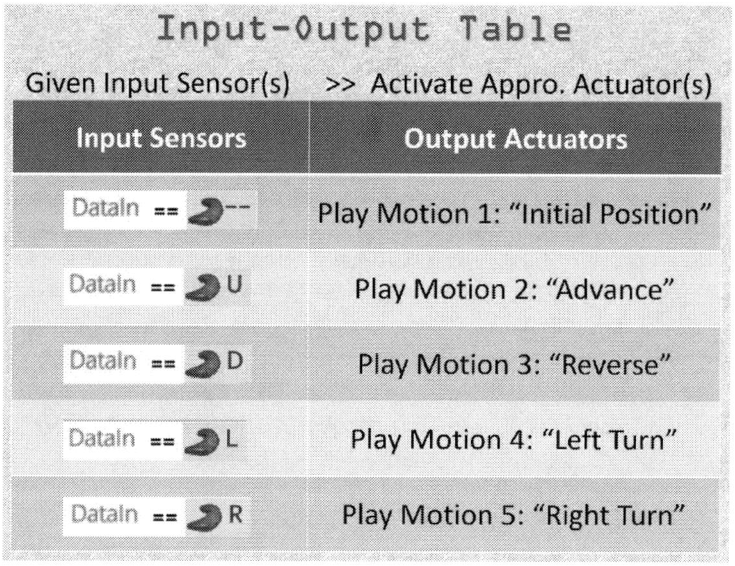
Input-Output Table

Given Input Sensor(s) >> Activate Appro. Actuator(s)

Input Sensors	Output Actuators
DataIn == --	Play Motion 1: "Initial Position"
DataIn == U	Play Motion 2: "Advance"
DataIn == D	Play Motion 3: "Reverse"
DataIn == L	Play Motion 4: "Left Turn"
DataIn == R	Play Motion 5: "Right Turn"

Fig. 6.20 Sensor-Actuator Pairs mirroring list of Event-Action Pairs shown in Fig. 6.18.

In this project, the author used the existing MOTION-LISTs/PAGEs provide by ROBOTIS, so the five MOTION-LISTs/PAGEs were already "prepared" in the MOTION-GROUP "RC Walk 1" (see Fig. 6.10). If he had the need to create new motions, he would have to create them at this point in time.

Please note that the previous procedures are independent of any computer programming language and the author highly recommends that beginner roboteers to always start with these procedures instead of jumping directly into coding or try to adapt someone else codes, as these procedures would help the roboteer refine in his/her mind the possible situations that the robot may encounter and the appropriate tasks to be performed by the robot, before committing them into hard codes, thus saving the programmer lots of time in the long run. **In the "Computational Thinking" practice, these steps would correspond to the "Decomposition" phase** (Beecher, 2017; Denning and Tedre, 2019).

In real practice, the user should not expect to come up with a complete list of Event-Action and Sensor-Actuator Pairs in the first trial - the author never did in his many years of practice! Thus, refinement steps are "normal"! Furthermore, in more complex situations, the Event-Action pairs may have to follow a hierarchy of "cascading" events as shown in more advanced projects later in this book.

6.3.6 Event-Programming Approach

Another benefit for using Event-Action and Sensor-Actuator Pairs is that they will help with the actual coding (as shown later in Section 6.3.7), as the TASK syntax would support practically a one-to-one correspondence of all Event-Action pairs A to Z (see Fig. 6.21) into similar "physical" locations within the final program, except for two aspects that had not been yet discussed:

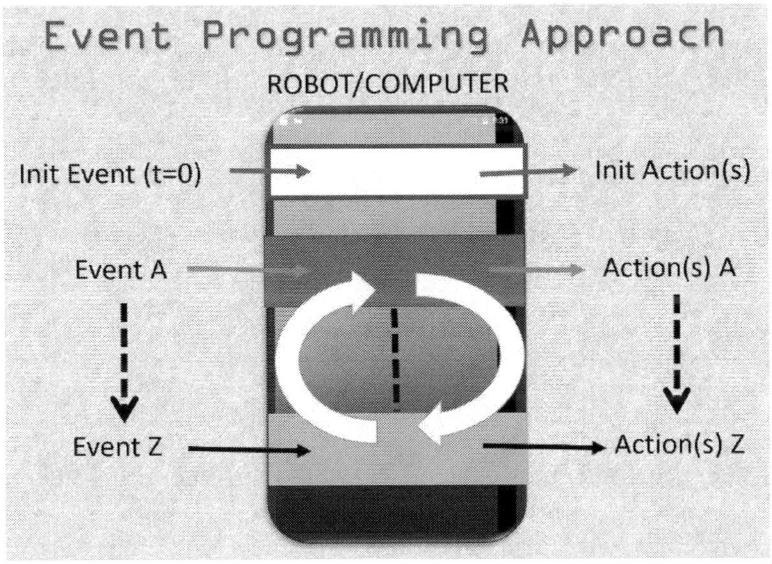

Fig. 6.21 Event-Action pairs matching locations in TASK code.

1) A main code looping structure that allows the controller to go through the list of possible Events A to Z that may occur during run time, as it listens to what the Sensors tell it about the current state of the World. This overall loop represents the Interactions Cycle as shown in Figs. 6.15, 6.16 and 6.17.

2) The Init Event which occurs once at time=0, i.e. when power is applied to the robot or when the TASK program starts running. The Init Event would usually result in a known state for the robot. For this project, we need to set all XL-320s into Joint Mode and at maximum Torque Limit and Goal Velocity levels. The MINI should also be set in the Initial Position, standing ready for actions, i.e. to receive commands from the VRC.

In "Computational Thinking" practice, these steps would correspond to the "Pattern Recognition" phase.

6.3.7 Application to Basic RC Walker

Recalling the previous discussions made in Section 6.3.3, this project deals with **mutually exclusive Event-Action Pairs**, and TASK has a CONDITION structure that will function in this manner. It is called the IF-ELSE-IF structure which can be laid out in pseudo code below:

If (Event 1 is TRUE) then (Play Robot Motion 1)

Else If (Event 2 is TRUE) then (Play Robot Motion 2)

....

Else If (Event 5 is TRUE) then (Play Robot Motion 5)

In a way, readers can consider this IF-ELSE-IF structure as a listing of all possible Event-Action Pairs that the robot may encounter at any point in time when this program is executed. There is a specific way that the Hardware Controller (HC) will execute this IF-ELSE-IF structure at runtime:

- **Event 1** is always checked first, and if it happens to be **TRUE**, then Motion 1 is played, and HC jumps out of this IF-ELSE-IF structure from this spot and goes on executing the rest of the TASK program. If Event 1 happens to be **FALSE**, then HC goes on to **check Event 2 next**.
- If **Event 2** happens to be **TRUE**, then Motion 2 is played, and HC jumps out of this IF-ELSE-IF structure from this spot and goes on executing the rest of the TASK code. If Event 2 happens to be **FALSE**, then HC goes on to **check Event 3 next**.
- The same steps as described above apply to **Event 3**, **Event 4** and **Event 5**.

In "Computational Thinking" practice, these steps would also correspond to the "Pattern Recognition" phase.

A sharp reader may have noticed that the appropriate action for Event 1 (no button is pushed) is started in a shorter time as compared to the start of the appropriate action for Event 5 (R button is pushed) which has to wait for all Events 1 through 4 to have tested out FALSE. However, the time differences are in the microseconds for the HC, so the human user would not experience any lag in the response of the robot when pushing different buttons at runtime.

Fig. 6.22 displays the Init Event which was implemented using the Function Init:

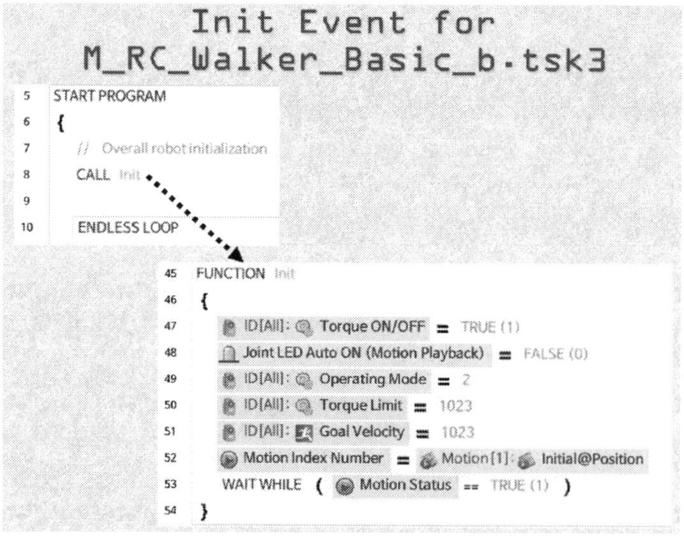

Fig. 6.22 Init Event in "M_RC_Walker_Basic_b.tsk3".

- When the HC starts to execute this program, it "calls" Function Init (Line 8), meaning that it shifts its Execution Counter to Line 45 to enter Function Init.
- Next, it executes the sequence of statements from Line 46 through 54:
  - Line 47 enables "Torque", i.e. electrical power, to all 17 servos.
  - Line 48 turns OFF the servos' LEDs during motion playback (to save some batteries power).
  - Line 49 sets all servos to Joint Mode.
  - Line 50 sets all servos' Torque Limit to the maximum value of 1023.
  - Line 51 sets all servos' Goal Velocity to the maximum value of 1023. (*Note: this setting is ignored while a MOTION-LIST/PAGE is performed, as the key-frame timing and playing speed used respectively in its MOTION-UNITS and MOTION-LIST/PAGES override this overall Goal Velocity setting)*.
  - Line 52 plays the MOTION-LIST/PAGE 1, i.e. "Initial Position", to set the MINI in a ready pose by using the "Play Motion Index Number" command.
  - Line 53 is always used with a "Play Motion Index Number" command to wait for a started MOTION-LIST/PAGE to finish and the robot to stop moving, i.e. when the flag "Play Motion Status" is set to FALSE.
  - Line 54 denotes the end of Function Init, so HC's Execution Counter is now restored back to the main program at Line 9 which a blank statement, so it goes to the next executable statement Line 10 and therefore enters the Main Endless Loop.

In "Computational Thinking" practice, these steps would also correspond to the "Abstraction" phase.

Fig. 6.23 contains the **essential algorithm** used in this project (i.e. sequential steps for HC to execute):

1. To check for any new data input from the VRC (Line 13) and if there is new data input, read it in and save it in Parameter "DataIn" (Line 15). If not, HC just waits at Line 13.
2. Depending on the actual value contained in "DataIn" at runtime, implement the correct motion for the robot to make using an IF-ELSE-IF control structure (Lines 16 through 40).
3. Wait for the started motion to finish (Line 41), i.e. for the robot to stop moving, before going through this algorithm once more.

In this algorithm, the author chose to implement the "Motion Status" check (Line 41) outside the IF-ELSE-IF structure (i.e. only once) instead of doing it right away after each "Play Motion Index Number" command as used in Function Init (i.e. for 5 times), so as to avoid obvious redundant statements and subsequently a smaller memory storage size for this TASK program.

In "Computational Thinking" practice, these steps would correspond to the "Algorithm" phase.

Main Endless Loop for M_RC_Walker_Basic_b.tsk3

```
ENDLESS LOOP
{
    // Wait if no new Remocon Data Received
    WAIT WHILE ( Remocon Data Arrived == FALSE (0) )
    // Read in Remocon Input
    DataIn = Remocon RXD
    IF ( DataIn == -- )
    {
        // Ready Pose
        Motion Index Number = Motion[1]: Initial@Position
    }
    ELSE IF ( DataIn == U )
    {
        // Go forward
        Motion Index Number = Motion[2]: Advance
    }
    ELSE IF ( DataIn == D )
    {
        // Go backward
        Motion Index Number = Motion[3]: Reverse
    }
    ELSE IF ( DataIn == L )
    {
        // Turn left
        Motion Index Number = Motion[4]: Left@Turn
    }
    ELSE IF ( DataIn == R )
    {
        // Turn right
        Motion Index Number = Motion[5]: Right@Turn
    }
    WAIT WHILE ( Motion Status == TRUE (1) )
}
}
```

Fig. 6.23 Main Endless Loop in "M_RC_Walker_Basic_b.tsk3".

By now, readers must have realized that this Version B is not quite the same as the "previously advertised" Version A (see Fig. 6.12 in Section 6.2). This was made on purpose to show that it is perfectly normal to have 2 TASK programs that would yield the same runtime performance for a robot as evidenced in Video 6.7. The difference in program size between the two versions was only 6 bytes to Version A's advantage.

But it may be educational for the readers to know the reasons why the author created Version A:

1. As Function Init already put the robot in Initial Position at the start of runtime, there really was no need to check for the case of DataIn == -- as the robot is already doing what it is

supposed to do for the case when there is no new data input coming in from the VRC. Also, it is because the other 4 motions always end with the Initial Position.

2. **Using the IF check on "Play Motion Status" at the beginning of the Endless Loop in Version A (Line 13 in Fig. 6.12) would allow the HC to do other actions** - if needed – in the Main Endless Loop such as reading in sensors data and/or perform more "meaningful" actions for the robot, **instead of just wasting controller cycles with the WAIT WHILE construct (Line 41 in Version B)**.

In "Computational Thinking" practice, these steps would also correspond to the "Algorithm" (Refining) phase.

In practice, diligent users would soon realize that backtracking various steps of this procedure for improving one's TASK program will be the "norm" for any meaningful robotic project.

6.3.8 TASK Support Utilities

The TASK Sub-tool of TASK 3 also provide several support utilities during the programming and debugging phases of the user's robotics projects:

1. In its Programming Menu (F2) (see Fig. 6.24):
    a. The usual editing functions are implemented – Cut, Copy, Paste, Redo & Undo.
    b. The user can export a TSK3 program into several PNG files depending on the length of the TSK3 program or into a single ASCII text file.
    c. The user can also choose the appropriate COM port before downloading one's TASK code.
2. In its Debugging Menu (F3) (see Fig. 6.25):
    a. The user can start or stop the Program Output Monitor to see print outputs at runtime, if the user's program uses PRINT or PRINTLN commands.
    b. If the user's program uses the Remote Controller RC-100 features, then the Virtual Remote Controller facility is accessible here. The user needs to remember that the arrow keys (U-D-L-R) and the "number" keys near the top of the keyboard can also be used instead on left-clicking with the mouse into the "buttons" shown in Fig. 6.25.

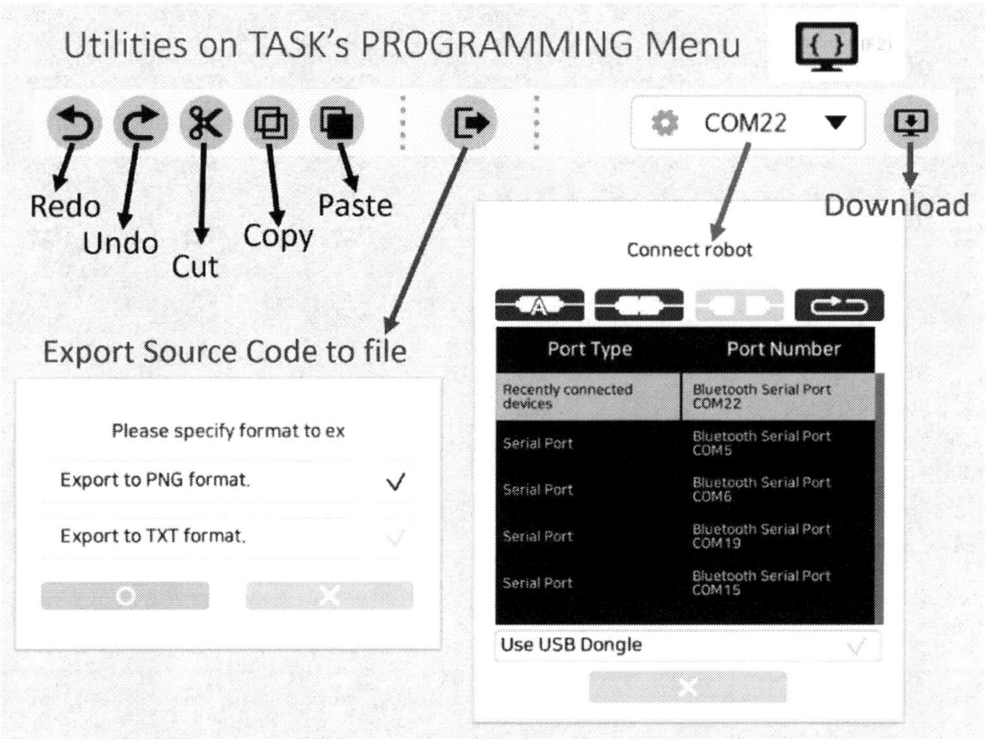

Fig. 6.24 Utilities found in TASK's Programming Menu (F2).

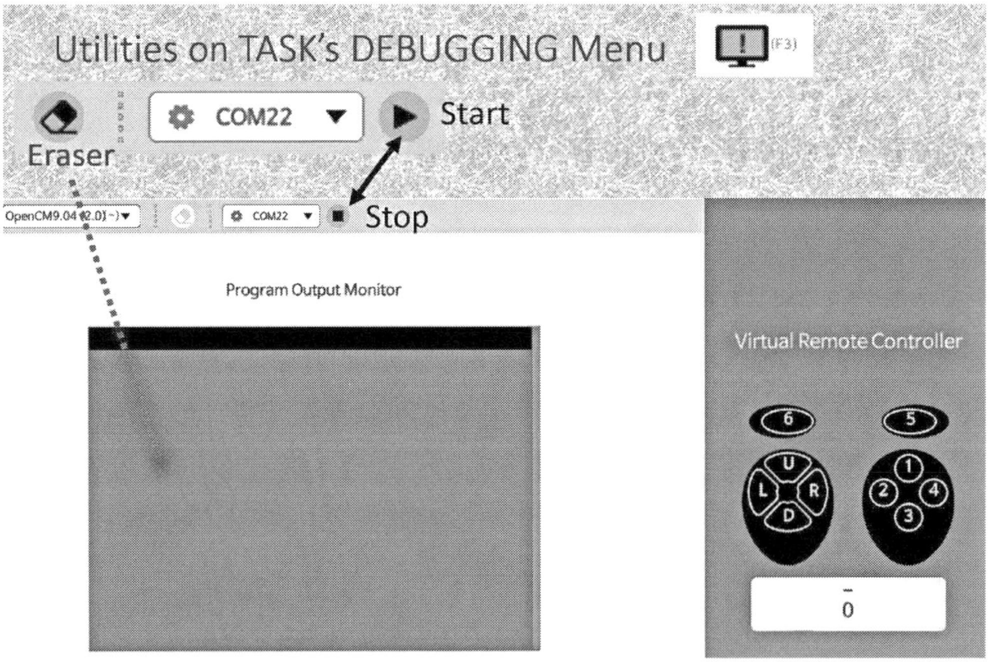

Fig. 6.25 Utilities found in TASK's Debugging Menu (F3).

When writing a TSK3 program, the user mostly uses the "Instructions" Tab (F9) of the Right Sidebar (see Fig. 6.26) to choose the various computer statements implemented in TASK. But there are other tabs that are also very useful:

1. The "Variables" Tab (F10) can be used to find how many variables are used in the user's program as well as which line numbers was a chosen Variable used at. The Editing tool can be used to change a Variable's name throughput the whole TASK program.
2. The "Functions" Tab (F11) (see Fig. 6.27) can be used in a similar way as for the "Variables" Tab, but of course only for FUNCTIONs used in the user's program.
3. The "Search" Tab (Ctrl+F) can search for any text string in the user's program.
4. Finally, the "Compile" Tab to find syntax errors that may be present in the user's program, including the line number where they occur (see Fig. 6.28).

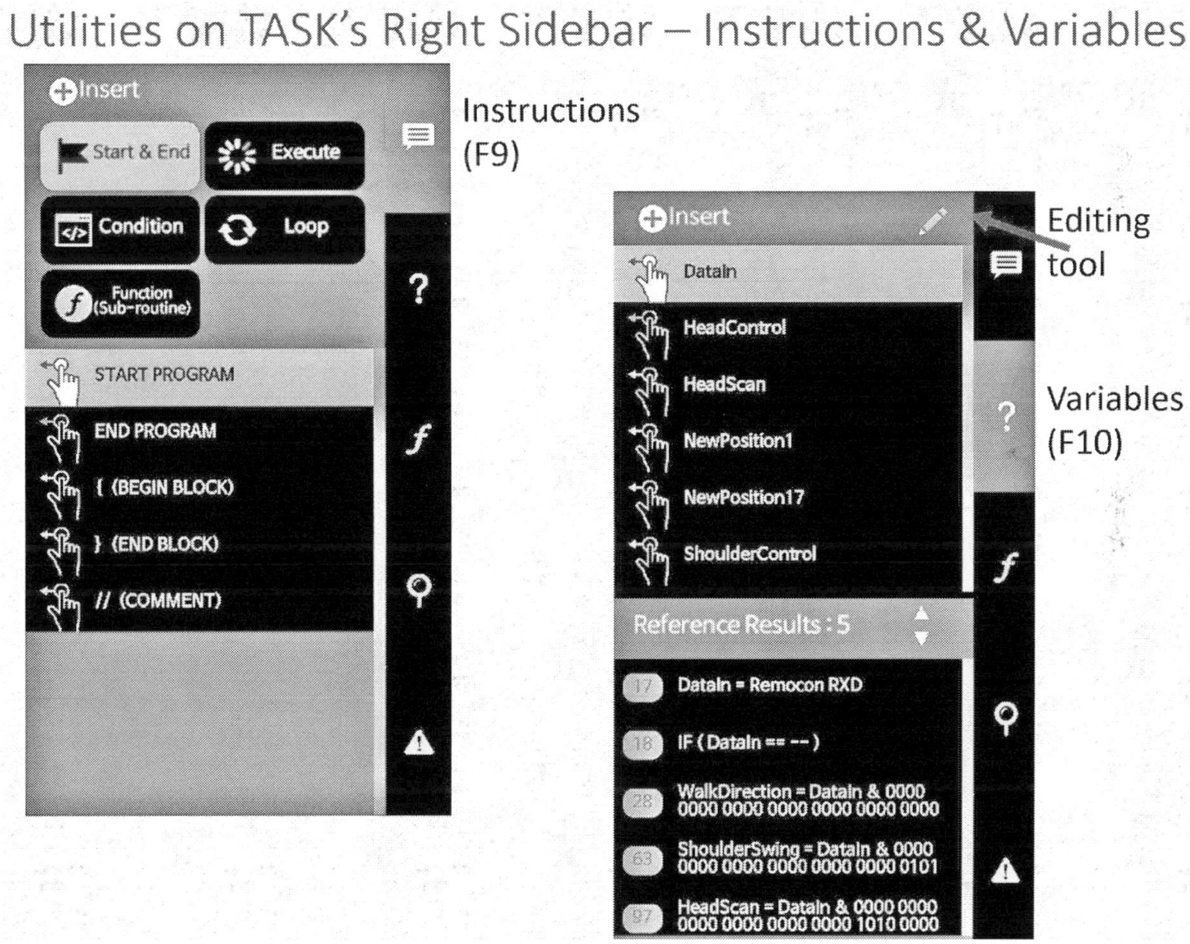

Fig. 6.26 Utilities found in TASK's Right Sidebar – "Instructions" & "Variables".

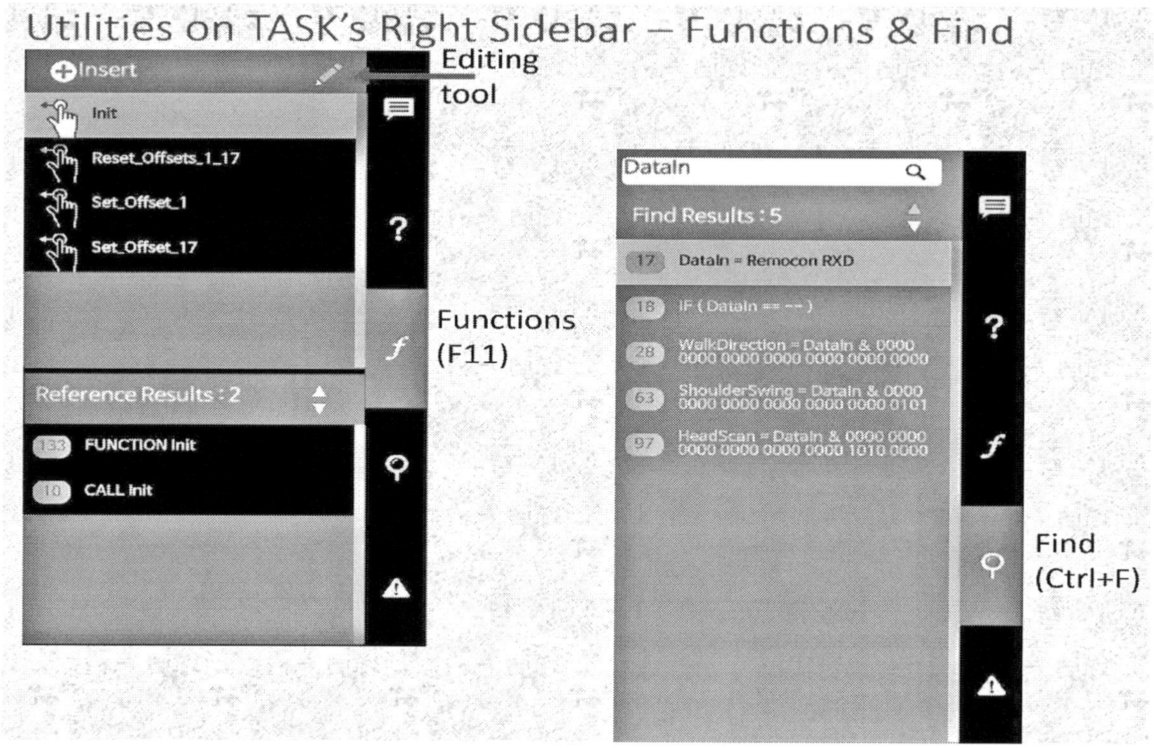

Fig. 6.27 Utilities found in TASK's Right Sidebar – "Functions" & "Find".

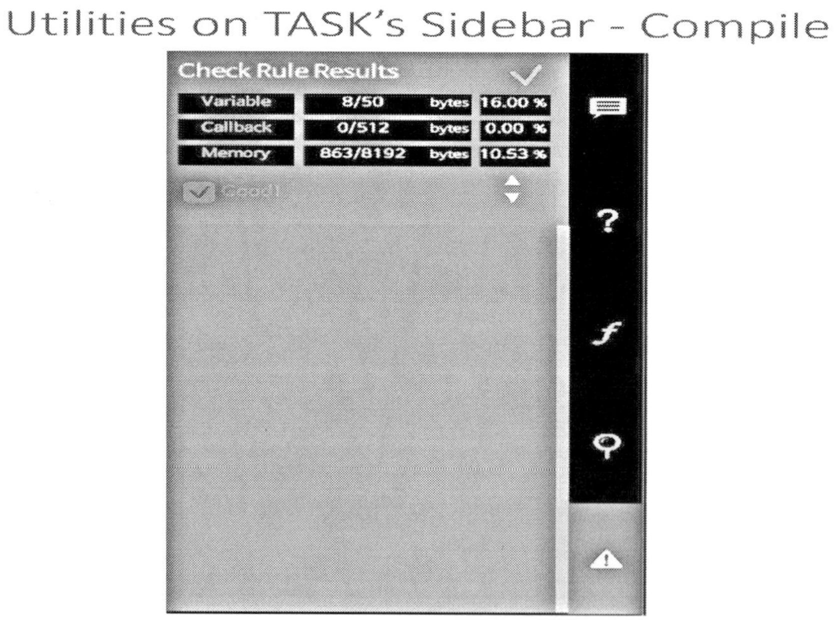

Fig. 6.28 "Compile" Utility found in TASK's Right Sidebar.

Within this Section 6.3, the author hopes to have provided a step-by-step general procedure so that beginner roboteers can start from goals/tasks defined for a robot system and end with a functioning TASK program that can be further refined with more advanced concepts to be provided in later sections and chapters.

6.4 RC Walker with Independent Servos Control

Expanding on the previous project "Basic RC Walker", let's add a new capability to our MINI. Instead of using only 1 RC button at any time to control the robot at runtime, let's now use multiple buttons simultaneously on the RC-100 (Physical or Virtual) to accomplish these tasks:

1. We want to use "U-D-L-R" buttons to serve as a **Walk Direction** (only one direction is allowed at any time).
2. The buttons "1" and "3" are to be used when we want to **raise ("1") OR lower ("3")** the **right arm** of the MINI via **direct control of Servo ID=1 (i.e. right shoulder joint) while the robot is walking in any direction or standing still**.
3. The buttons "2" and "4" are to be used when we want to make the MINI's **head scan left ("2") OR scan right ("4")** via **direct control of Servo ID=17 while the robot is walking in any direction or standing still**.

This new capability will require us to resolve two issues:

1. How to separate the information from the different RC buttons from the single message (e.g. "DataIn") that our program would receive from the VRC when the user pushes on several RC buttons at the same time (see Section 6.4.1).
2. From Chapter 4, we know that when a MOTION-LIST/PAGE is being played all 17 servos are controlled by this MOTION-LIST/PAGE, thus how can a TASK program be written to control only specific servos among those 17 servos? (see Section 6.4.2).

6.4.1 Filtering a RemoCon Message

So far (Sections 6.1.5 and 6.3), the reader has been shown the basic usage of messages sent by the VRC, let's now look closer at the structure of these "RemoCon" messages.

A full "RemoCon" packet is based on a 16-bit user-designed message wrapped up in a 6-byte packet (http://emanual.robotis.com/docs/en/parts/communication/rc-100/#communication-packet). For this Section 6.4, we'll restrict ourselves to using only the lower 10 bits of this 16-bit message (in Section 6.7, the reader will be shown how to use all 16 bits).

Each of the 10 RC Buttons "U-D-L-R-1-2-3-4-5-6" is represented by 1 bit in a 10-bit number/message, with the "U" button using the bit-0 position and the "6" button using the bit-9 position (see Fig. 6.29).

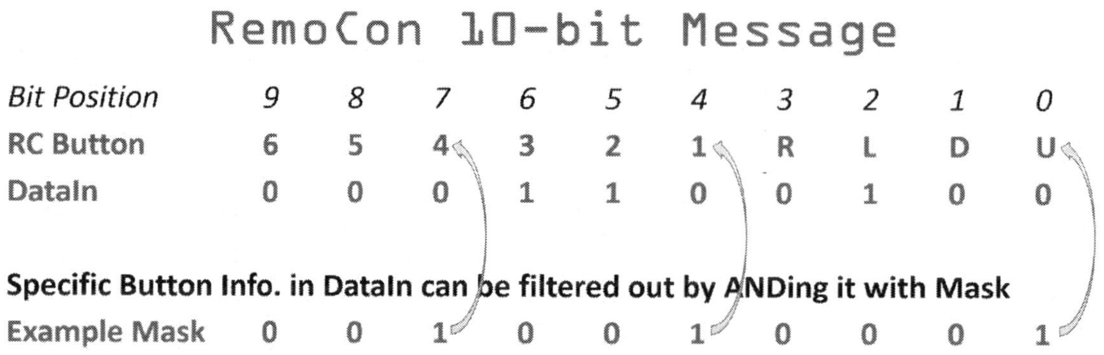

Fig. 6.29 Structure of a "RemoCon" 10-bit message.

Whenever one or several buttons are pushed by the user, this 10-bit message is sent from the Physical and Virtual RC-100 to the robot (i.e. to the TASK program and saved in Parameter "DataIn"). The "DataIn" row in Fig. 6.29 shows that RC Buttons "L", "2" and "3" are pushed as these bits are set to "1". All the "un-pushed" buttons of course have their bits set to 0.

The author needs to point out a unique feature with the ROBOTIS RC-100 system:

- Whenever the user releases all buttons, the Physical and Virtual RC-100 would send out the **message "0"** (i.e. all bits set to 0) **ONCE**. Other robotics systems may not implement the same scheme. In a "bad" BlueTooth environment, sometimes that "0" message, although sent, may not be "received" by the BT-210 or BT-410 receivers and subsequently not by the TASK program also.

The main issues in using the RC-100 are to know how many buttons are being pushed, and how to sort out which buttons are being pushed, as **information about all 10 buttons are sent over**, inside each transmit-ted RC message.

First, the issue of "how many" can be controlled by the programmer by using a "Mask" constant in binary format whereas a "1" is used for any Bit Position(s) that the programmer is interested in. For example, in Fig. 6.29, the programmer is only concerned with Bit Positions 0, 4 and 7 which correspond respectively to RC Button "U", "1" and "4".

In the current project, we are interested in detecting 3 events:

1. **When either U, D, L or R Button is pushed**, thus a mask constant with the binary value of 0000001111 would be used to filter any "DataIn" value. When a bitwise AND operation is made between "DataIn" with this "U+D+L+R" Mask, only the information regarding the "U-D-L-R" Buttons will be remain in the Filtered Result (see Fig. 6.30 for an example). In this example, the reader can see that only input from the "L" Button is retained. Inputs from Button "2" and "3" are cleared out to "0", basically "ignored". The interested reader can use the Truth Table for "A & B" at the bottom of Fig. 6.30 to verify these results. Please note that this filtering process does not prevent the RC operator from pushing more than one UDLR button at once. This issue needs to be addressed by other logical constructs later inside the TASK program.
2. **When either 1 or 3 Button is pushed**, so a mask constant with the binary value of 0001010000 would be needed to filter any "DataIn" value. Next, if a bitwise AND operation is made between "DataIn" with this "1+3" Mask, then only the information regarding the "1" or "3"

Buttons will be retained in the Filtered Result (see Fig. 6.31 for an example). In this example, the reader can see that only input from the "3" Button is retained and inputs from Buttons "L" and "2" are "ignored". Similarly, the issue of the RC operator pushing on both "1" and "3" Buttons together will be resolved with other TASK programming constructs.

AND Operation with Mask U+D+L+R

Bit Position	*9*	*8*	*7*	*6*	*5*	*4*	*3*	*2*	*1*	*0*
RC Button	6	5	4	3	2	1	R	L	D	U
DataIn	0	0	0	1	1	0	0	1	0	0
						&				
Mask U+D+L+R	**0**	**0**	**0**	**0**	**0**	**0**	**1**	**1**	**1**	**1**
Filtered Result	***0***	***0***	***0***	***0***	***0***	***0***	***0***	***1***	***0***	***0***

A	B	A & B
1	0	0
1	1	1
0	0	0
0	1	0

Fig. 6.30 Example of filtering of "DataIn" by a "U+D+L+R" mask.

AND Operation with Mask 1+3

Bit Position	*9*	*8*	*7*	*6*	*5*	*4*	*3*	*2*	*1*	*0*
RC Button	6	5	4	3	2	1	R	L	D	U
DataIn	0	0	0	1	1	0	0	1	0	0
						&				
Mask 1+3	**0**	**0**	**0**	**1**	**0**	**1**	**0**	**0**	**0**	**0**
Filtered Result	***0***	***0***	***0***	***1***	***0***	***0***	***0***	***0***	***0***	***0***

Fig. 6.31 Example of filtering of "DataIn" by a "1+3" mask.

3. **When either 2 or 4 Button is pushed**, so a mask constant with the binary value of 0010100000 would be needed to filter any "DataIn" value. Then, if a bitwise AND operation is made between "DataIn" with this "2+4" Mask, only the information regarding the "2" or "4" Buttons will be retained in the Filtered Result (see Fig. 6.32 for an example). In this example, the reader can see that only input from the "2" Button is retained and inputs from Buttons "L" and "3" are

"ignored". Similarly, the issue of the RC operator pushing on both "2" and "4" Buttons together will be resolved using other TASK programming constructs.

AND Operation with Mask 2+4

Bit Position	*9*	*8*	*7*	*6*	*5*	*4*	*3*	*2*	*1*	*0*
RC Button	6	5	4	3	2	1	R	L	D	U
DataIn	0	0	0	1	1	0	0	1	0	0
						&				
Mask 2+4	0	0	1	0	1	0	0	0	0	0
Filtered Result	***0***	***0***	***0***	***0***	***1***	***0***	***0***	***0***	***0***	***0***

Fig. 6.32 Example of filtering of "DataIn" by a "2+4" mask.

Next, we need to take on the issue of how to isolate specific servos from Position Control commands issued by a MOTION-LIST/PAGE being played.

6.4.2 Servo Joint Offset Special Value "1024"

The last time that we used Servo Joint Offset parameters was in Section 4.4 regarding the OFFSET-POSE being used to adjust for variations in the geometric framing configuration of a typical "physical" MINI robot, as compared to the ideal one represented by the "graphical" MINI robot. These values were small, actually they are restricted to the range from -255 to 255 (please see more details at this web link http://emanual.robotis.com/docs/en/software/rplus1/task/programming_02/#joint-offset). This link also mentions that a special value of "1024" can be used to spare specific servo(s) from the effect of a robot motion being performed.

The reader may/should remember that an XL-320 actuator can only take on Position Control values between 0 and 1023 (see Section 1.1.2 and Chapter 4). Thus, adding an offset value of 1024 just make this value way "out of range", and the XL-320 firmware is written to ignore this "out-of-range" situation for those servos that got set with a Joint Offset value of "1024". This means that "during a motion play" these targeted servos are not "affected" by data contained in pre-designed MOTION-UNITs. During this "unconstrained" time period, the programmer can then "sneek in" some "legal" GOAL POSITION commands to these targeted servos to get them to do his/her bidding! (*in a way converting a "bug" into a "feature"*).

Recapping the previous discussions, the proper way to use the special value of "1024" for Joint Offset is as follows:

1. Set 1024 to the Joint Offset parameter(s) for selected servo(s) that the programmer wants to isolate from the effect of a given MOTION-LIST/PAGE. This parameter can be found in the TASK commands menu as shown in Fig 6.33.
2. The key idea to remember that the "1024 magic" only happens during the playing of any MOTION-LIST/PAGE that uses previously set actuators, **thus the MOTION-STATUS flag needs to be monitored.**

3. GOAL POSITION commands do not use JOINT OFFSETs, and thus they need to comply with the [0-1023] range. **GOAL POSITION commands are effective for all servos when MOTION-STATUS is FALSE, but they are effective for only those servos with JOINT-OFFSETs set to 1024 when MOTION-STATUS is TRUE.**

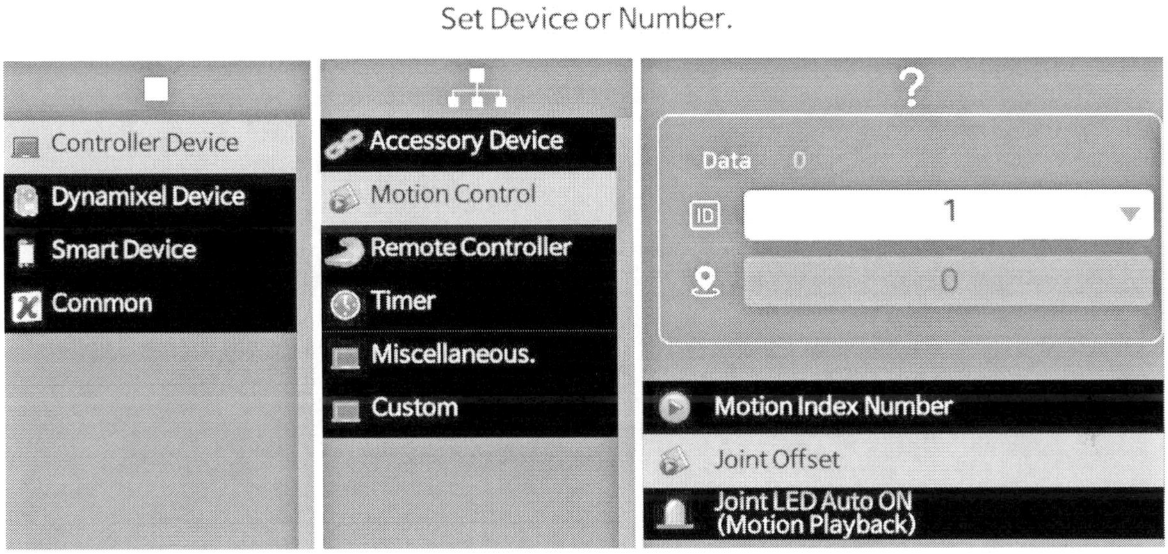

Fig. 6.33 Finding the location of the Joint Offset parameter in the TASK commands menu.

We are now ready to find a solution to the current project "RC Walker with Independent Servos Control".

6.4.3 Construction of "M_RC_Walker_IS.tsk3"

As this is another RC application, the general algorithm developed for "M_RC_Walker_Basic_b.tsk3" should still be applicable. In the "Basic" application, only the "Walk Direction" was needed: i.e. we need only to know which U-D-L-R Button is pushed and then play the appropriate MOTION-LIST/PAGE to make the MINI walk in the commanded direction.

In this "Independent Servos" (IS) application, we need to figure out a similar "U-D-L-R" solution, and also whether Buttons 1 or 3 are pushed so that we can swing up or down the right arm, as well as whether Buttons 2 or 4 are pushed so that we can scan the robot's head piece left or right. Thus, the **essential algorithm** used in this project should have 5 main steps each time that the Main Endless Loop is iterated during runtime:

1. To check for any new data input from the VRC: if no new data, HC just waits by doing nothing and if there is new data input, read it in and save it in Parameter "DataIn" and HC goes on to the next step.
2. Once "DataIn" is acquired, filter out "DataIn" for the first time with Mask "U+D+L+R" to obtain Parameter "WalkDirection" and implement the correct MOTION-LIST/PAGE for the robot to perform using an IF-ELSE-IF control structure based on "WalkDirection".

3. Filter out "DataIn" for a second time but with Mask "1+3" to obtain Parameter "Button_1_3" and implement the correct GOAL-POSITION commands on Servo ID=1 to swing the robot's right arm up or down using an IF-ELSE-IF control structure based on "Button_1_3".
4. Filter out "DataIn" for the third time but with Mask "2+4" to obtain Parameter "Button_2_4" and implement the correct GOAL-POSITION commands on Servo ID=17 to scan the robot's head piece left or right using an IF-ELSE-IF control structure based on "Button_2_4".
5. In the special case of (DataIn == 0) which means that the user just releases all RC buttons, reset all JOINT-OFFSETs to zero and put the MINI into its Ready Pose.

Using the essential algorithm described above, the author started to create the TASK program named "M_RC_Walker_IS.tsk3" illustrated by Figs. 6.34 through 6.39. Needless to say that some "adjustments" had to be made "along the way", but overall the overall ideas spelled out in the original algorithm were sound enough.

Fig. 6.34 shows the start of the program "M_RC_Walker_IS.tsk3" which calls the Function "Init" which has the same statements as before for "M_RC_Walker_Basic_b.tsk3", except for Lines 140 and 141:

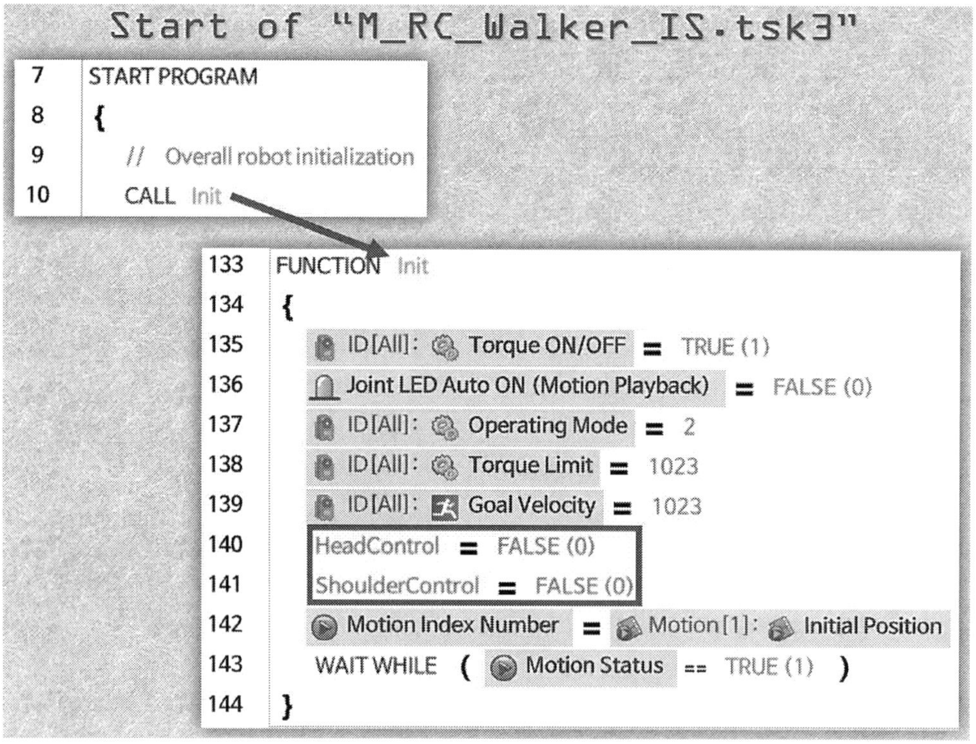

Fig. 6.34 Start of program "M_RC_Walker_IS.tsk3".

- Line 140 initializes a logical Parameter named "HeadControl" to FALSE. The intent is to set "HeadControl" to TRUE whenever the program detects the FIRST action by the user to move the head piece, i.e. when the user FIRST pushes on Button 2 or 4.
- Line 141 initializes a logical Parameter named "ShoulderControl" to FALSE. The intent is to set "ShoulderControl" to TRUE whenever the program detects the FIRST action by the user to swing the right arm, i.e. when the user FIRST pushes on Button 1 or 3.

- In general, the author had found that it is useful to create logical flags to correspond to various "states" that the robot may be at any given time. In this "IS" project, the MINI can be viewed as to be able to take on 4 possible states:
  - a "Ready Pose" state defined by (DataIn == 0), i.e. when no button is pushed.
  - A "Walking" state defined by (MOTION-STATUS == TRUE) or FALSE depending on the programmer's point of view. The author did not see the need to drill down further to U-D-L-R status.
  - A "HeadControl" state, whereas when it is TRUE, it is important to set Joint Offset for Servo 17 to 1024 before performing any MOTION-LIST/PAGE and/or to use appropriate GOAL-POSITION commands. And when it is FALSE, its JOINT-OFFSET should be reset to 0.
  - A "ShoulderControl" state, whereas when it is TRUE, it is important to set Joint Offset for Servo 1 to 1024 before performing any MOTION-LIST/PAGE and/or to use appropriate GOAL-POSITION commands. And when it is FALSE, its JOINT-OFFSET should be reset to 0.

Fig. 6.34 also shows that when the HC returns to the Main Program from Function Init, the robot is its Ready Pose (Lines 142-143).

Fig. 6.35 shows Part 1 of the Main Endless Loop in "M_RC_Walker_IS.tsk3":

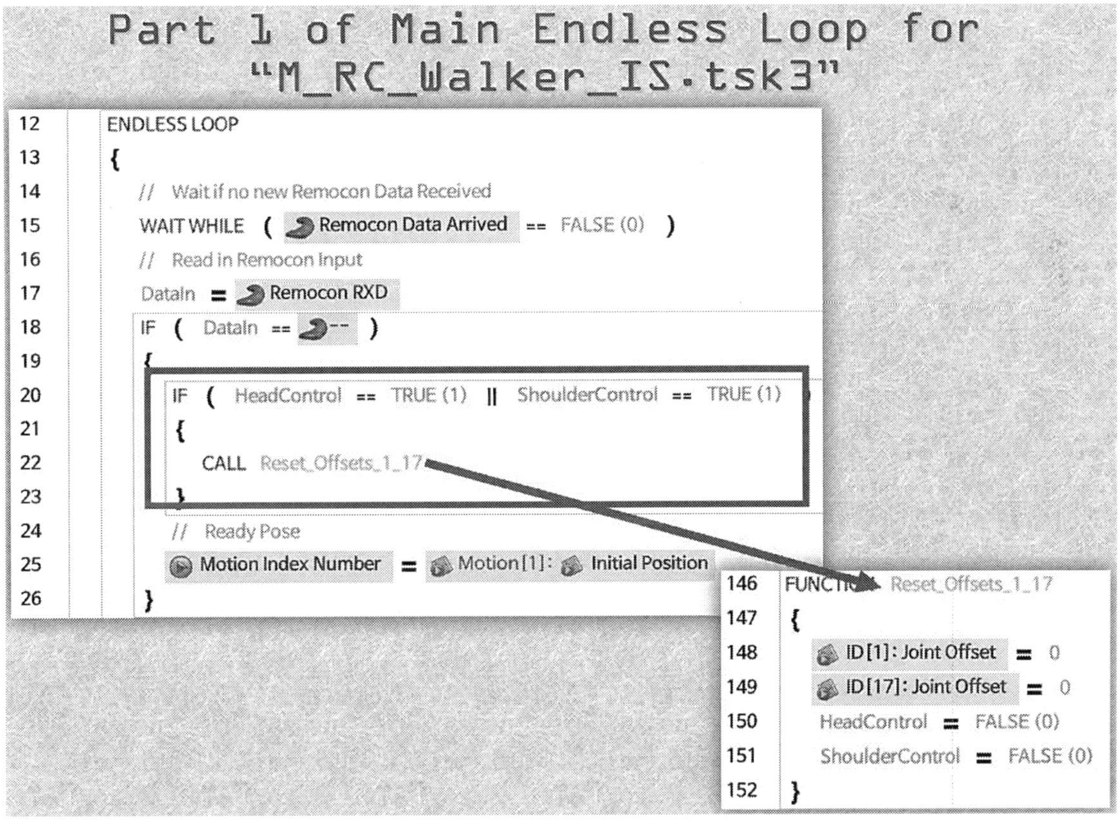

Fig. 6.35 Part 1 of Main Endless Loop in "M_RC_Walker_IS.tsk3".

- Lines 15 and 17 represent "Step 1" of the "Essential Algorithm" proposed previously.
- Lines 18-26 represent "Step 5" of the proposed "Essential Algorithm" whereas DataIn is found to be equal to 0 (i.e. no RC button is pushed) then, like before for the "Basic" solution, the robot is set to perform "Ready Pose" (Lines 25). But only **after** the HC checks to see if the user had tried to move the Head or Right Shoulder of the robot previously, via the logical Parameters "HeadControl" and "ShoulderControl". If these situations are found to be TRUE, then HC will call on Function "Reset_Offsets_1_17" to reset JOINT-OFFSETs for Servos 1 and 17 back to 0, and to reset Parameters "HeadControl" and "ShoulderControl" back to FALSE.
- The curious reader can move this group of statements down to the end of the Main Endless Loop to see it would generate a different runtime behavior for the robot, in keeping with its being "Step 5" of the "Essential Algorithm".
- The curious reader can also move the IF block to be after the "Play Motion Index Number" statement to see any changes in the behavior of the robot.

Fig. 6.36 shows Part 2 of the Main Endless Loop in "M_RC_Walker_IS.tsk3":

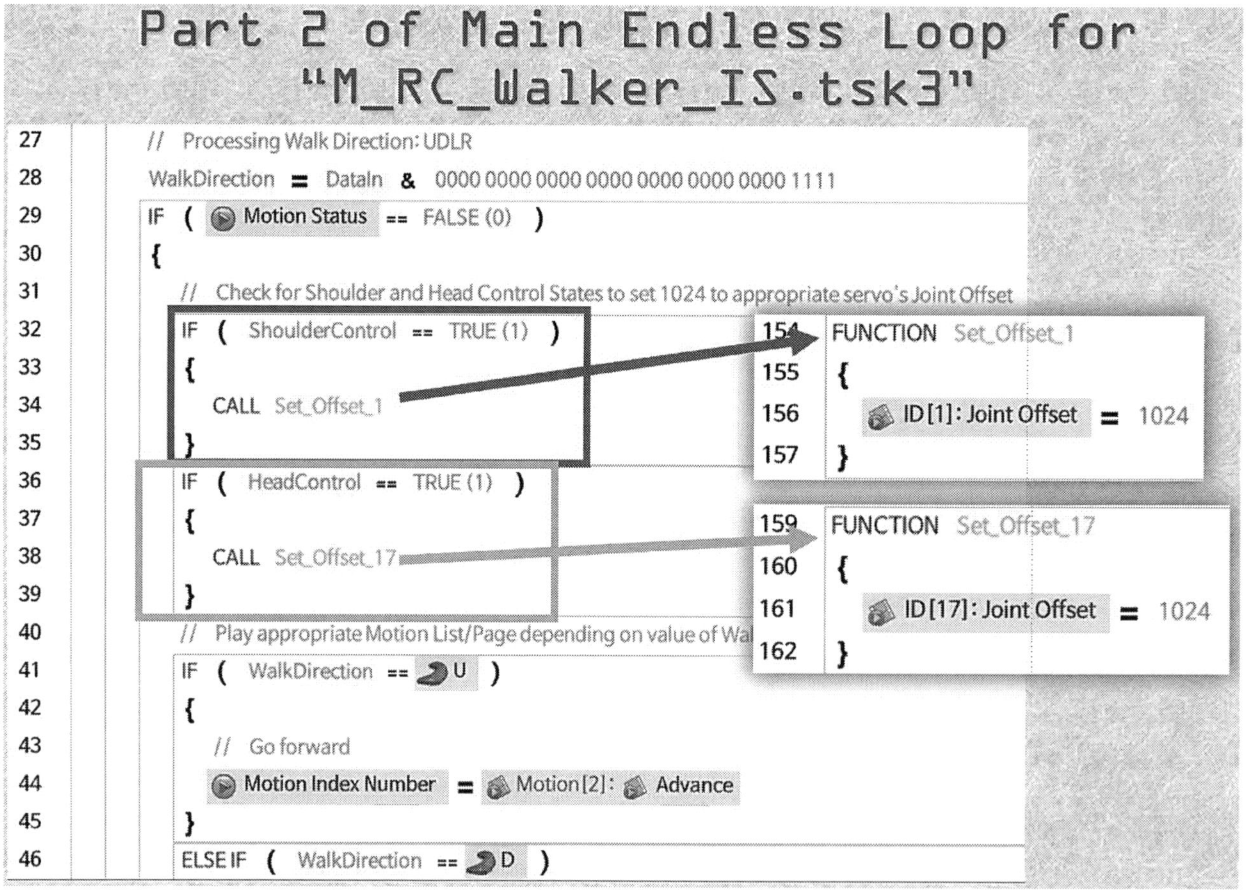

Fig. 6.36 Part 2 of Main Endless Loop in "M_RC_Walker_IS.tsk3".

- Line 28 filters out “WalkDirection” from “DataIn” using the Mask “U+D+L+R”.
- Line 29 checks to see if the robot is NOT performing any MOTION-LIST/PAGE via Parameter “Motion-Status” being equal to FALSE. And most importantly, this check was done with an IF block instead of a WAIT-WHILE loop: that is using the approach used in “M_RC_Walker_Basic_a.tsk3” instead of the approach used in “M_RC_Walker_Basic_b.tsk3”. The author leaves the readers to figure out the reason(s) why on their own here!
- So, if Line 29 evaluates to TRUE, i.e. the robot is not performing or just finished performing a Motion, then a big block of statements is entered, starting from Line 30 in Fig. 6.36 to Line 61 in Fig. 6.37.
- Thus, the IF block at Line 32 is executed to check whether the “ShoulderControl” state is active and if this state is active then the HC sets JOINT-OFFSET for Servo 1 to 1024 as required.
- Next, the IF block starting at Line 36 does a similar checking procedure for the “HeadControl” state and when appropriate would set the JOINT-OFFSET for Servo 17 to 1024 as required.
- An IF-ELSE-IF structure (from Line 41 of Fig. 6.36 to Line 61 of Fig. 6.37 is executed next to make the robot Advance, Reverse, Turn Left or Turn Right depending on the value contained in Parameter “WalkDirection”.

Part 3 of Main Endless Loop for “M_RC_Walker_IS.tsk3”

```
46        ELSE IF ( WalkDirection == D )
47        {
48            // Go backward
49            Motion Index Number = Motion[3]: Reverse
50        }
51        ELSE IF ( WalkDirection == L )
52        {
53            // Turn left
54            Motion Index Number = Motion[4]: Left Turn
55        }
56        ELSE IF ( WalkDirection == R )
57        {
58            // Turn right
59            Motion Index Number = Motion[5]: Right Turn
60        }
61    }
```

Fig. 6.37 Part 3 of Main Endless Loop in “M_RC_Walker_IS.tsk3”.

Here the author would pose a question to the readers, without providing an answer:

- What would the robot do if the RC operator pushes on several U-D-L-R buttons together at runtime? *(Hint: it has something to do with the IF-ELSE-IF structure used).*

Fig. 6.36 and 6.37 represent "Step 2" of the proposed "Essential Algorithm" with some logic regarding the "HeadControl" and "ShoulderControl" states added in to make sure that JOINT-OFFSETs for Servo 1 and 17 are appropriately set before making the robot move around (please remember that JOINT-OFFSET are effective only during the performance of MOTION-LIST/PAGEs).

Fig. 6.38 illustrates "Step 3" of the proposed "Essential Algorithm" when Buttons 1 and 3 are checked to see if they were pushed by the RC operator who wanted to move the Shoulder Joint (i.e. Servo 1):

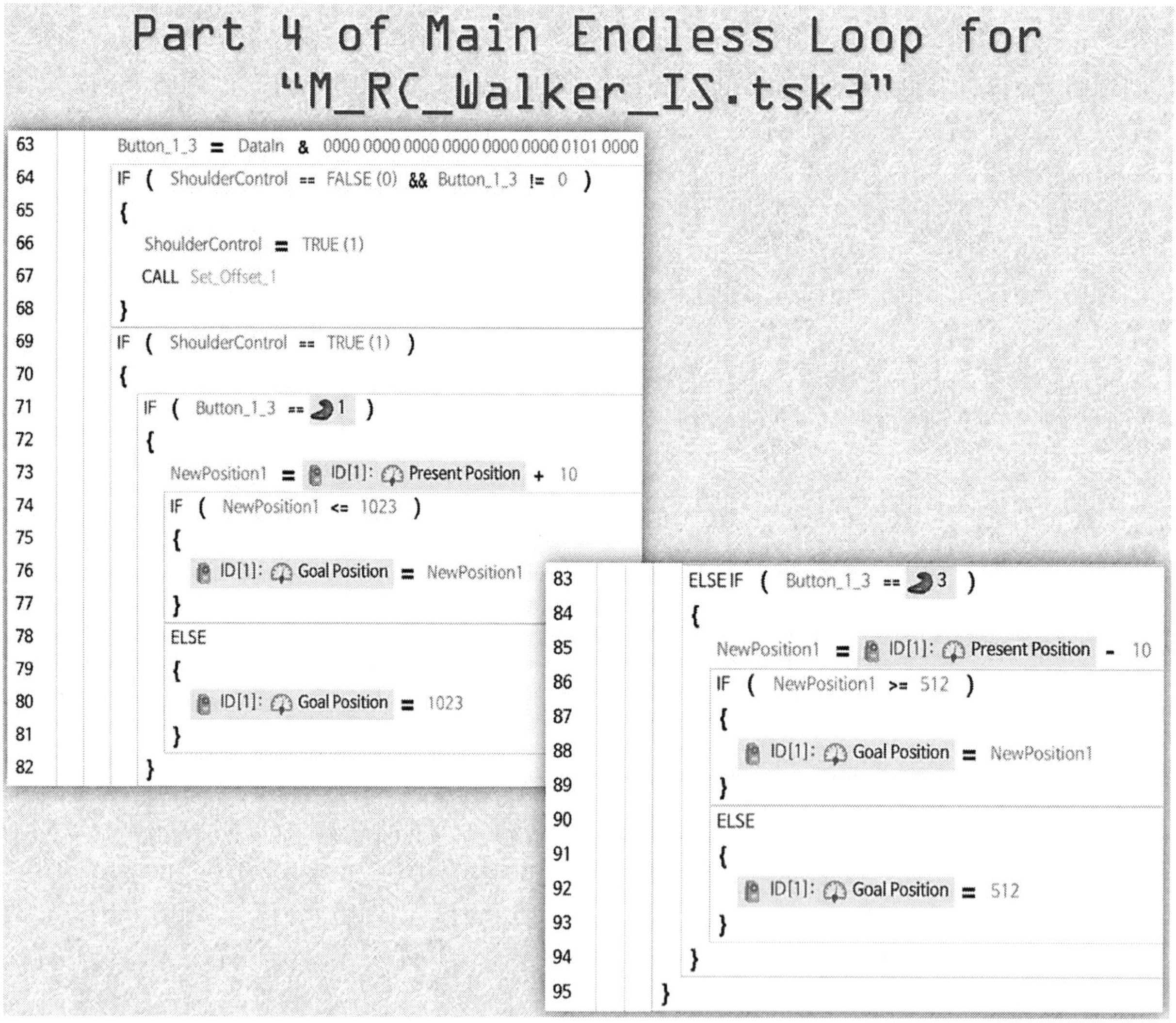

Fig. 6.38 Part 4 of Main Endless Loop in "M_RC_Walker_IS.tsk3".

- Line 63 filters out the states of Buttons 1 and 3 by ANDing “DataIn” with Mask “1+3” and by saving their values in Parameter “Button_1_3”.
- The IF structure comprising Lines 64 through 68 was designed to be a detection for the event when the RC operator first pushes on either Button 1 or 3: i.e. when “ShoulderControl” was still set to its initial value of FALSE and when the value for “Button_1_3” just turned NON-ZERO because the RC operator had just pushed on either Button 1 or 3, or both. When this type of event occurs, “ShoulderControl” is set to TRUE (Line 66), therefore making sure that this IF structure (64-68) won’t be triggered again within the same “game session”. “ShoulderControl” can be reset to FALSE only by another IF structure defined between Line 18 and 26 in Fig. 6.35. This second IF structure (18-26) is triggered only when the RC operator releases all buttons, making the condition (DataIn == 0) to become TRUE. It is during the execution of this second IF structure (18-26) that the Function “Reset_Offsets_1_17” is called and “ShoulderControl” is reset to FALSE by Line 151 (see Fig. 6.35).
- The IF structure defined between Lines 64 and 68 also sets the JOINT-OFFSET for Servo 1 to 1024 in preparation for the next IF structure defined between Lines 69 and 95 to be instantiated.
- The “69-95” IF structure is designed to check which Button 1 or 3 is being pushed by the RC operator. Please note that “Button_1_3” is used in conjunction with an IF-ELSE-IF structure to create an EITHER-OR situation between Button 1 (Line 71) and Button 3 (Line 83). Furthermore, this IF-ELSE-IF ensures that if the RC operator pushes on Buttons 1 and 3 together, the robot would not respond at all to this situation as no Event-Action Pair had been defined for it.
- If “1” was pushed, the IF structure between Lines 71 and 82 comes into play, i.e. it raises the right arm up by adding 10 Position units to its Present Position (Line 73) and by assigning this result to a GOAL-POSITION command for each time that it detects that “1” is pushed. The reader may remember that Servo 1 is at Position 512 when the robot is in Ready Pose, so the right arm is raised up from Position 512 towards Position 1023 which is an Upper Limit. To comply with this constraint, the IF-ELSE structure between Lines 74 and 81 is used to clamp the Servo 1’s Goal Position to the value of 1023 when needed.
- If “3” was pushed instead, the ELSE-IF structure between Lines 83 and 94 comes into play, i.e. it lowers the right arm down by subtracting 10 Position units from its Present Position (Line 85) and by assigning this result to a GOAL-POSITION command for each time that it detects that “3” is pushed. So, the right arm would be lowered from a Position near 1023 towards Position 512 being the Lower Limit. To comply with this constraint, the IF-ELSE structure between Lines 86 and 93 is used to clamp the Servo 1’s Goal Position to the value of 512 when needed.

Fig. 6.39 illustrates "Step 4" of the proposed "Essential Algorithm" when Buttons 2 and 4 are checked to see if they were pushed by the RC operator who wanted to move the Neck Joint (i.e. Servo 17) which is restricted to Servo Positions between 350 to 700, due to the author's construction choices:

Part 5 of Main Endless Loop for "M_RC_Walker_IS.tsk3"

```
97   Button_2_4 = DataIn & 0000 0000 0000 0000 0000 0000 1010 0000
98   IF ( HeadControl == FALSE (0) && Button_2_4 != 0 )
99   {
100      HeadControl = TRUE (1)
101      CALL Set_Offset_17
102  }
103  IF ( HeadControl == TRUE (1) )
104  {
105      IF ( Button_2_4 == 2 )
106      {
107          NewPosition17 = ID[17]: Present Position + 10
108          IF ( NewPosition17 <= 700 )
109          {
110              ID[17]: Goal Position = NewPosition17
111          }
112          ELSE
113          {
114              ID[17]: Goal Position = 700
115          }
116      }
117      ELSE IF ( Button_2_4 == 4 )
118      {
119          NewPosition17 = ID[17]: Present Position - 10
120          IF ( NewPosition17 >= 350 )
121          {
122              ID[17]: Goal Position = NewPosition17
123          }
124          ELSE
125          {
126              ID[17]: Goal Position = 350
127          }
128      }
129  }
130  }
131  }
```

Fig. 6.39 Part 5 of Main Endless Loop in "M_RC_Walker_IS.tsk3".

- The logic for "Step 4" is identical to "Step 3" except that different Parameters had to be used such as "HeadControl" and "Button_2_4", and different constants such as 700 and 350.
- Line 97 filters out the states of Buttons 2 and 4 by ANDing "DataIn" with Mask "2+4" and by saving their values in Parameter "Button_2_4".
- The IF structure comprising Lines 98 through 102 was designed to be a detection for the event when the RC operator first pushes on either Button 2 or 4: i.e. when "HeadControl" was still set to its initial FALSE value and when the value for "Button_2_4" just turned NON-ZERO because the RC operator had pushed on either Button 2 or 4, or both. When this kind of event occurs, "HeadControl" is set to TRUE (Line 100), therefore making sure that this IF structure

(98-102) is instantiated only once within the same "game session". "HeadControl" can be reset to FALSE only by the IF structure defined between Line 18 and 26 in Fig. 6.35. This "18-26" IF structure is triggered only by the RC operator releasing all RC buttons thus allowing the condition (DataIn == 0) to become TRUE. It is during the execution of this (18-26) IF structure that Function "Reset_Offsets_1_17" is called and "HeadControl" is reset to FALSE by Line 150 (see Fig. 6.35).

- The IF structure defined between Lines 98 and 102 also sets the JOINT-OFFSET for Servo 17 to 1024 in preparation for the next IF structure defined between Lines 103 and 129 to be executed.
- The "103-129" IF structure is designed to check which Button 2 or 4 is being pushed by the RC operator. Please note that "Button_2_4" is used in conjunction with an IF-ELSE-IF structure to create an EITHER-OR situation between Button 2 (Line 105) and Button 4 (Line 117), and to ignore all other situations.
- If Button "2" is pushed, the IF structure between Lines 105 and 116 gets instantiated, i.e. it rotates the robot's head to the left by adding 10 Position units to its Present Position (Line 107) and by assigning this result to a GOAL-POSITION command every time that it detects a push on Button "2". The reader may remember that Servo 17 is at Position 512 at start and Position 700 is the Left Upper Limit for Servo 17 resulting from the author's mechanical construction choices.
- If Button "4" is pushed instead, the ELSE-IF structure between Lines 117 and 129 gets instantiated, i.e. it rotates the robot's head (i.e. Servo 17) to the right by 10 Position units each time that it detects a push on Button "4" (Line 119), with Position 350 being the Right Upper Limit for Servo 17.

Video 6.8 shows the runtime performance of the TASK program "M_RC_Walker_IS.tsk3" where the reader can see that "Head Scanning" and "Arm Raising" can be performed with or without the walking behaviors. This capability was implemented using the special value of 1024 for selected servos' JOINT-OFFSET parameters which can keep them from being affected by a MOTION-LIST/PAGE that is being performed by the MINI,

This "Independent Servo Control" project also illustrates different ways to use IF, ELSE and ELSE-IF structures to check for events that are critical to the robot's behaviors at runtime and the main challenge was how to manage those structures within the Main Endless Loop. The "management" approach that the author used was to change the focus from "trees" to "forest" and vice-versa to make sure that the "Sense-Think-Act" Paradigm is adhered to at all levels of abstraction, along with the use of Flag/State Parameters to allow the programmer to keep track of the different "robot-states" that may be involved.

6.5 RC Smart Avoider

In this project "M_RC_Smart_Avoider.tsk3", Remote Control techniques are combined with Autonomous Behavior based on Sensor Input via two NIR distance sensors (IRSS-10) mounted on either side of the MINI's head piece (see Fig. 1.4).

The new goal is to add an Autonomous Behavior to the Basic Walker algorithm used in the previous project "M_RC_Walker_Basic_a.tsk3" such that:

- While under the RC control of an operator, the MINI will monitor its two IRSS-10 sensors for any potential obstacle in its front and/or on its two sides (i.e. within the detection zones for the two IRSS-10 sensors as mounted in the configuration shown in Fig. 1.4).
- If no obstacle is detected, the TASK code should allow and process the various Walk commands from the human operator as normal.
- If an obstacle is detected to be "close" enough, the robot should respond by performing "autonomously" and "appropriately" avoidance maneuvers to steer the MINI away from the detected obstacle. Additionally, during this "avoiding" time period, the robot should "ignore" all RC commands from the operator until it determines that it is clear from the detected obstacle. At that time, the robot will go back to its normal RC behavior, accepting RC commands from the operator – that is until a new obstacle is detected!

As this project requires fast responses from the IRSS-10 sensors as well as from MOTION-LIST/PAGEs being performed, there are two new concepts that readers need to master: programmed stoppage of a current MOTION-LIST/PAGE being performed and appropriate use of the CALLBACK Function.

6.5.1 "Programmed Stoppage" of Motion-List/Page

In the MOTION-GROUP "RC_Walk_1" used so far (see Fig. 6.10), each of its MOTION-LIST/PAGE performs 3 repetitions of the chosen robot movement such as Advance, Reverse, Turn Left or Turn Right for each time that it is invoked. For this "Smart Avoider" project where a quick response to an obstacle just detected is of paramount importance, we will need to use a different MOTION-GROUP where robot movements are shorter in time duration and performed only once when invoked. Thus the "Smart Avoider" MOTION-GROUP was created (see Fig. 6.40).

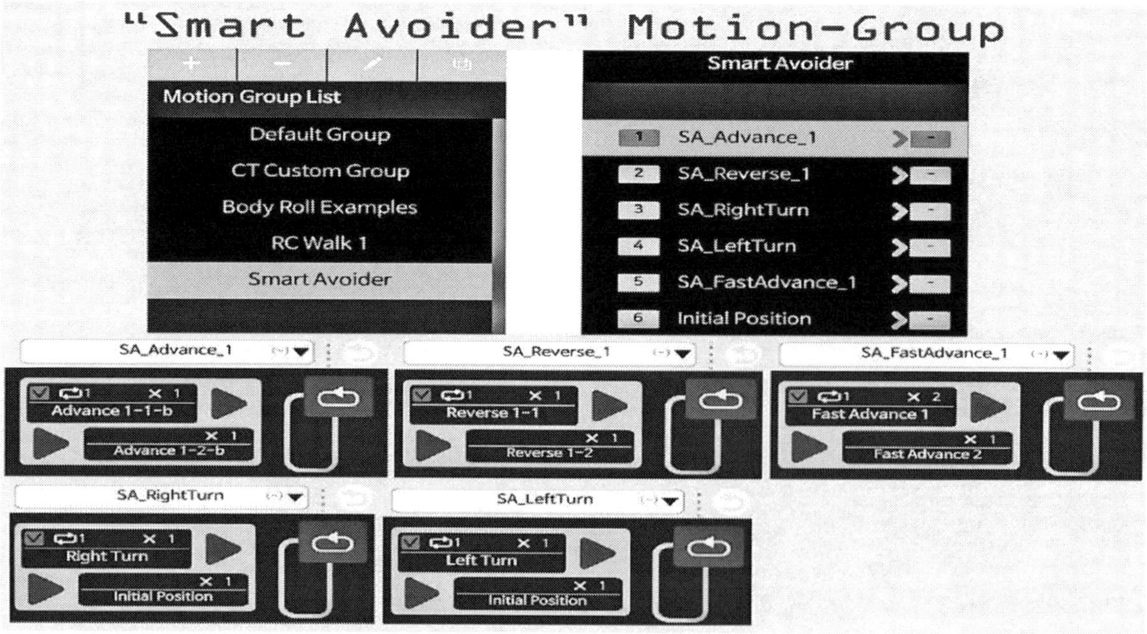

Fig. 6.40 MOTION-GROUP "Smart Avoider".

The "Smart Avoider" MOTION-GROUP has 6 MOTION-LIST/PAGEs:

1. "SA_Advance_1" uses MOTION-UNIT (MU) "Advance 1-1-b" (437 ms) as LINKED MU and "Advance 1-2-b" (218 ms) as EXIT MU.
2. "SA_Reverse_1" uses MU "Reverse 1-1" (421 ms) as LINKED MU and "Reverse 1-2" (218 ms) as EXIT MU.
3. "SA_RightTurn" uses MU "RightTurn" (281 ms) as LINKED MU and "InitialPosition" (390 ms) as EXIT MU.
4. "SA_LeftTurn" uses MU "LeftTurn" (281 ms) as LINKED MU and "InitialPosition" (390 ms) as EXIT MU.
5. "SA_FastAdvance_1" uses MU "Fast Advance 1" (515 ms at double-speed so equivalent to 258 ms) as LINKED MU and "Fast Advance 2" (210 ms) as EXIT MU.
6. "Initial Position" uses MU "Initial Position" for both LINKED MU and EXIT MU.

EXIT MOTION-UNITs are needed in this project because IRSS-10 sensors data can reach the critical threshold values for obstacle detection at any time during the performing of the current MOTION-LIST/PAGE. When this event happens, the TASK program will need to trigger "REGULAR STOP" or "EMERGENCY STOP" procedures equivalent to the ones provided by the MOTION Sub-Tool in Section 4.2.2 (e.g. keys "F8" and "ESC" as shown in Fig. 4.20).

In TASK 3, to trigger an EMERGENCY STOP, i.e. to make the MINI stop after performing the LAST KEY-FRAME of the CURRENT LINKED MU, the programmer needs to issue the LOAD command "**Motion Index Number = 0**". To trigger a REGULAR STOP, i.e. to make the MINI switch from the CURRENT LINKED MU to the corresponding EXIT MU and perform this EXIT MU until its LAST KEY-FRAME, the programmer needs to issue the LOAD command "**Motion Index Number = -1**". For more information please see web link (http://emanual.robotis.com/docs/en/software/rplustask3/task_parameters/#motion-index-number).

6.5.2 CALLBACK Function

The TASK Sub-Tool offers a special Function named CALLBACK which is executed independently of the Main Program and its User-Defined Functions. It executes itself every 8 ms (7.81 ms, to be exact) which corresponds to the smallest time period allowed between KEY-FRAMES defined in a MOTION-UNIT (see Section 4.1.2). Because of the short execution time of 8 ms, the CALLBACK Function has some restrictions on its size and the types of commands that can be used in it:

- Its size cannot be more than 512 bytes.
- LOOP, LABEL, JUMP and CALL types of commands are not allowed.
- A maximum of 2 External Device Calls (e.g. access to Sensors, RC-100 buttons or Dynamixels) are recommended. More programming tips for accessing Dynamixels are presented in Section 6.8.4.

For more information, the reader can visit this web link (http://emanual.robotis.com/docs/en/software/rplustask3/task_instructions/#callback).

In this project, the acquisition of "current" data from the IRSS-10 sensors, mounted on Ports 1 and 4, is performed by the CALLBACK Function (see Fig. 6.41) and their "latest" data (within 8 ms) are recorded in

Variables "NIR_Left" and "NIR_Right" and available to the programmer to use whenever and wherever inside the Main Program and its User-Defined Functions.

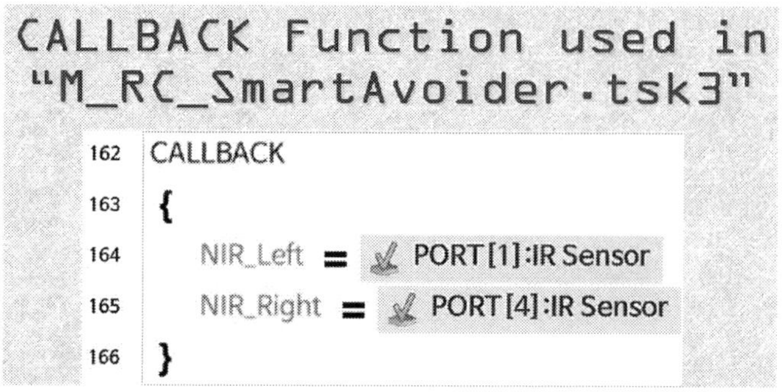

Fig. 6.41 CALLBACK Function used in "M_RC_SmartAvoider.tsk3".

6.5.3 Construction of "M_RC_SmartAvoider.tsk3"

For this project, the "Sense-Think-Act" still works well for us, however the Event-Action pairs become more complex in structure and intertwined as the robot's tasks get more sophisticated. The author's approach had been to create the **Essential Algorithm** first (i.e. the "forest" view) and to deal with the **Linkages** as they arose (i.e. the "tree" view).

Incorporating the "Smart Avoider" goal described at the beginning of Section 6.5 with the previous RC algorithm for the Basic Walker, the following Essential Algorithm (i.e. Main Endless Loop) was laid out as the first approach:

Main Endless Loop {

When new RC data comes in, save it in DataIn

If (DataIn == 0) then (Play robot's Initial Position)

Else { (i.e. U-D-L-R data coming in)

If (No Obstacle within NIR detection ranges) then (Execute RC command as normal)

Else { (i.e. Obstacle within NIR detection ranges)

Stop the robot

Figure out whether Obstacle is Frontal, on Left or on Right of robot and activate appropriate robot's avoidance moves <u>until obstacle out of NIR detection ranges</u>

*Use LEDs to signal robot's successful avoidance of obstacle and **<u>put robot in Initial Position</u>***

Wait until Operator releases last RC button (which made robot walk into obstacle)

} (End of Inner Else)

} (End of Outer Else)

} (End of Main Endless Loop, so loop back to wait for new RC data)

Please note that it is the author's preference to resolve any emergency issue within the same cycle that it arose within the Main Endless Loop, as this makes the program logic easier to understand.

Fig 6.42 illustrates the Initialization Steps used in this project "M_RC_SmartAvoider.tsk3":

- Four new Variables are initialized: Threshold_1, Threshold_2, Threshold_3 and Obstacle (Lines 71-74).

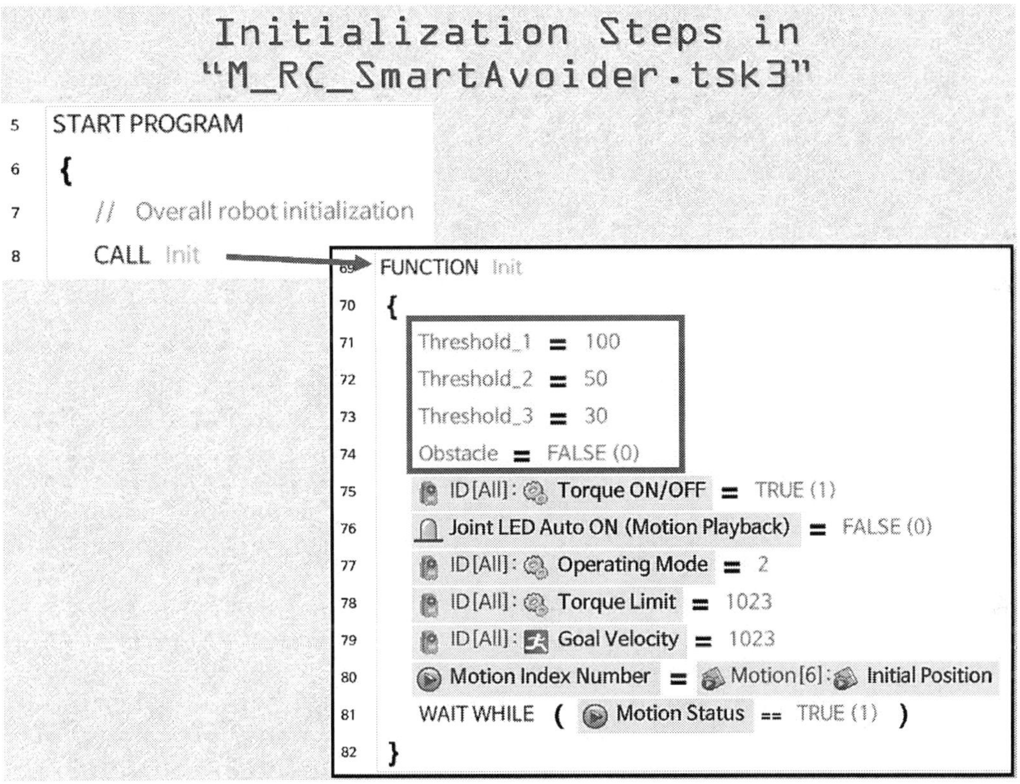

Fig. 6.42 Initialization Steps in "M_RC_SmartAvoider.tsk3".

- The remaining initialization steps are the same ones used the Basic Walker project (see Section 6.3).

Fig. 6.43 describes Part 1 of the Main Endless Loop used in this project:

- The usual procedure of waiting for new RC data, then to pick it up and to save it in Variable "DataIn" is shown in Lines 13 and 14.
- If (DataIn == 0), then the 5th new Variable "Direction" is set to "0" and Function "Walk_Execute" is called (Lines 108-133).

- Function “Walk_Execute” has the same control structure used in the Basic Walker project (Version A), except that “Direction” (1-2-3-4) is used instead of the various RC constants (U-D-L-R) – just to show the reader another approach.
- Please note Line 128 illustrating the special case for (Direction == 0), whereas the MINI is put into “Initial Position” only when the Current Motion Index Number (i.e. the last played motion) is different from “Initial Position”). This is to avoid redundancy in performing “Initial Position” again when the robot was in “Initial Position” already.
- Line 128 also shows that “Motion Index Number” can be “read-in” and used as a Variable, while Line 130 shows its more common use as a “Command” whereas a Constant is “written” into it.

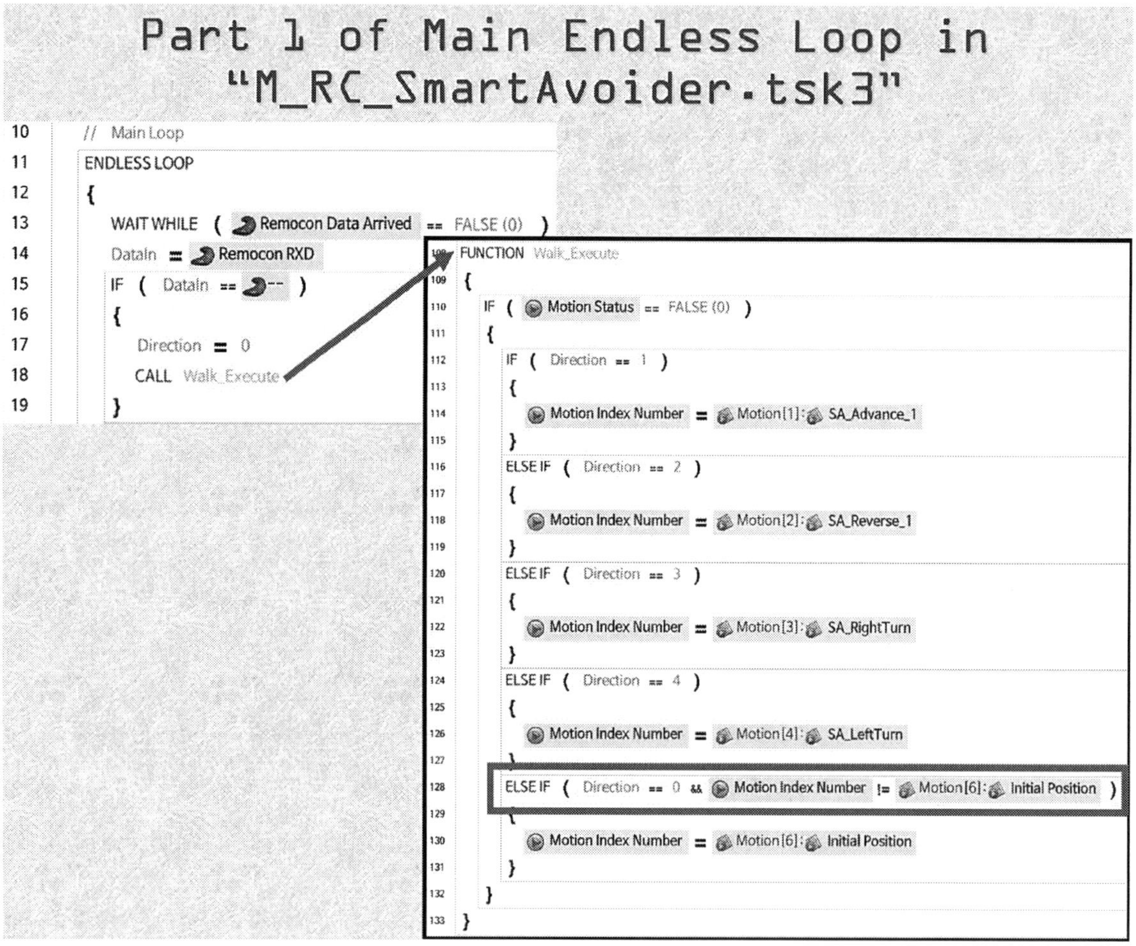

Fig. 6.43 Part 1 of Main Endless Loop in “M_RC_SmartAvoider.tsk3”.

Part 2 of the Main Endless Loop corresponds to the situation when an **RC Button** is pushed (U-D-L-R) **and** that there is **no obstacle** within the **user-given range** of the IRSS-10 sensors. In this case, it is OK for the robot to perform the commanded moves (see Fig. 6.44):

- When the ELSE block begins at Line 20 in Fig. 6.44, some "non-zero" **RC Buttons** had been pushed.
- Additionally, the IF condition in Line 23 is checked to see if both "NIR_Left" and "NIR_Right" have values less than "Threshold_1" (i.e. "100", the reader can also try other higher or lower values to see different resulting behaviors in the MINI). If this condition evaluates to TRUE, meaning that there is no obstacle detected by "NIR_Left" and "NIR_Right", then the Hardware Controller calls Function "Set_Direction" (Line 25) to obtain a value (1/2/3/4) for Parameter "Direction" which depends on the value (U/D/L/R) contained in Parameter "DataIn" (see Lines 86 to 105 in right panel of Fig. 6.44).
- Next, Function "Walk_Execute" is called to make the MINI perform the appropriate MOTION-LIST/PAGE corresponding to the current value of "Direction" (see right panel in Fig 6.43).

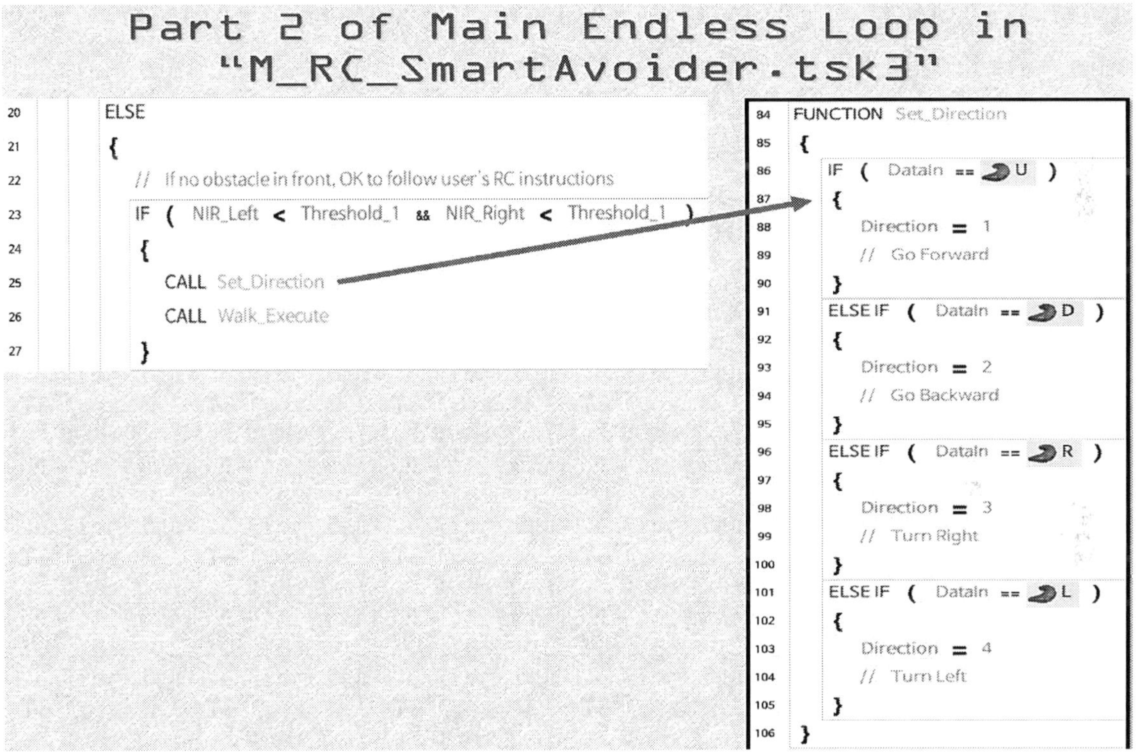

Fig. 6.44 Part 2 of Main Endless Loop in "M_RC_SmartAvoider.tsk3".

Parts 3, 4 and 5 of the Main Endless Loop correspond to the situation when an **RC Button** is pushed (U-D-L-R) **and** that there is an **obstacle detected** within the **user-given range** of the IRSS-10 sensors. In this situation, the robot must perform an avoidance procedure which has three phases:

1. Stop the robot, via a Regular Stop or an Emergency Stop (programmer choice). This is Part 3 of the Main Endless Loop and it is described in Fig. 6.45.
2. Next, the robot needs to figure out whether the obstacle is straight up front or is it more on the left or more on the right of the robot, and then perform the appropriate maneuver (Walk Backward, Turn Right or Turn Left) to get it clear from the obstacle. This is Part 4 of the Main Endless Loop and it is described in Fig. 6.46.

3. Once clear from the obstacle, the robot needs to flash its LEDs to let the RC Operator know of this new situation, and then it waits for the RC Operator to release the current RC Button that brought on this crisis in the first place. This is Part 5 of the Main Endless Loop and it is described in Fig. 6.47.

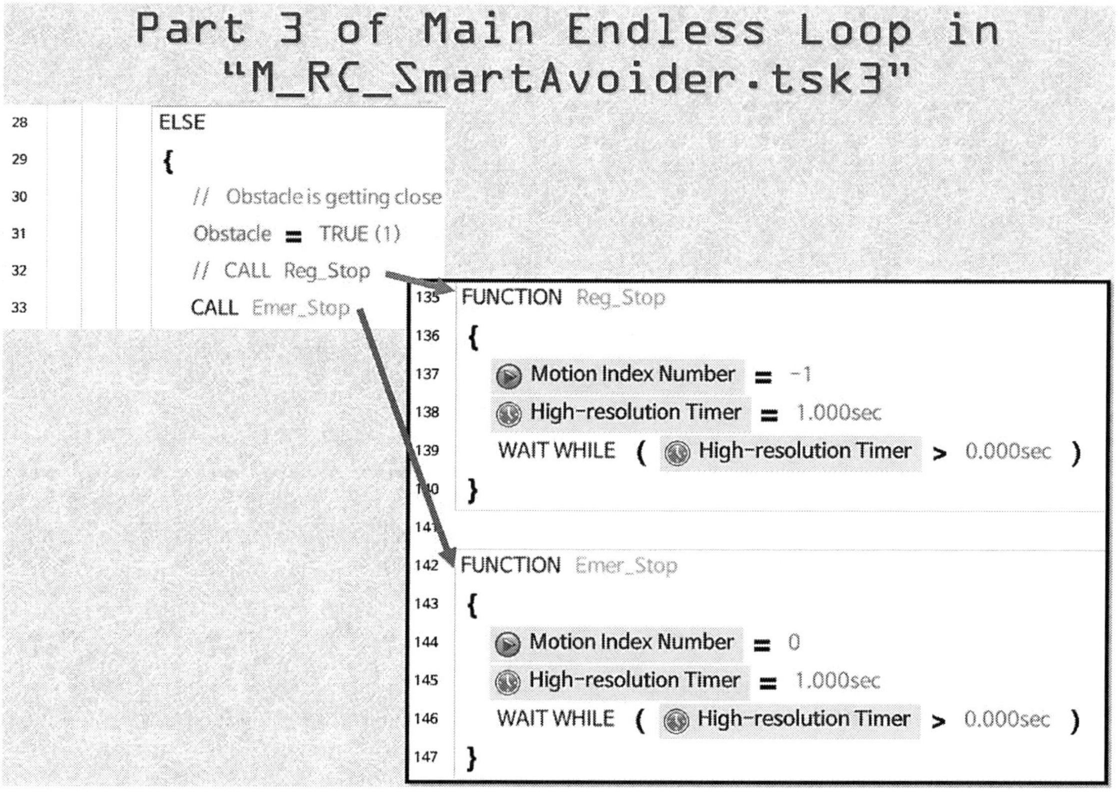

Fig. 6.45 Part 3 of Main Endless Loop in "M_RC_SmartAvoider.tsk3".

Let's look at the details shown in Fig. 6.45 Part 3 of the Main Endless Loop:

- The ELSE on Line 28 is quite important as it represents the negation of the logical statement of the IF structure used on Line 23 of Fig. 6.44. Using De Morgan's rules/laws (https://en.wikipedia.org/wiki/De_Morgan%27s_laws), ELSE would then corresponds to the situation when this condition (**(NIR_Left >= Threshold_1) .OR. (NIR_Right >= Threshold_1)**) is TRUE. This information is used later to design the main LOOP-WHILE controlling the Obstacle Avoidance algorithm used in Part 4 of the Main Endless Loop (see Line 34 in Fig. 6.46).
- Next, Parameter "Obstacle" is set to TRUE at Line 31.
- Then the programmer can use either Function "Reg_Stop" or "Emer_Stop" by invoking either special Motion_Index_Number's values of "**-1**" (Line 137 in right panel of Fig. 6.45) or "**0**" (Line 144 in right panel of Fig. 6.45). These special values were previously discussed in Section 6.5.1 as they would result in a different pose for the MINI after the Motion Stoppage is

performed, i.e. last pose of current MOTION-LIST/PAGE or last pose of corresponding EXIT MOTION-LIST/PAGE.

Part 4 of the Main Endless Loop lays out the steps used to find where the obstacle is located so that appropriate obstacle-avoiding moves can be executed next (see Fig. 6.46):

Part 4 of Main Endless Loop in "M_RC_SmartAvoider.tsk3"

```
34  LOOP WHILE ( NIR_Left > Threshold_2 || NIR_Right > Threshold_2 )
35  {
36     NIR_Dif = NIR_Left - NIR_Right
37     IF ( NIR_Dif <= 10 && NIR_Dif >= -10 )
38     {
39        // Go Backward
40        Direction = 2
41        CALL Walk_Execute
42        WAIT WHILE ( Motion Status == TRUE (1) )
43     }
44     LOOP WHILE ( NIR_Left > Threshold_3 && NIR_Right < NIR_Left )
45     {
46        // Turn Right
47        Direction = 3
48        CALL Walk_Execute
49     }
50     LOOP WHILE ( NIR_Right > Threshold_3 && NIR_Left < NIR_Right )
51     {
52        // Turn Left
53        Direction = 4
54        CALL Walk_Execute
55     }
56  }
```


Fig. 6.46 Part 4 of Main Endless Loop in "M_RC_SmartAvoider.tsk3".

- The LOOP-WHILE at Lines 34-56 is designed as a "**capture-loop**" because at this point in the program, the two Variables "NIR_Left" and "NIR_Right" should have values that correspond to this TRUE condition (**(NIR_Left >= Threshold_1) .OR. (NIR_Right >= Threshold_1)**) (as per previous discussions regarding the ELSE on Line 28). As Threshold_1 (=100) is larger than Threshold_2 (=50) (see Fig. 6.42), **"MATHEMATICAL LOGIC" guarantees that the**

LOOP-WHILE starting at Line 34 will always be entered when an obstacle is detected in the physical world (as per the ELSE branch on Line 28).

- This LOOP-WHILE contains three Event-Action pairs that correspond respectively to the three possible locations of the obstacle with respect to the MINI robot:
  - If the obstacle is in "front" of the MINI, it should make one step back, then keep on turning right or left until the obstacle is no longer detected.
  - If the obstacle is on the "left" of the MINI, it should keep on turning right until the obstacle is no longer detected.
  - If the obstacle is on the "right" of the MINI, it should keep on turning left until the obstacle is no longer detected.
  - The reader should have noticed that the "actions" used to deal with a frontal obstacle include the "actions" used to deal with an obstacle on the left or on the right also. Therefore, we are not dealing with the "standard" situation of "mutually exclusive" Event-Actions Pairs that had been encountered so far, consequently the "standard" IF-ELSE-IF structure would not be applicable here.
- Lines 36-37 are used to determine if the **obstacle is in "front" of the MINI** because Variables "NIR_Left" and "NIR_Right" would then be **within ± 10 units** of each other. If this condition is TRUE (Line 37), then the MINI needs to step backwards **once** (Lines 40-42). Please note that an IF structure is used in Line 37 (as per standard "Event Programming Approach", see Section 6.3.7), but an IF-ELSE-IF structure is not used afterwards because of the discussions in the previous paragraph *(also see more discussions in the "Review Questions" Section)*.
- Please also note the WAIT-WHILE on (MOTION-STATUS == TRUE) used in Line 42: this construct is used to make sure that the robot finishes its ONE backward step, before it takes a right or left turn next, as Function WALK_EXECUTE happens not to use such a WAIT-WHILE in this project.
- The next two Event-Action Pairs ("obstacle on left" or "obstacle on right") are implemented using LOOP-WHILE instead of the usual IF or ELSE-IF because of 2 reasons:
  - The author wanted to resolve the obstacle issue within the same iteration cycle of the Main Endless Loop that the obstacle was detected in.
  - The reader had probably already noticed that the "Left-Turn" and "Right-Turn" MOTION-LIST/PAGEs only rotated the robot "slightly" when they are performed once (i.e. when using IF or ELSE-IF), and a LOOP-FOR is also not applicable here as the number of repeats is not known in advance.
- The Event-Action Pair for "Obstacle on Left" (Lines 44-49) is implemented before the Event-Action Pair for "Obstacle on Right" (Lines 50-55): this was just an arbitrary choice made by the author, it could have been the other way around and the overall behavior of the MINI would remain the same:
  - Please note that both inner LOOP-WHILEs used Parameter "Threshold_3" (= 30) which is smaller than "Threshold_2" (= 50) which is used in the outer LOOP-WHILE defined at Line 34 (see Fig. 6.42). This means that once the two inner LOOP-WHILEs are exited "properly" (i.e. the robot clears the obstacle), the Condition used for the outer LOOP-WHILE would evaluate to FALSE also and would then exit, i.e. the outer LOOP-WHILE was used only once *(also see more discussions in the "Review Questions" Section)*.
  - Please also note that both inner LOOP-WHILEs do not use a WAIT-WHILE on (MOTION-STATUS == TRUE) after FUNCTION "Walk_Execute" is called (Lines 48 and 54), as in the case for the Frontal-Obstacle's IF (Line 42), because this WAIT-

WHILE would NOT be necessary as the combination of these LOOP-WHILEs and the IF statement using (MOTION-STATUS == FALSE) (Line 110 of Fig. 6.43) inside FUNCTION "Walk_Execute" provides the same results.

- In summary:
    - A "Frontal Obstacle" would trigger the MINI to perform 1 BACKWARD move and then to "most likely" perform as many RIGHT-TURN moves as needed to clear the obstacle.
    - An "Obstacle on the Left" would trigger the MINI to perform as many RIGHT-TURN moves as needed to clear the obstacle.
    - An "Obstacle on the Right" would trigger the MINI to perform as many LEFT-TURN moves as needed to clear the obstacle.

When the HC finishes with Part 4 of the Main Endless Loop, the MINI clears the obstacle (or at least the two IRSS-10 sensors cannot detect anything within the set range of less than 30 units). At this point, the author made a conceptual premise that the RC operator most likely still have his/her finger on the RC Button that made the MINI walk into the obstacle in the first place. Therefore, the next important programming feature to implement is to somehow make the HC ignore any "current" RC Button input and keep the MINI at the Ready Pose until he/she releases the "culprit" RC Button. This is the main job for Part 5 of the Main Endless Loop.

Fig. 6.47 illustrates the controlling procedures used in Part 5 of the Main Endless Loop:

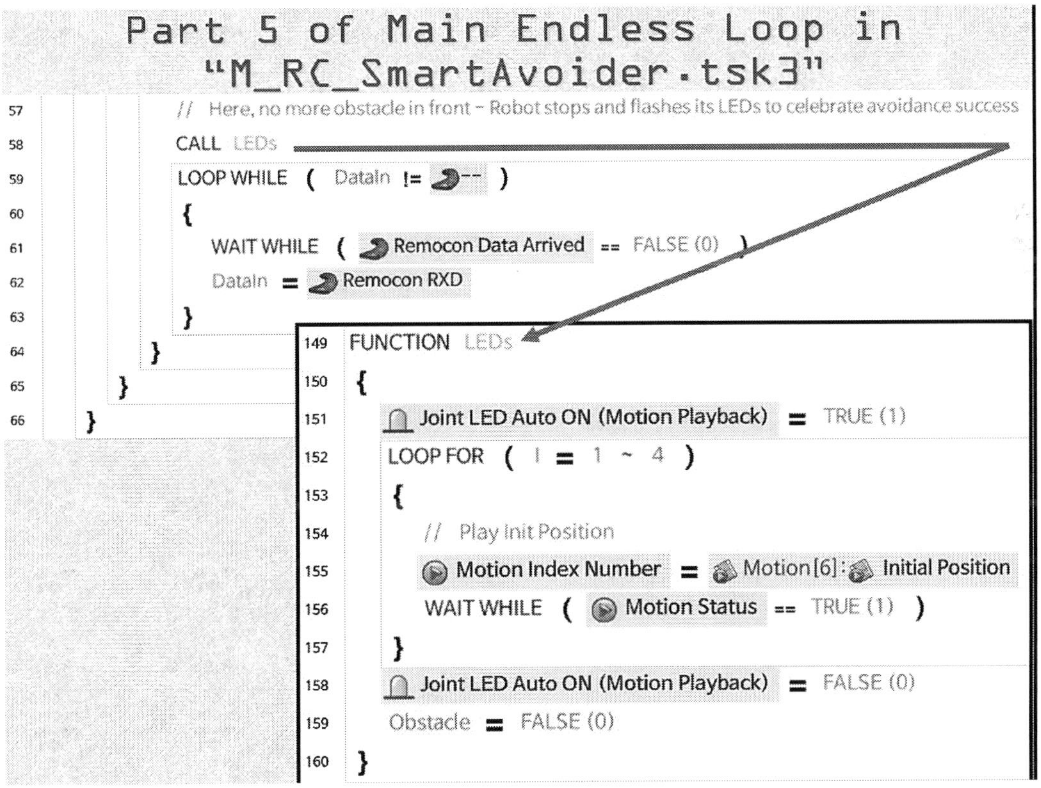

Fig. 6.47 Part 5 of Main Endless Loop in "M_RC_SmartAvoider.tsk3".

- The first procedure in Part 5 (Line 58) is to call Function "LEDs" (Lines 149-160) which:
    - First, sets to TRUE the Parameter "Joint LED Auto ON (Motion Playback)" so that from this point on, the LEDs on the XL-320s will flash during the playing of any MOTION-LIST/PAGE (Line 151).
    - Second, uses a LOOP-FOR (Lines 152-157) to put the MINI into the "Initial Position" Pose **4 times**, just so that the RC operator can see the MINI flashing the LEDs of the XL-320s four times.
    - Third, resets to FALSE the Parameter "Joint LED Auto ON (Motion Playback)" so that from now on the XL-320s won't flash their LEDs when performing a MOTION-LIST/PAGE (to save on battery power).
    - Finally, resets the important Parameter "Obstacle" to FALSE (Line 159), meaning that the "Obstacle Detection" algorithm is again active.
- The second procedure in Part 5 (Lines 59-63) uses a LOOP-WHILE to ignore the "current/culprit" RC Button input produced by the RC operator until he/she releases this Button (which could be any of the U-D-L-R Buttons):
    - The LOOP-WHILE on Line 59 uses a "double-negative" in its LOOP-CONDITION which is always hard to explain to readers! This WHILE loop keeps on looping whenever the logical statement **("DataIn" NOT EQUAL to "No Button")** evaluates to TRUE, which is equivalent to **("DataIn" EQUAL to "Any Button")** in everyday language.
    - Let's assume that the operator still presses on the "culprit" Button (i.e. "Any Button"), even though he/she saw the robot LEDs flash 4 times, then this LOOP-WHILE is active, and Lines 61-62 are executed saving the "new" RC Button into Parameter "DataIn". Please note that Parameter "DataIn" would always have a non-zero value for as long as any RC Button is pushed. When the operator releases the "culprit" Button, e.g. "No Button" is pushed, the RC-100 will send the value "—" (i.e. "0") ONCE which would make this LOOP-WHILE stop, and the HC finally reaches the end of the Main Endless Loop and it loops back to Line 11 of Fig. 6.43. The MINI is then ready to handle any RC input from the operator again!

Video 6.9 shows a run-time performance of the program "M_RC_SmartAvoider.tsk3" proving that this algorithm does work overall, but for some reason unknown to the author, Line 158 would not execute reliably at times (i.e. occasionally, the LEDs won't turn off after the obstacle was cleared).

6.5.4 Alternate Solution for Smart Avoider

There is an alternate algorithm/solution to the Main Endless Loop for the "Smart Avoider" project as shown below:

Main Endless Loop {

If (NIR Sensors detect no obstacle) then { *// Take "RC" Path*

When new RC data comes in, save it in DataIn

If (DataIn == 0) then (Play robot's Initial Position)

Else { (i.e. U-D-L-R data coming in)

Execute RC command as normal

} // End of processing for U-D-L-R buttons

} // End of RC Path

Else { // Obstacle within NIR detection ranges so take "Autonomous" Path

Stop the robot

Figure out whether Obstacle is Frontal, on Left or on Right of robot and activate appropriate robot's avoidance moves *<u>until obstacle out of NIR detection ranges</u>*

Use LEDs to signal robot's successful obstacle avoidance & *<u>put robot in Initial Position</u>*

Wait until Operator releases last RC button (which made robot walk into obstacle)

} // End of autonomous obstacle avoidance section

} (End of Main Endless Loop, so loop back to check NIR Sensors)

This alternate approach is implemented in the program "M_RC_SmartAvoider_alt.tsk3" (see Fig. 6.48). Some readers may prefer this logic flow as it corresponds visually to the "Sense-Think-Act" diagram drawn earlier in Fig. 6.15. In Fig. 6.48, Line 15 could be "conceptually" located at the branching out of the "RC" and "Autonomous" Paths, but we all know that there is no TASK code in the "cloud" – at least not yet!

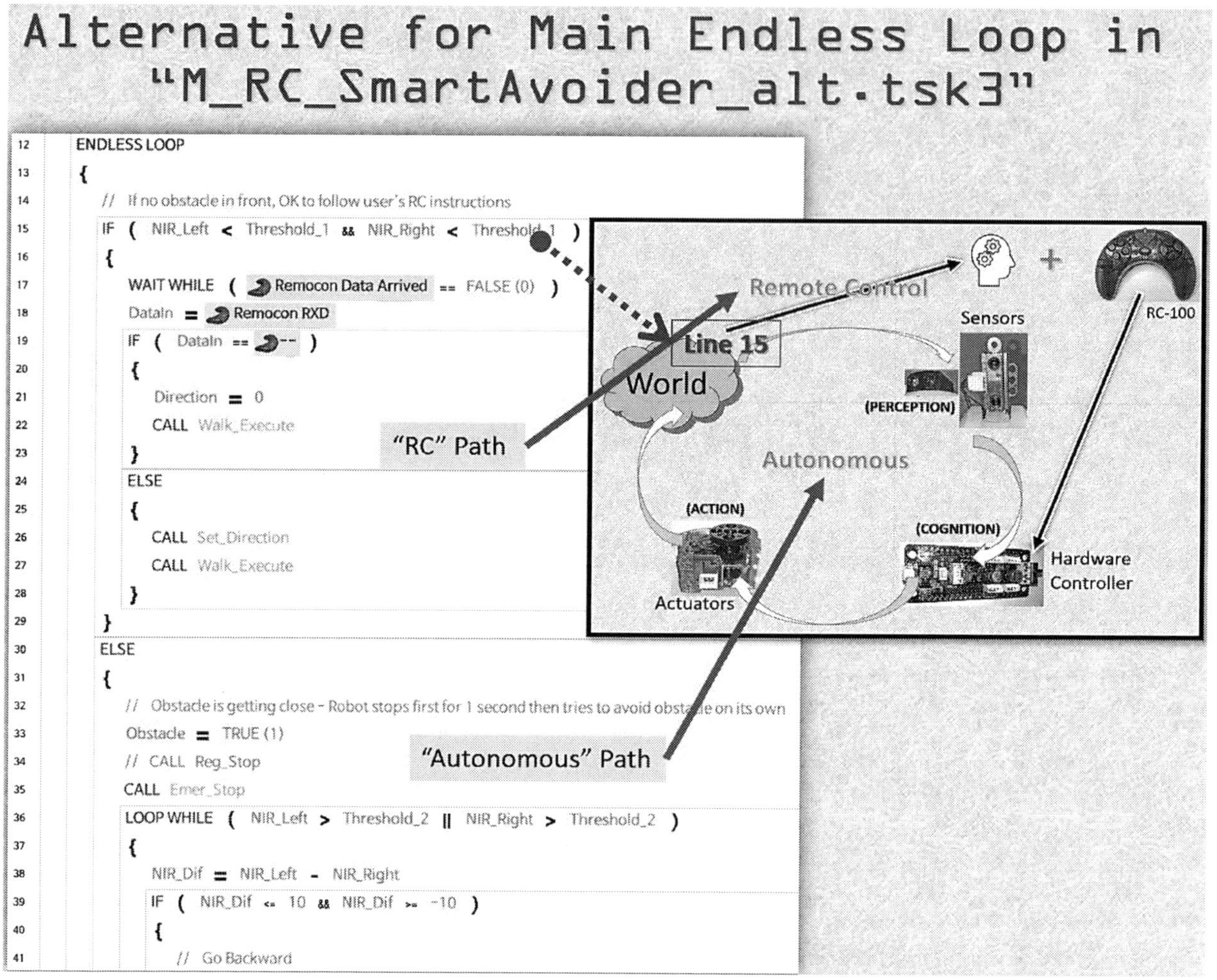

Fig. 6.48 Alternate Solution to Main Endless Loop in "M_RC_SmartAvoider_alt.tsk3".

There is no difference in run-time performances between the two algorithms that the author could discern.

The "Smart Avoider" project showcased a mixed "RC/Autonomous" approach to control the MINI, in the next project, we'll go "full-autonomous".

6.6 Autonomous Random Walker

This "Random Walker" project is similar to the "Smart Avoider" project in the robot's behavior for avoiding an obstacle, but instead of accepting its Walk Direction inputs from the RC operator like before, the "Random Walker" would decide on its Walking Direction and the Number of Repeats (for its resulting MOTION-LIST/PAGE) from a random number generator which is already built-in in the OpenCM9.04-C's firmware.

6.6.1 Using Built-In Random Number Generator

The OpenCM9.04-C's random-number generator can provide the TASK programmer with a random stream of integer values from 0 to 255 pulled out from a uniform probability distribution function (see Line 75 in Fig. 6.49).

Function Set_Direction in "M_RandomWalker.tsk3"

```
73  FUNCTION Set_Direction
74  {
75      Random = Random Number
76      IF ( Random >= 0 && Random <= 63 )
77      {
78          Direction = 1                            (25%)
79          // Go Forward
80      }
81      ELSE IF ( Random >= 64 && Random <= 127 )
82      {
83          Direction = 2                            (25%)
84          // Go Backward
85      }
86      ELSE IF ( Random >= 128 && Random <= 191 )
87      {
88          Direction = 3                            (25%)
89          // Turn Right
90      }
91      ELSE IF ( Random >= 192 && Random <= 255 )
92      {
93          Direction = 4                            (25%)
94          // Turn Left
95      }
96      Print Screen = Direction
97  }
```

Fig. 6.49 Function "Set_Direction" in "M_RandomWalker.tsk3".

For the random "Walk Direction" feature, the author wants each possible moving direction U-D-L-R to have equal probability: i.e. each for 25% of the time – thus the division of the full range [0-255] into 4 smaller ranges [0-63], [64-127], [128-191] and [192-255] (see Fig. 6.49).

For the random "Number of Repeats" feature, the author wants to "favor" the smaller Number of Repeats: i.e. 40% for 2 repeats, 39% for 3 repeats and 21% for 6 repeats (see Fig. 6.50).

Function Set_Motion_Repeats in "M_RandomWalker.tsk3"

```
FUNCTION Set_Motion_Repeats
{
    Random = 2 Random Number
    IF ( Random >= 0 && Random <= 100 )
    {
        Repeats = 2                              (40%)
    }
    ELSE IF ( Random >= 101 && Random <= 200 )
    {
        Repeats = 3                              (39%)
    }
    ELSE IF ( Random >= 201 && Random <= 255 )
    {
        Repeats = 6                              (21%)
    }
}
```

Fig. 6.50 Function "Set_Motion_Repeats" in "M_RandomWalker.tsk3".

6.6.2 New "Walk Execute" Approach

The "Essential Algorithm" for the "Random Walker" project is a modification of the Main Endless Loop used in "M_RC_SmartAvoider_alt.tsk3" and it is shown below:

Main Endless Loop {

If (NIR Sensors detect no obstacle) then { // Take "Random Walk" Path

Set a random Walk Direction

Set a random Number of Repeats for Motion-List/Page corr. to Current Walk Direction

Call Function Walk_Execute

} // End of Random Walk Path

Else { // Obstacle within NIR detection ranges so take "Obstacle Avoidance" Path

Stop the robot

Figure out whether Obstacle is Frontal, on Left or on Right of robot and activate appropriate robot's avoidance moves until obstacle out of NIR detection ranges

Use LEDs to signal robot's successful Obstacle Avoidance & put robot in Initial Position

} // End of obstacle avoidance section

} (End of Main Endless Loop, so loop back to check NIR Sensors)

Part 1 of the above Main Endless Loop lays out the procedures to perform when no obstacle is detected (see Fig. 6.51):

Part 1 of Main Endless Loop in "M_RandomWalker.tsk3"

```
ENDLESS LOOP
{
    // If no obstacle in front, OK to random walk
    IF ( NIR_Left < Threshold_1 && NIR_Right < Threshold_1 )
    {
        CALL Set_Direction
        CALL Set_Motion_Repeats
        CALL Walk_Execute
    }
```

Fig. 6.51 Part 1 of Main Endless Loop in "M_RandomWalker.tsk3".

- Call Function "Set_Direction" (see Fig. 6.49).
- Call Function "Set_Motion_Repeats" (see Fig. 6.50).
- Call Function Walk_Execute (see Figs. 6.52 and 6.53).

Function "Walk_Execute" (Lines 116-150) in the "Random Walker" project is substantially changed from the ones used in the previous projects (see Figs. 6.12, 6.23 and 6.43) as a LOOP-FOR is needed to handle the Number of Repeats of the Current MOTION-LIST/PAGE as shown in Line 120 in Fig. 6.52. Furthermore, as Parameter "Repeats" could take on the maximum value of "6" (see Line 112 in Fig. 6.50), a **new issue** arises as it is possible that the MINI gets too close to the obstacle before the LOOP-FOR has finished its programmed number of iterations. Thus, the author found the need for a **new "Obstacle Detection" procedure to be activated** right after Part 1 of Function "Walk_Execute" (see Fig. 6.53), i.e. **within each iteration of the LOOP-FOR and right after the MINI starts the commanded MOTION-LIST/PAGE**.

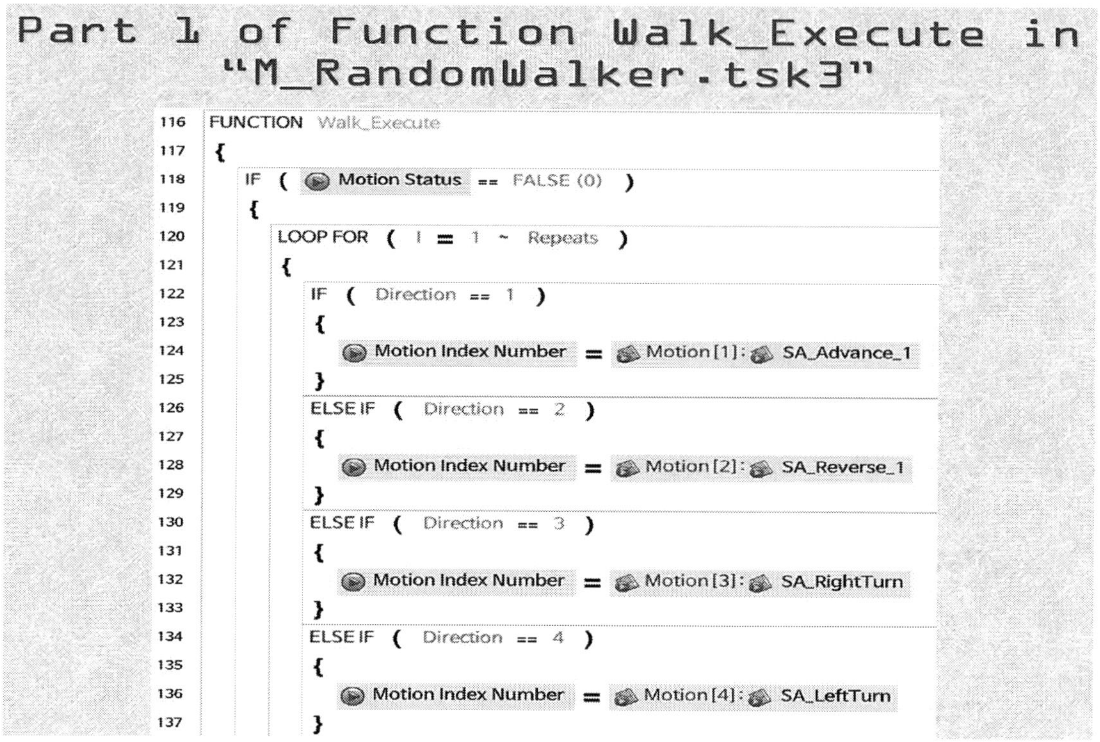

Fig. 6.52 Part 1 of Function "Walk_Execute" in "M_RandomWalker.tsk3".

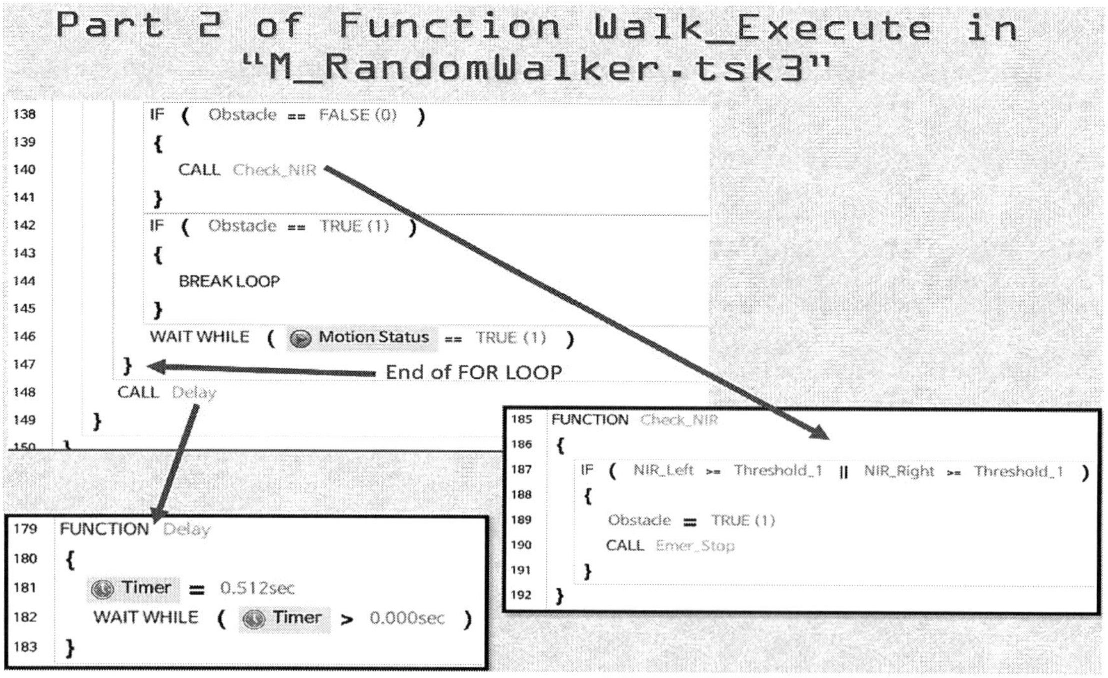

Fig. 6.53 Part 2 of Function "Walk_Execute" in "M_RandomWalker.tsk3".

This new "Obstacle Detection" procedure relies on two IF structures at Lines 138-141 and at Lines 142-145:

- The first IF structure (Lines 138-141) is only activated when there is "no obstacle detected", which should happen most of time (please note that Parameter "Obstacle" was initialized to FALSE in Function "Init" at start – see Line 61 in Fig. 6.54). Thus, when there is "no obstacle detected", Function "Check_NIR" at Line 140 is called (defined in Lines 185-192 of Fig. 6.53). And when Function "Check_NIR" is executed, two possible outcomes can present themselves:
    - "No obstacle is detected" as per the IF condition used on Line 187, then this Function is quickly exited at Line 192, and Parameter "Obstacle" retains its FALSE value.
    - "An obstacle is detected" as per the IF condition on Line 187, then Parameter "Obstacle" is set to TRUE, and Function "Emer_Stop" is called to stop all robot movements using the Special Motion Index Number "0" (see Fig. 6.55 and Section 6.5.1).

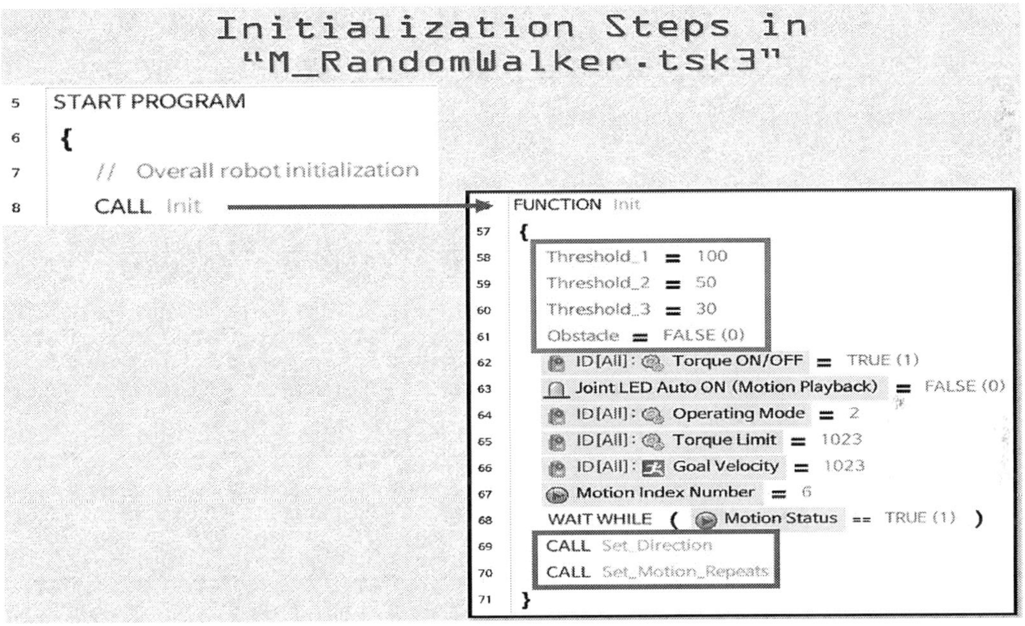

Fig. 6.54 Function "Init" in "M_RandomWalker.tsk3".

- The second IF structure (Lines 142-145) on the other hand is only activated when "an obstacle is detected", i.e. Parameter "Obstacle" is set to TRUE, and in this case, it will BREAK this LOOP-FOR and will jump next to Line 148.
- Thus, at run-time of this code, when the HC reaches Line 142 in Fig. 6.53, two possible outcomes can happen:
-
    - If "no obstacle was detected" via Function "Check_NIR" (in the First IF) then Parameter "Obstacle" was still set to FALSE and the HC just jumps to Line 146 and waits there for MOTION-STATUS to become FALSE. This event happens when the current MOTION-LIST/PAGE ends its performance. Next, the LOOP-FOR continues

its next iteration and so on until its last iteration, if no obstacle is encountered for the remaining iterations. Then a small delay at Line 148 is used for the MINI to let the frame's vibrations to dissipate (after such extensive movements) before exiting Function "Walk_Execute" as normal and the HC goes back to the Main Endless Loop at Line 19 then back to Line 12 of Fig. 6.51 (with "Obstacle" set to FALSE). In short, if this code is run on the MINI in an open space, the robot will just do "random" movements (i.e. it is on the "Random Walk" Path).

- However, if "an obstacle was detected" via Function "Check_NIR" (in the First IF) then Parameter "Obstacle" is set to TRUE, and so the IF condition on Line 142 also evaluates to TRUE, then the command "**BREAK LOOP**" is executed and the **LOOP-FOR is immediately aborted** (regardless of the current value of Counter I with respect to the value of "Repeats" in Line 120). Next, the HC jumps to Line 148 for a short delay (please remember that Function "Emer_Stop" was already activated inside Function "Check_NIR" so the robot was already standing still at this point). Then Function "Walk_Execute" is exited as normal and the HC goes back to the Main Endless Loop at Line 19 then Line 12 of Fig. 6.51, but this time "Obstacle" is set to TRUE and the IF condition in Line 14 will evaluate to FALSE (because its converse already evaluated to TRUE at Line 187 of Fig. 6.53), so Parts 2, 3 and 4 of the Main Endless Loop will now become active (i.e. "Obstacle Avoidance" Path).

```
Functions Reg_Stop & Emer_Stop in
"M_RandomWalker.tsk3"

FUNCTION Reg_Stop
{
    Motion Index Number = -1
    High-resolution Timer = 1.000sec
    WAIT WHILE ( High-resolution Timer > 0.000sec )
}

FUNCTION Emer_Stop
{
    Motion Index Number = 0
    High-resolution Timer = 1.000sec
    WAIT WHILE ( High-resolution Timer > 0.000sec )
}
```

Fig. 6.55 Functions "Reg_Stop" and "Emer_Stop" in "M_RandomWalker.tsk3".

Parts 2, 3 and 4 of the Main Endless Loop are defined by the big ELSE structure between Line 20 and Line 52 (see Figs. 6.56, 6.57 and 6.58).

The reader may recall that Function "Check_NIR" can "detect an obstacle" by way of the IF condition in Line 187 of Fig. 6.53 – this is the **first** way to detect obstacles in this project's code, i.e. **while the robot is performing required MOTION-LIST/PAGEs**. Part 2 of the Main Endless Loop is designed to handle the **second** way that an "obstacle can be detected" which is by way of the IF condition in Line 14 of Fig. 6.51 (to be exact when this condition evaluates to FALSE). At this point, the "Walk_Execute" Function just exited normally and Parameter "Obstacle" was set to "FALSE" (basically the robot was momentarily stopped and its frame vibrations may not have completely dissipated yet), but then an obstacle comes within the ranges of detection by the two NIR sensors, thus sending the HC to line 20 to begin the big ELSE structure corresponding to the "Obstacle Avoidance" Path.

The details of Part 2 of the Main Endless Loop are listed in Fig. 6.56:

- If "Obstacle" was set to FALSE, then it is set to TRUE and Function "Reg_Stop" is called (see Fig. 6.55 and Section 6.5.1), i.e. the Special Motion Index Number "-1" is used to stop the robot as an extra cautionary step, as the robot was already in the last pose of the last U-D-L-R moves which happens to be the "Initial Position" pose also.
- If "Obstacle" was set to TRUE (by Function "Check_NIR" in previous discussions), then Part 2 is skipped.

Part 2 of Main Endless Loop in "M_RandomWalker.tsk3"

```
20        ELSE
21        {
22          // Obstacle is getting close - Robot stops first for 1 second
23          IF ( Obstacle == FALSE (0) )
24          {
25            Obstacle = TRUE (1)
26            CALL Reg_Stop
27          }
```

Fig. 6.56 Part 2 of Main Endless Loop in "M_RandomWalker.tsk3".

Part 3 of the Main Endless Loop is designed to determine whether the obstacle is "Frontal" or on the "Left" or "Right" of the MINI (see Fig. 6.57) and then to perform the appropriate maneuvers to get the robot clear of the obstacle. It is the same algorithm used for Part 4 of the previous "Smart Avoider" project (see Fig. 6.46), so its explanations won't be repeated here.

Part 3 of Main Endless Loop in "M_RandomWalker.tsk3"

```
LOOP WHILE ( NIR_Left > Threshold_2 || NIR_Right > Threshold_2 )
{
   NIR_Dif = NIR_Left - NIR_Right
   IF ( NIR_Dif <= 10 && NIR_Dif >= -10 )
   {
      // Go Backward
      Direction = 2
      CALL Walk_Execute
   }
   LOOP WHILE ( NIR_Left > Threshold_3 && NIR_Right < NIR_Left )
   {
      // Turn Right
      Direction = 3
      CALL Walk_Execute
   }
   LOOP WHILE ( NIR_Right > Threshold_3 && NIR_Left < NIR_Right )
   {
      // Turn Left
      Direction = 4
      CALL Walk_Execute
   }
}
```

Frontal Obstacle

Obstacle on Left

Obstacle on Right

Fig. 6.57 Part 3 of Main Endless Loop in "M_RandomWalker.tsk3".

Part 4 of the Main Endless Loop is designed to flash the XL-320s LEDs four times to signal to viewers that the MINI had cleared the obstacle and then reset Parameter "Obstacle" to FALSE so that the robot can perform its random walks again (see Fig. 6.58).

Part 4 of Main Endless Loop in "M_RandomWalker.tsk3"

```
            // Here, no more obstacle in front - Robot stops and flashes its LEDs to celebrate avoidance success
            CALL LEDs
        } (End of Outer Else started at Line 20)
    } (End of Main Endless Loop started at Line 11)
}(End of Main Program)
```

```
FUNCTION LEDs
{
    Joint LED Auto ON (Motion Playback) = TRUE (1)
    LOOP FOR ( I = 1 ~ 4 )
    {
        // Play Init Position
        Motion Index Number = Motion[6]: Initial Position
        WAIT WHILE ( Motion Status == TRUE (1) )
    }
    Joint LED Auto ON (Motion Playback) = FALSE (0)
    Obstacle = FALSE (0)
}
```

Fig. 6.58 Part 4 of Main Endless Loop in "M_RandomWalker.tsk3".

Video 6.10 shows a runtime demonstration of the program "M_RandomWalker.tsk3".

6.6.3 Basic Concepts Review

If the reader compares the structures (laid out in pseudo-codes) of the Main Endless Loops between the "Smart Avoider" and "Random Walker" projects (see Sections 6.5.4 and 6.6.2 and of course the actual TASK codes), the reader can readily notice their similarities. **Thus, inside a TASK program, there are no conceptual differences between "RC", "Autonomous", "Random Walk" or "Obstacle Avoidance" Paths, just different behaviors or states that the programmer can put the robot in to satisfy some project requirements.**

Through Sections 6.3 to 6.6, the author hopes that the beginner in robotics can now appreciate the applicability of the "Sense-Think-Act" Paradigm and the derived tools such as "Event-Action" Pairs in any robotics project that the reader may encounter from now on.

The author also hopes that there were enough and varied examples for the beginner roboteer about how to use basic computing structures such as ENDLESS LOOP, IF, IF-ELSE-IF, Parallel/Sequential IFs, Parallel/Sequential LOOP-WHILEs (and their infinite combinations) in helping the reader create efficient solutions to various problems encountered as he/she "wades" into a robotics project.

We'll finish out this Chapter 6 with some more advanced MINI projects.

6.7 The "Mimicking MINIs"

Back in Section 6.4.1, the reader had been shown how to filter a "Remocon" Message to obtain only specific information about specific bits comprising the RC-100 message which has only 10 bits, each bit corresponding to each of the 10 Buttons U-D-L-R-1-2-3-4-5-6. However, a full "Remocon" message has 16 bits (http://emanual.robotis.com/docs/en/parts/communication/rc-100/#communication-packet), thus this "Mimicking MINIs" project will show the reader how to use the other 6 bits via a "Message Shaping" procedure.

For this project, **the reader does need to have access to two MINIs**, one will be set up as the Leader and the other as the Follower. The servos of the upper body of the Leader (i.e. Servo IDs from 1 to 6 and 17) will be disabled so that they can manipulated by hand. **The goal of this project is to design TASK codes to run on the Leader and Follower robots such that when the Leader's disabled servos are manipulated one-by-one or as a unit, the corresponding servos on the Follower will match exactly their movements.**

6.7.1 Master & Slave BT-210

First, the BT-210s on the Leader and Follower robots need to be modified so that **they communicate only to each other** (and not to the PC or Mobile Device as we had been using them). In wireless communications speak, the Leader's BT-210 needs to be set up as a Master and the Follower's BT-210 needs to be set up as a Slave. The ROBOTIS e-manual does carry how-to instructions for this procedure, along with a video, at this web link http://emanual.robotis.com/docs/en/parts/communication/bt-210/#communication-mode.

At this point, a sharp reader would raise the question about how to download the TASK codes from the development computer (PC or Mobile Device) to the MINIs now that they are set up to only talk to each other? There are several options: a) temporarily use the **same Slave BT-210** to download the special codes to the **Leader and Follower robots but separately**, or b) use an LN-101 if the reader has access to one.

6.7.2 "Message Shaping" Concept

For this project, the Leader robot needs to send to the Follower robot information about its own chosen XL-320 regarding its **ID** number (valued at 1 through 6 or at 17) and its corresponding **Present Position** (i.e. a number between 0 and 1023). When the Follower receives these items, it can use them to match its own servo **ID** and to set its **Goal Position** with the Leader's **Present Position**.

The highest value for Servo Position is "1023" in Decimal which translates to "0000 0011 1111 1111" in Binary (i.e. 10 bits are needed for Servo Position information). The highest value for Servo ID in this project is "17" in Decimal which translates to "0000 0000 0001 0001" in Binary (i.e. 5 bits are needed for Servo ID information). Thus, it is straightforward to come up with the following scheme to partition the 16-bit "Remocon" message into two sections:

- The lowest 10 bits will carry the information for Servo Position (in italics).
- The highest 5 bits will carry the information for Servo ID (in bold).
- The 11th bit is used as a "demarcation" bit – not carrying any vital information.
- Thus, a typical "Remocon" message for this project could look like this one:
    - "**1000 1**0*10 0000 0000*" which means that ID=17 and Servo Position = 512 (i.e. 10 0000 0000).

The crucial task will be how to shift the binary number "0000 0000 0001 0001" for ID=17 left-wise 11 bits so that it becomes "1000 1000 0000 0000", and then we can just **add this left-shifted ID** information to the **standard Servo Position** information "0000 0010 0000 0000" to yield the final message "1000 1010 0000 0000".

Now, most computer languages such as C++ and Python support left-shift "<<" and right-shift ">>" operators but TASK does not. However, we can use the multiplication "*" and division "/" operators to do the same respective jobs as the left-shift and right-shift operators, for example:

- ID = 17
- ShiftedID = (0000 1000 0000 0000) * ID // *"1000 0000 0000" is "2048" in decimal*
- ReconstructedID = ShiftedID / (0000 1000 0000 0000)

6.7.3 "Open-Loop" Solution for "Mimicking MINIs"

In the "Open Loop" solution, the Leader robot just sends its information over to the Follower robot which will strive to match the Leader's motions, whether the Follower can do the required motions or not.

The "Essential Algorithm" for the Leader in "M_MimickingMinis_**L**_OpenLoop.tsk3" is shown below:

Initialization of the Leader Robot & put it in Initial Position.

Disable Leader's Upper Body (i.e. Servos 1 through 6 and 17) by setting their Torque Limit and Goal Velocity to 0.

Main Endless Loop { // each Loop Iteration works on only 1 servo of the MINI's Upper Body

For ONE chosen servo:

Prepare message containing "ID" and "Present Position" of chosen servo.

Send message to Follower robot

Update ID to next servo ID in the list for Leader's Upper Body

} // End of Main Endless Loop

While the "Essential Algorithm" for the Follower in "M_MimickingMinis_**F**_OpenLoop.tsk3" is shown below:

Initialization of the Follower Robot & put it in Initial Position.

Main Endless Loop { // each Loop Iteration works on only 1 servo of the MINI's Upper Body

Wait for a new Remocon packet to arrive

Decode received message into the "ID" component and "Target Position" component

If ("Target Position" is Valid) then

*Set **Follower's Servo** with (received) "ID" to a Goal Position = (received) Target Position*

} // End of Main Endless Loop

Fig. 6.59 shows the initialization steps for the Leader robot:

- Function "Init" first sets all servos to Max Torque and Velocity (Lines 38-41) so that it can set the Leader into the Ready pose (i.e. MOTION 6) at Lines 43-44.
- Function "SetUpperBody" then "disables" Servo 17 (Lines 49-50) and Servos 1-6 so that the operator can move them by hand. A LOOP-FOR, using counter "I", is used for the Servo Group with ID 1 through 6 (Lines 51-55). Please note the Custom Commands used for "Goal Velocity" (i.e. Address 32 at Line 53) and for "Torque Limit" (i.e. Address 35 at Line 54). This type of information can be found in the Control Table for the XL-320 available at this web link (http://emanual.robotis.com/docs/en/dxl/x/xl320/#control-table-of-ram-area).

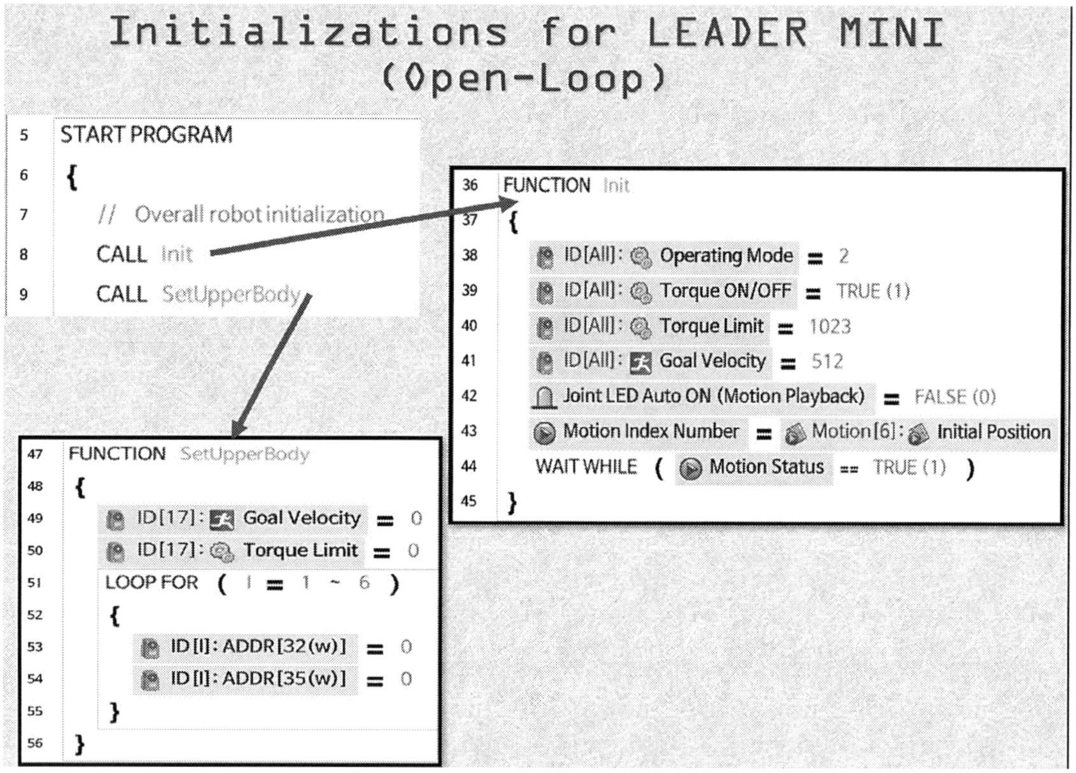

Fig. 6.59 Initialization Steps for Leader robot in "M_MimickingMinis_**L**_OpenLoop.tsk3".

Fig. 6.60 shows the Main program for "M_MimickingMinis_**L**_OpenLoop.tsk3" for the Leader robot:

- Please note that Parameter ID is initialized to 1 on Line 12, before entering the Endless Loop at Line 13.
- Line 15 shows how to shift "ID" left-wise 11 bits by Multiplying it with the binary constant "1000 0000 0000" (i.e. "2048" in decimal).
- Line 16 shows how to combine this shifted-ID's information with the value of "Present Position" (i.e. Address 37 in the XL-320 Control Table) by using the Addition operator.

```
Main Program for LEADER MINI
(Open-Loop)
12  ID = 1
13  ENDLESS LOOP
14  {
15      MessageSent = 0000 0000 0000 0000 0000 1000 0000 0000 * ID
16      MessageSent = MessageSent + ID[ID]: ADDR[37(w)]
17      Remocon TXD = MessageSent
18
19      IF ( ID < 6 )
20      {
21          ID = ID + 1
22      }
23      ELSE IF ( ID == 6 )
24      {
25          ID = 17
26      }
27      ELSE IF ( ID == 17 )
28      {
29          ID = 1
30      }
31      High-resolution Timer = 0.100sec
32      WAIT WHILE ( High-resolution Timer > 0.000sec )
33  }
```

Fig. 6.60 Main program in "M_MimickingMinis_**L**_OpenLoop.tsk3".

- Line 17 sends this combined message to the Follower's Controller to be processed in program "M_MimickingMinis_**F**_OpenLoop.tsk3".
- Lines 19 through 30 are used to update ID to its next value and to keep it within the cycle 1-2-3-4-5-6-17.
- Lines 31-32 is just a WAIT of 100 ms, but it is quite important to prevent Remocon buffer overflow on the Follower's side. The user can adjust it for optimal performance of the robots.

Fig. 6.61 shows the Main program for "M_MimickingMinis_**F**_OpenLoop.tsk3" for the Follower:

- Line 8 calls Function Init which is the same one as for the Leader code.
- Once a new Message from the Leader is saved into "MessageReceived" (Line 14). The Follower begins its extraction of the ID part and of the Servo Position part using the bit-wise "&" operator and the Division operator.
- Line 16 shows how the "&" operator extracts the highest 5 bits of the 16-bit "MessageReceived" to obtain the "shifted-ID" component (which is still in an incorrect bit position to use).
- Line 17 shows how to shift "ID" right-wise 11 bits by Dividing it with the binary constant "1000 0000 0000" (i.e. "2048" in decimal). The "ID" information is now correct to use by the Follower's Controller and saved into Parameter "ServoID".
- Line 18 does another extraction with the "&" operator and "MessageReceived" to get the lowest 11 bits (10 bits would have been OK too). This result is saved into Parameter "TargetPosition".

- Lines 19-22 use an IF structure to make sure that "TargetPosition" has the correct numerical range [0-1023] before using it to set the Follower's servo (with the same ID as the Leader's) to a new Goal Position (i.e. Address 30 on Line 21).

Main Program for FOLLOWER MINI (Open-Loop)

```
8   CALL Init
9
10  // Main Loop
11  ENDLESS LOOP
12  {
13      WAIT WHILE ( Remocon Data Arrived == FALSE (0) )
14      MessageReceived = Remocon RXD
15      // Decode Message Received
16      ServoID = MessageReceived & 0000 0000 0000 0000 1111 1000 0000 0000
17      ServoID = ServoID / 0000 0000 0000 0000 1000 0000 0000
18      TargetPosition = MessageReceived & 0000 0000 0000 0000 0000 0111 1111 1111
19      IF ( TargetPosition <= 1023 )
20      {
21          ID[ServoID]: ADDR[30(w)] = TargetPosition
22      }
23  }
```

Fig. 6.61 Main program in "M_MimickingMinis_**F**_OpenLoop.tsk3".

Video 6.11 shows the runtime performance of two MINIs running the Leader and Follower codes for the "Open Loop" solution.

6.7.4 "Closed-Loop" Solution for "Mimicking MINIs"

In the "Closed Loop" solution, the Leader robot still sends its information to the Follower robot which will strive to match the Leader's motions, however if some Follower's servo experiences a Torque Overload situation (because it cannot go where it is supposed to go, for example it gets stuck by some obstacle), then the Follower would send information (to the Leader) about the overloaded servo's ID and its Servo Position where the overload is occurring. This means that "Remocon" messages need to go both ways between the Leader & Follower robots. Thus, both Leader and Follower codes for the "Closed Loop" solution will now have a "Sending" section and a "Receiving" section.

Fig. 6.62 shows Part 1 of the Leader program "M_MimickingMinis_**L**_ClosedLoop.tsk3", which is the same for the Open Loop case (see Figs. 6.59 and 6.60):

Part 1 of Main for LEADER MINI (Closed-Loop)

```
CALL Init
CALL SetUpperBody

// Main Loop
ID = 1
ENDLESS LOOP
{
    MessageSent = 0000 0000 0000 0000 0000 1000 0000 0000 * ID
    MessageSent = MessageSent + ID[ID]: ADDR[37(w)]
    Remocon TXD = MessageSent

    IF ( ID < 6 )
    {
        ID = ID + 1
    }
    ELSE IF ( ID == 6 )
    {
        ID = 17
    }
    ELSE IF ( ID == 17 )
    {
        ID = 1
    }
```

Fig. 6.62 Part 1 of Main program in "M_MimickingMinis_**L**_ClosedLoop.tsk3".

Fig. 6.63 shows Part 1 of the Follower program "M_MimickingMinis_**F**_ClosedLoop.tsk3", which is the same for the Open Loop case (see Fig. 6.61), except for Line 12 (outside of the Main Endless Loop) initializing Parameter ID to 1:

Part 1 of Main for FOLLOWER MINI (Closed-Loop)

```
9   CALL Init
10
11  // Main Loop
12  ID = 1
13  ENDLESS LOOP
14  {
15      WAIT WHILE ( Remocon Data Arrived == FALSE (0) )
16      MessageReceived = Remocon RXD
17      // Decode Message Received
18      ServoID = MessageReceived & 0000 0000 0000 0000 1111 1000 0000 0000
19      ServoID = ServoID / 0000 0000 0000 0000 0000 1000 0000 0000
20      TargetPosition = MessageReceived & 0000 0000 0000 0000 0000 0111 1111 1111
21      IF ( TargetPosition <= 1023 )
22      {
23          ID[ServoID]: ADDR[30(w)] = TargetPosition
24      }
```

Fig. 6.63 Part 1 of Main program in "M_MimickingMinis_**F**_ClosedLoop.tsk3".

Fig. 6.64 shows Part 2 of the Follower program “M_MimickingMinis_**F**_ClosedLoop.tsk3” which is “new” code:

```
Part 2 of Main for FOLLOWER MINI
(Closed-Loop)
26      MessageSent = 0000 0000 0000 0000 0000 1000 0000 0000 * ID
27      LoadData = ID[ID]: ADDR[41(w)] & 0000 0000 0000 0000 0000 0011 1111 1111
28      LoadLimit = 512
29      IF ( LoadData > LoadLimit )
30      {
31          MessageSent = MessageSent + ID[ID]: ADDR[37(w)]
32          Remocon TXD = MessageSent
33      }
34
35      IF ( ID < 6 )
36      {
37          ID = ID + 1
38      }
39      ELSE IF ( ID == 6 )
40      {
41          ID = 17
42      }
43      ELSE IF ( ID == 17 )
44      {
45          ID = 1
46      }
```

Fig. 6.64 Part 2 of Main program in “M_MimickingMinis_**F**_ClosedLoop.tsk3”.

- Line 26 left-shifts “ID” by 11 bits (same as for the Leader) as the first step of preparing Parameter “MessageSent”.
- Line 27 defines Parameter “LoadData” using the lowest 10 bits of the “Present Load” (i.e. Address 41) of Servo identified by Parameter “ID” (initially set to “1” by Line 12 of Fig. 6.63, then updated subsequently by the Lines 35-46 in Fig. 6.64).
- Line 28 sets “LoadLimit” to “512” (the user can change this value as needed).
- The IF condition in Line 29 checks to see if “LoadData” is over “LoadLimit”. If this condition is “TRUE”, Line 31 adds the “Present Position” (Address 37) of this “overloaded” Servo to the previously set “MessageSent” (see Line 26). Finally, Line 32 uses “Remocon TXD” to send this message from the Follower robot to the Leader robot.
- Lines 35-46 does the job of updating “ID” with the proper numerical value (i.e. 1-2-3-4-5-6-17) for the next iteration of the Main Endless Loop.

Essentially, Part 2 will inform the Leader robot about any “Follower” servo that happens to be “overloaded” (i.e. “Present Load” greater than “LoadLimit”) and this message contains this Servo’s ID and its Present Position where the overloading is occurring. When the Leader robot receives this message, it will decode it and makes its “corresponding” Servo “twitch” at the “corresponding” Servo Position: this is the job of Part 2 of the Main program in “M_MimickingMinis_**L**_ClosedLoop.tsk3” (see Fig. 6.65):

- Lines 33-34 makes the Leader's Controller "do nothing" for 0.05 s. This Time Delay is quite important as without it, the Leader would be sending its packets "too fast" for the Follower's communications buffer to handle. The user can vary this Time Delay to see its effect of the Follower's response.

```
Part 2 of Main for LEADER MINI
(Closed-Loop)
33        High-resolution Timer = 0.050sec
34        WAIT WHILE ( High-resolution Timer > 0.000sec )
35
36        IF ( Remocon Data Arrived == TRUE (1) )
37        {
38            MessageReceived = Remocon RXD
39            MotorID = MessageReceived & 0000 0000 0000 0000 1111 1000 0000 0000
40            MotorID = MotorID / 0000 0000 0000 0000 0000 1000 0000 0000
41            TargetPosition = MessageReceived & 0000 0000 0000 0000 0000 0011 1111 1111
42            ID[MotorID]: ADDR[35(w)] = 512
43            ID[MotorID]: ADDR[30(w)] = TargetPosition
44            High-resolution Timer = 0.250sec
45            WAIT WHILE ( High-resolution Timer > 0.000sec )
46            ID[MotorID]: ADDR[35(w)] = 0
47        }
48    }
```

Fig. 6.65 Part 2 of Main program in "M_MimickingMinis_**L**_ClosedLoop.tsk3".

- Line 36 checks to see if the Follower robot had sent over any new message to the Leader robot. If this is TRUE, Line 38 saves this message in Parameter "MessageReceived".
- Line 39 filters out the "MotorID" part and Line 40 right-shifts "MotorID" by 11 bits.
- Line 41 extracts the Follower's "overloaded" Servo Position and saves it into Parameter "TargetPosition".
- Using the values extracted for "MotorID" and "TargetPosition", the Leader increases its corresponding Servo's "Torque Limit" (i.e. Address 35) to 512 (normally set to 0), and commands this Servo to go to Goal Position (i.e. Address 30) set by the value of "TargetPosition".
- This Servo's Torque Limit is then held at "512" for 0.25 s (Lines 44-45) before it is reset back to its normal value of "0" (Line 46). Resulting from this Hi-Lo setting sequence for the Torque Limit, the human operator should "feel" (in theory) a "twitching" motion by this corresponding Leader's Servo at the "overloaded" Servo Position. However, in practice, the gear-train of the XL-320 would only mechanically yield in a "higher" resistance to manipulation felt by the operator. On another similar application using the AX-12A on GERWALKs, the author did feel this "twitching" motion

(https://www.youtube.com/watch?v=OLbS3RvBR0c&list=PLtix7rPAJwqz85vK-9Dl1nNKnD0xFB8J9&index=54&t=0s).

Video 6.12 shows the runtime performance of two MINIs running the Leader and Follower codes for the "Closed Loop" solution.

6.8 Using MINI's Arms as Gripper

Back in Section 4.4, we encountered the concept of JOINT-OFFSET for a typical Dynamixel XL-320 in the MOTION Sub-Tool and in context of the Calibration Pose for the MINI robot, there we used it as **positive or negative** angular rotation values (i.e. in **degrees**). Then in Section 6.4.2, we used it in the TASK Sub-Tool with the special value of "**1024**" to isolate targeted Dynamixels from the effect of a MOTION-LIST/PAGE being played, in order to use GOAL-POSITION commands directly on these Dynamixels during Motion Play (please remember that "legal" GOAL-POSITION values are from 0 to 1023 – see Fig. 1.3). In this section, the reader will see a third way to use JOINT-OFFSET within the numerical range from -255 to 255, and in concert with a MOTION-LIST/PAGE being played. For more information regarding JOINT-OFFSET, please visit this web link http://emanual.robotis.com/docs/en/software/rplustask3/task_parameters/#joint-offset. **Essentially, a JOINT-OFFSET is "added" to the SERVO POSITION defined in a MOTION-UNIT to become the "final" value to be used as GOAL-POSITION for a targeted servo.**

6.8.1 "MOTION from STILLNESS" using Variable Joint-Offset

In this mini-project "M_Head_JointOffset.tsk3", we are using the "Arms Gripper" MOTION-GROUP and the MINI is programmed to play "continuously" the MOTION-LIST/PAGE named "Initial Position", i.e. the Ready Pose. In this "Initial Position" MOTION-UNIT, there is only 1 KEY-FRAME where the Head Servo (#17) is set to go GOAL POSITION "512" (i.e. head facing forward), thus playing "Initial Position" continuously means to keep the robot in STILLNESS. However, by varying the JOINT-OFFSET of Servo #17 appropriately, we can get the Head Servo into MOTION, thus the Zen-like notion of "MOTION from STILLNESS"!

The TASK program "M_Head_JointOffset.tsk3" starts by calling the usual Function "Init" (see Fig. 6.66) where Line 56 calls the "Initial Position" into play.

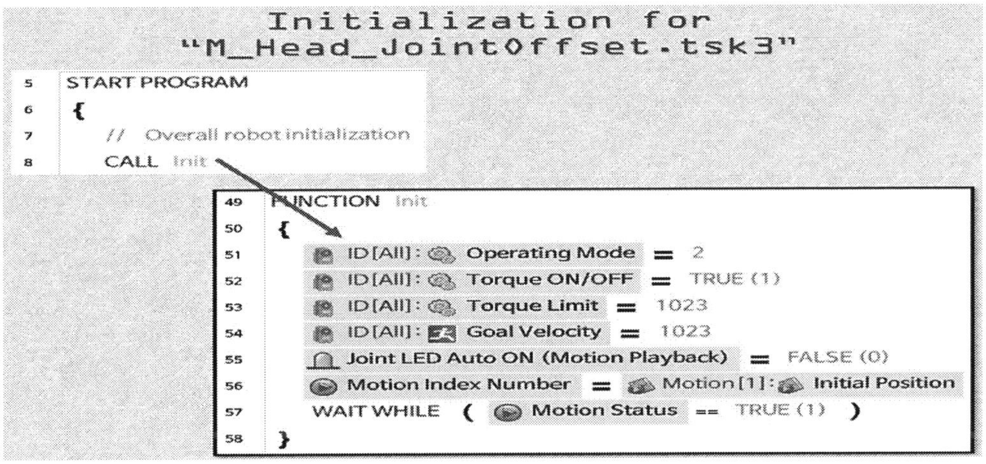

Fig. 6.66 Initialization of Main program in "M_Head_JointOffset.tsk3".

In Part 1 of the Main Endless Loop (Fig. 6.67), the IF structure from Line 12 to Line 35 picks up any "new" RC Button being sent from the RC-100 Remote Controller (Line 12) and records it into Parameter "ReceiveData" (Line 14). Depending on the value received for "ReceiveData": Button "1", Button "R" or Button "L", an inner IF-ELSE-IF structure is used to assign the proper value to Parameter "CurrentJointOffset":

- **If** Button "1" is pushed (Line 15), then "CurrentJointOffset" is set to "0" (Line 17).
- **Else If** Button "R" is pushed (Line 19), then "CurrentJointOffset" is **decremented** by 1 (Line 21). However, as the lower limit for a legal JOINT-OFFSET value in this type of application is "-255", the IF structure (Lines 22-25) makes sure to limit "CurrentJointOffset" to "-255" if necessary.
- **Else If** Button "L" is pushed (Line 19), then "CurrentJointOffset" is **incremented** by 1 (Line 29). However, as the upper limit for a legal JOINT-OFFSET value in this type of application is "+255", the IF structure (Lines 22-25) makes sure to clamp "CurrentJointOffset" to "+255" if necessary.
- Please note that Parameter "CurrentJointOffset", whether positive or negative in value, will be added to the Initial Value of Servo #17 which is "512".

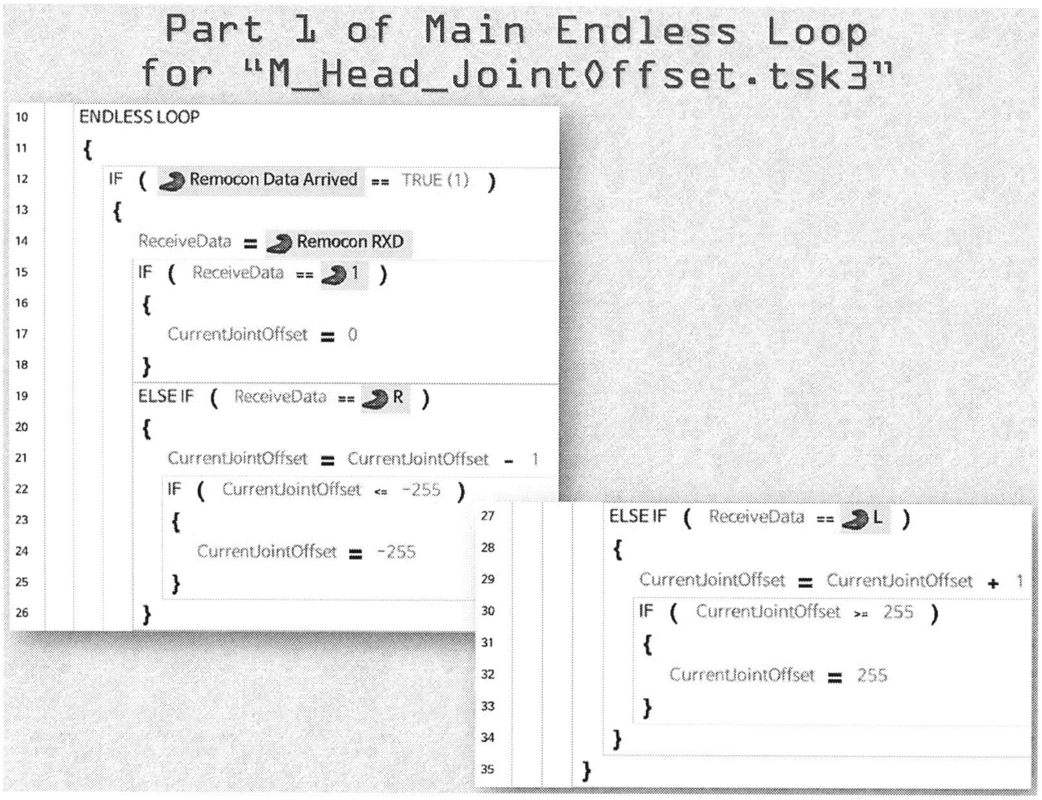

Fig. 6.67 Part 1 of Main Endless Loop in "M_Head_JointOffset.tsk3".

Part 2 of the Main Endless Loop accomplishes three tasks (see Fig. 6.68):

- If "Motion Status" is FALSE (Line 36), i.e. a MOTION-LIST/PAGE is **not being played**, then assign "CurrentJointOffset" to the "Joint Offset" of Servo 17 (Line 38) and play "Initial Position" (Line 39), in order to modify the Initial Servo 17 Position of "512" and move the Head Servo to the left or right of the MINI as commanded by the operator.
- Lines 41-43 print out Servo 17's "Joint Offset", "Goal Position" and "Present Position". When this program is running, the reader should see "Goal Position" being equal to "512 + Joint Offset". However a "negative" value of "Joint Offset" will be printed as its 2-complement version (see link https://www.programminglogic.com/how-computers-represent-negative-binary-numbers/ for a review). Unfortunately, the reader will also see that "Present Position" is usually "off" from "Goal Position" by 2 to 4 units depending on the age of the XL-320 servos used.
- Lastly a short delay of 0.1 sec is used so that the operator can view a "not-too-fast" scrolling of the above data printout on the Debugging Screen inside the TASK Sub-Tool.

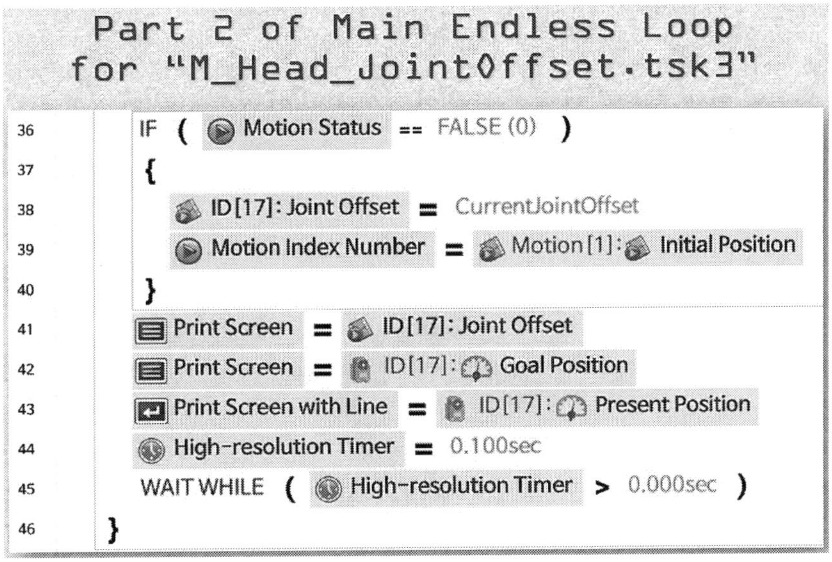

Fig. 6.68 Part 2 of Main Endless Loop in "M_Head_JointOffset.tsk3".

Video 6.13 shows the runtime performance of "M_Head_JointOffset.tsk3". Please remember that "Joint Offset #17", when "positive-valued" (i.e. when the Head is rotating Left), is printed as "normal positive" integers. However, when it takes on "negative" values (i.e. when the Head is turning Right), it is printed out as its 2-complement version (for example, "65535" stands for "-1", and "65533" stands for "-3").

6.8.2 Application to Adaptive Grasping

We are now ready to apply the above "Joint Offset" technique to the two arms of the MINI to make them function as a gripper that can adapt to different sizes of the object being grasped.

Let's first look at the "Arms Gripper" MOTION GROUP more closely (it is located in the file "MINI_Custom.mtn3"). It has 3 MOTION-LISTs/PAGEs: the usual "Initial Position", and two new ones named "AG_OpenArms" and "AG_CloseArms":

- "AG_OpenArms" contains only the MOTION-UNIT "Open Arms" which also contains only one robot pose (see Fig. 6.69-A)
- Similarly, "AG_CloseArms" contains the MOTION-UNIT "Close Arms" which contains also one robot pose (see Fig. 6.69-B).

The goal of the "Arms Gripper" project is to allow the MINI's two arms to grip objects of varying sizes that can fit between the MINI's "Open Arms" and "Close Arms" poses by playing continuously the MOTION-LIST/PAGE "AG_CloseArms", and by adjusting the JOINT-OFFSETS of Servos 3, 4, 5 and 6 "on the fly", depending on the "current" and "respective" PRESENT-LOAD data received from the same Servos.

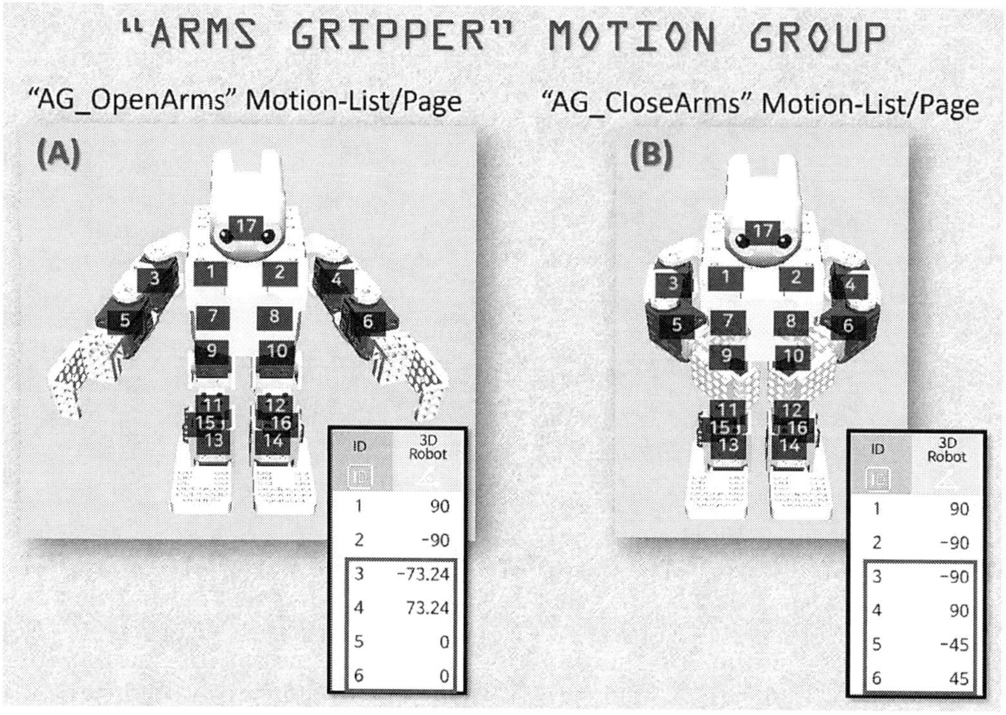

Fig. 6.69 MOTION-UNITs "Open Arms" **(A)** and "Close Arms **(B)** in "MINI_Custom.mtn3".

Please note that JOINT-OFFSET values are of the same unit type as the servo's GOAL-POSITION Parameter (e.g. [0-1023] and please review Fig. 1.3 if needed). Thus, we'll have to convert the meaning of the changes in "degree settings" for Servos 3-6 as shown in Fig. 6.69 (A) and (B) to the needed changes in the JOINT-OFFSET values for the same servos:

- Please notice that when the Left Arm closes, Servos 3 and 5 take on more "negative" degree-values ("-73.24° to -90°" and "0° to -45°) which correspond to a shift towards the "raw value" "0" of the GOAL-POSITION Parameter. Thus to "differentially open" the Left Arm when it is "squeezing too hard" on the grasped object, the JOINT-OFFSET Parameters for Servos 3 and 5 will need to be "positive-valued" because we want to "increase" their final GOAL-POSITION Parameters towards the "raw value" "1023" Position.
- Conversely, when the Right Arm closes, Servos 4 and 6 take on more "positive" degree-values ("73.24° to 90°" and "0° to 45°") which correspond to a shift towards the "1023" value of the GOAL-POSITION Parameter. Thus to "differentially open" the Right Arm when it is "squeezing too hard" on the grasped object, the JOINT-OFFSET Parameters for Servos 4 and 6 will need to be "negative-valued" because we want to "decrease" their final GOAL-POSITION Parameters towards the "0" Position.

The author implemented two versions of the TASK codes for this "Arms Gripper" project: one uses the Standard Function protocol ("M_ArmsGripper_F.tsk3") while the other uses the CALLBACK Function protocol ("M_ArmsGripper_CB.tsk3").

6.8.3 Standard Function (SF) Solution

Fig. 6.70 shows Part 1 of the Main Program for "M_ArmsGripper_F.tsk3" using the usual initializations in Function "Init".

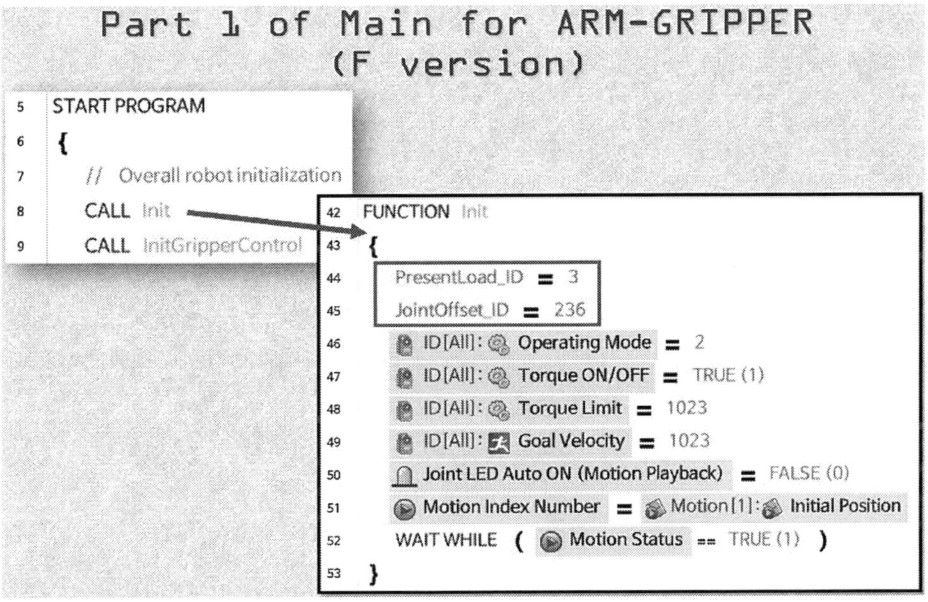

Fig. 6.70 Part 1 of the Main Program for "M_ArmsGripper_F.tsk3" and Function "Init".

Part 1 also uses Function "InitGripperControl" which implements "Custom" commands using "Direct Addressing" for various Dynamixel Parameters and introduces a new Dynamixel Parameter named "Punch" (see Fig. 6.71):

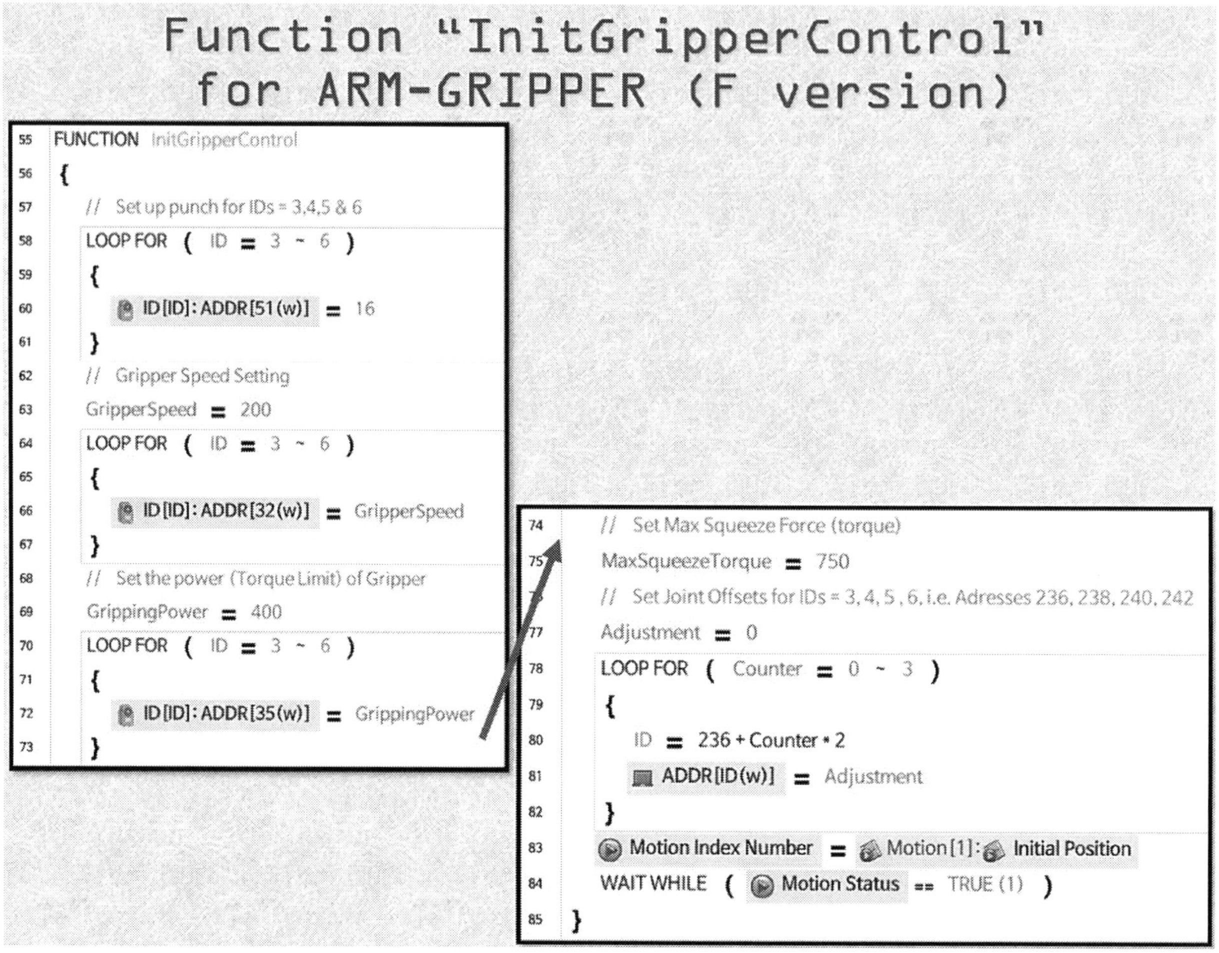

Fig. 6.71 Function "InitGripperControl" for "M_ArmsGripper_F.tsk3".

- For Servos 3-6, the Loop structure (Lines 58-61) uses Address 51 which corresponds to the Parameter named "Punch" for XL-320s (http://emanual.robotis.com/docs/en/dxl/x/xl320/#punch51). "Punch" is the minimum electrical current used to drive the motor from a full-stop to overcome static friction and inertia of the XL-320 drive train. By default, it is set to "32" but we'll reduce it to "16" for this project for smoother transition when continuously "opening" and "closing" the MINI arms to adjust for the size of the object being grasped.
- Lines 63 to 67 re-adjust the Speed (Address 32) of Servos 3-6 to a low value of 200.
- Lines 69 to 73 re-adjust the Torque Limit (Address 35) of Servos 3-6 to a low value of 400 for an initial "softer" squeeze on the grasped object.
- Line 75 sets Parameter "MaxSqueezeTorque" to "750" which is used later to adjust the gripper's hold on the object being grasped (via JOINT-OFFSET).
- Lines 77-82 initialize to "0" the corresponding JOINT-OFFSET Parameters (2 bytes) for each Servo (ID 3 through 6), i.e. from Addresses 236 through 242, using the "brand-new" POLYNOMIAL" statement from TASK 3 (Line 80).
- Lines 83-84 put the MINI into its Initial Position.

Fig. 6.72 shows Part 2 of the Main Program for "M_ArmsGripper_F.tsk3" which is the usual Main Endless Loop:

Part 2 of Main for ARM-GRIPPER (F version)

```
11  ENDLESS LOOP
12  {
13      IF ( Remocon Data Arrived == TRUE (1) )
14      {
15          ReceiveData = Remocon RXD
16          IF ( ReceiveData == 1 )
17          {
18              GripState = 1
19              CALL InitGripperControl
20          }
21          ELSE IF ( ReceiveData == U )
22          {
23              GripState = 2
24              CALL OpenGripper
25          }
26          ELSE IF ( ReceiveData == D )
27          {
28              // Close Gripper
29              GripState = 3
30          }
31      }
32      IF ( GripState == 3 )
33      {
34          Motion Index Number = Motion[3]: AG_CloseArms
35          CALL AdjustJointOffsets
36      }
37      Timer = 0.256sec
38      WAIT WHILE ( Timer > 0.000sec )
39  }
```

Fig. 6.72 Part 2 (Main Endless Loop) of the Main Program for "M_ArmsGripper_F.tsk3".

- The first part of the Main Endless Loop is controlled by an outer IF structure (Lines 13-31) checking on whether a new Remocon message has arrived (Line 13). If this is TRUE, the new message is saved into Parameter "ReceiveData". Next, an inner IF-ELSE-IF structure is used to deal with specific cases:
    - **If** "Button 1" is received (Line 15), then "GripState" is set to "1" and Function "InitGripperControl" is called (Lines 18-19). The operator would use this option when he/she wants to reset the MINI to its "Initial Position" and to open the Arms Gripper.
    - **Else If** "Button U" is received (Line 21), then "GripState" is set to "2" and Function "OpenGripper" is called (Lines 23-24). The operator would use this option when he/she only wants to open the Arms Gripper. Fig 6.73 shows the details of Function "OpenGripper":

- The first FOR LOOP is used to reset all JOINT-OFFSETs back to "0" (Lines 89-93).
- The second FOR LOOP is used to set the servos (3 to 6) Speed to "GripperSpeed".
- Lastly, the MOTION-LIST/PAGE "AG_OpenArms" is played (Lines 98-99).

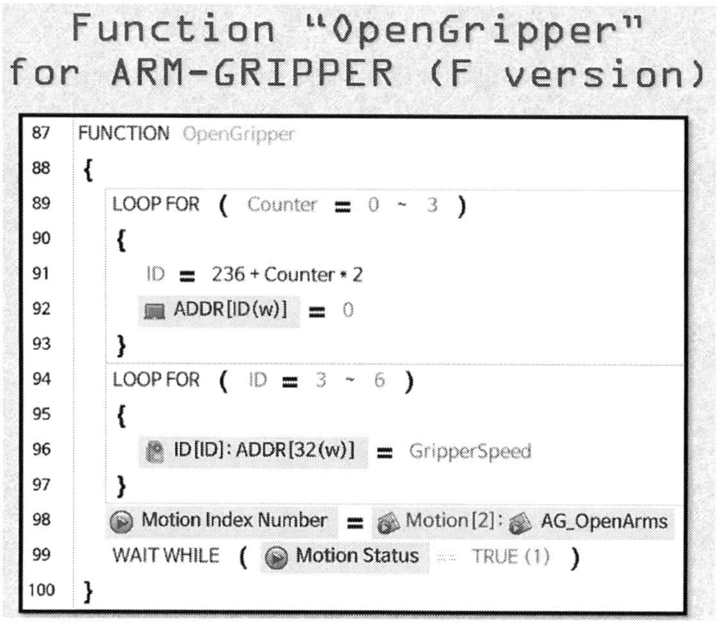

Fig. 6.73 Function "OpenGripper" for "M_ArmsGripper_F.tsk3".

- **Else If** "Button U" is received (Line 21 in Fig. 6.72), then "GripState" is set to "3" and both the inner IF-ELSE-IF and outer IF structures are exited.

- The second part of the Main Endless Loop is implemented using an IF structure (Lines 32-36) checking on whether (GripState == 3). If this condition is TRUE, the MOTION-LIST/PAGE "AG_CloseArms" is played and the Function "AdjustJointOffsets" is called (Lines 34-35). In other words, once "GripState" is set to "3", the "AG_CloseArms" MOTION-LIST/PAGE will be played every time that the Main Endless Loop iterates and it will be up to the operator to push either Button "1" or Button "U" to get out of this GripState 3.
- The last part of the Main Endless Loop is a 0.256 second time delay (adjustable by the user) implemented to see how fast or slow can this adaptive grasping algorithm perform in the physical world.

Function "AdjustJointOffsets" is implemented as a big FOR LOOP iterating on Parameter "PresentLoad_ID" varying from 3 to 6 (i.e. Servo ID from 3 to 6). This FOR LOOP starts on Line 106 and ends on Line 182 in the source code "M_ArmsGripper_F.tsk3". Due to lack of space, only the Beginning

Section of this Function is described in Fig. 6.74 (Lines 102-128) and its Ending Section (Lines 176-182) is described in Fig. 6.75.

Fig. 6.74 shows the Beginning Section of Function "AdjustJointOffsets":

Function "AdjustJointOffsets" (1) for ARM-GRIPPER (F version)

(Beginning Section)

```
102 FUNCTION AdjustJointOffsets
103 {
104     // Present Load Address is 41 for each servo ID = 3,4,5,6
105     // Joint Offset Addresses are 236, 238, 240, 242 for Servo IDs 3, 4, 5, 6
106     LOOP FOR ( PresentLoad_ID = 3 ~ 6 )
107     {
108         IF ( PresentLoad_ID == 3 )
109         {
110             CurrentLoad3 = ID[3]: Present Load
111             IF ( CurrentLoad3 <= 128 )
112             {
113                 Adjustment3 = 0
114             }
115             ELSE IF ( CurrentLoad3 > MaxSqueezeTorque )
116             {
117                 Adjustment3 = Adjustment3 + 1
118                 IF ( Adjustment3 > 255 )
119                 {
120                     Adjustment3 = 255
121                 }
122             }
123             ID [3]: Joint Offset = 0 + Adjustment3
124         }
125         ELSE IF ( PresentLoad_ID == 4 )
126         {
127             CurrentLoad4 = ID[4]: Present Load
128             IF ( CurrentLoad4 <= 128 )
```

Fig. 6.74 Beginning Section of Function "AdjustJointOffsets" for "M_ArmsGripper_F.tsk3".

- Line 106 sets up the FOR LOOP to iterate on the values 3-4-5-6 of the Counter "PresentLoad_ID".
- As the TASK Sub-Tool does not currently support arrays, a rather awkward-looking nested IF-ELSE-IF structure had to be used, along with a series of "similar" Parameters had to be implemented: from "CurrentLoad3" to "CurrentLoad6" and from "Adjustment3" to "Adjustment6".
- For example, if (PresentLoad_ID == 3), the IF structure from Line 108 to Line 124 is entered, and the "Present Load" of Servo ID 3 is read and saved into "CurrentLoad3" (Line 110). Next,

an inner IF-ELSE-IF structure is used to check "CurrentLoad3" against the Constants "128" and "750". **IF** it is less or equal to 128 (Line 111), then "Adjustment3" is set to 0 (Line 113), i.e. the Right Arm has not encountered the object yet, or it is just starting to squeeze on the object. **ELSE IF** "CurrentLoad3" is greater than "MaxSqueezeTorque" (previously set to 750) (Line 115), then "Adjustment3" is incremented by 1 (Line 117), i.e. Servo 3 is squeezing "too hard" on the object, so Servo 3 needs to "back away" so as to "open up" the Right Arm some. The next-inner IF structure (Lines 118-121) makes sure that "Adjustment3" won't go over its Upper Limit of "255". Lastly, this appropriately-adjusted "Adjustment3" Parameter is then "added" to Servo 3's JOINT-OFFSET Parameter (Line 123).

- When the user reads the complete source code, he/she would note that a similar technique of incrementing "Adjustment5" is used for Servo 5 of the Right Arm, while the reverse technique (i.e. decrementing "Adjustment4" and "Adjustment6") is used for Servos 4 and 6 of the Left Arm. Of course, all this "switching" is needed because the Right and Left Arms happen to be "mirror-image" of each other.

Function "AdjustJointOffsets" (2) for ARM-GRIPPER (F version)

(Ending Section)

```
176         // Print Servo ID, Present Position, Present Load, Joint Offset
177         Print Screen = PresentLoad_ID
178         Print Screen = ID[PresentLoad_ID]: ADDR[37(w)]
179         Print Screen = ID[PresentLoad_ID]: ADDR[41(w)]
180         JointOffset_ID = 236 + (PresentLoad_ID - 3) * 2
181         Print Screen with Line = ADDR[JointOffset_ID(w)]
182     }
183 }
```

Fig. 6.75 Ending Section of Function "AdjustJointOffsets" for "M_ArmsGripper_F.tsk3".

Fig. 6.75 shows the Ending Section of Function "AdjustJointOffsets" which is used to print out each set of servo data on the same display line, per iteration of the Main Endless Loop:

- Current Servo ID (Line 177).
- Present Position (Address 37) of Current Servo (Line 178).
- Present Load (Address 41) of Current Servo (Line 179).
- Joint Offset (appropriate Address computed by Line 180) of Current Servo (Line 181).

Video 6.14 shows the runtime performance of the TASK program "M_ArmsGripper_F.tsk3".

6.8.4 CALLBACK (CB) Function Solution

In two previous projects, "RC Smart Avoider" (Section 6.5) and "Autonomous Random Walker" (Section 6.6), we used the CALLBACK Function **to read two IRSS-10 Sensors** which were directly connected to I/O ports 1 and 4 of the OpenCM9.04 Hardware Controller STM32F103CB (see Fig. 6.41). As it took only about 0.02 ms to obtain each sensor reading, as compared to the 8 ms cycle of the CALLBACK Function, we did not have to modify the way we programmed the Main Program and the Standard Functions that were accessing these sensors.

However, with this "Arms Gripper" project, we will need to access 4 XL-320s that are connected to their own Dynamixel network which can be considered as **external** to the STM32F103CB Controller. Furthermore, as the preceding TASK code "M_ArmsGripper_F.tsk3" indicated, the Main Program and Standard Functions would be now accessing the same Dynamixels that the CALLBACK Function would also access, possibly at the same time. Thus, the "real-time" communications time delays between various controllers and the potential network packets collision issues would need to be dealt with, in this CALLBACK solution.

From private communications with ROBOTIS Technical Support, the author obtained the following information (with the Dynamixel Network set at its default baud rate of 1 Mbps):

- When an XL-320 is connected properly without any communication issue, the STM32F103CB Controller should expect a time delay of 0.5 to 0.7 ms for a Dynamixel Read instruction to be completed.
- When an XL-320 is connected improperly, the maximum wait time for the STM32F103CB Controller issuing a Dynamixel Read instruction can be from 3 to 4 ms which is then a substantial chunk of the 8 ms cycle time of the CALLBACK Function.
- This is the reason why ROBOTIS recommends **not to program more than 2 Sensor or Dynamixel accesses** within the CALLBACK Function, as exceeding its 8 ms cycle time will yield unusual and unpredictable errors.

For this "CB Arms Gripper" project "M_ArmsGripper_CB.tsk3", we also need to worry about the Main Program or any of the Regular Functions issuing their own Dynamixel Read/Write instructions while CALLBACK is accessing the same targeted Dynamixel(s) with its own Read/Write instructions. For this issue, we'll use a simple solution via a Global Flag Parameter named "CallBackStart" which would be set to TRUE when we want CALLBACK to implement its programmed steps. When we want the Main Program or the Regular Functions to access these targeted Dynamixels, "CallBackStart" would be set to FALSE to make CALLBACK skip its programmed steps.

The CALLBACK code for "M_ArmsGripper_CB.tsk3" is shown in Figs. 6.76 and 6.77 which show that the same programming steps for Function "AdjustJointOffsets" are used for this CALLBACK Function except for 2 aspects:

- Line 132 uses an IF condition to check on the current value of Parameter "CallBackStart". If it is set to TRUE (by Main or other Functions), then all the CALLBACK programming steps from Line 133 to Line 216 would be executed. Otherwise, if "CallBackStart" was found to be set to FALSE (also by Main or other Functions) then CALLBACK is sent to Line 217 to terminate immediately.

CALLBACK Function (1) for ARM-GRIPPER (CB version)

(Beginning Section)

```
CALLBACK
{
    // Present Load Address is 41 for each servo ID = 3,4,5,6
    // Joint Offset Addresses are 236, 238, 240, 242 for Servo IDs 3, 4, 5, 6
    IF ( CallBackStart == TRUE (1) )
    {
        IF ( PresentLoad_ID == 3 )
        {
            CurrentLoad3 = ID[3]: Present Load
            IF ( CurrentLoad3 <= 128 )
            {
                Adjustment3 = 0
            }
            ELSE IF ( CurrentLoad3 > MaxSqueezeTorque )
            {
                Adjustment3 = Adjustment3 + 1
                IF ( Adjustment3 > 255 )
                {
                    Adjustment3 = 255
                }
            }
            ID [3]: Joint Offset = 0 + Adjustment3
        }
        ELSE IF ( PresentLoad_ID == 4 )
        {
            CurrentLoad4 = ID[4]: Present Load
```

Fig. 6.76 Beginning Section of CALLBACK Function in "M_ArmsGripper_CB.tsk3".

- For this CALLBACK Function, a FOR LOOP is not used as for Function "AdjustJointOffsets" to handle all 4 servos in each CALLBACK iteration, as this may be "too-heavy" a computational load for an 8 ms cycle time. Thus, each CALLBACK cycle would handle only ONE servo identified by the current value of Parameter "PresentLoad_ID" which is then designed to cycle from 3 to 6, at the end of CALLBACK, by the IF-ELSE structure shown in Fig. 6.77 (Lines 208 to 210).
- The print-out commands from Line 211 to 215 in Fig. 6.77 should not be used for optimal performance of this Arms Gripper project, but they are provided for the reader to try out as they would produce an unplanned but interesting slow-motion effect on the closing of the gripper (as the Print-Screen data would be sent to the PC at a slower 57 Kbps baud rate instead of the 1 Mbps within the Dynamixel network). If the reader wants to use this section of code, please comment out Lines 36 to 42 as shown in Fig. 6.83, as Function "PrintData" is then not needed.

CALLBACK Function (2) for ARM-GRIPPER (CB version) (Ending Section)

```
        // Update PresentLoad_ID for next iteration of CALLBACK
        IF ( PresentLoad_ID == 6 )
        {
            PresentLoad_ID = 3
        }
        ELSE
        {
            PresentLoad_ID = PresentLoad_ID + 1
        }
        // Print Screen = PresentLoad_ID
        // Print Screen = ID[PresentLoad_ID]: ADDR[37(w)]
        // Print Screen = ID[PresentLoad_ID]: ADDR[41(w)]
        // JointOffset_ID = 236 + (PresentLoad_ID - 3) * 2
        // Print Screen with Line = ADDR[JointOffset_ID(w)]
    }
}
```

Fig. 6.77 Ending Section of CALLBACK Function in "M_ArmsGripper_CB.tsk3".

Fig. 6.78 shows Part 1 of the Main Program for "M_ArmsGripper_CB.tsk3" which is mostly identical to Parts 1 and 2 for the Main Program in "M_ArmsGripper_F.tsk3" (see Figs. 6.70 and 6.72), except:

- There is no call to Function "AdjustJointOffsets" because it is no longer used.
- The 0.256 second Wait Loop would be replaced by a **counter-based** Wait Loop to be shown in Fig. 6.83.

Part 1 of MAIN for ARM-GRIPPER (CB version)

```
CALL Init
CALL InitGripperControl
// Main Loop
ENDLESS LOOP
{
    IF ( Remocon Data Arrived == TRUE (1) )
    {
        ReceiveData = Remocon RXD
        IF ( ReceiveData == 1 )
        {
            GripState = 1
            CALL InitGripperControl
        }
        ELSE IF ( ReceiveData == U )
        {
            GripState = 2
            CALL OpenGripper
        }
        ELSE IF ( ReceiveData == D )
        {
            // Close Gripper
            GripState = 3
        }
    }
    IF ( GripState == 3 )
    {
        Motion Index Number = Motion[3]: AG_CloseArms
    }
```

Fig. 6.78 Part 1 of Main Program in "M_ArmsGripper_CB.tsk3".

Figs. 6.79 to 6.82 show that all Standard Functions now **start by setting Parameter "CallBackStart" to FALSE and end by setting "CallBackStart" to TRUE** so that they don't interfere with CALLBACK Function regarding issuing Read/Write commands to Servos 3 through 6.

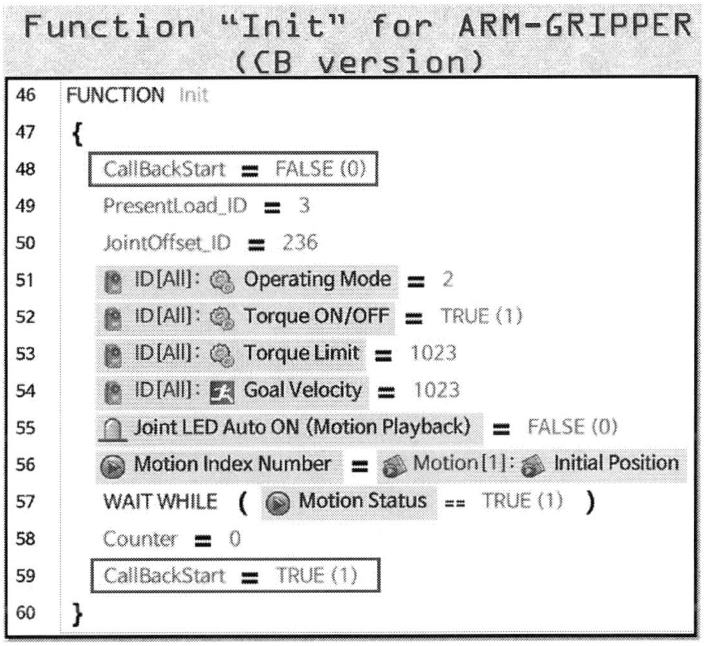

Fig. 6.79 Function "Init" in "M_ArmsGripper_CB.tsk3".

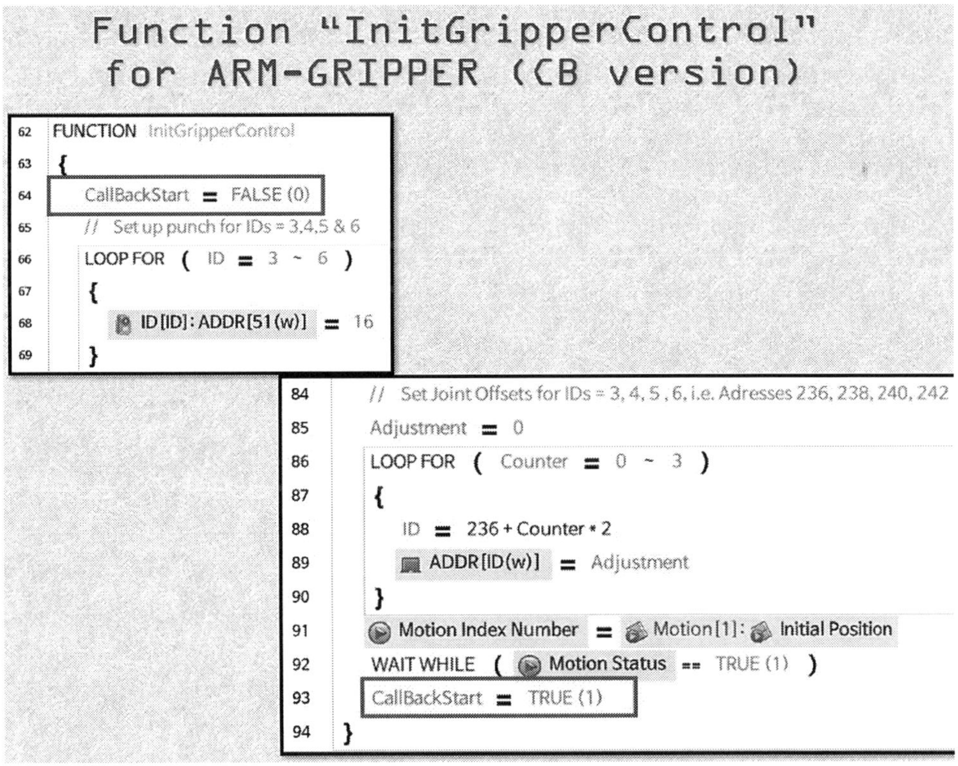

Fig. 6.80 Function "InitGripperControl" in "M_ArmsGripper_CB.tsk3".

Function "OpenGripper" for ARM-GRIPPER (CB version)

```
FUNCTION OpenGripper
{
  CallBackStart = FALSE (0)
  LOOP FOR ( Counter = 0 ~ 3 )
  {
    ID = 236 + Counter * 2
    ADDR[ID(w)] = 0
  }
  LOOP FOR ( ID = 3 ~ 6 )
  {
    ID[ID]: ADDR[32(w)] = GripperSpeed
  }
  Motion Index Number = Motion[2]: AG_OpenArms
  WAIT WHILE ( Motion Status == TRUE (1) )
  CallBackStart = TRUE (1)
}
```

Fig. 6.81 Function "OpenGripper" in "M_ArmsGripper_CB.tsk3".

Function "PrintData" for ARM-GRIPPER (CB version)

```
FUNCTION PrintData
{
  CallBackStart = FALSE (0)
  // Print Servo ID, Present Position, Present Load, Joint Offset
  LOOP FOR ( PresentLoad_ID = 3 ~ 6 )
  {
    Print Screen = PresentLoad_ID
    Print Screen = ID[PresentLoad_ID]: ADDR[37(w)]
    Print Screen = ID[PresentLoad_ID]: ADDR[41(w)]
    JointOffset_ID = 236 + (PresentLoad_ID - 3) * 2
    Print Screen with Line = ADDR[JointOffset_ID(w)]
  }
  CallBackStart = TRUE (1)
}
```

Fig. 6.82 Function "PrintData" in "M_ArmsGripper_CB.tsk3".

Fig. 6.83 describes Part 2 of the Main Program for "M_ArmsGripper_CB.tsk3" which shows a counter-based scheme for printing servos data, but **only for every 500 iterations** of the Main Endless Loop (just to show the reader an alternate approach to using the time-based WAIT-WHILE Loop normally used for this kind of task).

Part 2 of MAIN for ARM-GRIPPER

```
    Counter = Counter + 1
    IF ( Counter == 500 )
    {
        // Print Present Position, Present Load, Joint Offset
        CALL PrintData
        Counter = 0
    }
}
```

Fig. 6.83 Part 2 of Main Program in "M_ArmsGripper_CB.tsk3".

Video 6.15 shows the runtime performance of the TASK program "M_ArmsGripper_CB.tsk3".

Recapping the various uses of servo's JOINT-OFFSET, the author proposes that readers consider these issues to decide **when and how** to use JOINT-OFFSETs of targeted servos which need to be isolated from **MOTION-LIST/PAGEs being played**:

1. If the user's project happens to work well using **absolute** Servo Position values [**0-1023**] **at all times** for those targeted servos, then it is appropriate to set their JOINT-OFFSETs to **1024** and use **GOAL-POSITION** commands directly to affect these servos while playing with MOTION-LIST/PAGEs.
2. If the **timing and/or servo's position** (i.e. **when/wher**e the isolation task is needed) is **not known in advance**, then the **differential** use of JOINT-OFFSETs should be used, i.e. **use the range [-255, 255]** for them while playing related MOTION-LIST/PAGEs.
3. It is quite conceivable that the user would need to deal with a mixture of these situations, then care should be exercised when to set JOINT-OFFSET to 0, or 1024, or when to use them within the [-255, 255] range. The current project "Arms Gripper" is about alternating between the uses of the "0" value and the range [-255, 255], while the "RC Walker with Independent Servos Control" project was about alternating between the uses of the "0" and "1024" values (see Section 6.4). In another application for the BIOLOID system, the author had used a mixture of all 3 options for a standard robotic manipulator (see YouTube video https://www.youtube.com/watch?v=6QxEmptSjnE&list=PLtix7rPAJwqz85vK-9Dl1nNKnD0xFB8J9).

6.9 Autonomous Dowel Scanner using DMS-80

This last project of Chapter 6 uses the Smart Avoider Motion Group and uses the DMS-80 sensor (connected to I/O Port 2) to search for 2 white wood dowels placed at random in front of it, and next to command the MINI to autonomously approach the "closer" dowel first, and then to relocate the remaining dowel before approaching it as the final step.

Fig. 6.84 illustrates the physical layout for the "Dowel Scanner" program "M_DowelScanner.tsk3" as well as its constraints:

1. At the Start Position, the DMS-80 should be "looking" at empty space, as the proposed TASK program uses this condition to establish the "background" level as provided by the DMS-80 sensor readings.
2. At the Left Limit and Right Limit Positions, the DMS-80 should also be "looking" at empty space as a necessary assumption for the Dowel Scan algorithm to work.
3. The two white dowels can be located anywhere between the Left Limit and Right Limit Positions (but not in line with the Start Position). Essentially, "Dowel 1" should be on the robot's Left and "Dowel 2" should be on the robot's Right.

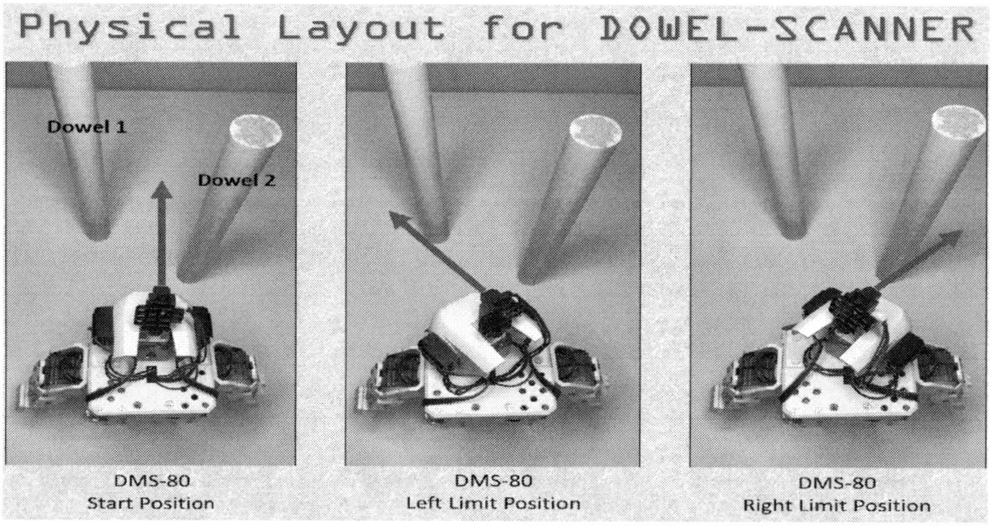

Fig. 6.84 Physical Layout and Contraints for Program "M_DowelScanner.tsk3".

The proposed overall Dowel Scanner procedure has 4 steps:

- **Step 1:** The MINI starts in its Ready Pose at the Start Position as shown in Fig. 6.84 (i.e. staring into empty space by setting Servo #17 to Position 512). Next, the robot collects and averages a series of DMS-80 sensor readings to establish the "Background NIR" level and set Parameter "Dowel Threshold" as somewhat higher than the "Background NIR" value.
- **Step 2:** The MINI swings the DMS-80 (via Servo #17) to its Left Limit Position and then swings it back slowly to the Right Limit Position while continuously collecting DMS-80 sensor data input. During this first scan for the Dowels, the goal is to determine which dowel, between "Dowel 1" and "Dowel 2", is the closer one to the robot according to DMS-80 sensor data

collected. This step will make use of the Parameters "Background NIR" and "Dowel Threshold" determined from Step 1.

- **Step 3:** At this point, the MINI resets and locks the DMS-80 to the Start Position (i.e. 512) and starts **rotating the whole robot** towards the "closer" dowel as determined by Step 2. Once the DMS-80 is "centered" on the "closer" dowel, the MINI will go forward towards this dowel to get closer to it within a given distance. The MINI would then stop and flash its LEDs waiting for the operator to remove this dowel.
- **Step 4:** The MINI now rotates its whole body left or right appropriately (thanks to Step 2) to search for the remaining dowel (which was the "farther" one). Once the DMS-80 is "centered" on this "farther" dowel, the MINI will go forward towards this dowel to get closer to it within a given distance. The MINI would then stop and flash its LEDs signaling that all its tasks are done.

An important point to mention before we get into the details of algorithms development is that the TASK program "M_DowelScanner.tsk3" uses the CALLBACK Function to get the latest DMS-80 Sensor reading and save it into Parameter "NIR_Center".

6.9.1 Scan Background & Set Dowel Threshold

Fig. 6.85 displays Part 1 of the Main Program for "M_DowelScanner.tsk3" (i.e. **Step 1**):

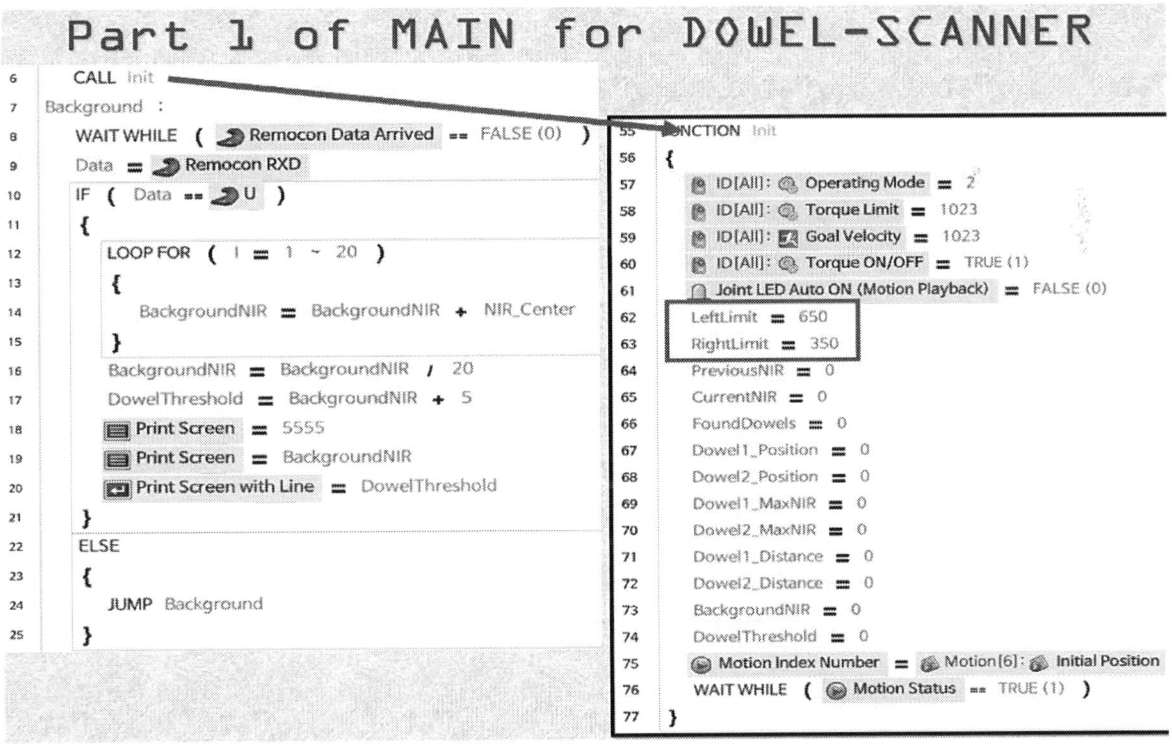

Fig. 6.85 Part 1 of Main Program in "M_DowelScanner.tsk3": Background NIR Determination.

- First, Function "Init" is called to initialize the servos in the usual manner (Lines 57-61). Function "Init" also sets "LeftLimit" to 650 and "RightLimit" to 350 (Lines 62-63). These values are just particular to the author's robot as he had constructed it. The reader's robot may have a different range of Servo Positions available to its Servo #17. Next, 11 Parameters are initialized to "0" (Lines 64-74), they will be used later in other code sections. Lastly, the MINI is set to its "Initial Position" (Lines 75-76).
- **Step 1** as proposed earlier is implemented between Lines 7 and 25. Essentially, the Controller waits for a new Button Input if any from the operator and this new input is saved into Parameter "Data" (Lines 8-9). If "Data" is the Button "U" (Line 10), then 20 sequential DMS-80 sensor data are collected, added together and averaged to obtain the final value for Parameter "BackgroundNIR" (Lines 12-16). Parameter "DowelThreshold" is set as (BackgroundNIR + 5) on Line 17. The reader's robot may need a value different than 5 depending on the performance of his/hers DMS-80 Sensor. Next, Lines 18-20 print out the results for "Background" and "DowelThreshold" for the operator's assessment of the validity of data collected and then the Controller goes on to Step 2 to scan for the dowels (see Fig. 6.86).
- If "Data" is not Button "U", then the Controller jumps back to Line 7 (Lines 22-25 of Fig. 6.85) and waits for the operator to eventually push Button "U".

6.9.2 Scan & Locate Both Dowels

Fig. 6.86 shows Part 2 of the Main Program for "M_DowelScanner.tsk3" (i.e. **Step 2**):

Part 2 of MAIN for DOWEL-SCANNER

```
26  ScanDowels :
27      WAIT WHILE ( Remocon Data Arrived == FALSE (0) )
28      Data = Remocon RXD
29      IF ( Data == 1 )
30      {
31          // Swing Head Servo (#17) from Left to Right and collect NIR data
32          CALL ServoScan
33      }
34      ELSE
35      {
36          JUMP ScanDowels
37      }
```

Fig. 6.86 Part 2 of Main Program in "M_DowelScanner.tsk3": Scan for Dowels 1 and 2.

- Similarly as for Part 1, the Controller waits for the operator to push on Button 1 for it to call Function "ServoScan" (Line 32) which implements a more complex algorithm involving other Functions to scan and locate which dowel, between Dowel 1 and Dowel 2, is the closer one to the robot as measured by the DMS-80. This algorithm is described via Figs. 6.87, 6.88, 6.89 and 6.90.

Fig. 6.87 lists the contents of Function "ServoScan" in "M_DowelScanner.tsk3":

Function "ServoScan" for DOWEL-SCANNER

```
79  FUNCTION ServoScan
80  {
81      PreviousNIR = 0
82      CurrentNIR = 0
83      FoundDowels = 0
84      Dowel1_Position = 0
85      Dowel2_Position = 0
86      Dowel1_MaxNIR = 0
87      Dowel2_MaxNIR = 0
88      Dowel1_Distance = 0
89      Dowel2_Distance = 0
90      ID[17]: Goal Velocity = 512
91      ID[17]: Goal Position = LeftLimit
92      WAIT WHILE ( ID[17]: Is Moving == TRUE (1) )
93      Timer = 1.024sec
94      WAIT WHILE ( Timer > 0 )
95      ID[17]: Goal Velocity = 64
96      ID[17]: Goal Position = RightLimit
97      CALL FindDowels
98  }
```

Fig. 6.87 Function "ServoScan" in "M_DowelScanner.tsk3".

- Lines 81-89 "zero out" 9 parameters that will be used for this First Scan algorithm.
- Lines 90-92 send Servo #17, along with the DMS-80, to the "LeftLimit" Servo Position at Speed 512.
- There it waits for 1 second for servo vibrations to subside (Lines 93-94).
- Next, it sets Servo 17's Speed to a low value of 64 (Line 95) and send Servo 17 to the "RightLimit" Servo Position (Line 96). In other words, the DMS-80 is starting a slow-scan to the right of the robot from its "LeftLimit" Position.
- The last step for Function "ServoScan" is to call Function 'FindDowels" which is described in Figs. 6.88 and 6.89.

Function "FindDowels" is quite important as it does the actual determination of which dowel is the closer one to the robot as measured by the DMS-80. Fig. 6.88 describes Part 1 of Function "FindDowels":

Part 1 of Function "FindDowels" for DOWEL-SCANNER

```
100 FUNCTION FindDowels
101 {
102     // Let IR Sensor scan "empty space" until IR Sensor sees edge of First Dowel
103     WAIT WHILE ( NIR_Center <= DowelThreshold )
104     Print Screen with Line = 6666
105     Print Screen = ID[17]: Present Position
106     Print Screen = NIR_Center
107     Print Screen with Line = DowelThreshold
108     // Scanning for First Dowel
109     PreviousNIR = BackgroundNIR
110     CurrentNIR = NIR_Center
111     LOOP WHILE ( FoundDowels == 0 && CurrentNIR >= PreviousNIR )
112     {
113         PreviousNIR = CurrentNIR
114         CurrentNIR = NIR_Center
115         Dowel1_Position = ID[17]: Present Position
116     }
117     Dowel1_Distance = CurrentNIR
118     Dowel1_MaxNIR = PreviousNIR
119     FoundDowels = FoundDowels + 1
120     CALL PrintData
121     // Let IR Sensor scan by the rest of Dowel 1 and until IR Sensor sees Background again
122     WAIT WHILE ( NIR_Center > BackgroundNIR )
```

Fig. 6.88 Part 1 of Function "FindDowels" in "M_DowelScanner.tsk3".

- When Function "FindDowels" starts at Line 103, the reader needs to remember that Servo #17 is starting to slowly rotate to the robot's right but Servo #17 is still close to the "LeftLimit" Position, therefore the DMS-80 should "see empty space" to the left of Dowel 1 (see Fig. 6.84). This means that Parameter "NIR_Center" (which is provided "live" by CALLBACK) would contain numerical values closer to "BackgroundNIR" and therefore "less or equal" than "DowelThreshold" which was set to be 5 units higher than "BackgroundNIR" (see Line 17 of Fig. 6.85). Thus, the WAIT-WHILE in Line 103 instructs the Controller to "do nothing" until the DMS-80 is starting to "see" the Left Edge of Dowel 1 (as shown in Fig. 6.84) which would then make "NIR_Center" greater than "DowelThreshold". When this event occurs, the WAIT-WHILE Loop exits to Line 104.
- Lines 104-107 prints out Servo 17's Present Position, the "latest" NIR_Center value which should be greater than "DowelThreshold" (because the DMS-80 should be scanning more towards the "interior" of Dowel 1 at this point in time). This is for the operator's usage to ascertain that sensor/actuator and algorithm are still working together OK!

- Here the reader needs to review Fig. 6.84 while performing a "thought-experiment" visualizing what the DMS-80 would "see" as its scans across a dowel from its left edge to its right edge:
    - The DMS-80 should see an increase in the value of reflected NIR (i.e. Parameter "NIR_Center" increasing) when it scans from the left edge to the middle line of the dowel due to the circular cross-sectional area of the dowel.
    - And for the same "cross-sectional" reason, the DMS-80 should provide a decrease in NIR_Center values as it scans from the middle line of the dowel to its right edge where it would see "empty space" again.
- To detect these changes in Parameter "NIR_Center", a pair of these "running values" are kept in Parameters "PreviousNIR" and "CurrentNIR". They are used in the mini-algorithm designed to locate the middle-line of Dowel 1 which is implemented from Line 109 to Line 119:
    - First, "PreviousNIR" is set equal to "BackgroundNIR" (Line 109) – most logical value to use since the DMS-80 was coming in from "empty space", and "CurrentNIR" is set equal to "NIR_Center" which represents the latest DMS-80 sensor reading provided by CALLBACK (Line 110). At this point in time, the DMS-80 is seeing the right edge area of Dowel 1, therefore "CurrentNIR" is greater than "PreviousNIR". These NIR values guarantee that the coming LOOP WHILE will be entered (Lines 111-116) as its Loop Condition is ((FoundDowel == 0) && (CurrentNIR >= PreviousNIR)) and FoundDowel was already set to "0" way back in Line 83 of Fig. 6.87 – inside Function ServoScan.
    - This LOOP WHILE contains statements designed to implement a "rolling" update of "PreviousNIR" and "CurrentNIR" (Lines 113-114) and to record the Present Position of Servo 17 into Parameter "Dowel1_Position" (Line 115).
    - Eventually, the DMS-80 passes over to the right of the middle line of Dowel 1 and "CurrentNIR" gets smaller than "PreviousNIR" forcing this LOOP WHILE to exit to Line 117 which saves "CurrentNIR" into "Dowel1_Distance".
    - Next, Line 118 saves "PreviousNIR" into "Dowel1_MaxNIR" and "FoundDowel" is incremented by 1 (from 0). These Parameters will be used later to determine the "closer' dowel between Dowel 1 and Dowel 2.
    - Function "DataPrint" is called next (Line 120 and see Fig. 6.90) for the operator's verification of data collected so far.
- At this point in time, the DMS-80 is scanning towards the right edge of Dowel 1, so the WAIT WHILE on Line 122 is used to instruct the Controller to "do nothing" until the DMS-80 starts to see the "empty space" between Dowel 1 and Dowel 2 (see Fig. 6.84) – that is when "NIR_Center" is no longer greater than "BackgroundNIR".

Now, it is time to locate Dowel 2 and to determine its "Distance" from the DMS-80 (see Fig. 6.89 for Part 2 of Function "FindDowels"):

- Part 2 of Function "FindDowels" follows the same algorithm developed in its Part 1, except for the slightly different LOOP WHILE's condition used on Line 134 to accommodate for Dowel 1 being already found (i.e. FoundDowels == 1) and the different Parameters used in Lines 138, 140, 141 and 142 to record Dowel 2's results.
- The last statement in Function "FindDowels" is to call Function "PrintData" on Line 143.

Part 2 of Function "FindDowels" for DOWEL-SCANNER

```
124     // Let IR Sensor scan "empty space" until IR Sensor sees edge of Second Dowel
125     WAIT WHILE ( NIR_Center <= DowelThreshold )
126     Print Screen with Line = 7777
127     Print Screen = ID[17]: Present Position
128     Print Screen = NIR_Center
129     Print Screen with Line = DowelThreshold
130
131     // Scanning Second Dowel
132     PreviousNIR = BackgroundNIR
133     CurrentNIR = NIR_Center
134     LOOP WHILE ( FoundDowels == 1 && CurrentNIR >= PreviousNIR )
135     {
136         PreviousNIR = CurrentNIR
137         CurrentNIR = NIR_Center
138         Dowel2_Position = ID[17]: Present Position
139     }
140     Dowel2_Distance = CurrentNIR
141     Dowel2_MaxNIR = PreviousNIR
142     FoundDowels = FoundDowels + 1
143     CALL PrintData
144 }
```

Fig. 6.89 Part 2 of Function "FindDowels" in "M_DowelScanner.tsk3".

Fig. 6.90 shows the details of Function “PrintData” which is self-explanatory.

Function "PrintData" for DOWEL-SCANNER

```
FUNCTION PrintData
{
    IF ( FoundDowels == 0 )
    {
        Print Screen with Line = 0
    }
    ELSE IF ( FoundDowels == 1 )
    {
        Print Screen with Line = 1111
    }
    ELSE IF ( FoundDowels == 2 )
    {
        Print Screen with Line = 2222
    }
    Print Screen = Dowel1_Position
    Print Screen = Dowel1_Distance
    Print Screen with Line = Dowel1_MaxNIR
    Print Screen = Dowel2_Position
    Print Screen = Dowel2_Distance
    Print Screen with Line = Dowel2_MaxNIR
}
```

Fig. 6.90 Function “PrintData” in “M_DowelScanner.tsk3”.

After printing data for Dowel 1 and Dowel 2, the Controller exits the Function “ServoScan” at Line 33 in Fig. 6.86. **The Controller now has all the information needed to decide whether Dowel 1 (on robot’s Left) or Dowel 2 (on robot’s Right) is the dowel closer to the robot.** The next task is to **relocate and approach** the **closer** dowel with the DMS-80 Sensor locked into the forward-looking mode.

6.9.3 Relocate & Approach Closer Dowel

Fig. 6.91 shows Part 3 of the Main Program for “M_DowelScanner.tsk3”:

- Similarly as for Part 1 and Part 2, the Controller waits for the operator to push on Button R (Lines 39-41) for it to set Servo 17 to Position 512 (Lines 44-45) and call Function “LocateApproach” (Lines 46) which implements the “Locate & Approach” algorithm involving Functions “LocateRight”, “LocateLeft”, “Approach” and “Stop” (see Figs. 6.92 and 6.93).

Part 3 of MAIN for DOWEL-SCANNER

```
LocateApproachDowels :
    WAIT WHILE ( Remocon Data Arrived == FALSE (0) )
    Data = Remocon RXD
    IF ( Data == R )
    {
        // Swing Head Servo (#17) to Front (to see Background)
        ID[17]: Goal Position = 512
        WAIT WHILE ( ID[17]: Is Moving == TRUE (1) )
        CALL LocateApproach
    }
    ELSE
    {
        JUMP LocateApproachDowels
    }
    // End of Main Program
}
```

Fig. 6.91 Part 3 of Main Program in "M_DowelScanner.tsk3".

Part 1 of Function "LocateApproach" for DOWEL-SCANNER

```
FUNCTION LocateApproach
{
    // When Dowel 2 is closer than Dowel 1
    IF ( Dowel2_MaxNIR > Dowel1_MaxNIR )
    {
        // Closest dowel is on the robot's right
        CALL LocateRight
        CALL Approach
        CALL Stop
        Print Screen with Line = 111111
        // Locate and approach 2nd dowel on the left
        CALL LocateLeft
        CALL Approach
        CALL Stop
        Print Screen with Line = 222222
    }
```

Fig. 6.92 Part 1 of Function "LocateApproach" in "M_DowelScanner.tsk3".

Part 2 of Function "LocateApproach" for DOWEL-SCANNER

```
184  ELSE
185      // When Dowel 1 is closer than Dowel 2
186    {
187      // Closest dowel is on the robot's left
188      CALL LocateLeft
189      CALL Approach
190      CALL Stop
191      Print Screen with Line = 111111
192      // Locate and approach 2nd dowel on the right
193      CALL LocateRight
194      CALL Approach
195      CALL Stop
196      Print Screen with Line = 222222
197    }
198  }
```

Fig. 6.93 Part 2 of Function "LocateApproach" in "M_DowelScanner.tsk3".

Figs 6.92 and 6.93 show that Function "LocateApproach" is designed to handle two possibilities:

- If Dowel 2 is closer to the MINI than Dowel 1 (Line 171 in Fig. 6.92), then the MINI rotates Right to locate Dowel 2 (Line 174), approaches it to a certain distance (Line 175) and stops (Line 176), waiting for the operator to remove Dowel 2. Next, the MINI rotates Left to locate Dowel 1 (Line 179), approaches it to a certain distance (Line 180) and performs the final stop (Line 181).
- If Dowel 1 is closer to the robot than Dowel 2, the ELSE branch (Line 184) is chosen by the Controller and two sets of mirrored tasks are performed: LocateLeft >> Approach >> Stop for Dowel 1, then LocateRight >> Approach >> Final Stop for Dowel 2.

Let's take the case when Dowel 2 is closer to the robot than Dowel 1 as an example to explain more programming details. This means that the condition (Dowel2_MaxNIR > Dowel_MaxNIR) on Line 171 in Fig. 6.92 evaluates to TRUE:

- Function "LocateRight" is then called (see Fig. 6.94):
  - A new Parameter "Locate_Threshold" is defined as the latest DMS-80 sensor data input "NIR_Center" with 20 more units added (Line 253). This assignment guarantees that the upcoming LOOP WHILE will be taken (Lines 254-261).
  - This LOOP WHILE is designed to keep on rotating the robot to the Right (by calling Function "Right" on Line 256) as long as the latest "NIR_Center" data input is less than "Locate_Threshold".

- Lines 257-259 print out the current values of "Local_Threshold" and "NIR_Center" to keep the operator aware of the current situation.
- Function Delay instructs the Controller to wait 0.5 second for the robot's frame vibrations to subside so that a more stable reading of the DMS-80 Sensor is achieved. The reader may have to adjust this time delay to suit his/her own situation.
- Eventually, these Right rotations would bring the DMS-80 to scan upon Dowel 2, then NIR_Center would be greater than "Locate_Threshold" and this LOOP WHILE and Function "LocateRight" terminate. Thus, the Controller is returned to Line 175 in Fig. 6.92, and Function "Approach" is called (see Fig. 6.95).

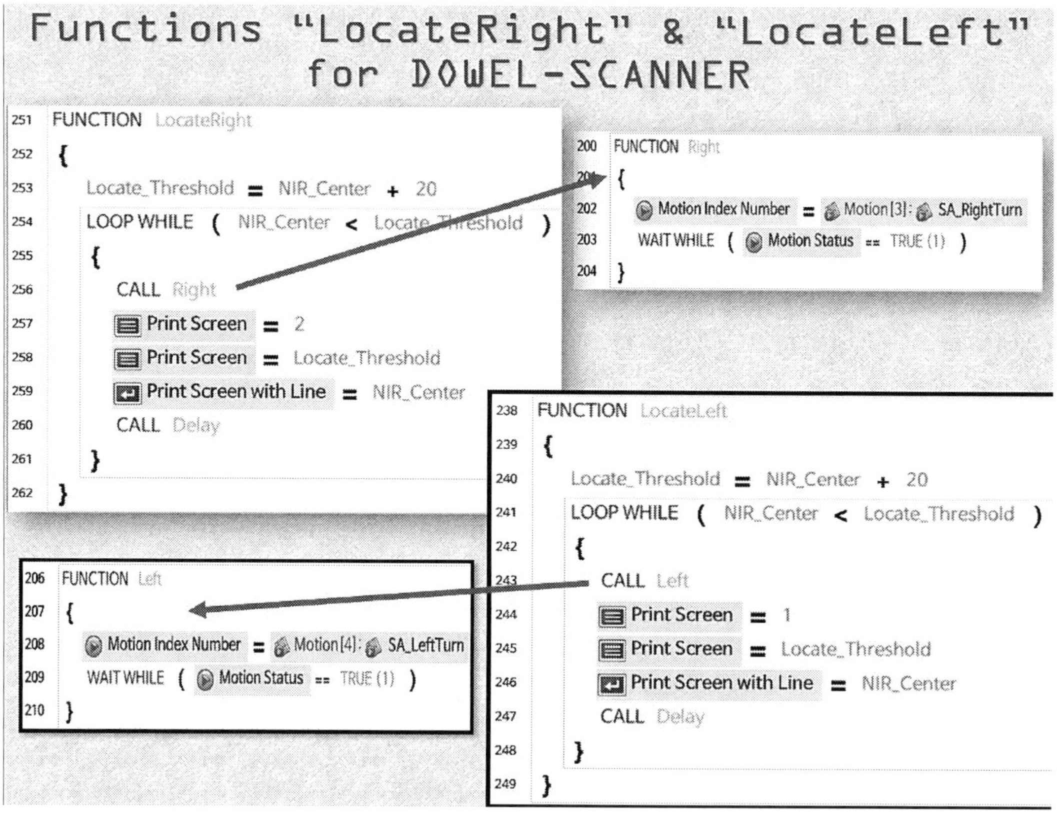

Fig. 6.94 Functions "LocateRight", "Right", "LocateLeft" and "Left" in "M_DowelScanner.tsk3".

Function "Approach" for DOWEL-SCANNER

```
264 FUNCTION Approach
265 {
266     PreviousNIR = NIR_Center
267     Stop_Limit = NIR_Center + 150
268     // Approaching Dowel and adjust right if needed
269     LOOP WHILE ( NIR_Center < Stop_Limit )
270     {
271         IF ( NIR_Center >= PreviousNIR )
272         {
273             PreviousNIR = NIR_Center
274             CALL Forward
275             CALL Delay
276         }
277         // Correction for author's robot which tends to drift left while walking forward
278         ELSE
279         {
280             PreviousNIR = NIR_Center
281             CALL Right
282             CALL Delay
283         }
284     }
285 }
```

Fig. 6.95 Function "Approach" in "M_DowelScanner.tsk3".

Fig. 6.95 lists the details of Function "Approach" which is designed to go forward towards the targeted dowel up to a certain distance controlled by Parameter "Stop_Limit":

- First, Line 266 saves the latest DMS-80 data input "NIR_Center" (from CALLBACK) into Parameter "PreviousNIR".
- Next, Parameter "Stop_Limit" is assigned with the sum of the latest DMS-80 data input "NIR_Center" and the Constant "150" (Line 267).
- Thus, Lines 266 and 267 assure that the coming LOOP WHILE (Lines 269-284) is executed.
- This LOOP WHILE is designed to move the robot forward if the condition (NIR_Center < Stop_Limit) remains TRUE. However, the author's MINI has a construction flaw which tends to drift this robot left while it is walking forward, thus an IF-ELSE structure had to be implemented inside this LOOP WHILE to correct for this left-drifting behavior:
    - Lines 271-276 check to see if the Current DMS-80 input **"NIR_Center" is greater or equal to "PreviousNIR"**. If this is TRUE, this means that in the physical world the MINI is approaching the targeted dowel more or less "straight on", so the robot should keep going forward (Line 274 and see Fig. 6.96 for Function "Forward"). Please note

that "PreviousNIR" is updated with the latest "NIR_Center" (Line 273) before making the robot move forward (Line 274). A 0.5 second time delay is also used for robot vibrations to subside (Line 275).
  - However, if the author's robot drifts too much to the left, then the **latest "NIR_Center" would be less than "PreviousNIR"** and the ELSE branch (Lines 279-283) is chosen and this robot would be commanded to turn Right (Line 281) to hopefully bring it back in line with the targeted dowel.
  - If the reader's robot moves forward straight, then the ELSE branch would never be chosen. However, if the reader's robot drifts right instead, then a "Call Left" should be used in Line 281.
- Eventually (and hopefully) the robot gets close enough to the targeted dowel so that "NIR_Center" is greater or equal to "Stop_Limit", then this LOOP WHILE terminates and Function "Approach" ends as well. Then the Controller returns to Line 176 of Fig. 6.82, and Function "Stop" is called (see Fig. 6.96).

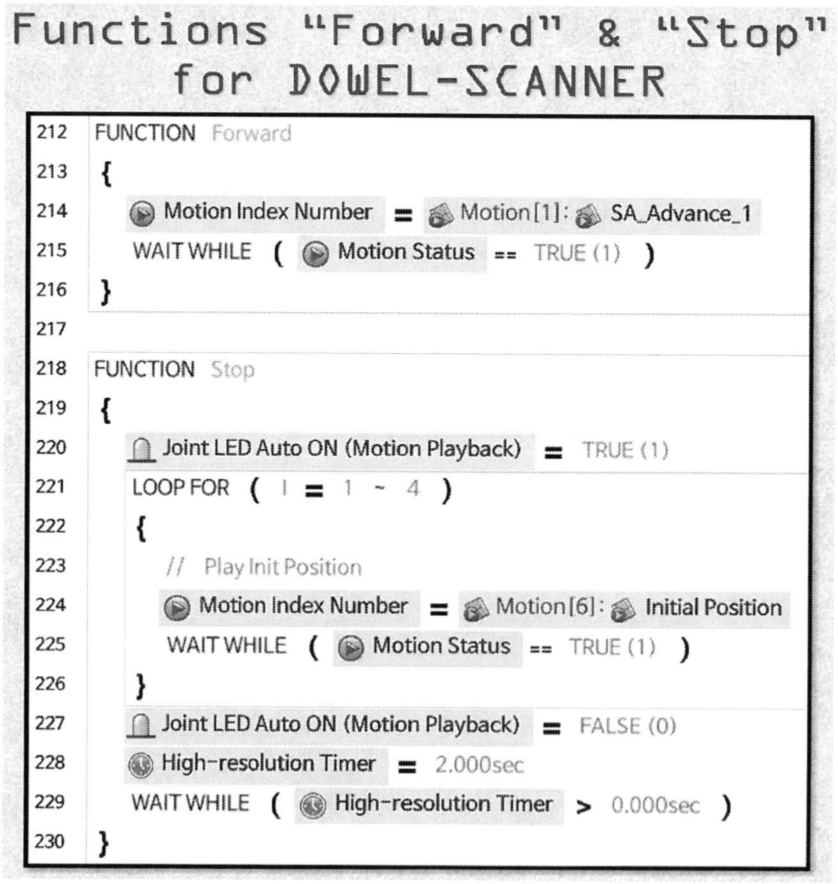

Fig. 6.96 Functions "Forward" and "Stop" in "M_DowelScanner.tsk3".

Fig. 6.96 shows the details of Function "Stop" which is designed to stop the robot and make its LED flash 4 times:

- Normally the robot LEDs are turned off during motion-plays, so they have to be turned on during the time that Function "Stop" is active, thus Line 220 is used to turn them on, and Line 227 is used to turn them off before exiting Function "Stop".
- The LOOP FOR structure is designed to play MOTION-LIST/PAGE "Initial Position" 4 times. In the physical world, when the operator can see the robot's LEDs flashing, it would be time to remove the targeted dowel. The 2-second wait construct (Lines 228-229) should give the operator enough time to perform this task.
- When Function "Stop" terminates, the Controller returns to Line 177 of Fig. 6.92 and prints out "111111" to show that the robot is done with its first dowel (Dowel 2 in this example). Next, the robot needs to deal with the "farther" dowel which is Dowel 1 in the current discussion.

6.9.4 Relocate & Approach Remaining Dowel

At this point in time, Dowel 1 would be on the robot's Left, thus Line 179 in Fig. 6.92 is invoked to put into play the Function "LocateLeft" (see Fig. 6.94) which uses a similar algorithm as for Function "LocateRight" which is already described in details in Section 6.9.3 – except that the robot would now be commanded to rotate Left until it scans upon Dowel 1.

Next, Line 180 in Fig. 6.92 becomes active and Function "Approach" is called to make the robot approach Dowel 1 in a similar manner as it did for Dowel 2 previously in Section 6.9.3. Line 181 executes next and flashed the robot's LEDs 4 times, then Line 192 prints out "222222" to mark the completion of all tasks for the robot!

Video 6.16 shows the runtime performance of the program "M_DowelScanner.tsk3".

6.10 Review Questions

1. How many types of ASSIGNMENT statements does TASK V.3 support?
2. Which LOOP type does TASK **not** support?
3. If used, how often is the CALLBACK function called upon?
4. What is the recommended number of External Device Calls for use in the CALLBACK function?
5. What is the maximum memory size allowable for the CALLBACK function?
6. How many TIMER device(s) is supported in TASK?
7. What are the two operating modes allowed for the XL-320 actuators used on the MINI?
8. In the example code "M_CountingLoopControl_Servo6.tsk3", try to print Parameter Count AFTER the LOOP FOR to see its final value. Is this value what the reader expected?
9. In the example code "M_ConditionalLoopControl_Servo6.tsk3", please modify it so that Button "U" is the one that will trigger the moves for Servo 6.
10. When All Buttons on the RC-100 are released, does this Remote Controller send any message back to the TASK program?
11. What is the TASK command to receive a Remocon message?
12. What is the TASK command to send a Remocon message?
13. How many bits does a Remocon message contain?
14. With which bit-wise operation can the programmer use to extract specific information from a Remocon message?
15. What is the number of bits used by TASK to internally store its Variables and Constants?
16. How can the programmer shift a TASK variable/constant for a given number of bits to the left?

17. How many ways can the JOINT-OFFSET setting of the XL-320 actuators be used?
18. Where are these JOINT-OFFSET settings are recorded? On the OpenCM9.04? Or on the respective XL-320 actuators?
19. What is the maximum number of Variables that can be used in a TASK program for the MINI robot?
20. What can the user do with the "Instructions" Tab of the TASK Sub-tool?
21. What can the user do with the "Variables" Tab of the TASK Sub-tool?
22. What can the user do with the "Functions" Tab of the TASK Sub-tool?
23. How can the user access the Virtual RC Controller in the TASK Sub-tool?
24. What are the steps used in "Computational Thinking" practice?
25. How can the programmer trigger an Exit Motion Unit?
26. How can the programmer monitor whether an actuator is in motion or not?
27. How can the programmer monitor whether a Motion List/Page is being played or not?
28. Are JOINT-OFFSET settings active when Motion-List/Pages are not being played?
29. What is the "Message Shaping" concept?
30. In Part 4 of the Main Endless Loop for "M_RC_SmartAvoider.tsk3" (see Fig. 6.46), change the IF structure used on Line 37 with a LOOP WHILE structure. Does the overall program work as before? If no, how can you modify the logic so that it works as before?
31. In Part 4 of the Main Endless Loop for "M_RC_SmartAvoider.tsk3" (see Fig. 6.46), change the LOOP WHILE structures used on Lines 44 and 50 into PARALLEL IF structures instead. Does the overall program work as before? If no, how can you modify the logic so that it works as before?

6.11 References

Beecher K (2017) Computational Thinking: A Beginner's Guide to Problem-Solving and Programming. BCS Learning & Development Ltd, Swindon, UK.

Ben-Ari M, Mondada F (2018) Elements of Robotics. Springer Publishers International, Cham, Switzerland.

Denning PJ, Tedre M (2019) Computational Thinking. The MIT Press, Cambridge.

Matarić MJ (2007) The robotics primer. The MIT Press, Cambridge.

Chapter 7: Using PLAY700 Mobile App

The Mobile App "R+m.PLAY700" was originally created to accompany the PLAY700 kit (http://www.robotis.us/software/play700/), but it also works with the OpenCM9.04-C using SMART DEVICE commands in conjunction with SMART CONSTANT parameters inside the TASK tool (see Fig. 7.1). Detailed information about these SMART functions and their corresponding arguments are contained in the file "SmartDeviceControlTable.PDF" accessible at www.cntrobotics.com/mini-book. This PDF file is a Chrome-translated English version of this Korean web page http://support.robotis.com/ko/software/mobile_app/r+smart/smanrt_manual.htm#Actuator_Address_0B3.

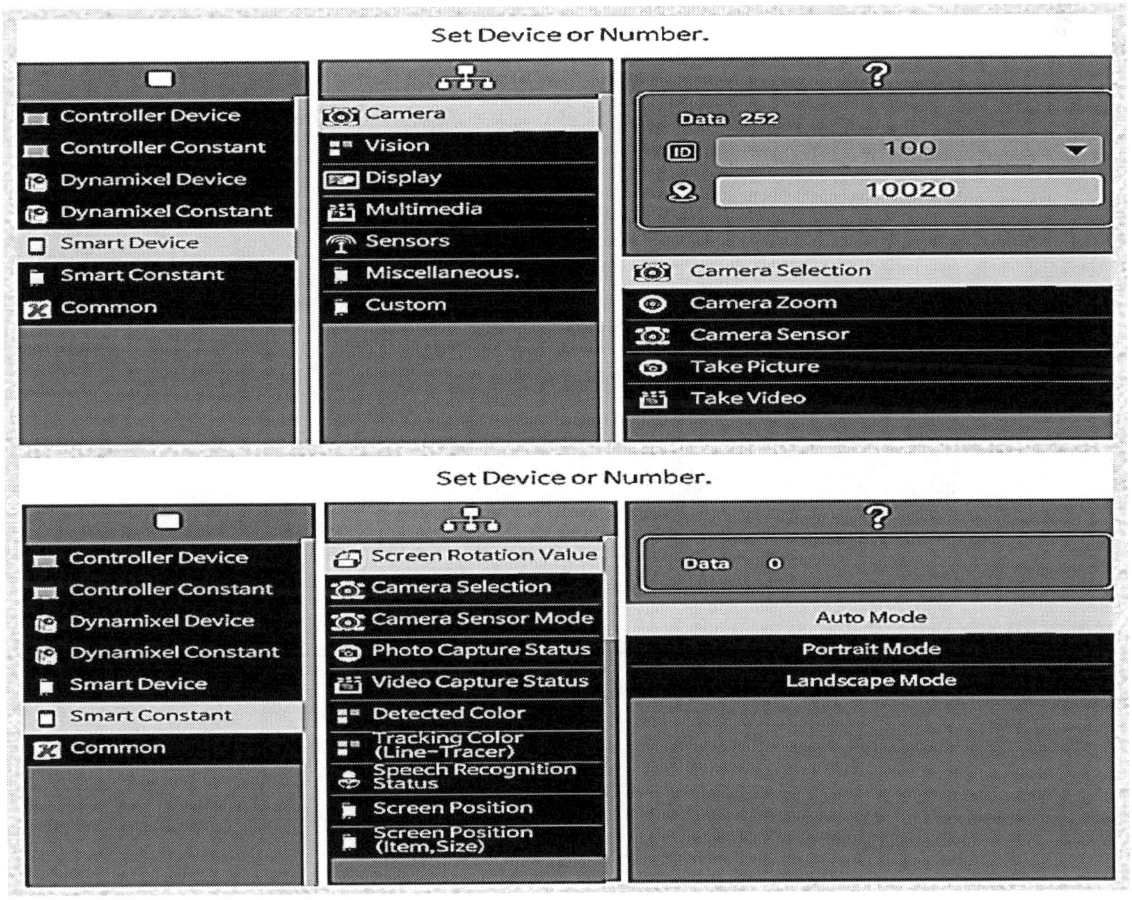

Fig. 7.1 SMART DEVICE & CONSTANT Options inside TASK Sub-Tool.

The R+m.PLAY700 App allows users to add more functionalities to standard TASK programs (see Chapter 6) by leveraging existing mobile services such as graphics, multimedia, video camera or speech recognition.

The goal of this Chapter is to present readers with a firm foundation in using the R+m.PLAY700 App with R+TASK V.3 via several targeted projects.

7.1 Installation of R+m.PLAY700

The reader can install the latest version of R+m.PLAY700 on iOS or Android devices at the appropriate web links (https://play.google.com/store/apps/details?id=com.robotis.play700&hl=en or https://itunes.apple.com/us/app/play700/id1156037721?mt=8).

7.1.1 Files Structure for PLAY700 App

For development work, the author uses a Windows 10 PC which interfaces well with Android devices, and all file and folder management can be done via Windows Explorer. Fig. 7.2 shows the folder/file structure for PLAY700 on an Android device.

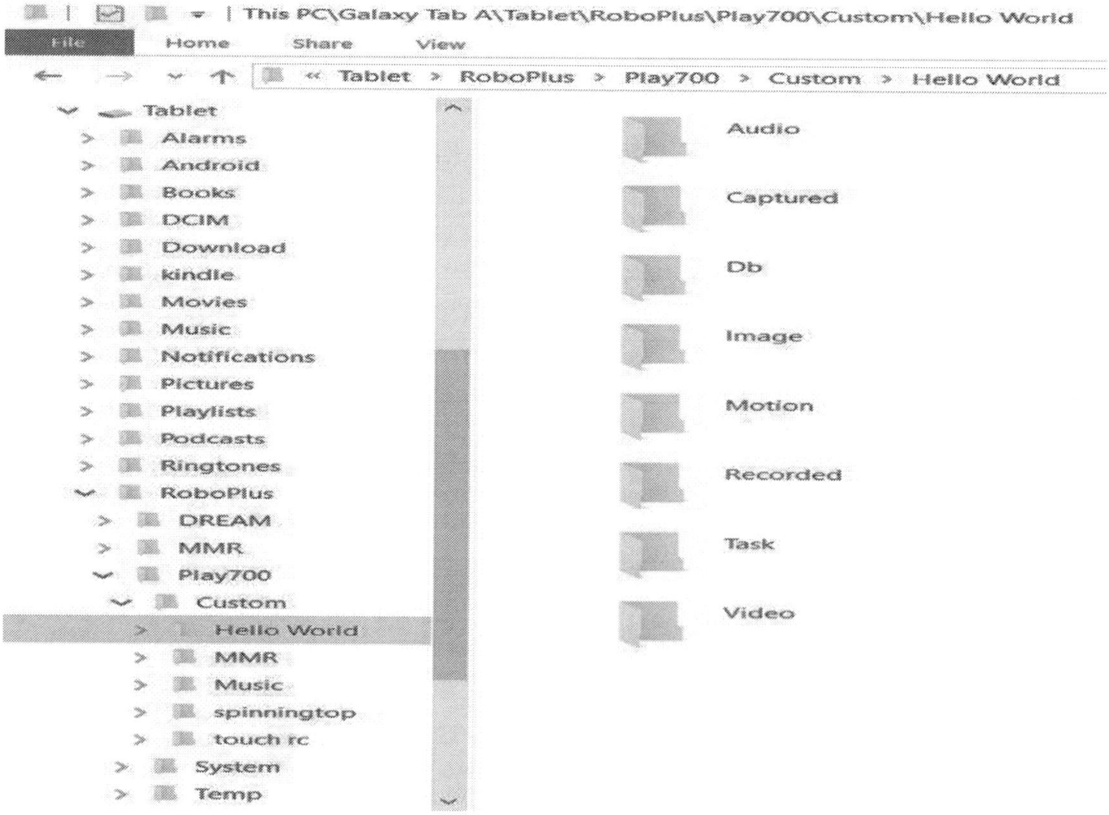

Fig. 7.2 File/Folder structure for PLAY700 App on an Android device.

For iOS devices, the author uses "iMazing™" to help with file/folder management for the PLAY700 App. Fig. 7.3 shows the folder/file structure for PLAY700 on an iOS device.

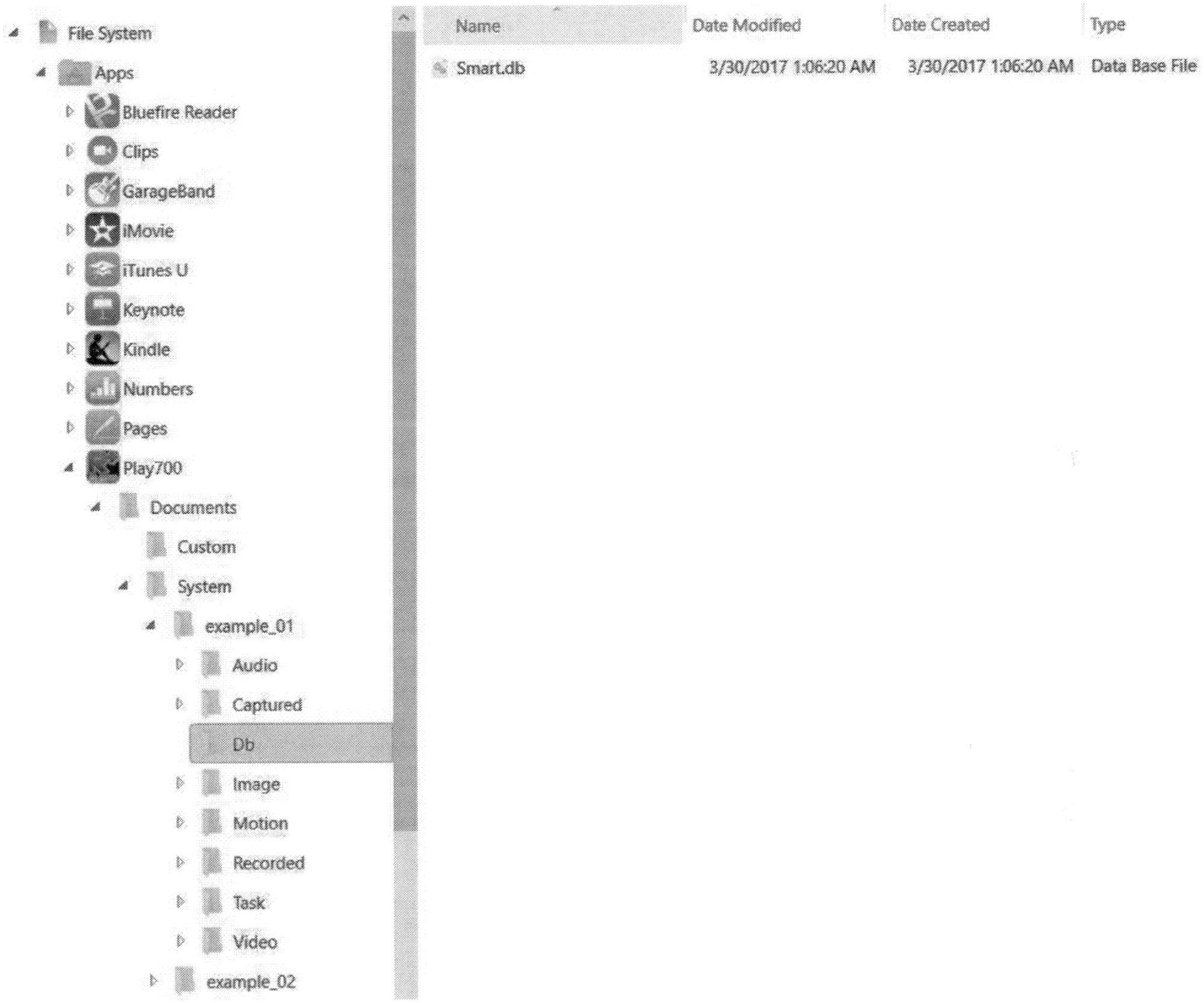

Fig. 7.3 File/Folder structure for PLAY700 App on an iOS device (via iMazing).

The reader can see that a PLAY700 project has many components represented by the folder structures shown in Figs. 7.2 and 7.3.

Fig. 7.2 shows that all ROBOTIS R+m apps are installed under a main folder named "RoboPlus", off the root directory of the main storage. As one drills into the PLAY700 folder, one can see three sub-folders "Custom", "System" and "Temp". The "System" sub-folder contains the 6 example projects provided by PLAY700. The "Custom" sub-folder contains the user-initiated projects, for example the "Hello World" project where the project's standard components are stored:

Audio – for audio files to be played by the mobile device but controlled by the TASK code at run time.

Captured – where pictures captured by the PLAY700 app via the mobile device's front and back cameras are kept.
Db – where SMART databases are stored, including "Text" and "Audio Input" lists used for Speech Recognition (these items can only be modified via the editing of the Project's Tools – see Section 7.2).
Image – where photos specifically used by a given project are first "registered" and then "stored" as background images (i.e. "**Bg**" sub-folder) or as foreground images (i.e. "**Fg**" sub-folder).
Motion – where MOTION files are stored.
Recorded - where videos captured by the PLAY700 app via the mobile device's front and back cameras are kept.
Task – where TASK files are stored if the R+m.TASK tool is used to create TASK codes on the mobile device.
Video – for video files to be played by the mobile device but controlled by the TASK code at run time.

The "Temp" folder is only used by the PLAY700 App for its own purposes.

For an iOS device, Fig. 7.3 shows that ROBOTIS R+m Apps are installed under a main folder named "Apps" off the root directory of the main storage. As one drills into the PLAY700 folder, one would encounter a "Documents" sub-folder which further branches out into two sub-folders "Custom" and "System" which contain similar sub-folders from Audio to Video as described in previous paragraphs.

7.1.2 Settings for PLAY700 App

Fig 7.4 shows a screenshot of the main menu window for the PLAY700 App where the reader can see a "gear' icon on the top right corner. Once this "gear" icon is tapped once, the "Settings" screen is shown as in Fig. 7.5.

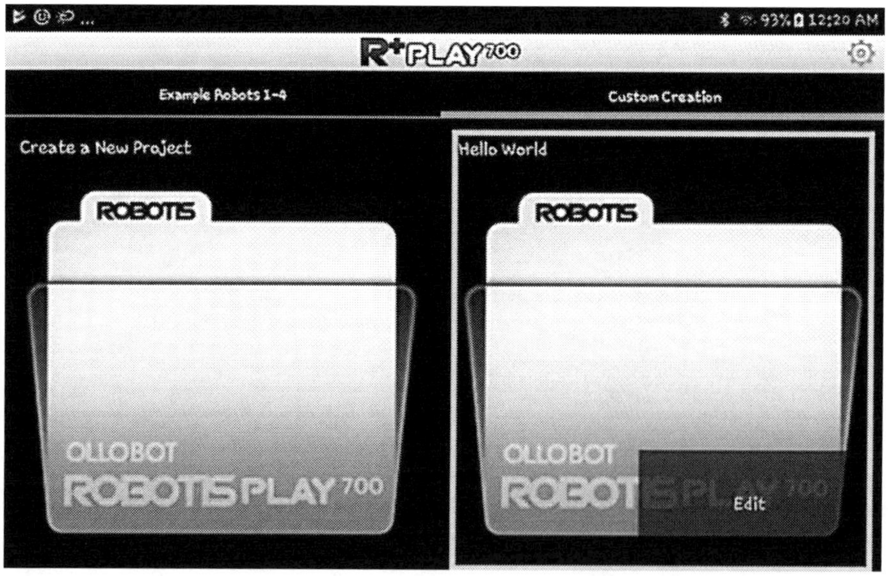

Fig. 7.4 Main Menu of PLAY700 app, opening on Custom Project "Hello World".

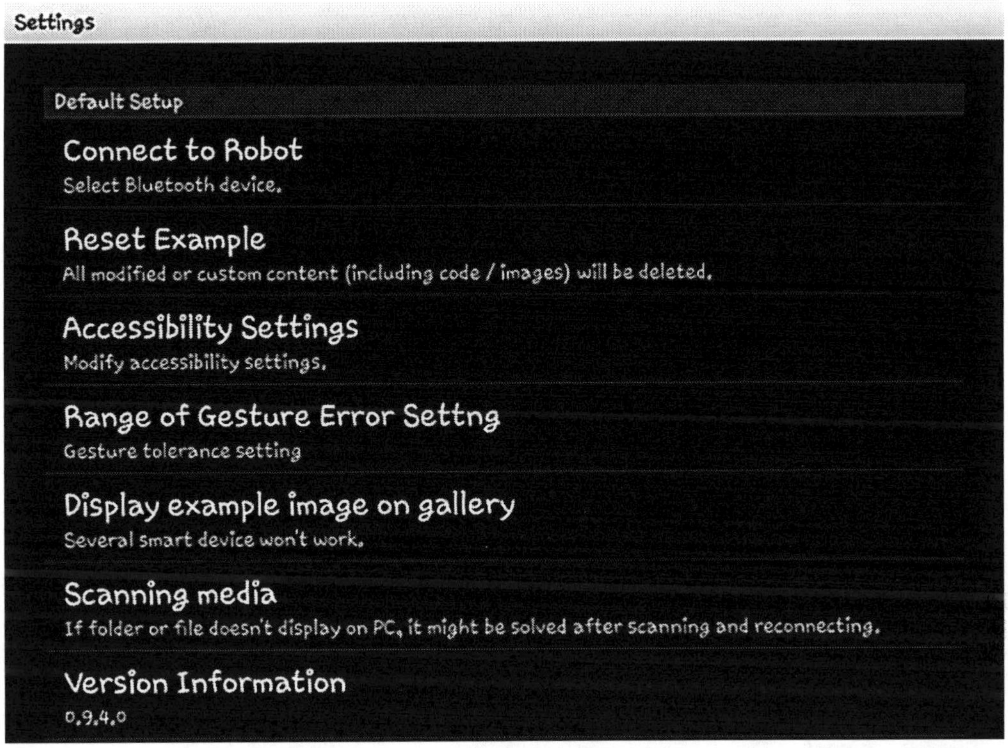

Fig. 7.5 App-level "Settings" for PLAY700 App.

The first item "Connect to Robot" allows the user to scan for new BT devices and pair them or to choose a specific BT device among those that were previously paired.

The second item "Reset example" can be used to reset all ROBOTIS example System projects to their originally installed states. This setting does not affect the Custom projects.

The third item "Accessibility Set-up" only pertains to Android devices where the user could set-reset many OS-level settings such as "Auto Rotate Screen" or "Screen Timeout". This "Accessibility" setting is not available for iOS devices.

The fourth item "Range of Gesture Error Setting" is used to set the accuracy of the gesture function (a number between 0 and 30 – available to both Android and iOS devices).

The fifth item "Display Example Image on Gallery" also pertains to Android devices only and does what its label implies.

The sixth item "Scanning Media" also pertains to Android devices only and removes the original "hidden" status of the folders and files inside the PLAY700 main folder so that Windows Explorer on a PC can "see" them.

7.2 PLAY700 Project Features

Going back to Fig. 7.4, when the user taps on the "Edit" button of a project, a screen as shown in Fig. 7.6 displays all the components/tools that can be used for this project. The current Android version (0.9.5.1) has all these tools operational, **but for the current iOS version (1.0.7), the following six tools are not yet functional**:

"Instrument" in the Multimedia group.
"Illumination" in the "Sensor" group.
"Received SMS", "Status Bar", "Vibration", "Application" in the "Other" group.

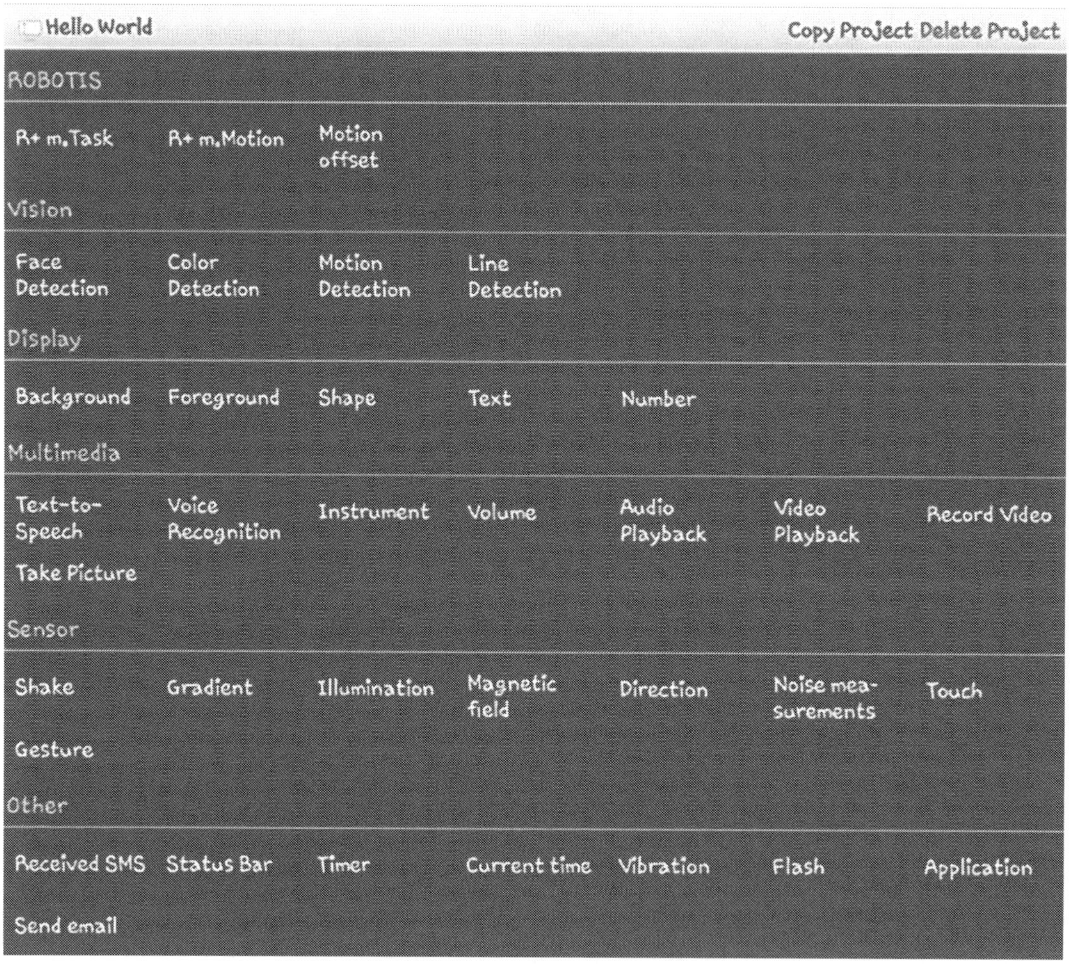

Fig. 7.6 Five groupings of services that a SMART Project can provide.

For example:

1) The "R+m.TASK" item allows the user to edit the TSKX files corresponding to this project.
2) The "Face Detection" item switches to the camera whereas the user can choose the front or back camera so that it can find a human face and then puts on a pair of sunglasses across the eyes, all in "real time".
3) The "Color Detection" item also kicks in the camera and shows a live image with a small centered region of interest where it would try to detect for 4 "colors": Black, Red, Green and Blue.

4) The "Display" section pertains to "Background" and "Foreground" images, and other items like "Text" and "Number". For example, the user can add and register Text Items if the user chooses the "Text" item in this "Display" section.
5) The "Multimedia" section can be used to record or play audio and video segments, to play musical instruments and for text-to-speech and speech recognition tasks.
6) The "Sensor" section pertains to sensors embedded in the mobile device such as its Touch, Gesture, Gyro and Tilt sensors.
7) The "Other" section applies to "Messages" and other "Applications" that can be linked to this project.

7.3 "Hello World" Project

There is an extensive list of "SMART DEVICE" commands as shown in Fig. 7.1 and this Section 7.3 should provide readers with a basic usage of the tool chain TASK/R+m.PLAY700. This web link http://emanual.robotis.com/docs/en/software/rplustask3/task_parameters/#smart-device has more information about the Smart Devices supported in TASK V.3.

This project will show how to display the text object "Hello World!" on the Mobile Device's Display Screen in different sizes and colors, and at randomized screen locations.

As far as TASK is concerned, the Mobile Device's Display Screen is divided into 25 zones arranged in a 5x5 grid in either Portrait or Landscape mode (see Fig. 7.7).

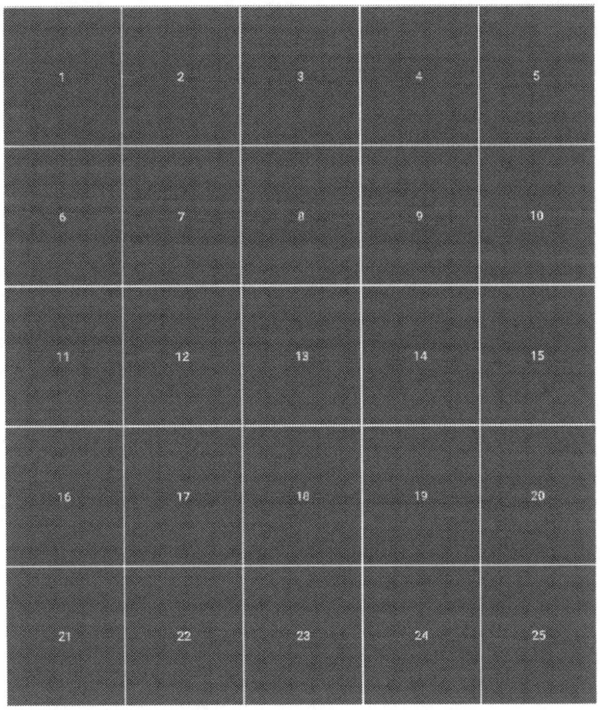

Fig. 7.7 Organization of the Display Screen on Mobile Device vis-à-vis TASK Tool.

When the Mobile Display is used as an Input Device, for example as a Touch Sensor (see more details in Section 7.3.2), the R+m.PLAY700 App reports a Zone Number between 1 and 25 (as shown in Fig. 7.7) to the TASK program at run time.

When the Mobile Display is used as an Output Device for a possible "Text", "Shape" or "Number" object (see Fig. 7.8), these Display Zones can be specified in two ways, depending on the programmer's needs, via a SMART CONSTANT Parameter which can have a fixed value or a variable value during run time:

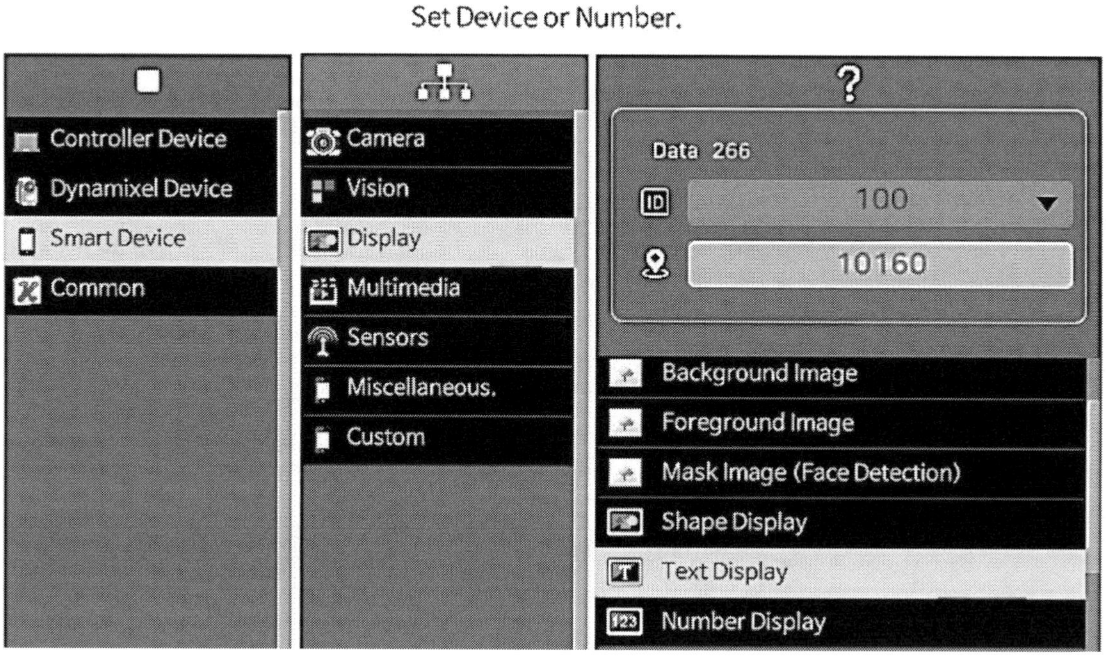

Fig. 7.8 Common Display Functions are for Shape, Text and Number in SMART DEVICE programming.

1) Fig. 7.9 illustrates the case where the parameters **Position** [3, 3] – i.e. Screen Center or Zone 13, **Item No.** [1], **Size** [100] and **Color** [White, i.e. 1] are all fixed values during run-time. This is the first way to specify the SMART CONSTANT Screen Position Parameter (SPP) which is a 4-byte number (equal to "23331085" in this case - see Data field at top of right panel in Fig. 7.9). The "SmartDeviceControlTable.PDF" file (at "address 10150") provides a detailed "formula" to compute the Screen Position Parameter (SPP) as follows:

SPP = **(Color * 16777216) + (Size * 65536) + (Item No * 256) + Position**

In this case, SPP = (1 * 16777216) + (100 * 65536) + (1 * 256) + 13 = **23331085**

In other words, "Position" forms the lowest byte of the 4 bytes used and "Item No" occupies the next higher byte (2nd byte). While "Size" is pushed into the 3rd byte, and "Color" is put into the highest 4th byte. **This formula applies to the display of Shape, Text and Number items**.

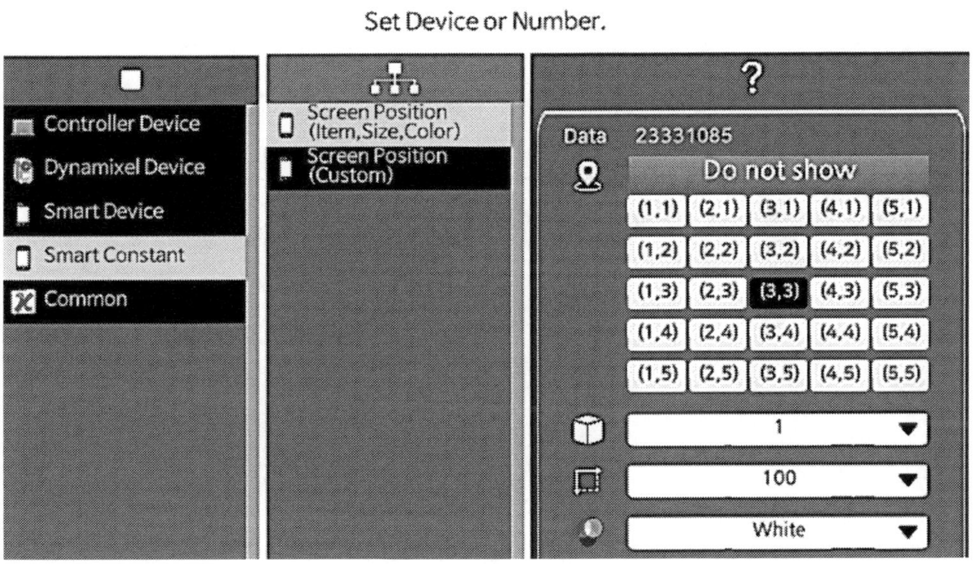

Fig. 7.9 Specification of a "fixed-value" Screen Position Parameter.

2) However, there may be times when the programmer needs to use a variable value for these parameters **Position, Item No, Size** or **Color** (i.e. variable arguments), then the programmer would need to use the "Custom" specification as illustrated in Fig. 7.10 where the "needed" Parameters are defined by 4 Variables named "PositionNo", "ItemNo", "SizeNo" and "ColorNo" respectively. Of course, these four Variables would have to be defined with the correct values before their use here: for example, "PositionNo" should be set to "13" if the Screen Center is to be used for Display.

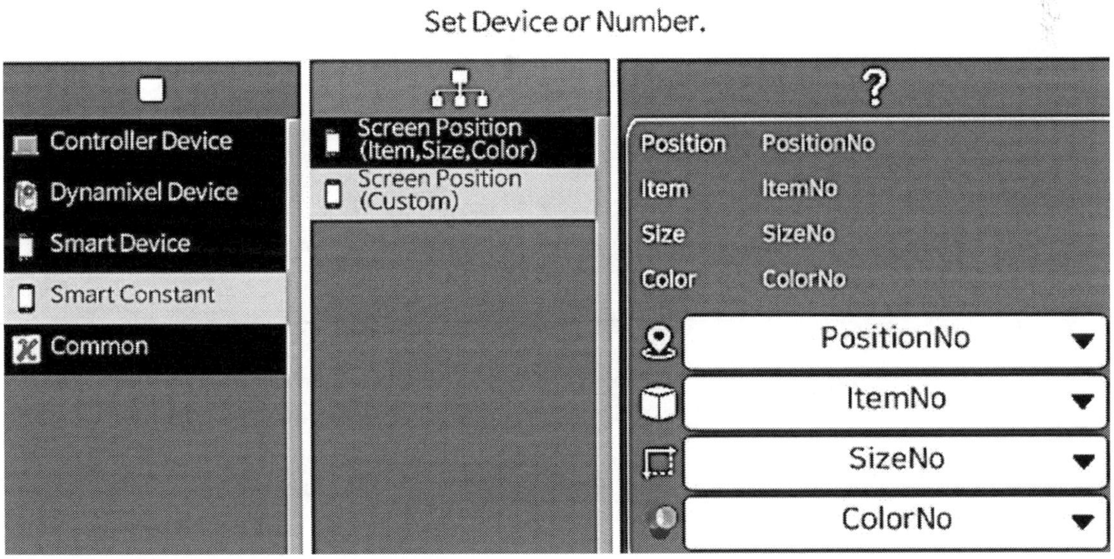

Fig. 7.10 Specification of a "variable-value" (i.e. Custom) Screen Position Parameter.

The program "M_HelloWorld_Random.tsk3" illustrates the use of a Custom Screen Position Parameter in conjunction with a Random Number (see Fig. 7.11):

1) Lines 6 to 9 initialize the values of Parameters PositionNo, ItemNo, SizeNo and ColorNo.

2) Line 10 waits on the user to make a loud enough noise to make the WAIT WHILE loop to exit to the next Line 11. This step is needed as Bluetooth connections between the Mobile Device and the MINI robot usually takes about 30-45 seconds to get established, so the TASK program must be designed to WAIT for this connection to occur.

3) Line 11 puts the mobile display screen in Portrait Mode and Line 12 clears the screen.

```
6   PositionNo = 0
7   ItemNo = 1
8   SizeNo = 0
9   ColorNo = 0
10  WAIT WHILE ( SMART: Noise (dB) < 50 )
11  SMART: Screen Rotation = Portrait Mode (1)
12  SMART: Text Display = 0
13
14  ENDLESS LOOP
15  {
16      Random = 2 Random Number
17      PositionNo = Random % 26
18      SizeNo = Random % 100
19      ColorNo = Random % 10
20      IF ( PositionNo > 0 && ColorNo > 0 && SizeNo > 20 )
21      {
22          SMART: Text Display = [Position:PositionNo],[Item:ItemNo],[Size:SizeNo],[Color:ColorNo]
23          High-resolution Timer = 1.000sec
24          WAIT WHILE ( High-resolution Timer > 0 )
25          SMART: Text Display = 0
26          High-resolution Timer = 0.250sec
27          WAIT WHILE ( High-resolution Timer > 0 )
28      }
29  }
```

Fig. 7.11 Using Custom Screen Position in "M_HelloWorld_Random.tsk3".

4) Then the Endless Loop (Lines 14-29) is entered whereas for each iteration, the following actions are performed:

a) A Random Number generated internally by the OpenCM9.04-C is saved into Parameter Random (Line 16).

b) Next, the Remainder Operator "%" is used to transform Parameter Random into an appropriate value for each of the Parameters PositionNo, SizeNo and ColorNo (Lines 17-19). The IF structure on Line 20 makes sure that various Parameters are in the proper ranges for reasonable display.

c) Line 22 combines the previous parameters into a Custom Screen Position to display Text Item 1 (i.e. "Hello World!") at a randomized screen position with randomized font sizes and colors.

d) Lines 23-24 allow the Text Display to stay on for 1 second, then Lines 25-27 clear the screen for 0.25 second.

Video 7.1 shows how to use the R+m.PLAY700 App to create a "Hello World" project and how to register the Text Item 1 "Hello World!", as well as the run time performance of "M_HelloWorld_Random.tsk3". The reader should try to comment out Lines 25-27 to obtain a "collage" effect at run time.

If the reader is interested in a **manual computation** of the SMART CONSTANT described above, please read the program "M_HelloWorld_Random_mc.tsk3" (included in the example codes package).

7.4 "Touch RC Walker" Project

This "Touch RC Walker" project's goal is to adapt the robot functionalities described in the "RC Walker with Independent Servos Control" project (i.e. "M_RC_Walker_IS.tsk3" in Section 6.4) using selected Smart Device features such as "Touch Area", "Text Display" and "Instrument Play" as well as a new operational feature which is the capability to change the Goal Velocity of Servo #17 of the MINI. This project uses the "RC Walk 1" MOTION GROUP from the file "MINI_Custom.mtn3".

7.4.1 "Touch Areas" in R+m.PLAY700 App

So far we have only used the RC-100 with its 10 buttons as a Remote Device for controlling the robot, and each Remocon message contains ON/OFF information regarding these 10 buttons "individually", so we have developed programming techniques to deal with that operational feature (see Section 6.4). However, the R+m.PLAY700 App handles the Mobile Device's Display Screen quite differently as it divides this Screen into 25 Touch Areas arranged in a 5x5 grid in either Portrait or Landscape mode (see left picture in Fig. 7.12). Furthermore, this App can only report to a TASK program about **2 Touch Areas** at any time during the runtime of a TASK/PLAY700 project. In other words, the programmer/operator can use only up to two fingers at a time when using the Mobile Display as a Remote Controller (as compared to all 10 buttons on the RC-100). But the programmer will be able to distinguish which "finger" touches the screen first as "Touch Area 1", and the later finger-press as "Touch Area 2". Inside the TASK Sub-tool, these "Touch Areas" are accessible via the "Smart Device" panel and in the "Sensors" category (see right and bottom pictures in Fig. 7.12). The Parameters "Touch1" and "Touch2" would contain a numerical value in the range from "1" to "25" depending on which Touch Area(s) on the Mobile Display get pressed, and they would contain "0" if all fingers are off the Mobile Display. Consequently, a different interface programming technique needs to be developed to handle Touch Areas (see Section 7.3.4).

Communications wise, the "SMART: Touch Area 1" and "SMART: Touch Area 2" commands are READ instructions that are sent from the MINI Controller to the Mobile Device through the 57 Kbps Bluetooth interface. Once the Mobile Device receives these instructions, it uses its own services to determine the current status of its Touch Areas and reports this information back to the MINI Controller via Bluetooth. Thus, this is a lengthy process as compared to the reading of Buttons on the RC-100.

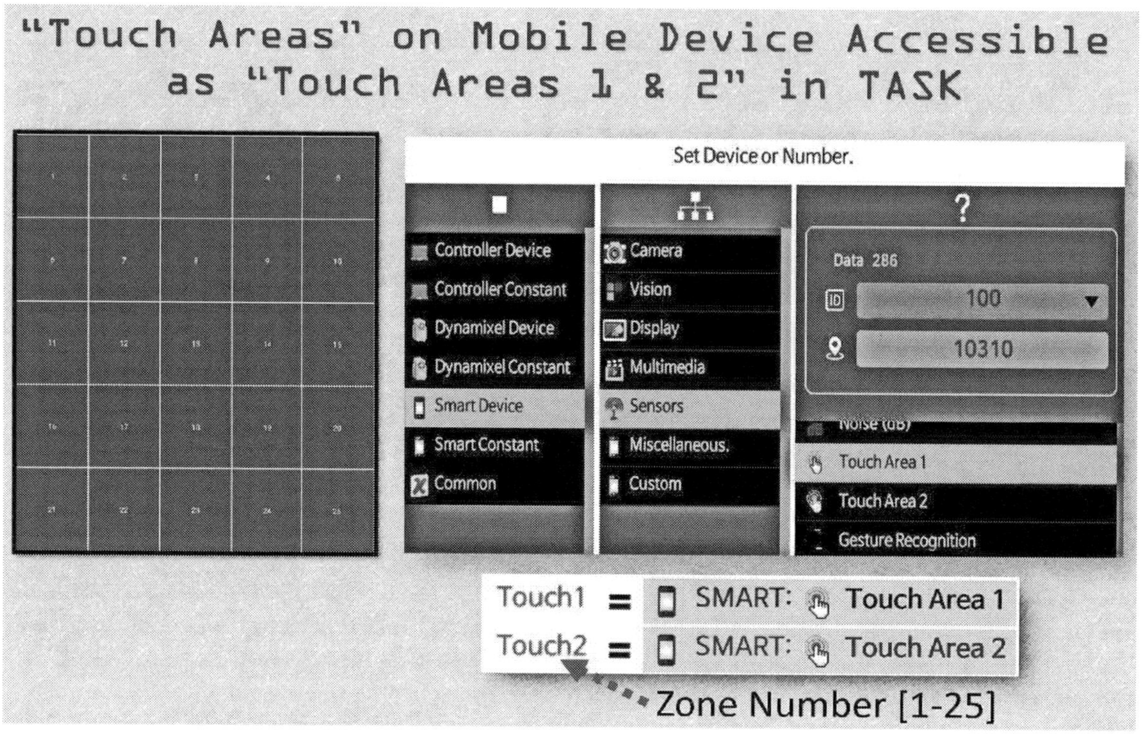

Fig. 7.12 Accessing & Programming Touch Areas on Mobile Display.

7.4.2 "Text" Items in R+m.PLAY700 App

For this Touch RC project, the first 10 "Text" Items as shown in the left picture in Fig. 7.13 are used. These "Text" Items need to be registered separately in an appropriate PLAY700 Project before using them in the TASK program (see Part 1 in Video 7.2).

Fig. 7.13 also show how to use the "SMART: Text Display" command to display a specific "Text" Item on the Mobile Display at runtime whereas the programmer needs to specify:

- Its display "Position" within the 5x5 grid of "Touch Areas" described earlier.
- Its numerical "Index" as registered on the PLAY700 App "Text" utility.
- Its display font "Size" [0-255].
- Its display "Color" from 10 choices: Unknown, White, Black, Red, Green, Blue, Yellow, Light Gray, Gray and Dark Gray.

These options are combined into a **32-bit/4-byte** integer value using the protocol described in Section 7.3 and then assigned to the "SMART: Text Display" utility which runs on the Mobile Device (see Fig. 7.14).

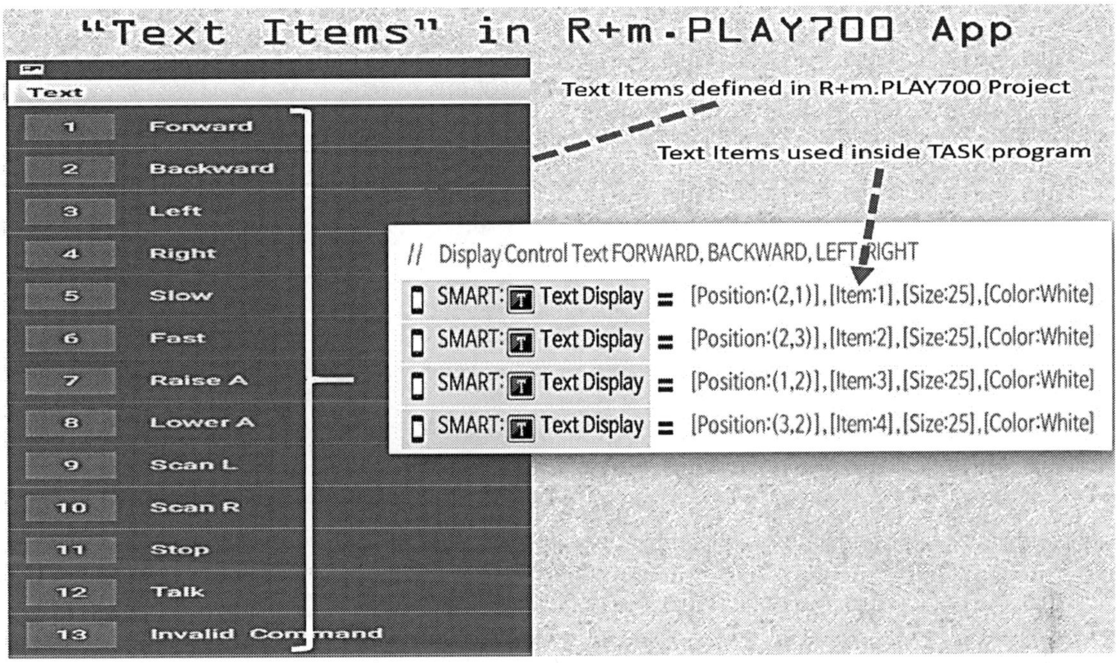

Fig. 7.13 "Text" Items used for Projects "M_TouchRC_Walker_IS_SD.tsk3" and "M_MixedRC_Walker_IS_SD.tsk3".

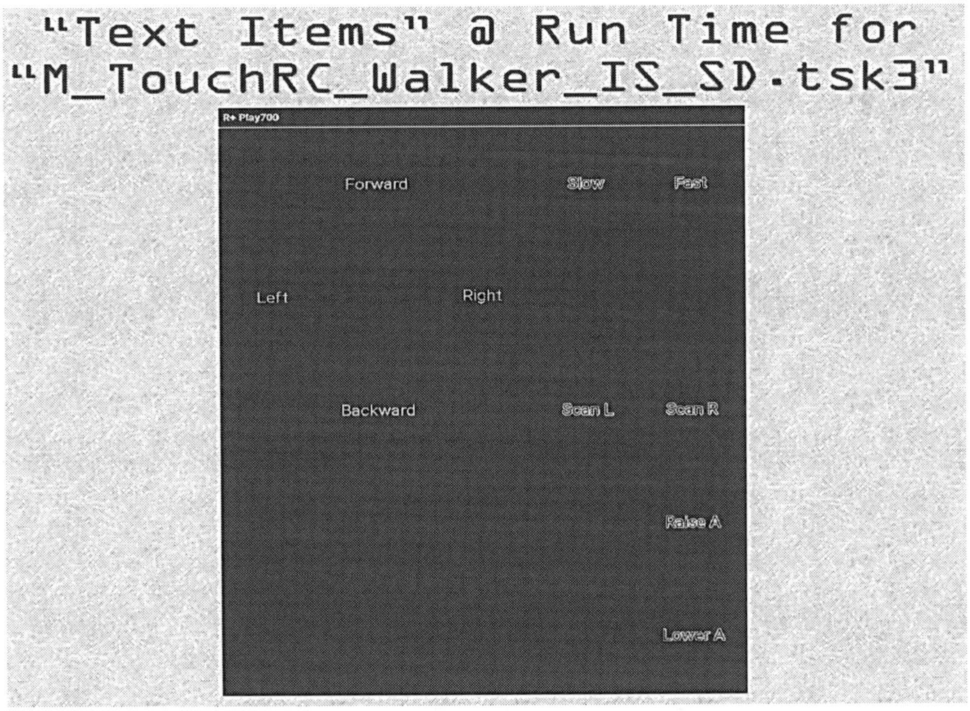

Fig. 7.14 "Text" Items displayed at run-time.

7.4.4 "Instruments Play" in R+m.PLAY700 App

The PLAY700 App also plays musical scales on 128 Musical Instruments (from "Acoustic Grand Piano" [1] to "Gunshot" [128]. For each instrument, there are 10 Octave settings [1-10] and 12 Musical Scales or Notes [1-12]: e.g. Note 1 is "Do", while Note 3 is "Re", and so forth until Note 12 which is "Shi". A **3-byte** SMART CONSTANT is associated with the Musical Instruments service where the lowest byte (Byte 1) contains the Note's value, while the next higher Byte (Byte 2) contains the Octave's value and Byte 3 contains the Instrument's value. The 4th byte of a typical SMART CONSTANT is not used for the Musical Instruments service and needs to be set to zero (see Fig. 7.15).

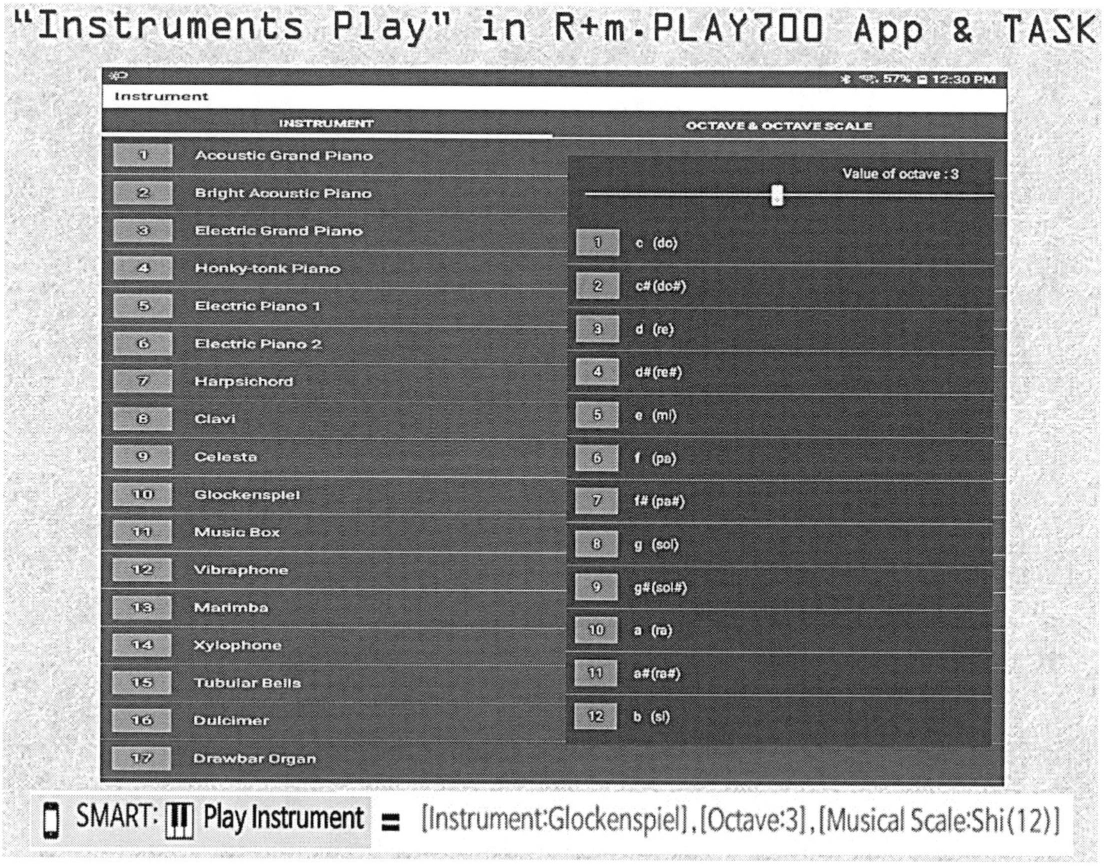

Fig. 7.15 Musical "Instruments" Service in R+m.PLAY700 App.

7.4.5 Construction of "M_TouchRC_Walker_IS_SD.tsk3"

This program "M_TouchRC_Walker_IS_SD.tsk3" uses the overall algorithm of the previous project "M_RC_Walker_IS.tsk3" which was described in details in Section 6.4.3. Thus, only relevant differences due to the use of SMART DEVICE functions will be emphasized in this section.

Fig. 7.16 describes the initialization steps done for servos in "M_TouchRC_Walker_IS_SD.tsk3" which are similar to the ones used in Fig. 6.34 for "M_RC_Walker_IS.tsk3", except for the setting of the Servo "Goal Velocity" Parameter:

- In Fig. 6.34, the "Goal Velocity" for All servos was set to a fixed constant value of "1023" (Line 139).
    - In Fig. 7.16, the "Goal Velocity" for All servos was set to a Variable named "Speed" (Line 150) which was initialized to "800" in the previous Line 149. However, this setting would be **ignored** while a MOTION-LIST/PAGE is performed, as the key-frame timing and playing speed, used respectively in its MOTION-UNITS and MOTION-LIST/PAGES, override this overall Goal Velocity setting. Once "HeadControl" or "ShoulderControl" is set to TRUE, i.e. when "Joint-Offset" is set to 1024 for Servos #1 or #17 in Functions "Set_Offset_1" or "Set_Offset_17", Parameter "Speed" can then be effective in controlling the "Goal Velocity" for Servos #1 and #17, **and only for those 2 servos** (see Fig. 7.20).

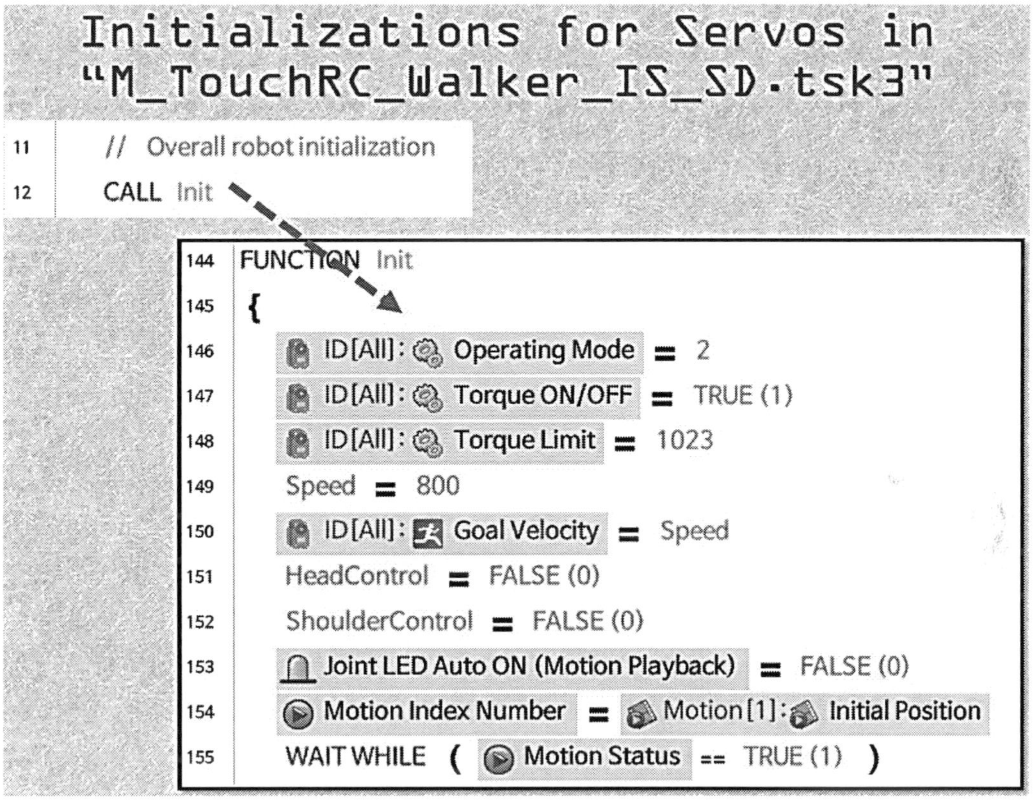

Fig. 7.16 Initializations for Servos in "M_TouchRC_Walker_IS_SD.tsk3".

Fig. 7.17 describes initialization steps done for the SMART DEVICE (i.e. R+m.PLAY700 App on the Mobile Device):

- Line 158 sets the Mobile Display in Portrait Mode (i.e. a SMART Write command).
- Line 159 clears the "Text" Display Screen by sending a "0" to the PLAY700 App (another SMART Write command).
- Line 161 makes the MINI Hardware Controller **wait until** a "SMART Noise" Parameter coming from the Mobile Device and with a dB value larger than 50 (usually the author whistles

loudly into the Mobile Device to achieve this result). **Line 161 is critical to a smooth run-time coordination between the TASK program running on the MINI and the corresponding PLAY700 App running on the Mobile Device**. The reader may remember from Section 1.3.4 that as soon as power is applied to the MINI, its embedded TASK program will immediately run. On the other hand, the human operator will always take some time to start the appropriate Custom Project on the PLAY700 App and the Mobile Device will always take several seconds to connect the Mobile Device Bluetooth port to the BT-210 running the MINI. Essentially Line 161 allows the PLAY700 App to catch up with the much faster TASK program. Without Line 161, all the SMART Text Display commands (Lines 164-176) would be sent over to the Mobile Device when it is not ready to receive them, and therefore would be "lost", i.e. the Mobile Device won't display the "Menu" as shown in Fig. 7.14. **In practice, it is best to make sure that the appropriate PLAY700 Project is activated before turning on the MINI to start the TASK program.** *A side note for the interested reader: on newer ROBOTIS controllers such as the CM-550, there is a "Check Smart-Device Connection" TASK command that can be used to check for the actual BT connection between the Mobile Device and the MINI. This command can be used before issuing such Lines as 158-159 for best synchronization between these two devices.*

Initializations for Smart Device in "M_TouchRC_Walker_IS_SD.tsk3"

```
157  // Getting remote device ready using SMART commands
158  SMART: Screen Rotation = Portrait Mode (1)
159  SMART: Text Display = 0
160  // User needs to make sure that robot is connected to mobile device. User whistles loudy when ready.
161  WAIT WHILE ( SMART: Noise (dB) < 50 )
162
163  // Display Control Text FORWARD, BACKWARD, LEFT, RIGHT
164  SMART: Text Display = [Position:(2,1)],[Item:1],[Size:25],[Color:White]
165  SMART: Text Display = [Position:(2,3)],[Item:2],[Size:25],[Color:White]
166  SMART: Text Display = [Position:(1,2)],[Item:3],[Size:25],[Color:White]
167  SMART: Text Display = [Position:(3,2)],[Item:4],[Size:25],[Color:White]
168  // Display Speed Control Text SLOW or FAST
169  SMART: Text Display = [Position:(4,1)],[Item:5],[Size:25],[Color:Red]
170  SMART: Text Display = [Position:(5,1)],[Item:6],[Size:25],[Color:Red]
171  // Display Shoulder Control Text RAISE A or LOWER A
172  SMART: Text Display = [Position:(5,4)],[Item:7],[Size:25],[Color:Yellow]
173  SMART: Text Display = [Position:(5,5)],[Item:8],[Size:25],[Color:Yellow]
174  // Display Head Control Text SCAN L or SCAN R
175  SMART: Text Display = [Position:(4,3)],[Item:9],[Size:25],[Color:Blue]
176  SMART: Text Display = [Position:(5,3)],[Item:10],[Size:25],[Color:Blue]
177  }
```

Fig. 7.17 Initializations for SMART DEVICE in "M_TouchRC_Walker_IS_SD.tsk3".

- Lines 164-176 produce the "Menu" as shown in Fig. 7.14. *A sharp reader may have noticed that certain Text Items were NOT "colored" correctly: "Slow", "Fast", "Raise A", "Lower A", "Scan L" and "Scan R". This was just because the author's Samsung Tab A 8" was not 100% compatible with the PLAY700 App at the time of writing of this book!*

Fig. 7.18 shows Part 1 of the Main Endless Loop whereas changes had to be made to accommodate the use of Touch Areas:

- Back in Fig. 6.35, Lines 15 and 17 were used to handle Button inputs from the RC-100, now Lines 16 and 17 are used to handle Touch Area inputs from the Mobile Device. Lines 16-17 record the First and Second Touch Area pressed by the user at run-time.
- The IF condition used in Line 18 makes sure that NO Touch Area had been activated before the MINI's Controller go on to Line 20 to check on whether Parameters "HeadControl" or "ShoulderControl" had been set to TRUE.
- Lines 19-26 in Fig. 7.18 correspond to Lines 19-26 of Fig. 6.35, regarding how to reset the OFFSETs for Servos 1 and 17, i.e. when no finger touches the mobile screen and when either or both Parameters "HeadControl" and "ShoulderControl" were previously set to TRUE.

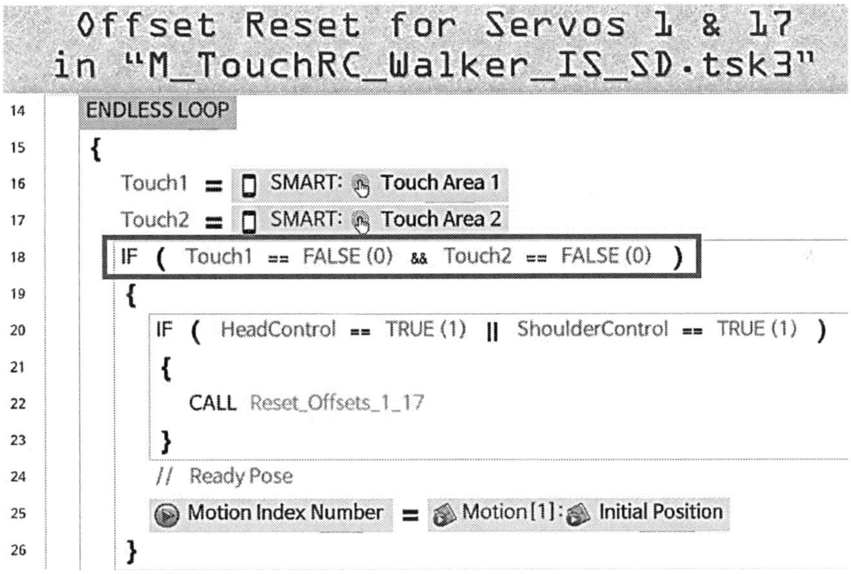

Fig. 7.18 Part 1 of Main Endless Loop in "M_TouchRC_Walker_IS_SD.tsk3".

Next, an IF-ELSE-IF structure is used to activate the appropriate robot Motions depending on the value recorded in Parameter Touch1 (see Fig. 7.19):

- Please note that as "**Touch1**" is the only parameter used in the various IF conditions used in this IF-ELSE-IF structure, this means that at run-time the operator is restricted to use his/her **FIRST** finger to control the Moving Directions for the robot in this project. In other words, pressing on the "Moving Direction" areas using his/her SECOND finger will not trigger any robot motion.

- The various SMART CONSTANTS (such as [Position:(2,1)]) correspond to the respective "zones" used to display "Text" Items (Forward, Backward, Left, Right) – see Fig. 7.17 and Lines 164-167.

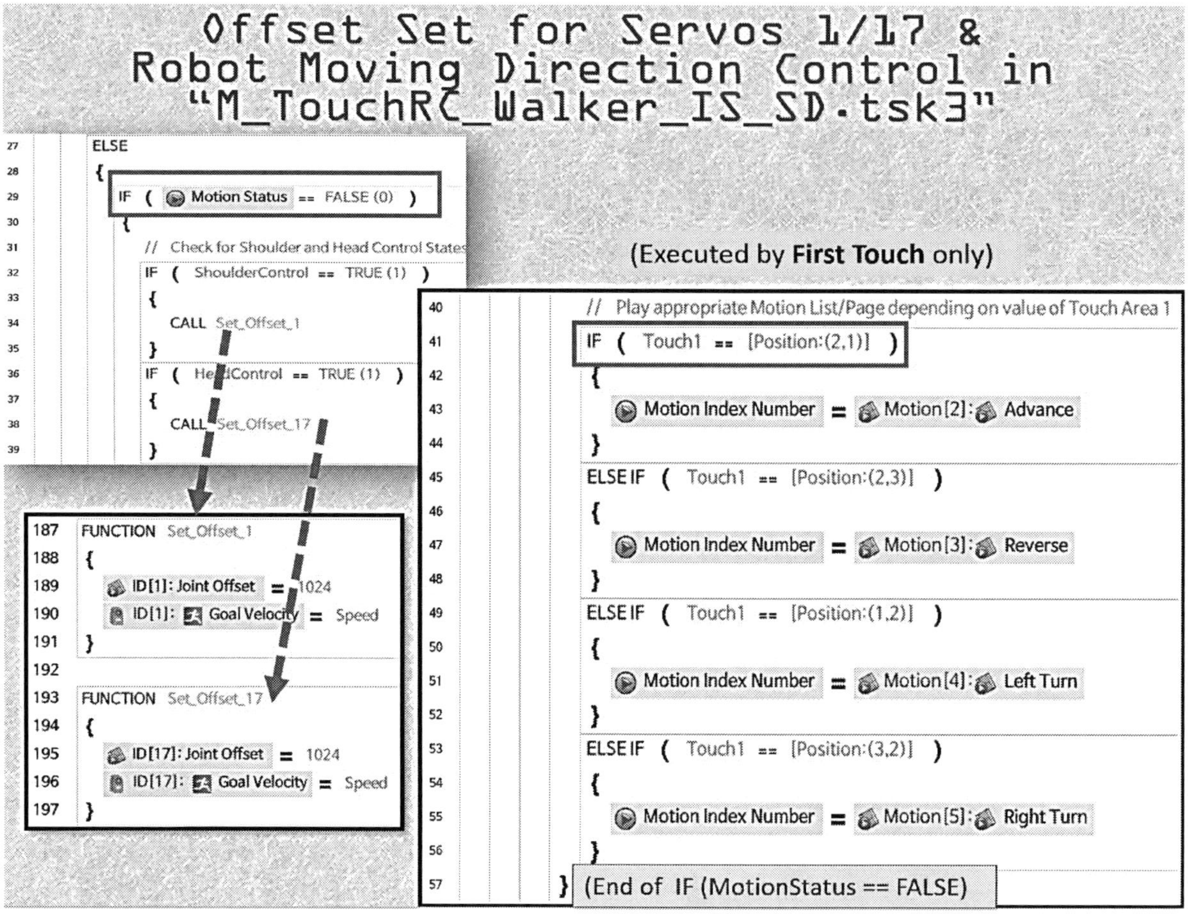

Fig. 7.19 Control of Robot Moving Directions in "M_TouchRC_Walker_IS_SD.tsk3".

However, a different scheme is used regarding the Touch Areas corresponding to the "Text" Items "Slow" (Zone "4", i.e. Position (4.1)) and "Fast" (Zone "5", i.e. Position (5,1)) which are used to decrement and increment the "Goal Velocity" Parameters for Servos #1 and #17 (see Fig. 7.20 and bottom left of Fig. 7.19):

- First, please note that ("Touch1" OR "Touch2") are implemented in the IF conditions used in Lines 59 and 63. This means that at run-time the operator can press on "Slow" or "Fast" using the FIRST finger (i.e. a "1-finger" press operation), or that the operator can press on a "Moving Direction" area with the FIRST finger and then presses on "Slow" or "Fast" with the SECOND finger. In everyday language, this means that the operator can use the "Slow" or "Fast" commands by themselves or in conjunction with a preceding "Moving Direction" command for the robot (i.e. while the robot is in motion).

- Second, a closer look at the Function "Increase_Speed" shown in Fig. 7.20 would reveal that a "new" user-interface programming technique is used for "Slow" and "Fast" as compared to the "Moving Direction" commands. Previously as implemented in Fig. 7.19, for as long as the user presses on the "Forward" Touch Area, the robot will move forward, and it will cease its forward motion when the operator lifts his/her finger off the "Forward" Touch Area. Fig. 7.20 shows that the "START" for activating Function "Increase_Speed" follows the same procedure (i.e. to press on Zone 5 or "Fast" to activate this Function – see Line 63).

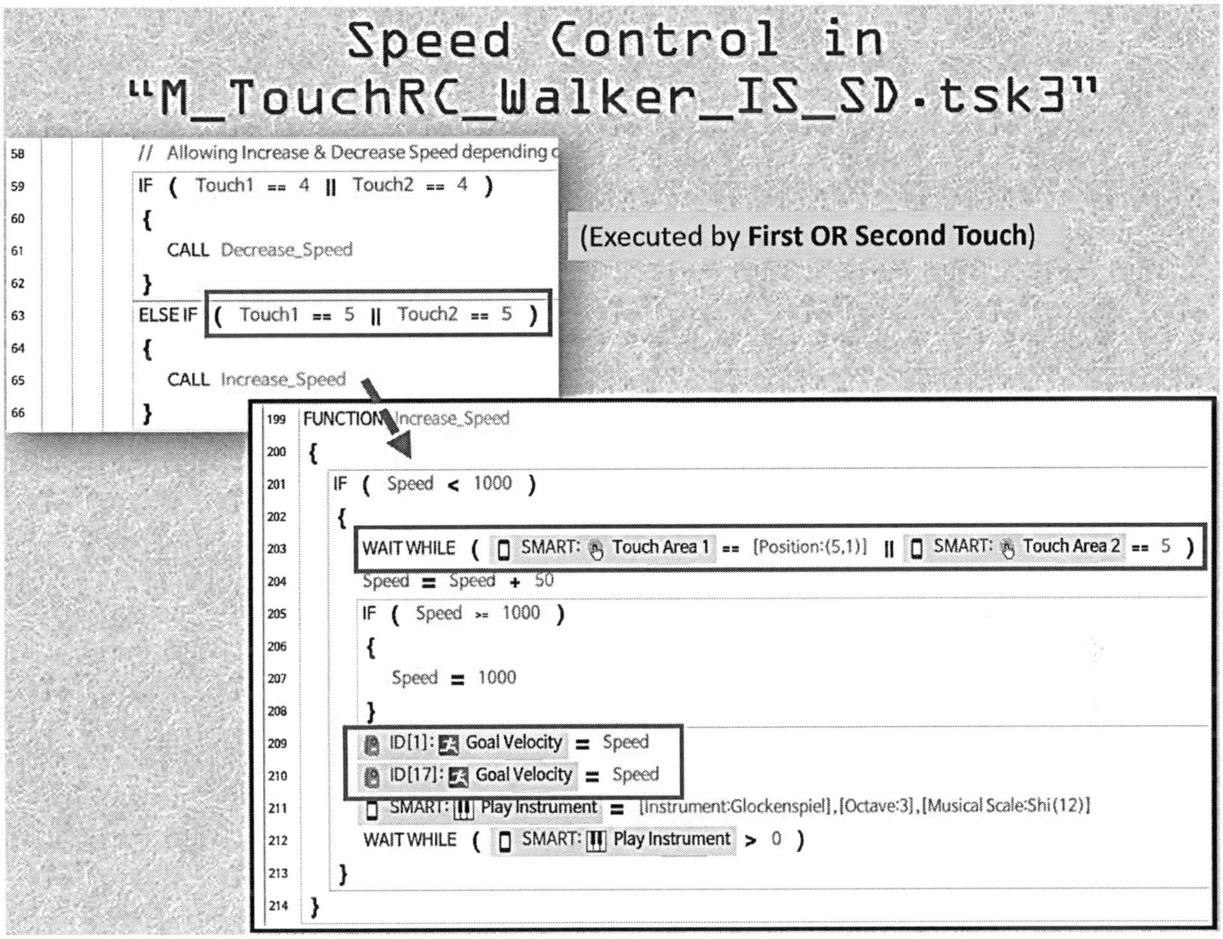

Fig. 7.20 Control of Servos' "Goal Velocity" in "M_TouchRC_Walker_IS_SD.tsk3".

- Next, Function "Increase_Speed" is "entered" and the IF condition in Line 201 shows that if "Speed" is greater or equal to 1000, this Function is immediately exited, meaning that this project observes an upper limit of 1000 for "Goal Velocity" although it can be set all the way to 1023 in previous projects.
- If the IF condition in Line 201 evaluates to TRUE (i.e. Speed < 1000), the MINI Controller executes Line 203 next, which tells it to WAIT as long as the LATEST input data for "Touch Area 1" or "Touch Area 2" show that the operator is still pressing on "Fast" (i.e. Position (5,1)

or Zone 5). Please make sure to distinguish between the use of Parameters "Touch1" and "Touch2" and commands "SMART: Touch Area 1" and "SMART: Touch Area 2":

    - Parameters "**Touch1**" and "**Touch2**" correspond to information about "Touch Area 1" and "Touch Area 2" as they were **recorded back on Lines 16 and 17**.
    - Commands "**SMART: Touch Area 1**" and "**SMART: Touch Area 2**" in Line 203 return the **latest information** from the Mobile Device regarding the FIRST and SECOND finger presses performed by the operator.
    - If the author had used Parameters "Touch1" and "Touch2" in this WAIT WHILE LOOP, a truly "FOREVER" loop would occur at run-time, and the robot would "seemingly" no longer correspond to any operator's command afterwards.

- Lastly, for the MINI's Controller to exit this WAIT WHILE LOOP, the operator would need to **lift his/her finger off Zone 5** (i.e. to make the WAIT WHILE LOOP's condition to be set to FALSE). This means that the operator can only increment "Goal Velocity" if and only if the operator "**presses then releases**" Zone 5 (i.e. "Text" Item "Fast"). This technique is used to guarantee that "Goal Velocity" is changing by an increment of "50" (Line 204) every time that Function "Increase_Speed" is called.
- Lines 205 to 208 make sure to clamp "Speed" to its upper limit of "1000".
- Lines 209-210 set Parameters "Goal Velocity" for Servos #1 and #17 to this new "Speed" value.
- Lines 211-212 play the "Glockenspiel" with "Shi" (Note 12). This is to provide the operator with an audio feedback to know that the "Increment Speed" operation is successful. When this Function "Increment_Speed" Function is called and "Speed" is already set to "1000", this Function is immediately exited at Line 214, so no "Shi" note would be heard by the operator.

Although not shown in Fig. 7.20, Function "Decrease_Speed" is similar in implementation but Zone 4 and a lower limit of "600" are used, along with a "Do" note.

Figs. 7.21 and 7.22 show that only "Touch2" is used for "Text" Items "Scan L" (Zone 14) and "Scan R" (Zone 15), meaning that they are also meant to be used as the "SECOND FINGER" press for the operator, i.e. in conjunction with a preceding "FIRST FINGER" press on one of the "Moving Direction" commands. In other words, "Scan L" and "Scan R" commands won't work by themselves, as previously implemented in "M_RC_Walker_IS.tsk3".

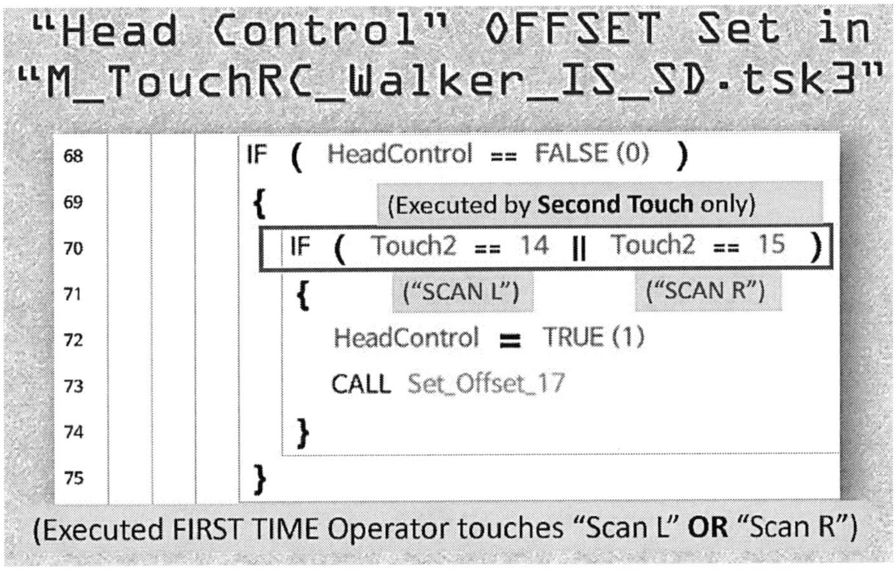

Fig. 7.21 Offset Setting for Servo 17 if "Head Control" is activated in "M_TouchRC_Walker_IS_SD.tsk3".

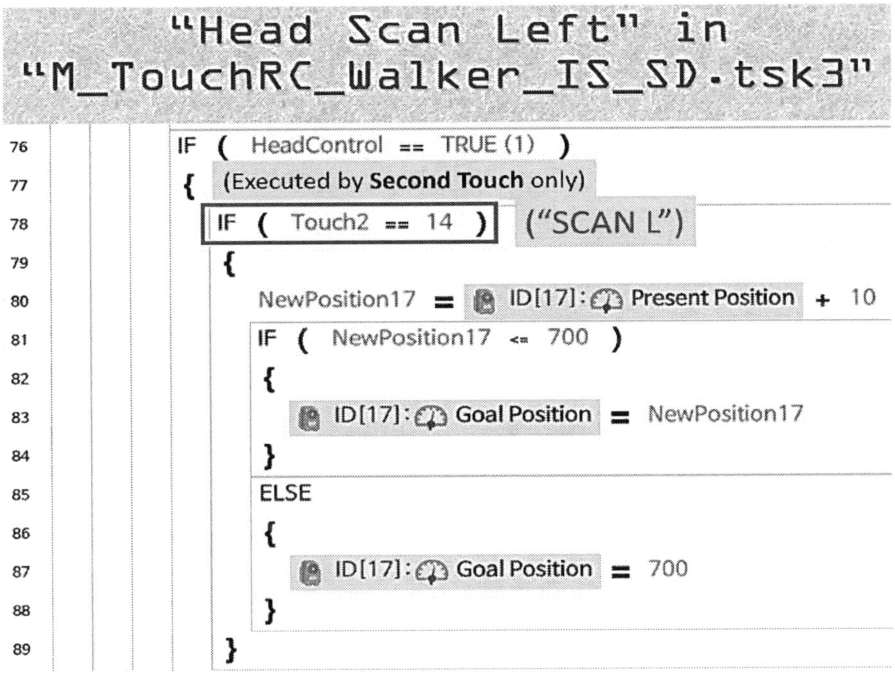

Fig. 7.22 "Head Scan Left" procedure in "M_TouchRC_Walker_IS_SD.tsk3".

Similarly, Fig. 7.23 shows that "Arm Raise" operations "Raise A" and "Lower A" (i.e. independent control of Servo 1) follows the same implementation approach as for "Scan L" and "Scan R" for Servo 17.

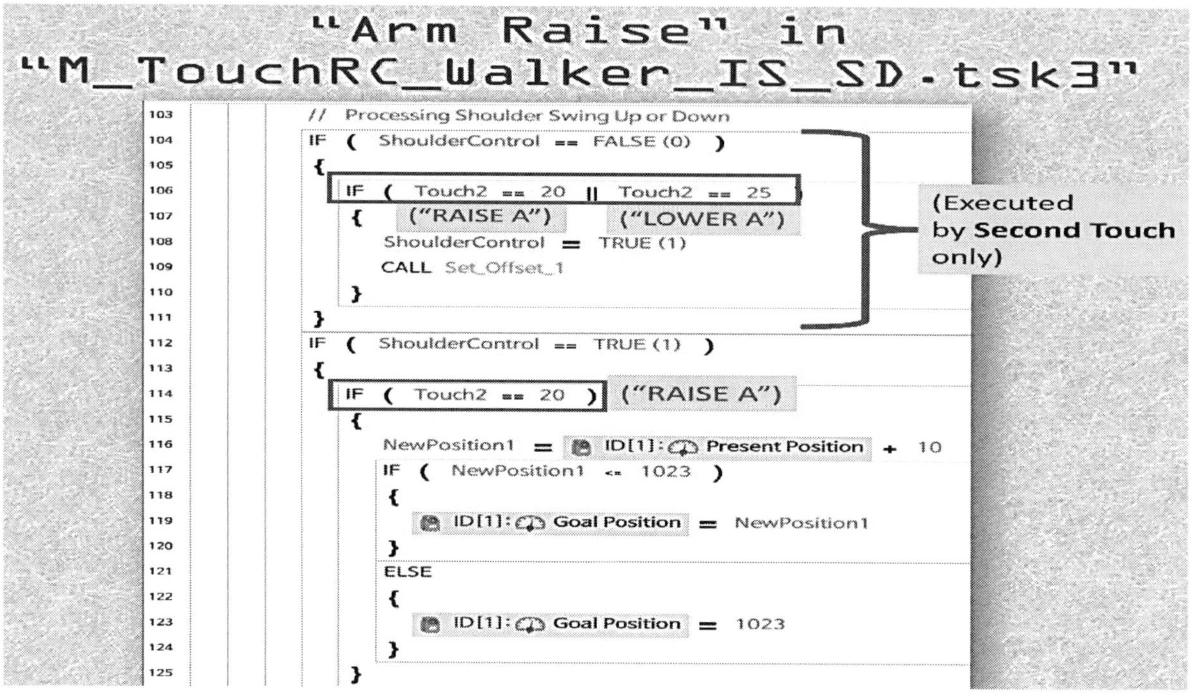

Fig. 7.23 "Arm Raise" procedure in "M_TouchRC_Walker_IS_SD.tsk3".

Video 7.2 shows the run-time performance of this project "M_TouchRC_Walker_IS_SD.tsk3".

7.5 "Mixed Control Walker" Project

For this "Mixed Control" project, the primary goal is to have two ways of controlling the Moving Directions of the robot and the Servos Goal Velocity setting using "**Touch Areas**" and "**Speech Recognition**". Secondary goals are to show the usage of the "**Text-to-Speech**" utility and the implementation of a Touch-based **Emergency Stop**. This project also uses the "RC Walk 1" MOTION GROUP from the author's file "MINI_Custom.mtn3".

7.5.1 "Speech Recognition" and "Text-to-Speech" in R+m.PLAY700 App

The "Speech Recognition" Function of the PLAY700 App relies on the web-based Google Speech Engine and it is a lengthy process. Thus, it is designed as a "FIRST FINGER" function.

Both "Speech Recognition" and "Text-to-Speech" functions rely on the same list of "Text" Items used for the "Display" function (see Figs. 7.24 and 7.25).

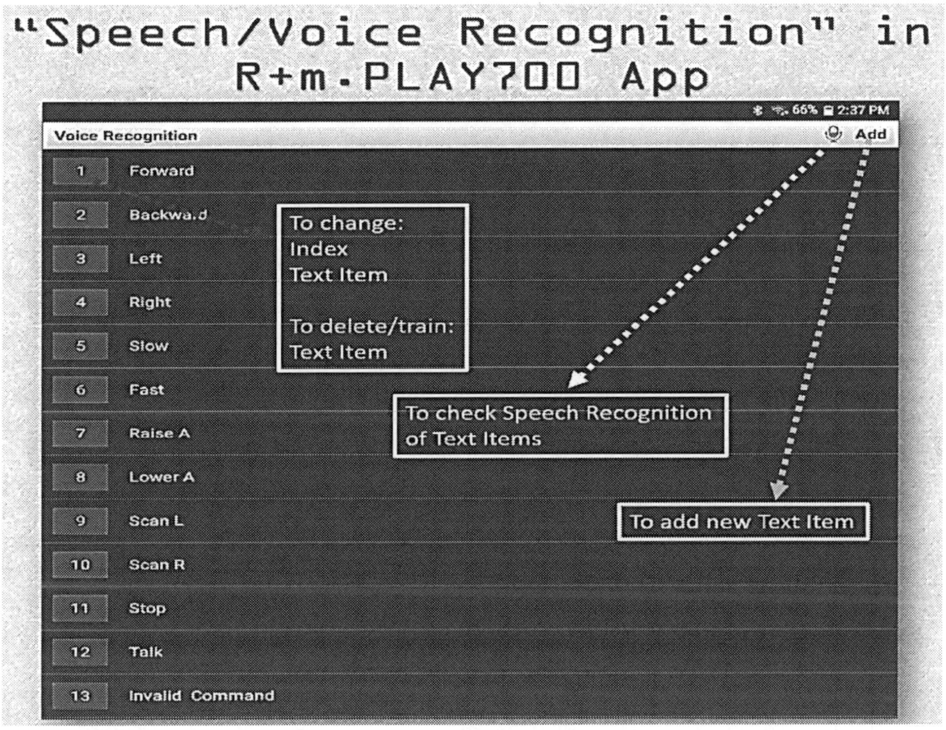

Fig. 7.24 "Speech/Voice Recognition" Utility in R+m.PLAY700 App.

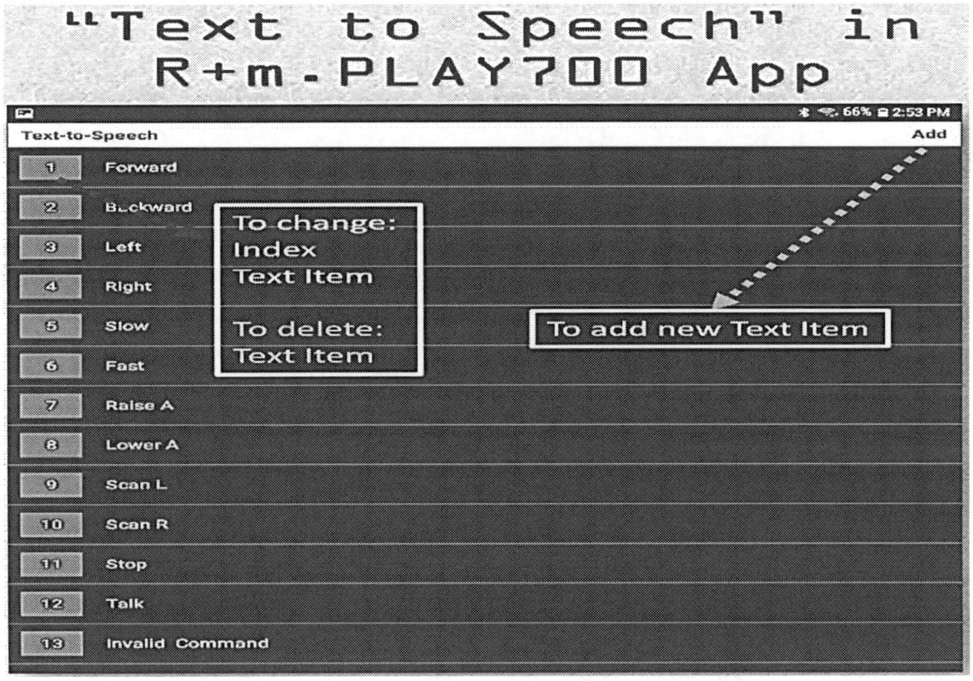

Fig. 7.25 "Text-to-Speech" Utility in R+m.PLAY700 App.

Fig. 7.24 shows that the user can edit the Item Index Number, change/delete the Text Item itself, and most importantly "test" and "train" a selected Text Item on one's personal speech patterns. Please see Part 1 of Video 7.3 for illustrations of such uses.

Fig. 7.25 shows that the user can similarly edit the Item Index Number, change/delete the Text Item itself. If the user taps on a selected "Text" Item, the user would hear the Item spoken with the default voice on the Mobile Device.

Of course, editorial changes in "Text" Items in **any utility** "Display", "Speech Recognition" or "Text-to-Speech" will affect **all three**.

7.5.2 Construction of "M_MixedRC_Walker_IS_SD.tsk3"

Fig. 7.26 shows the Function "Init" for the project "M_MixedRC_Walker_IS_SD.tsk3" which has 3 new statements:

- Line 196 initializes Parameter "InvalidCommand" to FALSE. This parameter is used to report on the success/failure of the Speech Recognition process.
- Line 197 initializes Parameter "VoiceCommand" to "Text" Item 11 which is "Stop" (see Fig. 7.24).

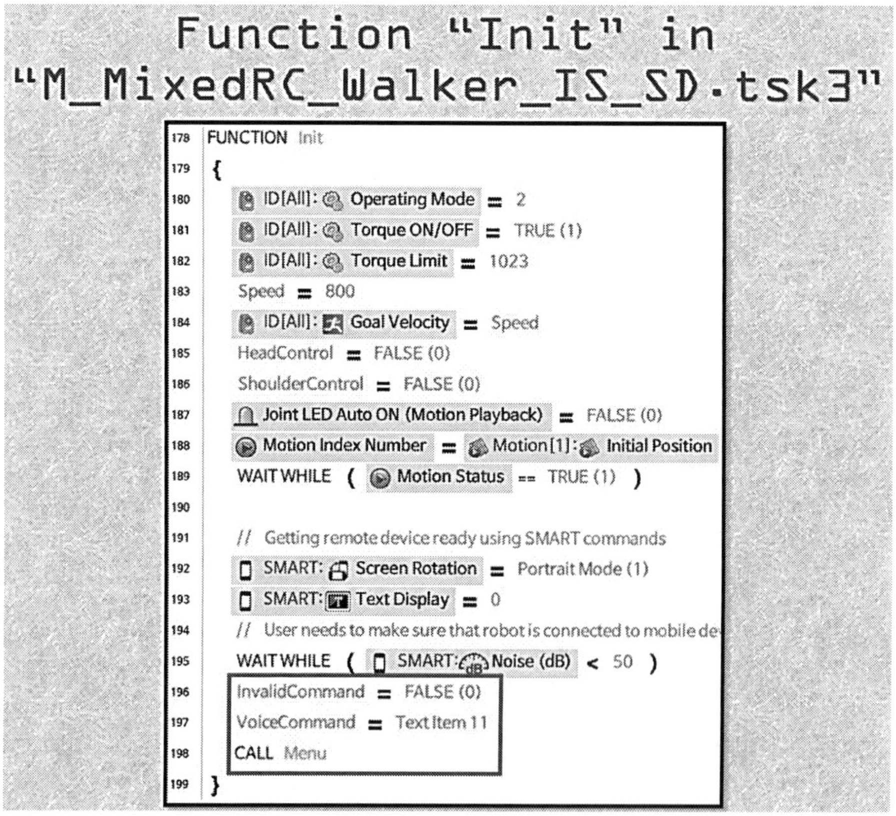

Fig. 7.26 Function "Init" in "M_MixedRC_Walker_IS_SD.tsk3".

- Line 198 calls the Function "Menu" which is listed in Fig. 7.27, showing the Display of two new "Text" Items "Stop" and "Talk" (Line 268 and 269 respectively).

Function "Menu" in "M_MixedRC_Walker_IS_SD.tsk3"

```
251 FUNCTION Menu
252 {
253     // Display Control Text FORWARD, BACKWARD, LEFT, RIGHT
254     SMART: Text Display = [Position:(2,1)],[Item:1],[Size:25],[Color:White]
255     SMART: Text Display = [Position:(2,3)],[Item:2],[Size:25],[Color:White]
256     SMART: Text Display = [Position:(1,2)],[Item:3],[Size:25],[Color:White]
257     SMART: Text Display = [Position:(3,2)],[Item:4],[Size:25],[Color:White]
258     // Display Speed Control Text SLOW or FAST
259     SMART: Text Display = [Position:(4,1)],[Item:5],[Size:25],[Color:Red]
260     SMART: Text Display = [Position:(5,1)],[Item:6],[Size:25],[Color:Red]
261     // Display Shoulder Control Text RAISE A or LOWER A
262     SMART: Text Display = [Position:(5,4)],[Item:7],[Size:25],[Color:Yellow]
263     SMART: Text Display = [Position:(5,5)],[Item:8],[Size:25],[Color:Yellow]
264     // Display Head Control Text SCAN L or SCAN R
265     SMART: Text Display = [Position:(4,3)],[Item:9],[Size:25],[Color:Blue]
266     SMART: Text Display = [Position:(5,3)],[Item:10],[Size:25],[Color:Blue]
267     // Display Stop & Talk
268     SMART: Text Display = [Position:(2,2)],[Item:11],[Size:25],[Color:Green]
269     SMART: Text Display = [Position:(3,5)],[Item:12],[Size:25],[Color:Green]
270 }
```

Fig. 7.27 Function "Menu" in "M_MixedRC_Walker_IS_SD.tsk3".

Fig. 7.28 shows that "Text-to-Speech" is used to "accompany" the MOTION PLAY of each Moving Direction performed by the robot when called upon.

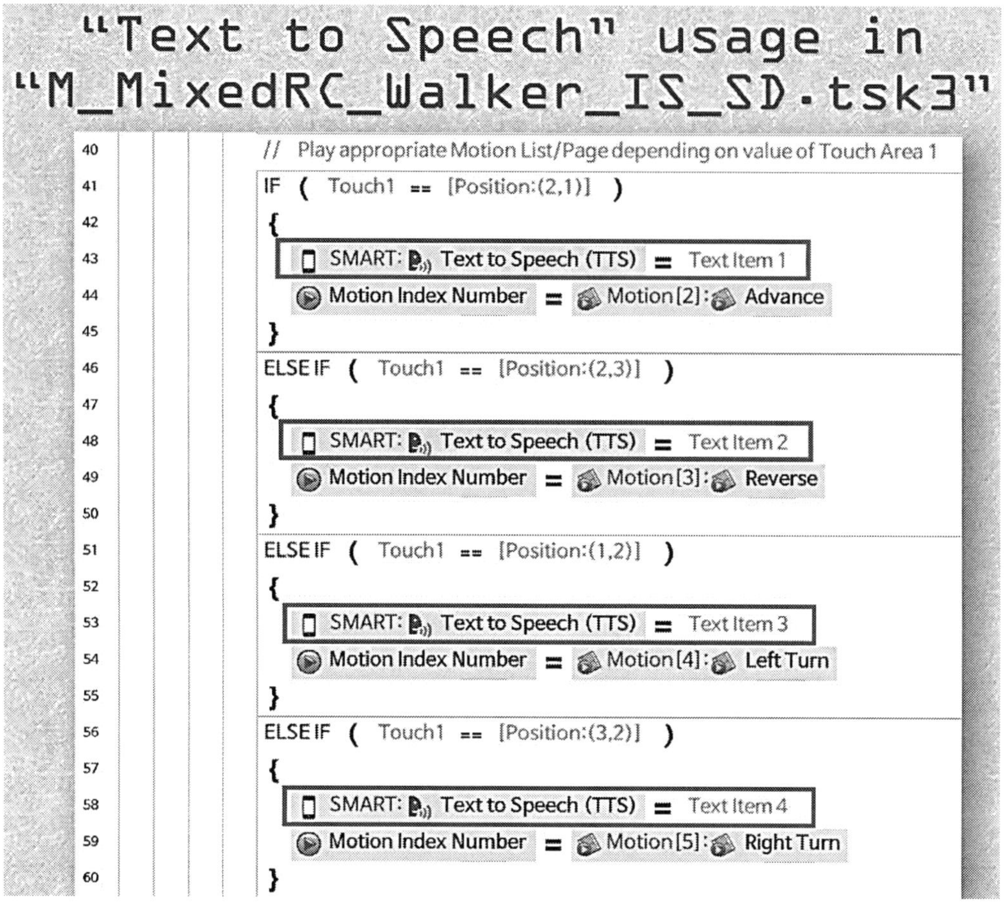

Fig. 7.28 "Text-to-Speech" usage to accompany MOTION PLAY in "M_MixedRC_Walker_IS_SD.tsk3".

The "Speech Recognition" utility (i.e. "Text" Item "Talk") is implemented as a "FIRST TOUCH" function, thus it is added as a new "ELSE-IF" block to the existing "IF-ELSE-IF" structure used for handling the Moving Directions of the robot (as previously shown in Fig. 7.19).

Fig. 7.29 shows the implementation details of "Speech Recognition" in the project "M_MixedRC_Walker_IS_SD.tsk3":

- Line 64 shows the usage of a WAIT WHILE LOOP based on "SMART: Touch Area 1", meaning that a "Press and Release" scheme has been implemented for "Talk" (same approach as used earlier for "Slow" and "Fast" in Section 7.3.5).
- The "Speech Recognition" process is a multi-step procedure illustrated by Lines 66, 67 and 69:
    - Line 66 starts the "Speech Recognition" process (i.e. send a "1" to PLAY700 App). At this point in time, the user would see the "Speak to Me" prompt appear on the Mobile Device. The user then would need to voice out (*as soon as possible and in the same tone of voice and speed as when doing speech-training*) the wanted verbal command among the following "Text" Items: Forward, Backward, Left, Right, Slow and Fast. This process has a built-in time-out period of about 5 seconds.

"Speech Recognition" usage in "M_MixedRC_Walker_IS_SD.tsk3"

```
61          // Play appropriate Motion List/Page depending on result of Speech Recognition
62          ELSE IF ( Touch1 == [Position:(3,5)] )  (Executed by First Touch only)
63          {
64              WAIT WHILE ( SMART: Touch Area 1 == [Position:(3,5)] )
65              // Start Speech Recognition Process
66              SMART: Speech Recognition = Start (1)
67              // Wait for SR to complete
68              WAIT WHILE ( SMART: Speech Recognition != 0 )
69              VoiceCommand = SMART: Result of Speech Recognition
70              IF ( VoiceCommand != -1 ) <==
71              {
72                  CALL ProcessCommand
73                  SMART: Result of Speech Recognition = 0
74              }
75              IF ( InvalidCommand == TRUE (1) )
76              {
77                  InvalidCommand = FALSE (0)
78                  VoiceCommand = Text Item 11
79              }
80          }
81      }
```

Fig. 7.29 "Speech Recognition" implementation in "M_MixedRC_Walker_IS_SD.tsk3".

- ○ The "Speech Recognition" tool would then pick up this verbal input and try to match it up with its web database for the actual text as previous created in the PLAY700 "Text" Item list. This process is quite lengthy thus Line 68 is used to wait for its termination. The "SMART: Speech Recognition" command is a READ instruction sent from the MINI to the Mobile Device. When the "Speech Recognition" process is concluded on the Mobile Device, a value of "0" is returned by the Mobile Device to the MINI Controller.
- ○ It is then OK to save the "SMART: Result of Speech Recognition" value (returned from PLAY700 App) into Parameter "VoiceCommand" (Line 69).

- If Parameter "VoiceCommand" is **valid**, i.e. **not equal to -1**, Function "ProcessCommand" is called (see details in Figs. 7.30 and 7.31) and the Parameter "Result of Speech Recognition" is cleared to 0 (Line 73) upon return from Function "ProcessCommand".
- Lines 75 to 79 handle the situation when an "InvalidCommand" is concluded while running Function "ProcessCommand". When this happens, Parameter "InvalidCommand" is reset to FALSE and "VoiceCommand" is reset to "Text" Item 11 (i.e. "Stop"). At this point, the MINI Controller exits the big IF-ELSE-IF structure handling the "Moving Directions" within the Main Endless Loop.

Let's now have a closer look at Fig. 7.30:

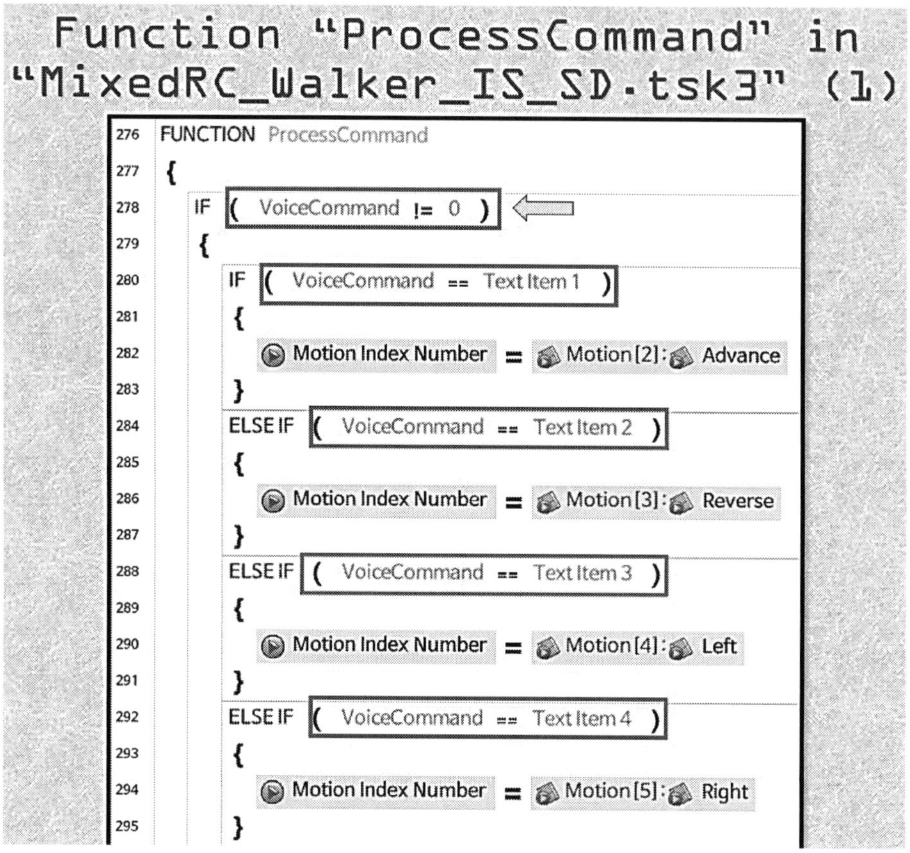

Fig. 7.30 Part 1 of Function "ProcessCommand" in "M_MixedRC_Walker_IS_SD.tsk3".

- From Line 70 in Fig. 7.29, we know that "VoiceCommand" is not equal to "-1" when Function "ProcessCommand" is entered. Line 278 in Fig. 7.30 further checks to see if "VoiceCommand" is not equal to "0". If this condition is FALSE, meaning that (VoiceCommand == 0), the MINI Controller is sent to Line 305 in Fig. 7.31 (basically we have an "Invalid Command" situation to deal with). However, if (VoiceCommand != 0), meaning that the "Speech Recognition" process concluded that the operator's verbal input was found to match with **one** "Text" Item among the list shown in Fig. 7.24, then the MINI Controller would tackle next the IF-ELSE-IF structure defined from Line 280 of Fig. 7.30 to Line 304 of Fig. 7.31.
- This IF-ELSE-IF structure is designed to MOTION PLAY the appropriate MOTION-LIST/PAGE that corresponds to the Moving Direction wanted by the operator ("Text" Item 1-4), or to decrease/increase Servos Goal Velocity if those corresponding "VoiceCommand" values were detected ("Text" Item 5 or 6). *Here, a sharp reader would wonder what would happen if he/she verbalized "Stop", then this IF-ELSE-IF structure would do "nothing" as the author did not program it to handle that situation!*

Let's next have a closer look at the situation when (VoiceCommand == 0), i.e. Line 305 in Fig. 7.31. This means that the "Speech Recognition" process had found no matching "Text" items from the list created by the user in the PLAY700 App:

- First, the MINI Controller puts the robot into its Initial Position (Line 307).
- Second, Parameter "InvalidCommand" is set to TRUE (Line 308).
- Then a group of SMART commands are executed in a specific order:
    - The Display Screen is cleared (Line 309).
    - "Text" Item 13 (i.e. "Invalid Command") is "spoken" (Line 310).
    - "Invalid Command" is also displayed at Position (3,3) (Line 311).
    - Line 312 is used to make sure that "Text" Item 13 is completely voiced out.
- Lines 313-314 make the MINI Controller "do nothing" for 0.5 second for the operator to take stock of the "Invalid Command" situation.
- Line 315 clears the Display Screen and Line 316 "re-displays" the original Menu.

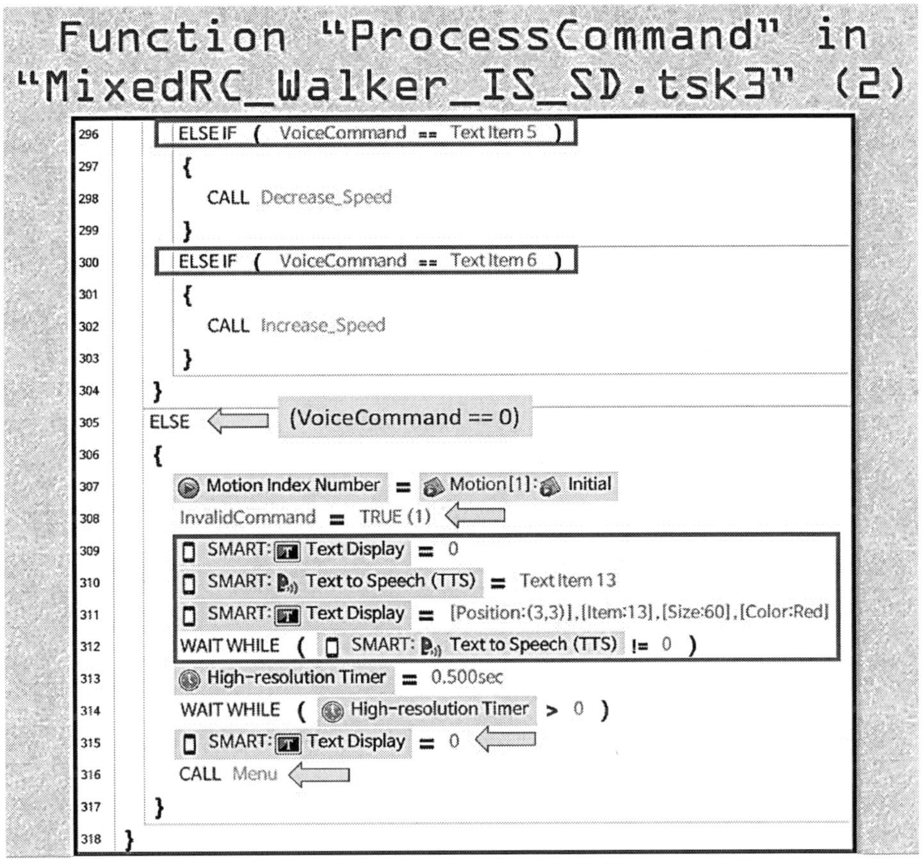

Fig. 7.31 Part 2 of Function "ProcessCommand" in "M_MixedRC_Walker_IS_SD.tsk3".

In this project, an "Emergency Stop" is programmed as a FIRST or SECOND TOUCH option for the operator (see Line 83 in Fig. 7.32):

- Lines 85-90 describe the steps executed next:

- Line 85 issues a MOTION (-1) command to **immediately stop** the robot current MOTION-LIST/PAGE from being played and have the robot perform the corresponding EXIT MOTION UNIT.
- Lines 86-90 use a LOOP WHILE to keep on playing "Shi" note on the "Instrument" FX 8, as long as the operator keeps pressing on the Display Screen with ONE or TWO fingers. The music will stop when all fingers are off the Mobile Screen.

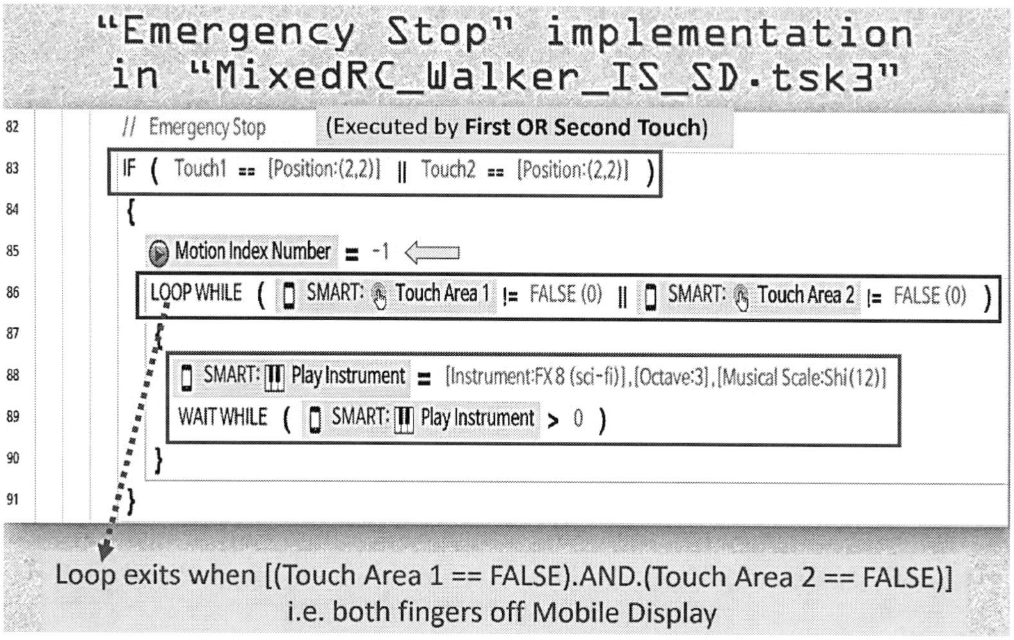

Fig. 7.32 "Emergency Stop" Implementation in "M_MixedRC_Walker_IS_SD.tsk3".

Video 7.3 shows the run-time performance of this project "M_MixedRC_Walker_IS_SD.tsk3".

7.6 "Camera Tracker" Project

The reader may remember from Section 4.4 that the author had no success in mounting a mobile phone in front of the MINI robot and maintaining any stable robot motion because of the location of the extra weight of the Mobile Phone. The next option is of course is to mount the Mobile Phone on the backside of the MINI (see Fig. 7.33) for this "Camera Tracker" project.

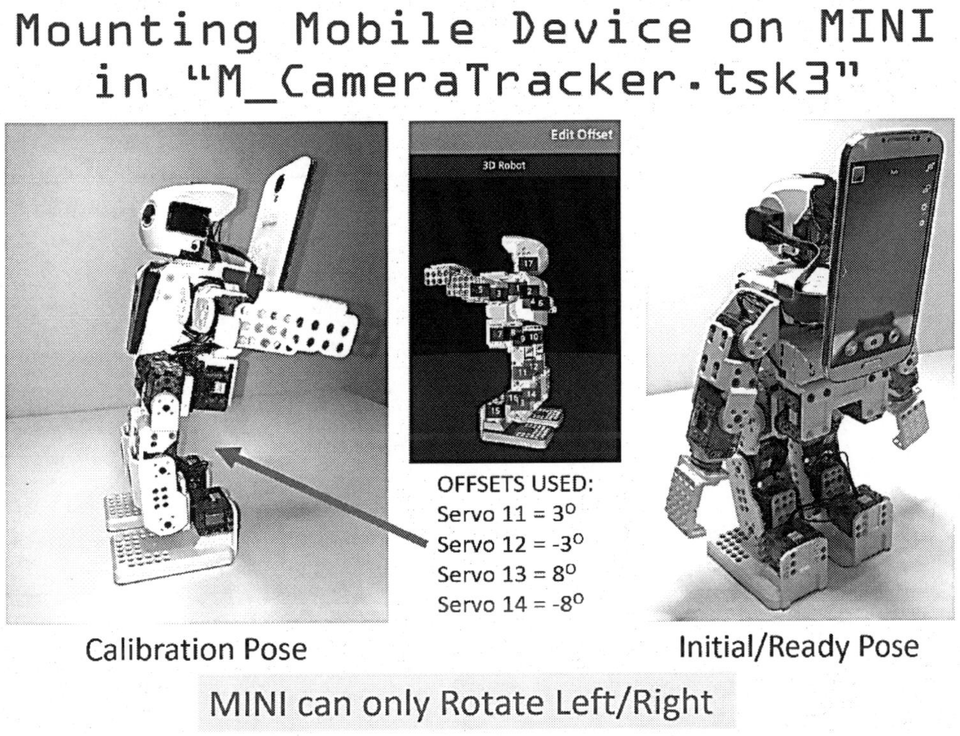

Fig. 7.33 "Back Mount" of Mobile Phone on Author's MINI.

Fig. 7.33 also shows the JOINT-OFFSETs to be applied to Servos 11, 12, 13 and 14 (essentially the robot's knee and ankle joints) to keep the overall Center of Gravity within the MINI's ground contact area for the Calibration Pose as well as for the Initial/Ready Pose.

With this mounting option, the author could make the robot turn Left and Right in a stable manner (however, still no stability obtained for walking Forward or Backward).

Besides taking/recording pictures and videos via the built-in cameras on the Mobile Device, the R+m.PLAY700 App also provides several rudimentary Image Processing modes (see Fig. 7.34):

- In the "Face Detection" mode, the SMART "Face Detection Area" command would return a number between 1 and 25 to the TASK program, representing the Zone Number where the "face" was detected.
- In the "Color Detection" mode, only pixels within Zone 13 (i.e. Screen Center or Position [3,3]) are evaluated for their color values to see if they fit, on the average, certain "fixed" internal numerical RGB criteria for 4 possible color values: 2 for Black, 3 for Red, 4 for Green and 5 for Blue.
- In the "Motion Detection" mode, the SMART "Motion Detection Area" command would return a number between 1 and 25 to the TASK program representing the Zone Number where "Motion" was detected.
- In the "Line Detection" mode, the user first needs to specify which color the App should be tracking for: 2 for Black, 3 for Red, 4 for Green and 5 for Blue. Furthermore, only the top 5 Screen Zones, i.e. Zone 1 through 5, are scanned by the App, thus the SMART "Line Detection Area" command would return a number between 1 and 5 to the TASK program if the user

chosen color is found among those zones at runtime. If no such-specified color is found, a value of "0" is returned to the TASK program.

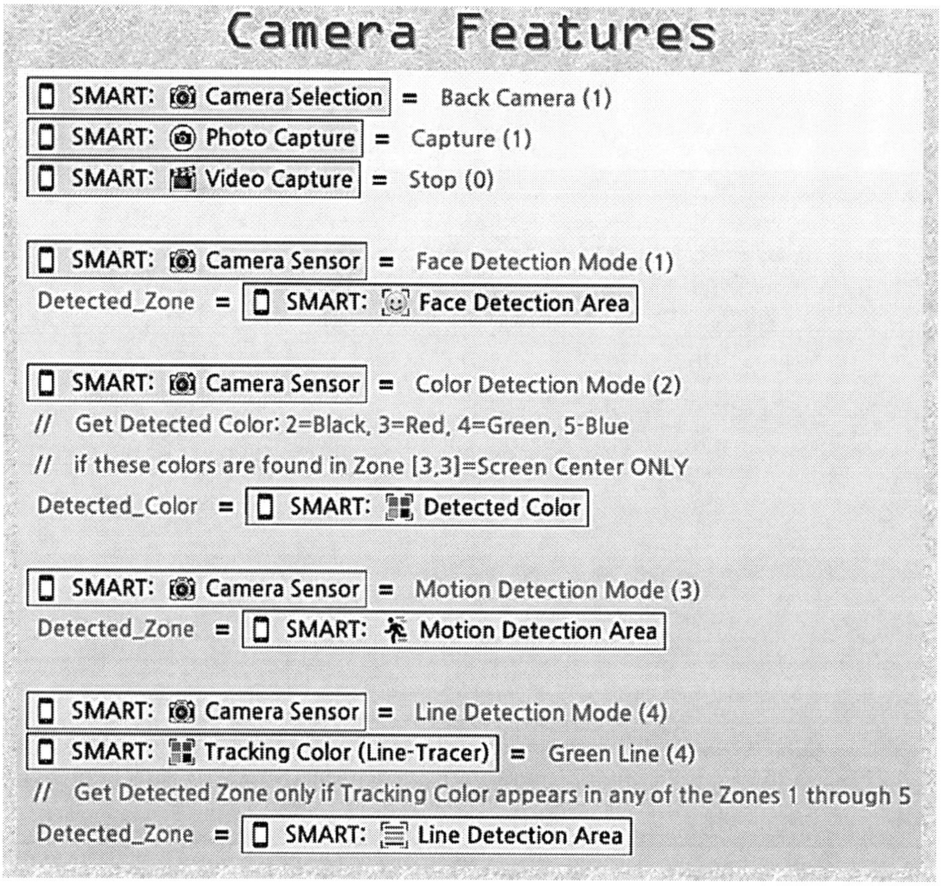

Fig. 7.34 Various Camera Functions programmable in TASK+PLAY700 App.

This "Camera Tracker" Project uses the "Line Detection" feature of the PLAY700 App. Fig. 7.35 shows the "Text" Items used in this project "M_CameraTracker.tsk3".

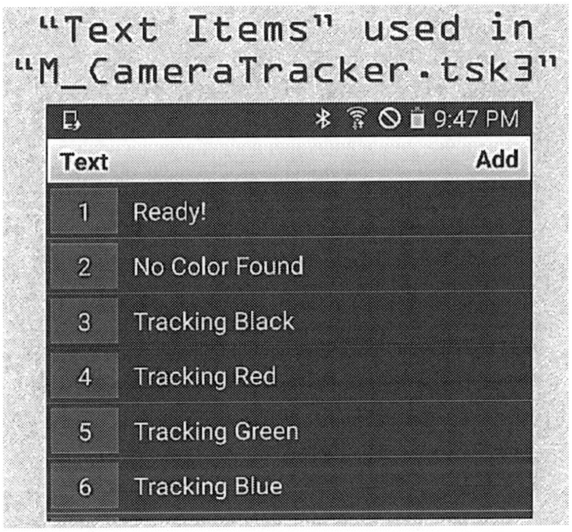

Fig. 7.35 "Text" Items used in "M_CameraTracker.tsk3".

Fig. 7.36 shows that only standard initializations are used for Servos in "M_CameraTracker.tsk3".

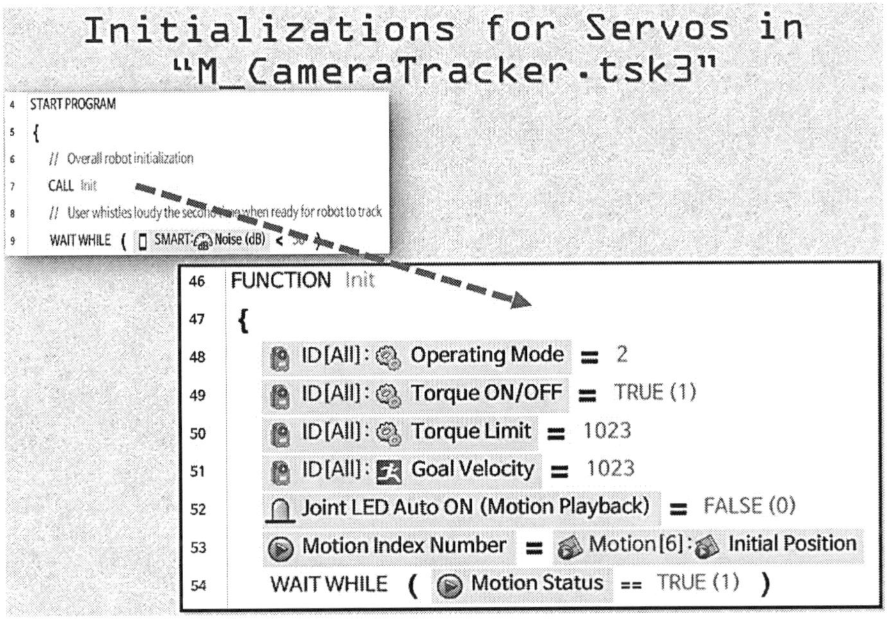

Fig. 7.36 "Initializations" for Servos in "M_CameraTracker.tsk3".

Fig. 7.37 shows the initializations for SMART DEVICE in "M_CameraTracker.tsk3":

- Lines 61-63 display "Ready!" ("Text" Item 1) for 2 seconds.
- Line 66 sets Parameter "LineColor" to "5" (i.e. Blue Line), to be used by the SMART Function "Tracking Color (Line-Tracer)" later at Line 75. If the user wants to track a different line color,

then a different value should be used instead (2, 3 or 4) for "LineColor" (see comment on Line 65).

- Next, Function "DisplayTrackedColor" is called (Line 67 and see its details in Fig. 7.38), and another 2-second wait is executed (Lines 68-69).
- Line 71 chooses the Back Camera and Line 72 sets its Zoom to 0.
- It is important to note that Lines 66, 74 and 75 form a triplet of statements that need to be executed in that order for a successful Image Processing procedure:
    - Set **Line Color** to be searched for in Zones 1 through 5 (Line 66).
    - Set the chosen "Camera Sensor" into "**Line Detection Mode**" (Line 74).
    - Start the "**continuous**" "SMART Tracking Color (Line-Tracer)" service by sending it the value contained in Parameter "LineColor" (i.e. Blue Line in this project). Line 75 can be re-issued with a different "LineColor" value if the programmer wants to switch the line-tracing task to a different color.
- In Line 76, Parameter "LinePosition" is initialized to "0" (meaning that no targeted color patch is found within Zones 1 through 5). Later, the actual Zone result, returned from the "SMART Line Detection Area" function, will be saved into "LinePosition" which then can be used to determine the proper moves (Left, Right or Stay Still) for the MINI to do to keep track of the targeted color patch.

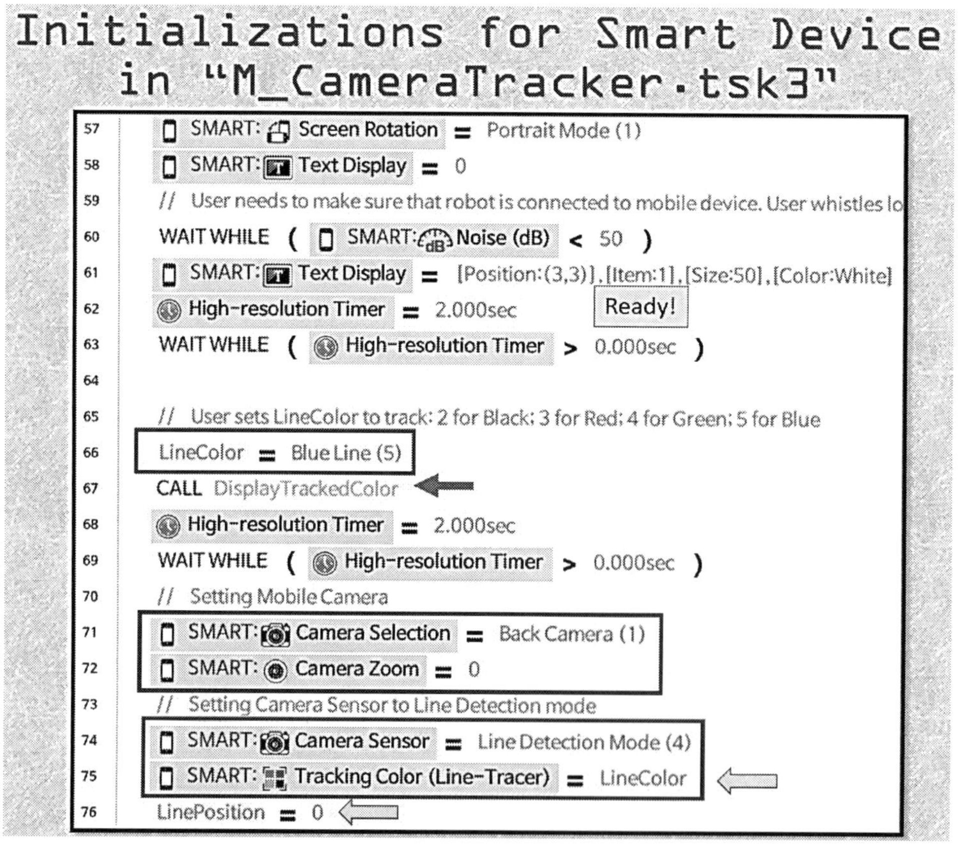

Fig. 7.37 "Initializations" for SMART DEVICE in "M_CameraTracker.tsk3".

The details of Function "DisplayTrackedColor" are listed in Fig. 7.38:

Function "DisplayTrackedColor" in "M_CameraTracker.tsk3"

```
79 FUNCTION DisplayTrackedColor
80 {
81     SMART: Text Display = 0
82     IF ( LineColor == Black Line (2) )
83     {
84         SMART: Text Display = [Position:(3,3)],[Item:3],[Size:50],[Color:White]
85     }                                        Tracking Black
86     ELSE IF ( LineColor == Red Line (3) )
87     {
88         SMART: Text Display = [Position:(3,3)],[Item:4],[Size:50],[Color:White]
89     }                                        Tracking Red
90     ELSE IF ( LineColor == Green Line (4) )
91     {
92         SMART: Text Display = [Position:(3,3)],[Item:5],[Size:50],[Color:White]
93     }                                        Tracking Green
94     ELSE IF ( LineColor == Blue Line (5) )
95     {
96         SMART: Text Display = [Position:(3,3)],[Item:6],[Size:50],[Color:White]
97     }                                        Tracking Blue
98 }
```

Fig. 7.38 Function "DisplayTrackedColor" in "M_CameraTracker.tsk3".

- Line 81 clears the Mobile Display.
- An IF-ELSE-IF structure is then used to properly display the "Text" Items 1 through 4 at the Center (i.e. Position (3,3)) of the Mobile Display (Lines 82 to 97).

Fig 7.39 displays Part 1 of the Main Endless Loop and Function "DisplayColorPosition":

- Parameter "LinePosition" is updated by issuing a READ command "SMART: Line Detection Area" to the Mobile Device which returns a Zone Number back to the TASK program and saved into "LinePosition" (Line 14).
- If "LinePosition" is non-zero (i.e. 1, 2, 3, 4 or 5 returned), i.e. a "Blue Line" is found within one of the 5 zones scanned, then the LOOP WHILE at Line 15 is entered.
- Line 17 is used to display "Text" Item 0 (which does not exist) in Black color into Position (3,4), essentially clearing out any "Text" Item previously displayed in that Position (3,4).

- Line 18 calls Function "DisplayColorPosition" which is also listed in Fig. 7.39. It uses SMART versions of "Print Screen" and "Print Screen with Line" to produce a scrolling line of data (in the Foreground of the Mobile Display) representing "LineColor" and "LinePosition" as a way to confirm data being received by the TASK program. Please note that this process is independent of the SMART Display Text function used so far.

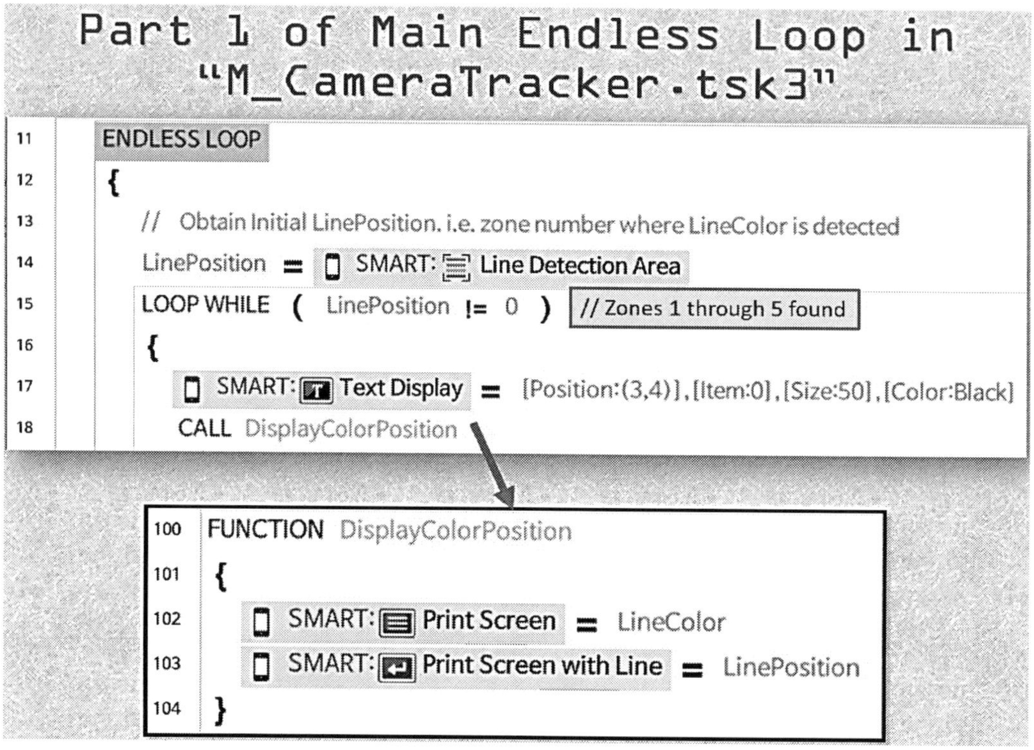

Fig. 7.39 Part 1 of Main Endless Loop and Function "DisplayColorPosition" in "M_CameraTracker.tsk3".

Part 2 of the Main Endless Loop is shown in Fig. 7.40:

- A familiar IF-ELSE-IF structure (Lines 19-33) is implemented to activate the proper MOTION-LIST/PAGE depending on the value recorded in Parameter "LinePosition":
  - **IF** (LinePosition equals 3), the MINI is facing the targeted color patch straight on, so it should stay still (i.e. Initial Position) (see Lines 19-23).
  - **ELSE-IF** (LinePosition equals 1 or 2), the targeted color patch is on the Left of the MINI, so it should rotate Left (see Lines 24-28).
  - **ELSE-IF** (LinePosition equals 4 or 5), the targeted color patch is on the Right of the MINI, so it should rotate Right (see Lines 24-28).
- Line 34 waits until the currently played MOTION-LIST/PAGE is concluded.
- Lines 35-36 wait for another 1 second to allow the operator to "mentally" confirm the MOTION just made by the robot.

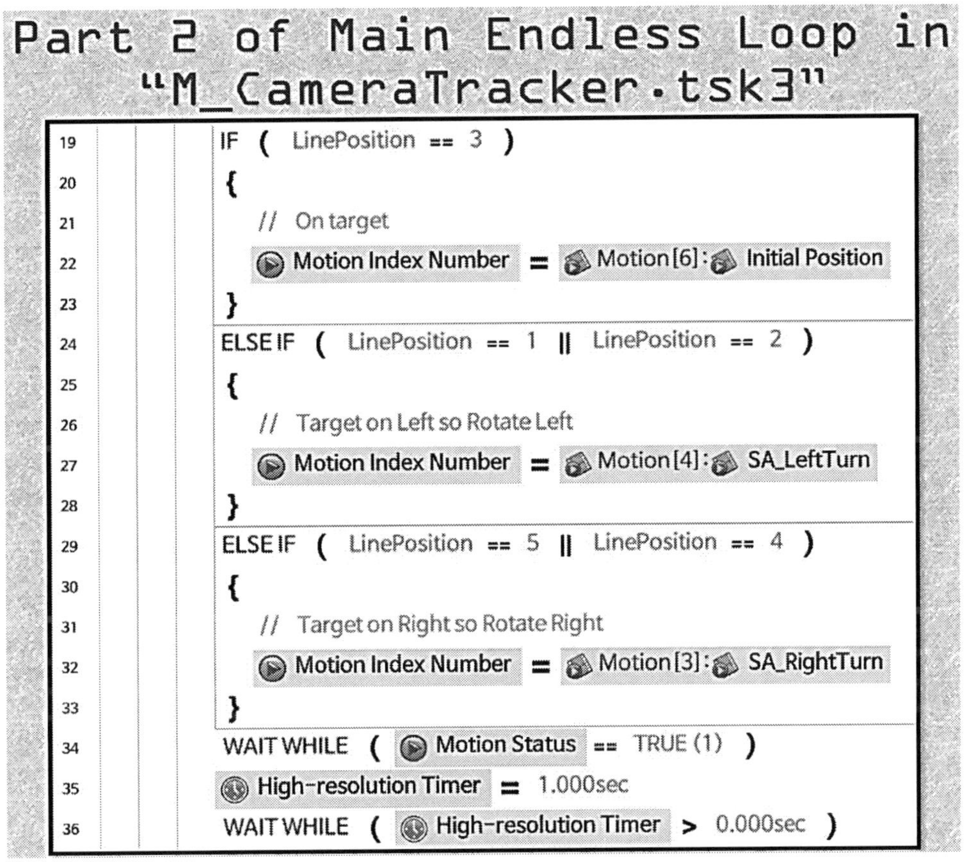

Fig. 7.40 Part 2 of Main Endless Loop in "M_CameraTracker.tsk3".

Part 3 of the Main Endless Loop is shown in Fig. 7.41:

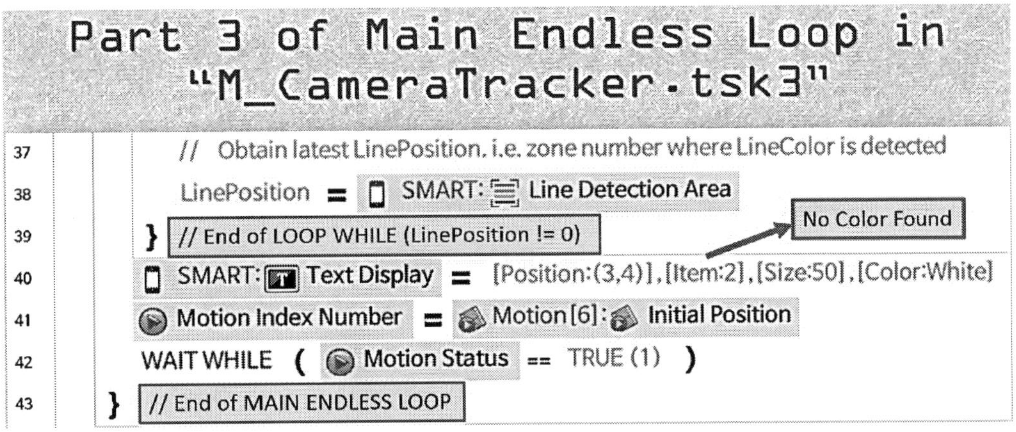

Fig. 7.41 Part 3 of Main Endless Loop in "M_CameraTracker.tsk3".

- The last action performed within the LOOP WHILE for (LinePosition != 0) is Line 37 to send a READ command "SMART: Line Detection Area" to the Mobile Device to retrieve the latest information for Parameter "LinePosition", before the LOOP WHILE reiterates itself.
- However, if LinePosition becomes 0, then the LOOP WHILE exits to Line 40 which displays "No Color Found" into Position (3,4).
- Lines 41 and 42 then stop the robot, and the Main Endless Loop reiterates to its Part 1 (Fig. 7.39).

Video 7.4 shows the run-time performance of this project "M_CameraTracker.tsk3".

In review, Chapter 7 shows how to add some MultiMedia services, ubiquitous on a typical Mobile Device, to work together with the TASK program without sacrificing the "local" access to Sensors and Dynamixels. The data throughput bottleneck once again is the BT connection at 57 Kbps. The "ideal" approach seems to be one adding a more general-purpose Controller such as the Raspberry Pi to the OpenCM9.04-C, and with a wired communication setup between the two Controllers. If the reader gets the opportunity to "upgrade" to the BIOLOID ENGINEER Kit, this is the configuration offered by ROBOTIS.

7.7 Review Questions

1. What is the name of the App that can provide Multimedia Services from a mobile Device to a TASK program running on the OpenCM9.04-C Controller?
2. List the types of services that this App can provide.
3. What are the 5 groupings of services that a SMART project can provide?
4. As far as TASK is concerned, how is the Mobile Display organized?
5. How many bytes does a SMART CONSTANT contain?
6. Describe the two ways that the Screen Position Parameter can be used.
7. How many Touch Area(s) can the Mobile App report to a TASK program?
8. How many Musical Instruments are available for a TASK program to use?
9. How can the programmer override the Speed Settings created by MOTION-UNITs and/or MOTION-LIST/PAGEs?
10. Where is the database for Speech Recognition located?
11. How many "colors" can the "Line Tracer" function from the Mobile App distinguish? What are they?

Chapter 8: Using EDBOT™ Tools

So far when we use ROBOTIS software tools, the robot run-time codes always reside on the MINI's OpenCM9.04-C Controller in a "compiled" binary machine code format which is about the best situation for a robot to perform in, along with local access to Dynamixels and Sensors at a communication rate of 1 Mbps (Chapter 6). However, we also saw the limitation of the MINI's firmware as we had to leverage in the PLAY700 App for Multimedia Services, albeit at 57 Kbps (Chapter 7).

In this Chapter 8, the robot control software runs instead on a general-purpose personal computer such as the Windows PC, Mac or Raspberry Pi and inside an "interpreting" shell, meaning that each line of code is "interpreted" one by one at run-time and converted into an appropriate command packet. This packet is then sent via the EDBOT tool to the OpenCM9.04-C through the BT communication hardware at a mere 57 Kbps. Then EDBOT waits for the MINI to respond with Dynamixels and Sensors data which are returned via the same 57 Kbps BT port (see Fig. 1.17). The EDBOT/SCRATCH wait time for Sensors data is around 100 ms, so data throughput is the main issue to consider when programming with EDBOT. However, this throughput issue is traded for the access to a much more powerful and faster computer with a variety of Multimedia resources to tap into. When using PYTHON with EDBOT, more sophisticated programming techniques such as multi-threading and access to software packages like OpenCV are then available to MINI programmers.

Circa 2015, the EDBOT software tool was first created as a server/client tool to be used within an actual classroom environment, along with teacher supervision of student access to individual ROBOTIS MINI robots (http://www.robotis.us/robotis-mini/), programmed with the MIT SCRATCH 2 Offline IDE (http://ed.bot/edbot/). It is available for purchase at http://ed.bot/order.

In January 2019, SCRATCH 3 (https://en.scratch-wiki.info/wiki/Scratch_3.0) was released with an online interface and a desktop interface (https://scratch.mit.edu/download). In March 2019, EDBOT V.5 was released allowing the additional usage of SCRATCH 3 via a web interface (http://scratch.ed.bot/) – a desktop version may be released in the future. Unfortunately, this web interface yields a slower communications speed with the MINI robot and makes autonomous behaviors harder to achieve, thus this chapter concentrates on the usage of the **SCRATCH 2 Offline Editor** instead.

This chapter uses EDBOT V.5.2.0.1523 which allows **multiple robot** programming via three main software interfaces: SCRATCH, PYTHON and JavaScript, and with additional languages such as Node.js, Java and C# and other platforms such as Chromebooks, Macs and Linux (http://support.ed.bot/edbot-software-index.html). This version also accommodates the 17-servo configuration of the MINI and the functionality of MOTION-LIST/PAGE "0" and "-1" to make use of the EXIT MOTION-UNITs (http://emanual.robotis.com/docs/en/software/rplustask3/task_parameters/#motion-index-number).

In this chapter, an individual user is assumed, i.e. the user's PC is used as the local host (Port 8080) and the EDBOT tool is used in its server mode.

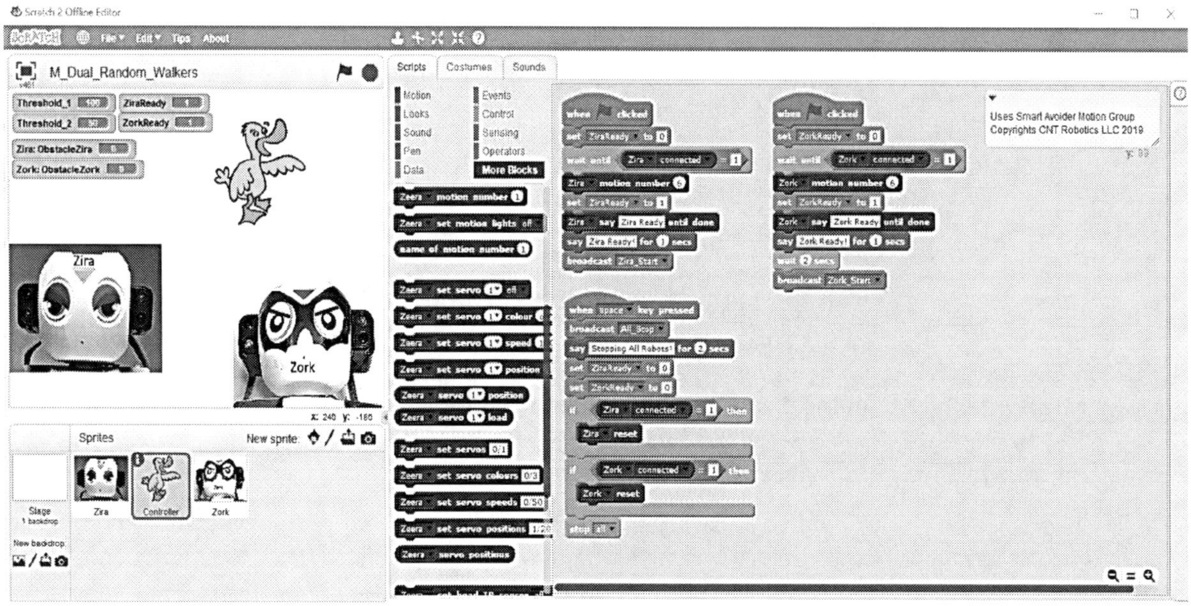

Fig. 8.1 Multi-robot application of the EDBOT tool with SCRATCH 2.

For this Chapter, the author assumes that the reader is already familiar with MIT SCRATCH 2 software. If not, the user is recommended to first read up on such works as Ford (2014), Warner (2015) or Vlieg (2016) to have a thorough understanding of the SCRATCH 2 language. In this Chapter, only selected SCRATCH 2 features, most applicable to robotics and unique to the workings of the EDBOT software, would be presented.

An important and necessary practice for the user is to make sure that he/she has created a blank/empty TASK program and downloaded it to the OpenCM9.04-C BEFORE downloading any SCRATCH 2 or PYTHON example project, as the author had noticed that an existing/working TASK code could interfere with EDBOT's operations.

8.1 EDBOT's Hardware and Software Installation Requirements

The user will need to purchase the appropriate EDBOT MINI kits (which has one IRSS-10 mounted on Port 1) and/or software product keys from Robots in School Ltd. (https://shop.ed.bot/collections/products). If the user has already obtained a MINI robot elsewhere, the user can just buy the EDBOT product key from Robots in School Ltd. (https://shop.ed.bot/collections/products/products/edbot-software) or from ROBOTIS USA (http://www.robotis.us/edbot-software-product-key/). **In this book, the author assumes that the user is using the EDBOT tool (Version 5.2.0.1523 or higher).**

Hardware wise, the EDBOT software works with the ROBOTIS BT-210 (http://www.robotis.us/bt-210/).

Please view Video 8.1 for the typical steps needed to install the EDBOT software on the user's PC. **Importantly, each software product key will be bound permanently to the specific BT-210s used, thus the user will need to plan out carefully his or her "robots" organization if the user wants to use the EDBOT software with multiple robots.** Fig. 8.2 shows the author's EDBOT set up for 5 robots: Zeeb is a DREAM robot, Zomby is a PLAY700 robot, while Zeera, Zira and Zork are MINI robots.

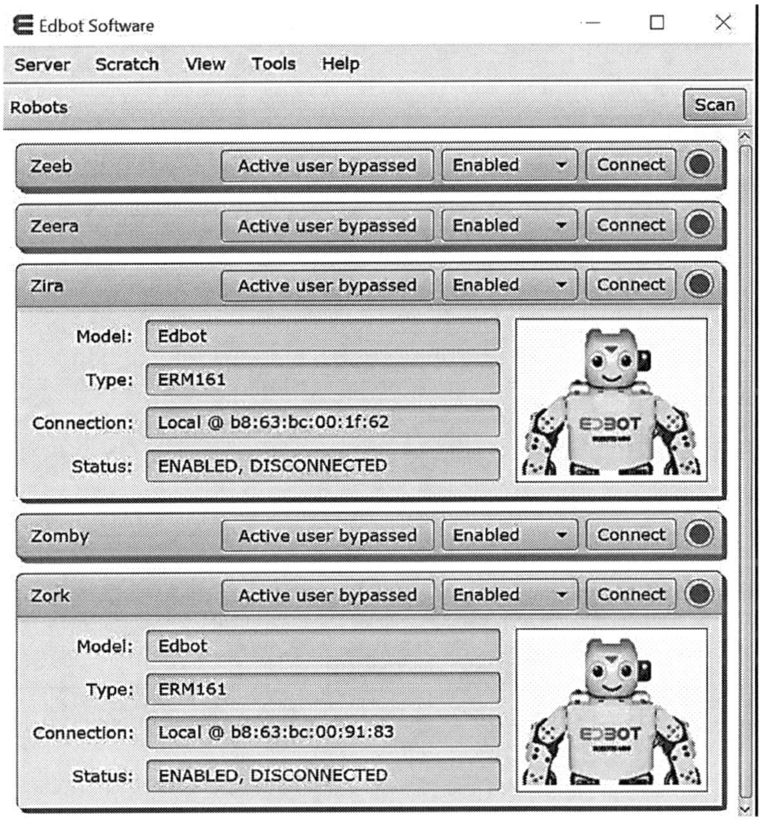

Fig. 8.2 Author's EDBOT Setup for his 2 MINI robots (Zira and Zork).

8.2 SCRATCH 2 Project File Conversion

The author used 2 MINI robots named "Zira" and "Zork" to create the SCRATCH 2 example codes presented in this Chapter, but the reader may use a different name for his/her robot(s), and thus will need to do an updating process for the SCRATCH SB2 files provided by the author.

Let's say that the user's robot name is "Spot" and that the user has put the author's sample codes in a folder named "Author" and has prepared a blank folder named "User" to receive the converted SB2 files whereas the "Zira" name is updated to "Spot". Next, let's assume that the user has installed the EDBOT software properly and has successfully registered his/her MINI robot to the name "Spot" via a specific BT module. Thus, when the user runs the EDBOT software, he or she would see a starting screen (EDBOT V.5.2) as shown in Fig. 8.3, when the item "Scratch" has been chosen from the Top Menu bar.

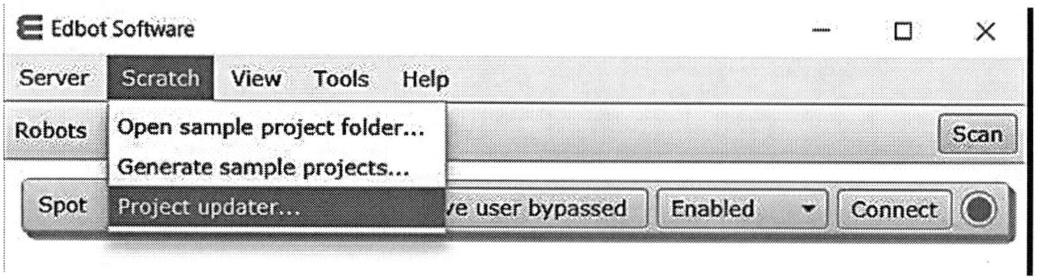

Fig. 8.3 Starting Screen of EDBOT (V.5.2) software with item "Scratch" chosen on Top Menu bar.

Next, the user should choose the item "Project Updater" and a new window pops up (Fig. 8.4) where the user can browse to select a given Author's SB2 file to be updated as "Source project". If the button "Different target project" is checked, then the user can specify a different file location and name for the "Target project". If not, the newly updated Target Project will overwrite the original Source Project/File.

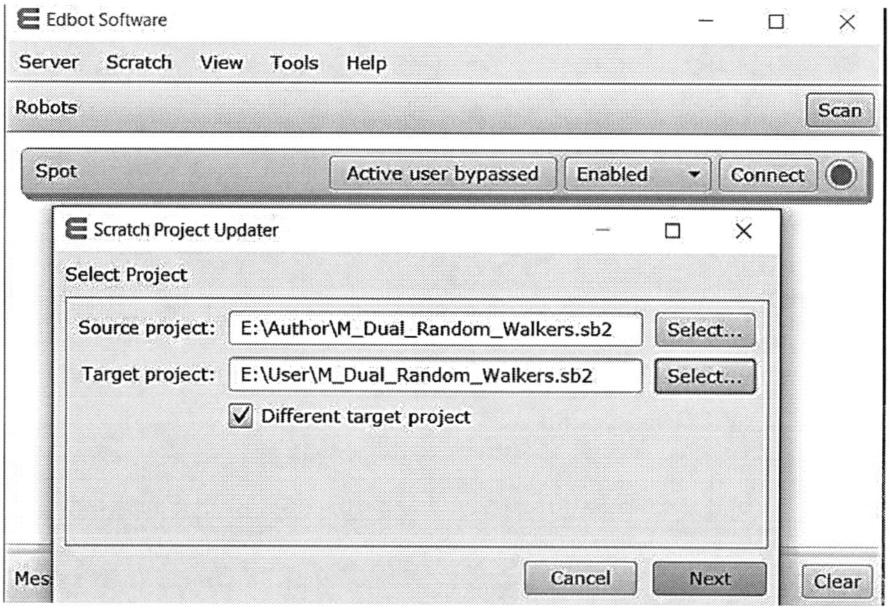

Fig. 8.4 Choosing "Source project" and "Target project" in "Scratch Project Updater".

When the user clicks on "Next" as shown in Fig. 8.4, the "Update Names" window pops up (see Fig. 8.5) where the user can verify that the Source Project's name "Zira" will be set to "Spot". If this does not happen, the user should click on the "Pull Down" menu under the column "Target project" and choose "Spot" from the choices provided. If the user has only 1 MINI robot, then the user needs to choose "[Remove]" for "Zeera" and "Zork" on the "Target project" column.

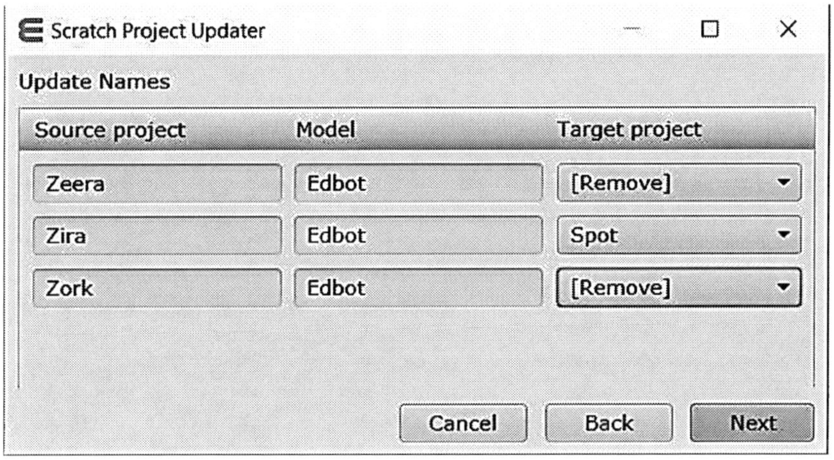

Fig. 8.5 "Update Names" pop-up window.

Finally, if the user clicks on "Next" as shown in Fig. 8.5, then a final "Summary" window pops up (see Fig. 8.6) where the user can click on "Apply" to affect the actual file update.

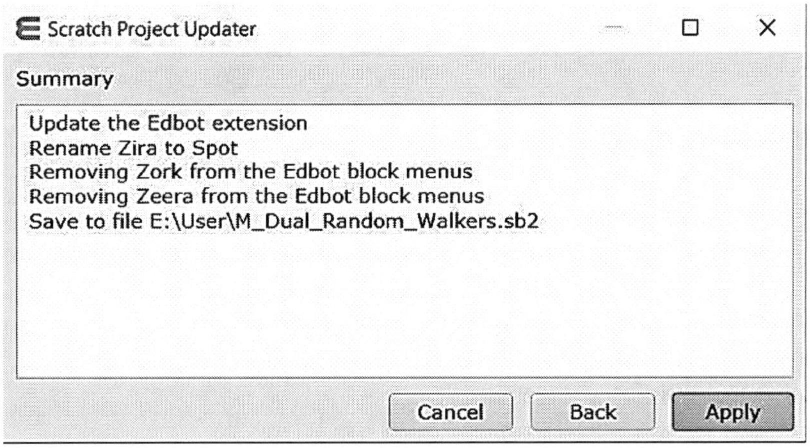

Fig. 8.6 "Summary" window before actual SCRATCH Project Update.

As the user can see, only ONE SB2 file can be converted each time that the "Project Updater" is used. Of course, only "EDBOT" related items are updated while SCRATCH constructs (such as graphics items) remain the same and need to be changed if needed (see Fig. 8.7).

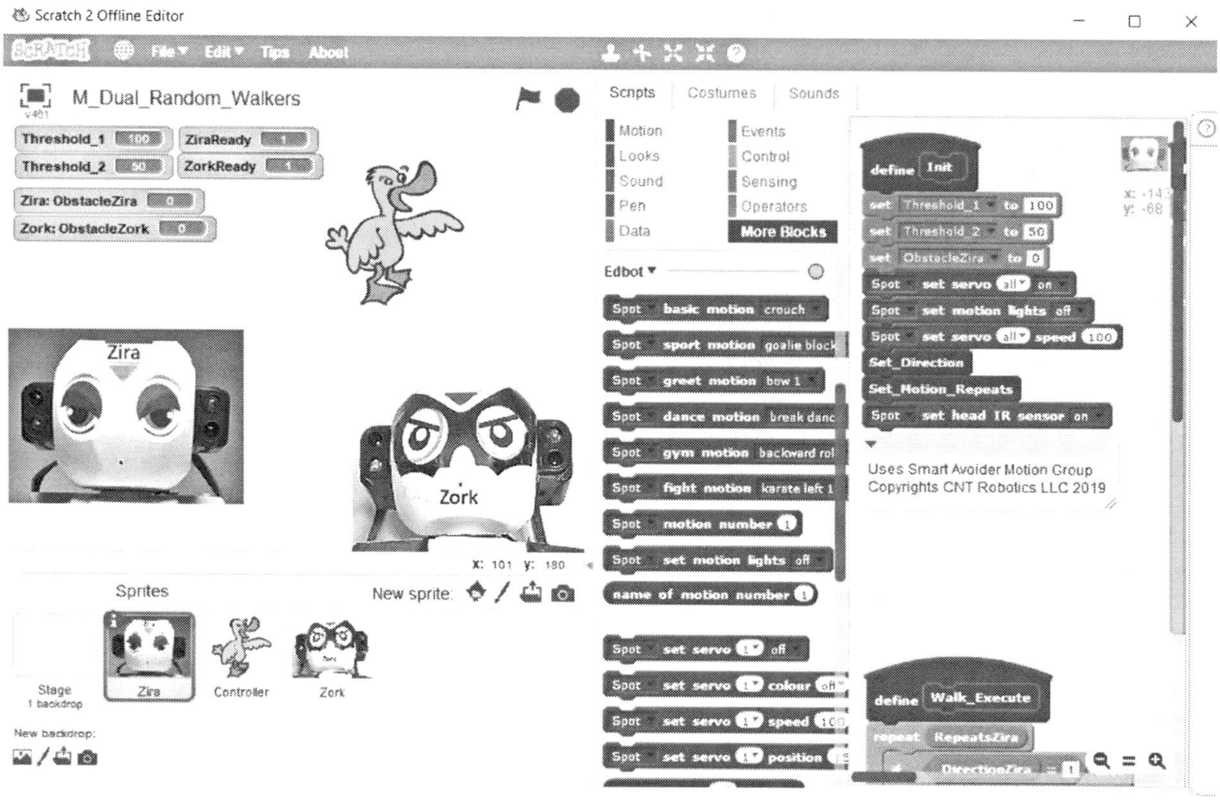

Fig. 8.7 SCRATCH project updated for "Spot" in EDBOT blocks.

Although SCRATCH 2 is designed for beginner programmers as young as 8 years of age, it has quite a sophisticated structure for its "project":

- Each SCRATCH project can have several actors called "SPRITES" (e.g. "Duck" Controller, Zira and Zork) listed in the bottom left panel of Fig. 8.7. The large area on the top left panel, where pictures of all Sprites are shown, is called the STAGE.
- Attached to each Sprite are SCRIPTS (actual codes), its COSTUMES (graphics components triggered by its SCRIPTS), and its SOUNDS (audio components also controlled by its SCRIPTS) – see top of mid-panel in Fig. 8.7.

8.3 SCRATCH 2 Blocks Provided with EDBOT Software

For the MINI, EDBOT provides 20 Stack Blocks and 7 Reporter Blocks which can be accessed under the "More Blocks" item of the SCRIPTS tab (see Fig. 8.8):

- Group 1 (see Fig. 8.9) includes Stack Blocks that work specifically with the MINI Default Motion Group so they are not used in this Chapter as we use Custom Motion Groups.
- Group 2 (see Fig. 8.10) are for general use, and the Motion Control blocks use the standard Motion Index Numbers as argument.

The EDBOT website has a support page at http://support.ed.bot/EDBOT-scratch2.html where the various Blocks are classified as: MOTION CONTROL, SINGLE SERVO, MULTI SERVO, SENSOR, SPEECH and GENERAL.

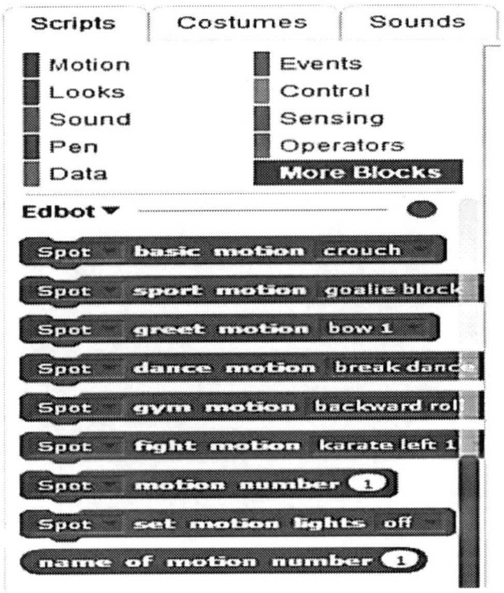

Fig. 8.8 EDBOT provides 20 Stack Blocks and 7 Reporter Blocks.

Fig. 8.8 also shows the standard SCRATCH Blocks which are grouped under different categories: **Motion, Looks, Sound, Pen, Data, Events, Control, Sensing and Operators**.

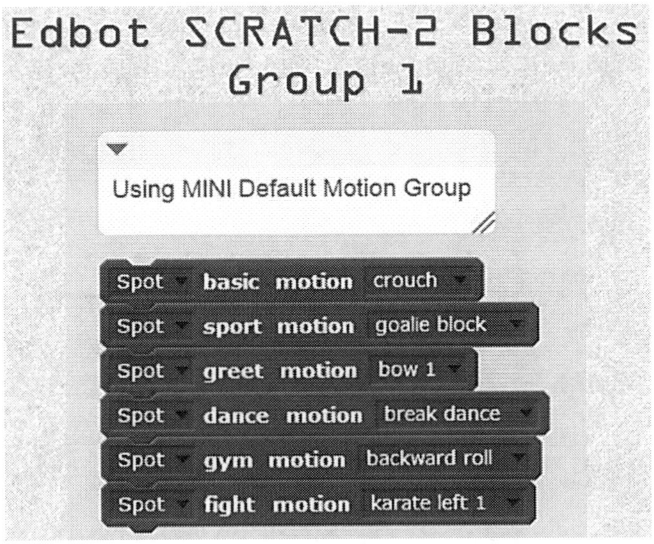

Fig. 8.9 Group 1 of EDBOT's Stack Blocks.

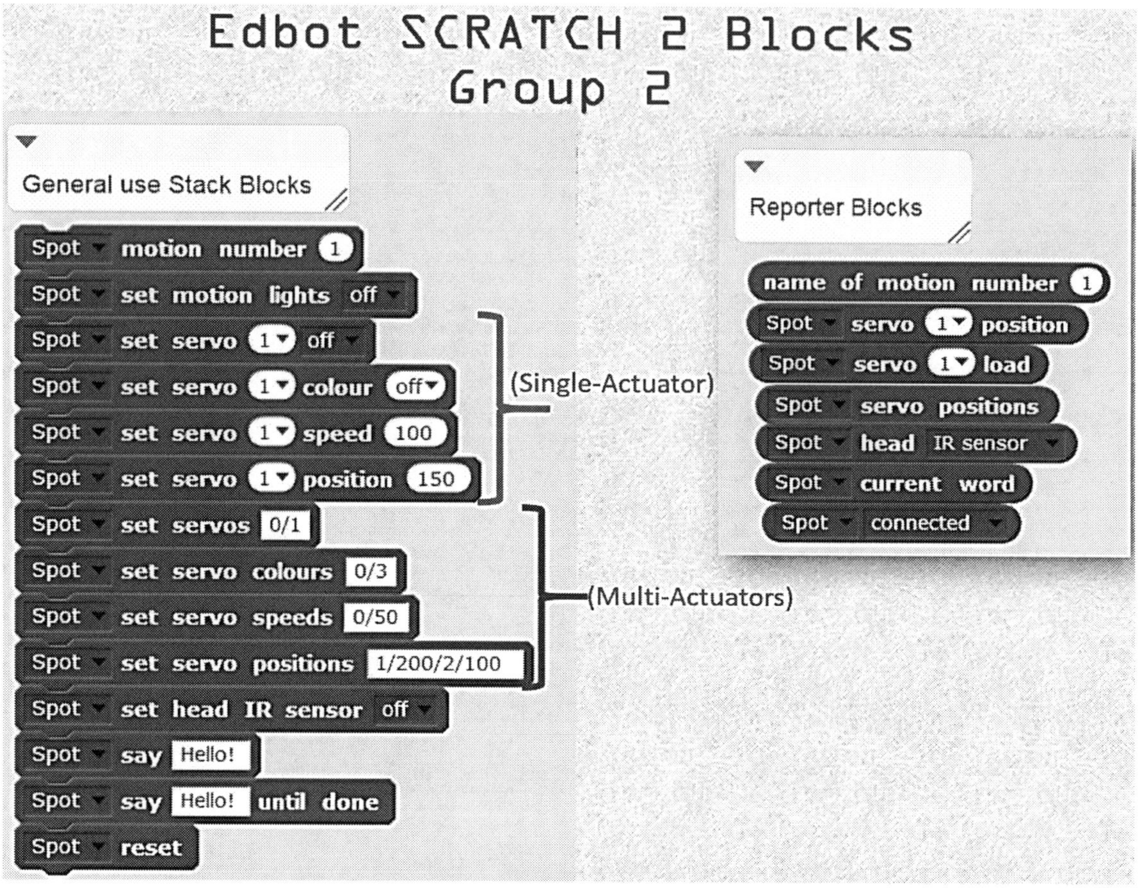

Fig. 8.10 Group 2 of EDBOT's Stack Blocks & Reporter Blocks.

8.3.1 MOTION CONTROL Blocks

Fig. 8.11 shows the currently supported EDBOT MOTION-related blocks:

- The "Name of Motion Number" Reporter Block is a unique EDBOT feature.
- The "Set Motion Lights" Stack Block is equivalent to the ROBOTIS "Joint LED Auto ON (Motion Playback)" command.

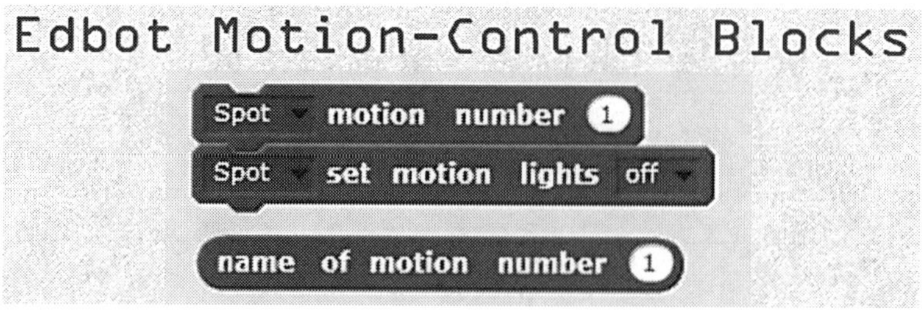

Fig. 8.11 EDBOT's Motion Control Blocks.

- The “Motion Number” Stack Block corresponds to the ROBOTIS “Play Motion Index Number” command. This EDBOT block does support MOTION (0) and MOTION (-1) so we can use EXIT MOTION UNITs in a similar manner as for ROBOTIS software (http://emanual.ROBOTIS.com/docs/en/software/rplustask3/task_parameters/#motion-index-number).
- However, there is an **important difference** that the user needs to be aware of when using this SCRATCH/EDBOT “Motion Number” Stack Block. Once this Stack Block is issued by the “SCRATCH+EDBOT” tool chain, the SCRATCH Script would “**wait at this step**” until the current MOTION-LIST/PAGE **has finished playing**, and EDBOT would also **not allow any other SCRATCH/EDBOT “Motion Number” Block from taking effect** during this “Motion Play” time period. Previously, when using the TASK Sub-tool, we had to “**programmatically**” watch for these situations ourselves, using the Conditional Expression (**MOTION STATUS == FALSE**). Perhaps, that is why there is no EDBOT Block corresponding to the ROBOTIS “Motion Status” Parameter that was much used in Chapters 6 and 7. Thus, when needed to, the author would devise a **work-around solution** (see Section 8.5).

8.3.2 SINGLE & MULTI SERVO Blocks

Fig. 8.12 shows the currently supported EDBOT SERVO-related blocks:

- The “Single-Servo” Blocks correspond the various ROBOTIS commands such “Goal Position”, “Goal Velocity”, etc… for individual servos, and they function in a similar manner but with slower data throughput.

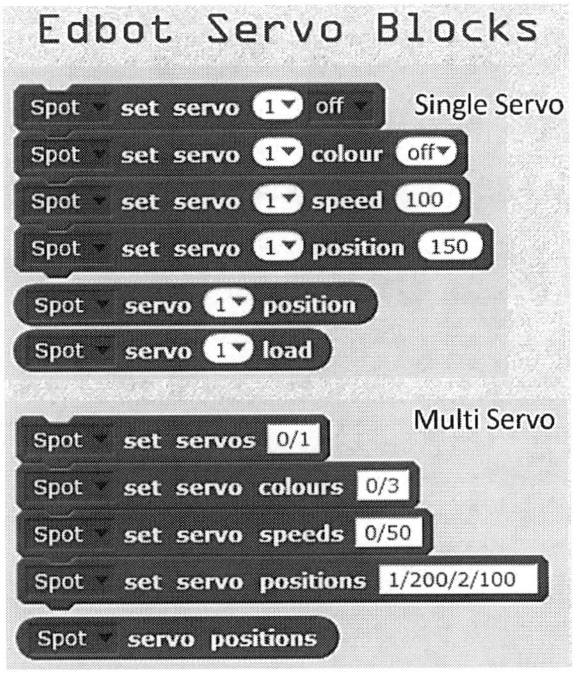

Fig. 8.12 EDBOT’s Single-Servo and Multi-Servo Blocks.

- The "Multi-Servo" Blocks allow the "near-simultaneous" setting of multiple servos via a "big" message sent from EDBOT to the MINI. Currently, there is no equivalent feature in ROBOTIS TASK programming for the MINI *(for newer controllers like the CM-550, the equivalent ROBOTIS feature is called "Synch-Write")*. For this Chapter, as we mostly control robot motions with MOTION-LIST/PAGEs, we "regretfully" do not need to use this SCRATCH/EDBOT powerful feature.

8.3.3 SENSOR Blocks

Sensor-wise, EDBOT currently assumes that there is only ONE IRSS-10 sensor connected to Port 1 of the MINI (see Fig. 8.13) – for its SCRATCH implementation. To access the full range of sensors and to use all four I/O Ports like in TASK previously, the user needs to switch to using PYTHON with EDBOT (see Section 8.7 and later).

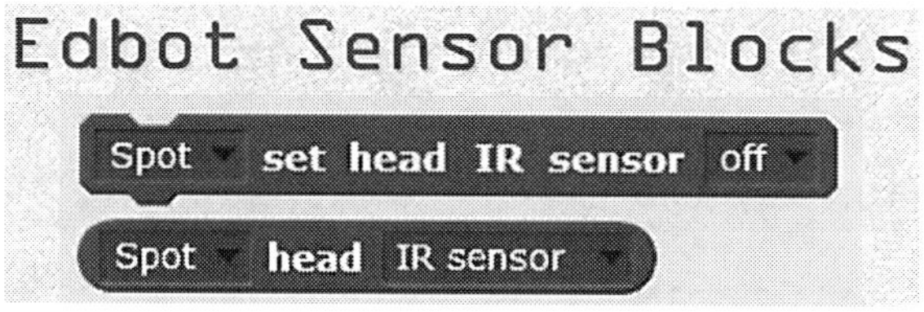

Fig. 8.13 EDBOT's Single NIR Sensor Blocks.

8.3.4 SPEECH Blocks

Fig. 8.14 shows the Speech Blocks implemented by EDBOT. These "audio" SAY Blocks are different from the standard SCRATCH's SAY Blocks which are "visual display" and "non-audio" outputs.

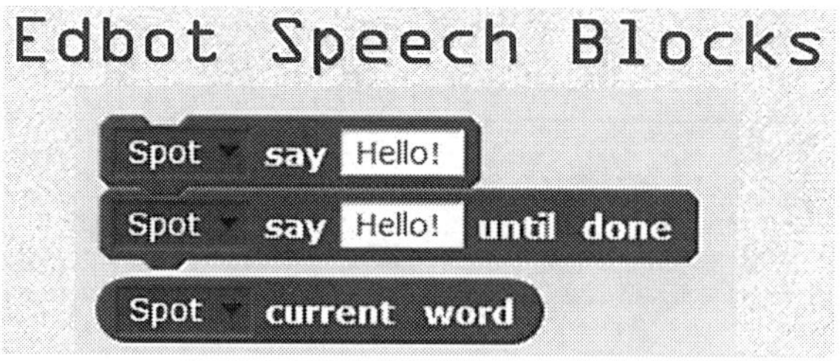

Fig. 8.14 EDBOT's Speech Blocks.

8.3.5 GENERAL Blocks

These GENERAL or SYSTEM blocks relate to the overall operation of the OpenCM9.04-C Controller: "Reset" is a Stack Block and "Status" is a Reporter Block (see Fig. 8.15). "Reset" will turn off all hardware components and stop all operations on the OpenCM9.04-C, but the SCRATCH code will still be active. The "Status" block is handy to check if your MINI robot is "Enabled" or "Connected" to the EDBOT software on the Desktop Computer or not, especially if Bluetooth happens to take a long time to connect properly when a SCRATCH project starts to run.

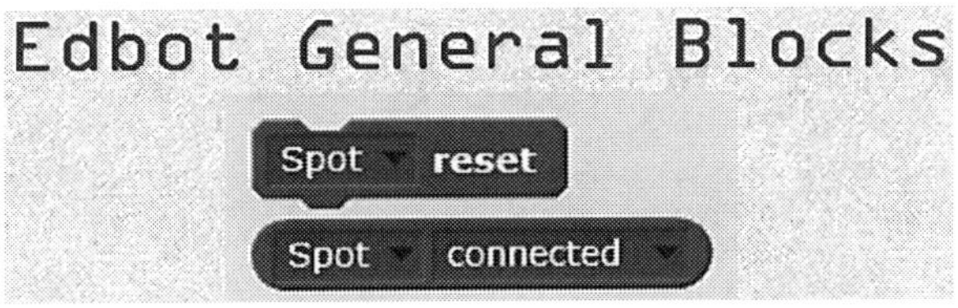

Fig. 8.15 EDBOT's General Blocks.

8.4 RC Walker with Independent Servo SCRATCH Project

In this section 8.4, the author's goal is to create a SCRATCH Project that is equivalent "as much as possible" to the "M_RC_Walker_IS.tsk3" (see Section 6.4), whereas Servos #1 (Left Shoulder) and #17 (Head) can be independently controlled from the effects of the currently performing MOTION-LIST/PAGE.

8.4.1 Differences between TASK & SCRATCH

Let's first discuss important differences between ROBOTIS TASK and SCRATCH+EDBOT:

- The RC-100 provides a 16-bit message that can tell the programmer about the status of all Buttons U-D-L-R-1-2-3-4-5-6 (see Section 6.4.1), while SCRATCH can access most keys of the standard computer keyboard but only as a "single" key being pressed. Fig. 8.16 shows the differences in handling "Button" or "Key" events between TASK and SCRATCH and it is obviously much simpler when using SCRATCH.

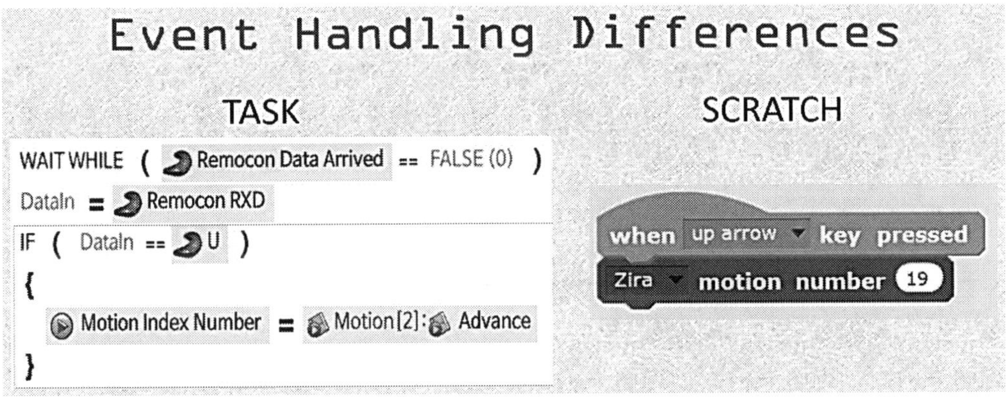

Fig. 8.16 Event Handling Differences between TASK and SCRATCH+EDBOT.

- Although the same MOTION-UNIT/PAGEs can be used by both TASK and SCRATCH+EDBOT, the "monitoring" of the MOTION-IN-PLAY is achieved differently between the two tools (see Section 8.3.1).

- There is no equivalent of JOINT-OFFSET commands in SCRATCH+EDBOT, therefore the independent control of the Shoulder Servo #1 cannot be implemented in SCRATCH. Regarding the Head Servo #17, the author's work-around solution is to use the default MOTIONS that are designed for the standard MINI with only **16 servos** (i.e. no head servo) and then use **direct single-servo commands** to move the **Head Servo #17**.

8.4.2 Keyboard Solution "M_RC_Walker16_IS.sb2"

This SCRATCH project uses the **Default Motion Group for the standard MINI with only 16 servos** and we'll use the following MOTION-LIST/PAGEs:

- MOTION-LIST/PAGE (1) is used for "Initial Position".
- MOTION-LIST/PAGE (19) is used for "Advance".
- MOTION-LIST/PAGE (20) is used for "Reverse".
- MOTION-LIST/PAGE (15) is used for "Right Turn".
- MOTION-LIST/PAGE (16) is used for "Left Turn".

Fig. 8.17 shows the Costumes used for robot Zira: "Idle" and "Ready".

Fig. 8.17 "Costumes" used for robot Zira.

Fig. 8.18 shows the MAIN SCRIPT for "M_RC_Walker16_IS:

- Block 1 is one of the "Event" type of Block and it would start whenever the user clicks on the FLAG Menu Item found on the Top-Right corner of the STAGE Area.
- Blocks 2 and 3 belong to the "Looks" category. Block 2 displays the costume "Zira_Idle_m" onto the STAGE. Block 3 displays a Text Balloon, also on the STAGE and usually located at the top right of the associated SPRITE Zira, which says "Zira Not Ready!" for 2 seconds then this Text Balloon would disappear.
- Block 4 combines a "WAIT UNTIL" SCRATCH Control Block with the EDBOT Reporter "Zira_Connected" to make the MAIN SCRIPT wait at this Block until "Zira_Connected" is

found to be set to "1" (i.e. until the BT connection is confirmed between EDBOT and the MINI).

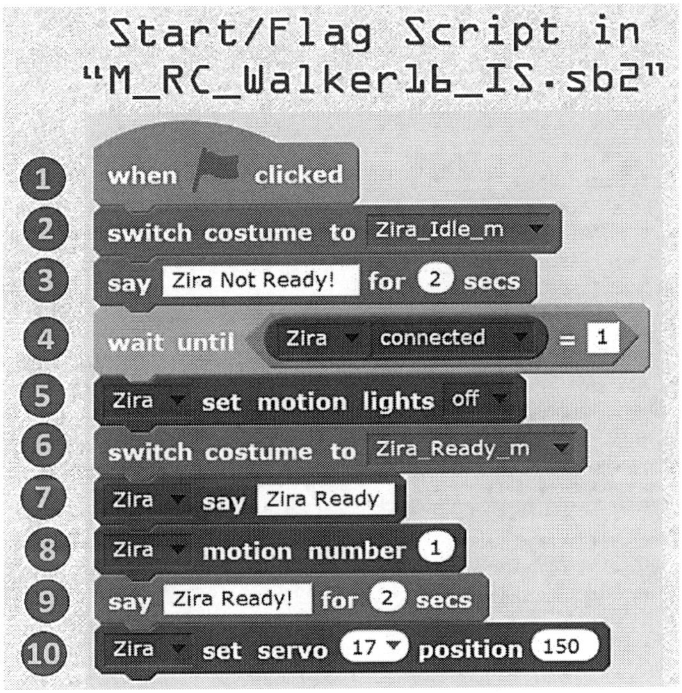

Fig. 8.18 MAIN SCRIPT for "M_RC_Walker16_IS.sb2".

- Block 5 is an EDBOT Block and it sets the "Motion Lights" option for Zira to be OFF, i.e. no LED blinking during MOTION PLAY.
- Block 6 switches Zira's Costume to "Zira_Ready_m".
- Block 7 makes Zira (i.e. EDBOT) speak via the Desktop speaker the words "Zira Ready".
- Block 8 commands the MINI to get into its Initial Position (i.e. MOTION (1)).
- Block 9 displays a Text Balloon showing "Zira Ready!" for 2 seconds before removing it.
- Block 10 makes sure to set Servo #17 (Head Servo) to the "150°" Goal Position (i.e. Goal Position "512" when using TASK).

Fig. 8.19 shows the SCRIPTS that correspond to various "Keyboard Input" Events:

- When the UP-ARROW key is pushed, MOTION (19) is played to make the MINI walk forward.
- When the DOWN-ARROW key is pushed, MOTION (20) is played to make the MINI walk backward.
- When the RIGHT-ARROW key is pushed, MOTION (15) is played to make the MINI turn right.
- When the LEFT-ARROW key is pushed, MOTION (16) is played to make the MINI turn left.
- When the "s" key is pushed, a custom set of movements are executed with Servo #17:

- Servo 17 Speed is set to 25%.
- Servo 17 is sent to Position 100°, i.e. to scan right.
- Next, this SCRIPT waits for 1 second.
- Servo 17 scans left by rotating towards Position 200°.
- Then, this SCRIPT waits for another 1 second.
- Lastly, Servo #17 is sent back to Position 150° (i.e. facing forward).

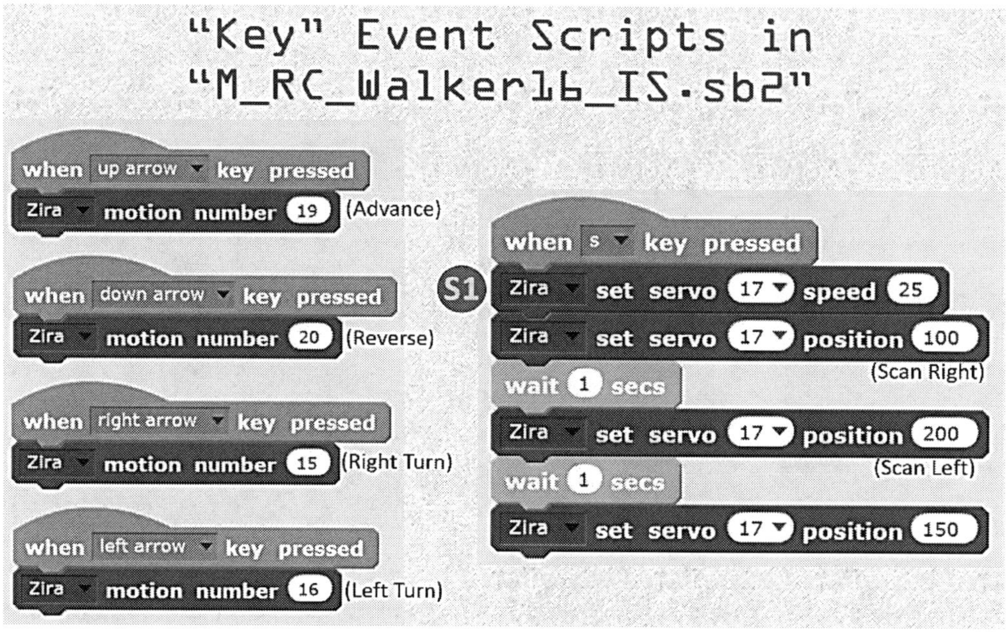

Fig. 8.19 Various "Keyboard Input" SCRIPTS in "M_RC_Walker16_IS.sb2".

One important point for the reader to notice is that an equivalent TASK solution would require the use of an IF-ELSE-IF structure to "glue" these 5 events together for them to work properly at run-time (see Figs. 6.36 and 6.37). But in SCRATCH, there is no need to do so because SCRATCH can execute/interprete "independent" EVENT SCRIPTS at run time.

Another minor issue to point out about the S1 Block in Fig. 8.19 which is used to set the Goal Velocity for Servo #17 to 25%:

> The first time that the author used it, he put it right after Block 10 of the MAIN SCRIPT as part of an initialization procedure for Servo #17 performed commonly in any TASK solution. However, the author found that Servo #17 got its speed reset to 100% every time that a MOTION-LIST/PAGE had been played. Thus, putting Block S1 in the "Key-s" SCRIPT was the **sure way** to avoid this issue.

So far, the reader can see that a SCRATCH solution requires a "parallel-thinking" approach instead of a "linear" approach used previously for TASK. In a way, the SCRATCH programmer work is done at the step for defining the "Event/Action" Pairs and he/she does not have to go through the "Algorithm Development" phase of the "Computational Thinking" paradigm as illustrated back in Section 6.3. This is because the Desktop Computer has a much more sophisticated OS than the much simpler firmware on the OpenCM9.04-C, allowing it to manage multiple threads of execution in real-time for SCRATCH.

Still, we do not yet know if this SCRATCH solution "M_RC_Walker16_IS.sb2" will produce run-time behaviors comparable to the equivalent TASK solution "M_RC_Walker_IS.tsk3". Video 8.2 shows that the run-time behaviors of the MINI were different between the 2 solutions because of the following implementation difference:

- The usage of Multiple RC Buttons is allowed in TASK, but this is not allowed in SCRATCH. For example, in Video 8.2, the operator pushed first on the "UP-ARROW" key and the MINI responded by walking forward. Next, while still pushing on "UP-ARROW", the operator also pushed on the "s" key, and Video 8.2 showed that the robot stopped walking forward and it started executing "only" its head scanning maneuver. Previously, in the TASK solution, this same robot was able to perform these 2 movements simultaneously.
- Further tests in Video 8.2 showed that Windows OS is feeding SCRATCH with a **sequence of single keys** as the operator typed them in on the keyboard (which is the "expected" way for a computer keyboard to operate anyhow!). Thus, the issue is not with SCRATCH per-se, but with the sequential sending of the typed keys by the keyboard via the PC's OS.

In retrospect, although the Virtual RC-100 Controller is using the same physical PC keyboard as SCRATCH, it treats/simulates the behavior of the UP/DOWN/LEFT/RIGHT Arrow Keys and the NUMBER Keys 1 through 6 VERY DIFFERENTLY. Consequently, **the way that a robot can receive its operator's inputs can profoundly affect the types of tasks that the robot can be programmed to do.**

8.5 Dual RC Walkers SCRATCH Project

Although EDBOT can support multiple robot connections, the results from Section 8.4.2 show us that there is no feasible way to control two MINIs independently from each other as SCRATCH can only detect a "single" keystroke at a time from the operator. But it should be feasible to **control two MINIs synchronously** to do the same motions as commanded by the operator's keystroke sequence.

In this "Dual RC Walkers" project, we'll combine keystroke events with MESSAGE BROADCAST techniques to synchronously control two MINIs. We'll also use 3 SPRITEs for this project: "Duck" Controller and the two robots Zira and Zork (see Fig. 8.20). The "Duck" Controller will have a supervisory role of the connections between EDBOT+SCRATCH and the robots. It also intercepts the incoming operator's keystrokes and broadcasts the appropriate messages to the 2 robots for them to perform the commanded motions synchronously.

Fig. 8.20 Various "Keyboard Input" SCRIPTS in "M_RC_Walker16_IS.sb2".

Two versions will be described: a Simplified Version having a bare-bone structure just enough to get the job done, and a Verbose Version that can help us see the interactions between key-strokes, messages and robot behaviors.

8.5.1 Simplified Version "M_Dual_RC_Walkers_M_S.sb2"

The project "M_Dual_RC_Walkers_M_S.sb2" uses the **MOTION-GROUP named Smart Avoider** from the file "MINI_Custom.mtn3".

The "Duck" Controller uses 3 types of SCRIPTs which are described in Figs. 8.21, 8.22 and 8.23.

Fig. 8.21 describes the Controller's FLAG Scripts, one for Zira and one for Zork (i.e. they are set **to run independently and simultaneously** when the operator clicks on the GREEN FLAG Icon). The reader can see that Zira and Zork share the same programming logic except for the use of different robot names in EDBOT-related Blocks. Thus, the author would use Zira's Script as demo code:

- Zira's FLAG Script starts by setting Parameter "ZiraReady" to "0".
- Next, SCRATCH waits until BT connection is established between SCRATCH and the MINI via EDBOT (i.e. when the condition (REPORTER "Zira – Connected" == 1 becomes TRUE).
- Zira's Motion LEDs are then set to OFF.
- Zira plays MOTION-LIST/PAGE (6), i.e. the Initial Position.
- Zira voices out the words "Zira Starting" through EDBOT Speech Utility (not SCRATCH's).
- Then Zira SPRITE (i.e. SCRATCH) displays a Text Balloon with the words "Zira Starting" for 2 seconds.
- Next, Parameter "ZiraReady" is set to "1".

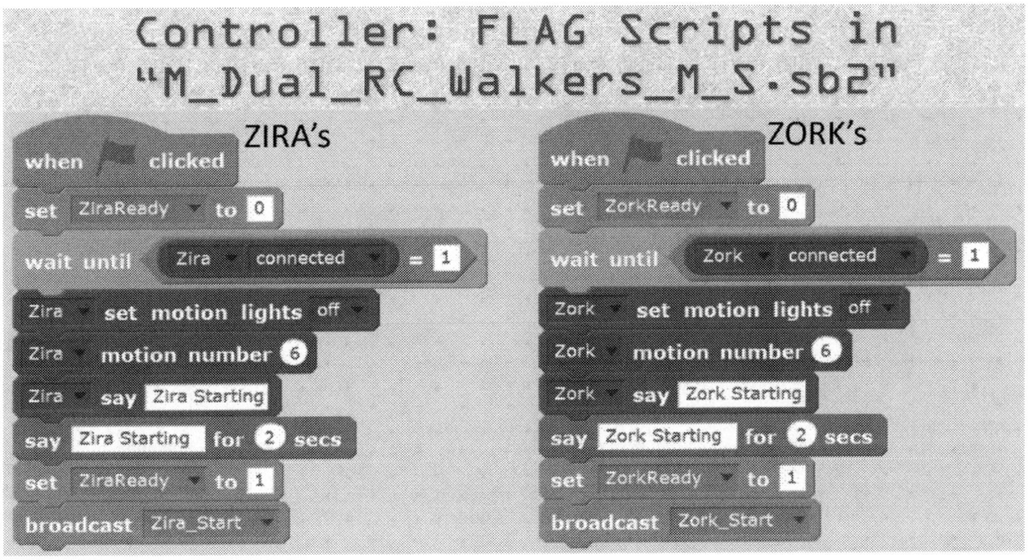

Fig. 8.21 Controller's FLAG SCRIPTS for Zira and Zork in "M_Dual_RC_Walkers_M_S.sb2".

- Lastly, the message "Zira_Start" is broadcasted and is picked up by Zira's SPRITE and SCRIPTS (see Fig. 8.25)

Please note that, as compared to TASK, SCRATCH allows a "physical" separation of the Scripts used for each separate "physical" robot, this feature is very useful to the programmer as it eases his/her ABSTRACTION process, and allows him/her to concentrate on one robot at a time during code creation. There is also a runtime benefit to have two separate FLAG Scripts because BT connections to the two robots via EDBOT sometimes do not happen for various reasons, so for example let's say that:

- Zira gets connected OK, so Zira's Script would continue beyond Zira's "Wait Until" Block, and Zira's subsequent Scripts would function as normal.
- But Zork did not get connected for some reasons, so Zork's Script would stop at his "Wait Until" Block. However, this SCRATCH project would not "crash" in this situation, and it would just only control Zira as previously programmed!

For the reader who may have only 1 MINI (let's say named "Spot") to work with, this "feature/bug" has a second benefit:

- Let's assume that the reader had done the appropriate Project Update step so "Zira" references got updated to "Spot" (see Section 8.2).
- Then, the reader would get a pleasant surprise when he/she sees that the Project "M_Dual_RC_Walkers_M_S.sb2" would work just fine with the single robot named "Spot" (*except for the various Text Displays and Text-to-Speech items in EDBOT and SCRATCH as the EDBOT Project Updater tool is not AI-enabled yet!*).

Fig. 8.22 displays the ARROW Scripts used by the Controller Sprite which are quite simple, as they just "broadcast" an "appropriate" message (Advance|Reverse|Turn Left|Turn Right) depending on which arrow key is pushed at runtime. When these messages are received by the robots SPRITEs, they are used to play the appropriate motions (Fig. 8.27).

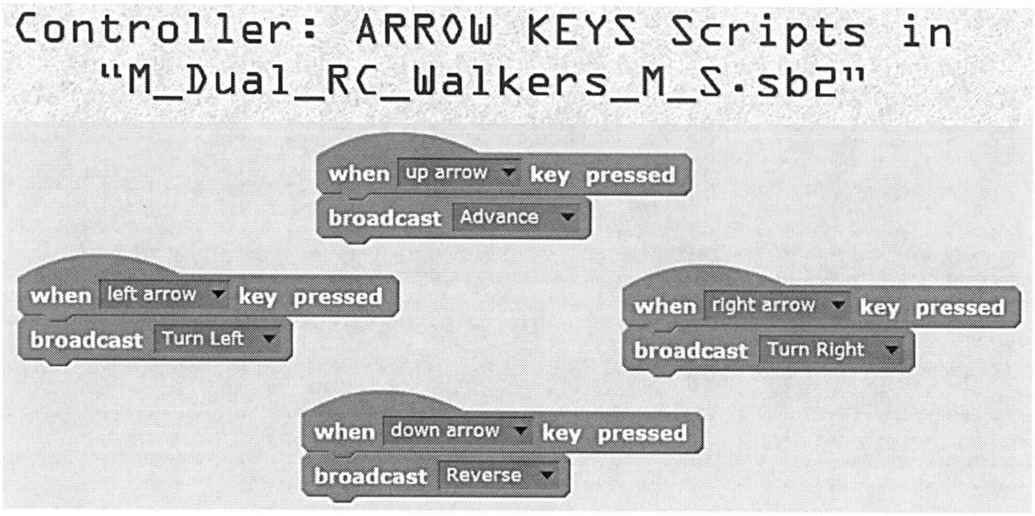

Fig. 8.22 Controller's ARROW Scripts in "M_Dual_RC_Walkers_M_S.sb2".

Once the robots are running their Scripts, the operator can issue an Emergency Stop by tapping on the Space Bar of the keyboard. Fig. 8.23 displays the SPACE BAR Script used by the Controller Sprite:

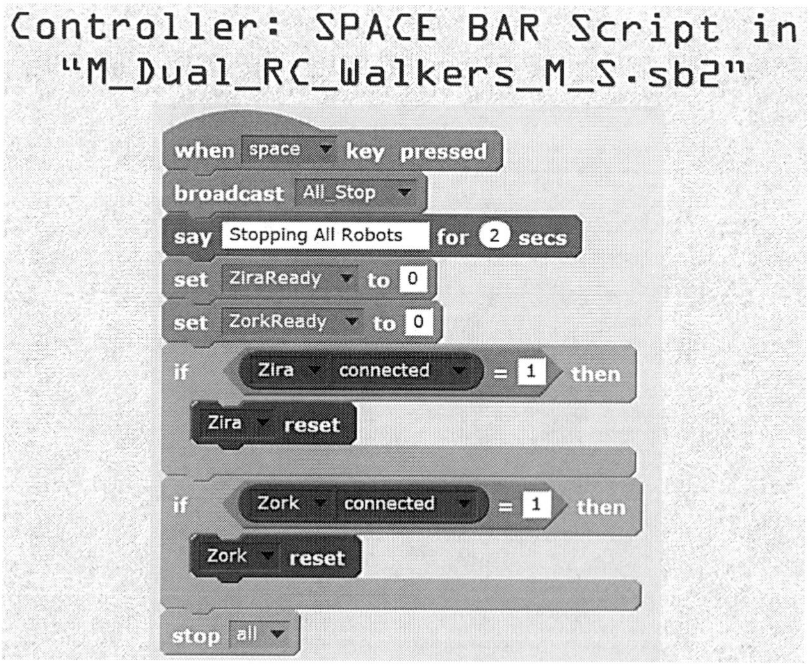

Fig. 8.23 Controller's SPACE BAR Script in "M_Dual_RC_Walkers_M_S.sb2".

- First the message, "All_Stop" is broadcasted to all robot Sprites.

- “Duck” Controller then pops up a Text Balloon “Stopping All Robot” for 2 seconds and resets Parameters “ZiraReady” and “ZorkReady” to “0”.
- The two IF structures are used to check which robot is connected so that an EDBOT “Reset” command is issued to the proper robot.
- Lastly, the SCRATCH Block “Stop All” terminates all Scripts in this project, effectively puts the SCRATCH tool back on stand-by.

Fig. 8.24 describes the FLAG Script for SPRITE Zira which would start at the same time as the other FLAG Scripts belonging to SPRITEs Controller and Zork:

- First, Zira’s Sprite switches its Costume to “Zira_Idle_m”.
- Next, Zira via EDBOT Text-to-Speech voices out “Zira Idle” and SCRATCH displays the “Zira Idle …” Text Balloon.

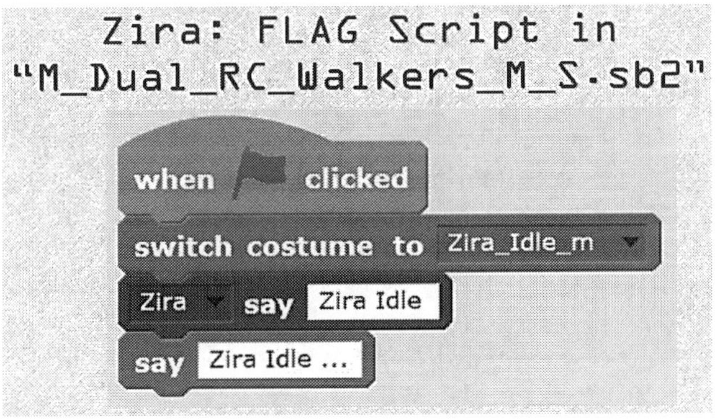

Fig. 8.24 Zira’s FLAG Script in “M_Dual_RC_Walkers_M_S.sb2”.

Fig. 8.25 lists the steps performed by Sprite Zira when it receives the message “Zira_Start”:

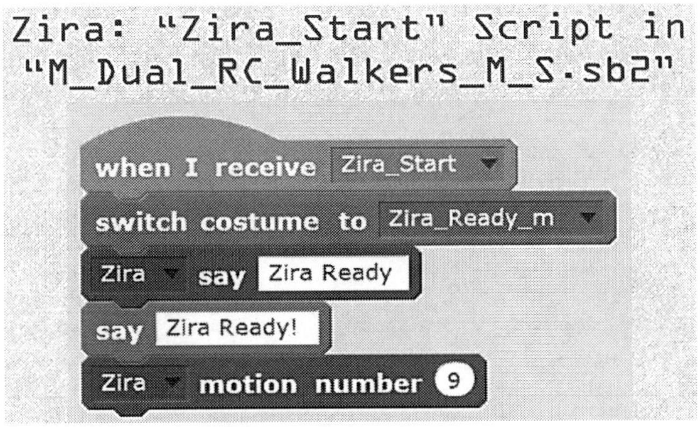

Fig. 8.25 Zira’s “Zira_Start” Script in “M_Dual_RC_Walkers_M_S.sb2”.

- Zira switches costume to “Zira_Ready_m”.

- EDBOT issues an audio message as "Zira Ready".
- SCRATCH displays a Text Balloon with the words "Zira Ready!".
- Lastly, Zira performs MOTION (9) which is a "Left-Hand Wave".

Fig. 8.26 lists the steps performed by Sprite Zira when it receives the message "All_Stop":

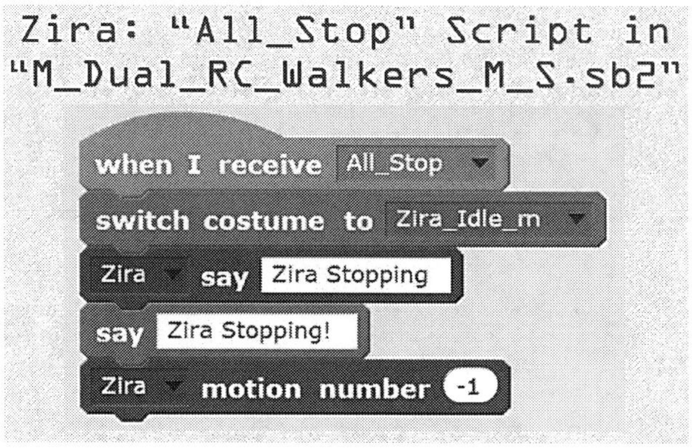

Fig. 8.26 Zira's "All_Stop" Script in "M_Dual_RC_Walkers_M_S.sb2".

- Zira's costume is switched back to its "Idle" version.
- Zira voices out and displays the message "Zira Stopping".
- Zira's last step is to perform a MOTION (-1) to stop its current Motion.

Fig. 8.27 lists the corresponding MOTIONs to be performed by Sprite Zira when it receives any of the Motion messages "Advance", "Reverse", "Turn Left" or "Turn Right".

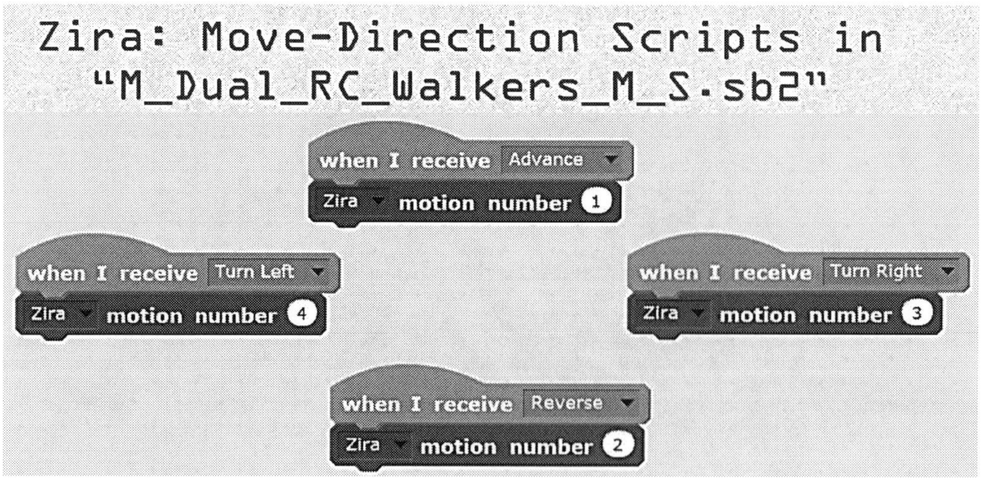

Fig. 8.27 Zira's "All_Stop" Script in "M_Dual_RC_Walkers_M_S.sb2".

Video 8.3 illustrates the runtime performances of "M_Dual_RC_Walkers_M_S.sb2":

- Synchronization of the two robots is quite good.
- If the operator types in a short list of moves for the robots in quick succession, the robots can be seen to perform that list in a delayed fashion, i.e. SCRATCH does keep a buffer of the keystrokes received.

8.5.2 Verbose Version "M_Dual_RC_Walkers_M_V.sb2"

The "Verbose" version "M_Dual_RC_Walkers_M_V.sb2" is designed with several goals:

- Implementation of SPRITE specific parameters "ZiraInMotion" and "ZorkInMotion" to obtain similar functionality to the TASK's MOTION-STATUS flags.
- All robots SPRITE Scripts are now MESSAGE-driven, while all FLAG Scripts are reserved for the Controller Sprite for a more hierarchical but also distributed control structure.
- New messaging tasks are used to show details of the progress of keystrokes and broadcasted messages during runtime.

Fig. 8.28 shows the FLAG Scripts used by the Controller Sprite to initialize Zira and Zork (let's use Zira's Script for demo):

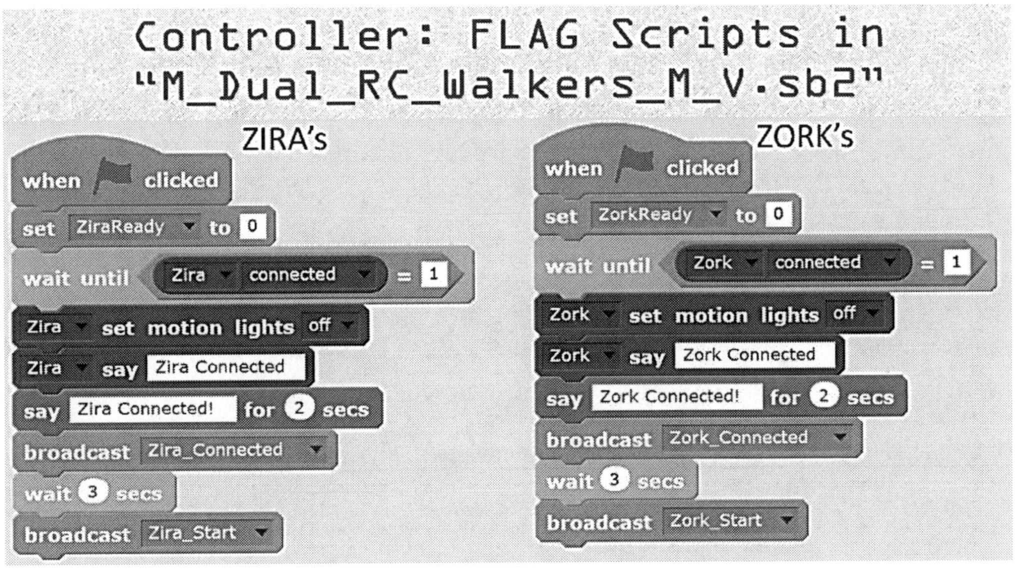

Fig. 8.28 Controller's FLAG SCRIPTS for Zira and Zork in "M_Dual_RC_Walkers_M_V.sb2".

- First, SCRATCH sets Parameter "ZiraReady" to "0" and waits for the BT connection to Zira to get established.
- Next, EDBOT sets Zira's MOTION LED setting to OFF and voices out "Zira Connected".
- SCRATCH then displays the Text Balloon "Zira Connected!" for 2 seconds.

- Next, SCRATCH broadcasts "Zira_Connected" and waits for 3 seconds (so that Zira can execute its "Zira_Connected" MESSAGE Script (Fig. 8.29).
- Lastly, SCRATCH broadcasts "Zira_Start".

The other Scripts for Controller Sprite remain un-changed, thus they are not repeated here.

Fig. 8.29 displays "Zira_Connected" Script showing Zira's steps after receiving the message "Zira_Connected":

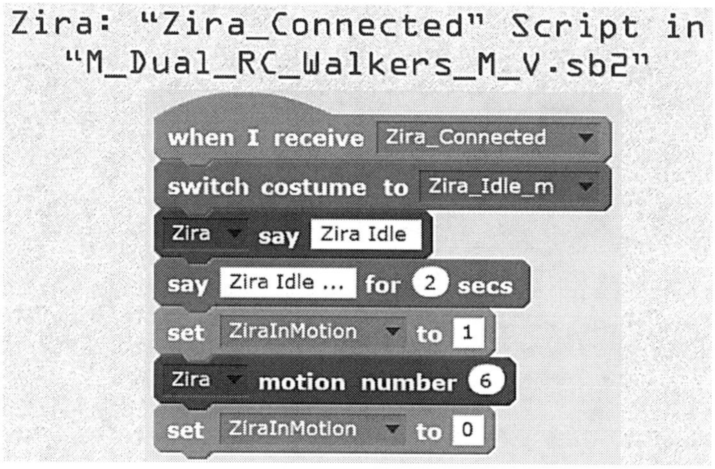

Fig. 8.29 Zira's "Zira_Connected" SCRIPT in "M_Dual_RC_Walkers_M_V.sb2".

- First, Zira's Sprite switches its Costume to "Zira_Idle_m".
- Next, Zira via EDBOT voices out "Zira Idle" and SCRATCH displays the "Zira Idle …" Text Balloon for 2 seconds.
- SCRATCH sets Parameter "ZiraInMotion" to "1", before commanding Zira to play MOTION (6) (i.e. Initial Position) via EDBOT. When this EDBOT Block is completed, SCRATCH resets Parameter "ZiraInMotion" to "0". This is an "explicit" implementation of the MOTION-STATUS Flag available in TASK programs. Parameter "ZiraInMotion" is used in other Zira's Scripts to check whether Zira is In-Play or Not when other actions need to be executed.

Fig. 8.30 displays "Zira_Start" Script showing Zira's steps after receiving the message "Zira_Start":

- Zira switches costume to "Zira_Ready_m".
- EDBOT issues the audio message "Zira Ready" and SCRATCH displays a Text Balloon with the words "Zira Ready!" for 2 seconds.
- Next, SCRATCH sets Parameter "ZiraInMotion" to "1", before commanding Zira to play MOTION (9) (i.e. Left Hand Wave) via EDBOT. When this EDBOT Block is completed, SCRATCH resets Parameter "ZiraInMotion" back to "0".
- Lastly, SCRATCH sets Parameter "ZiraReady" to "1".

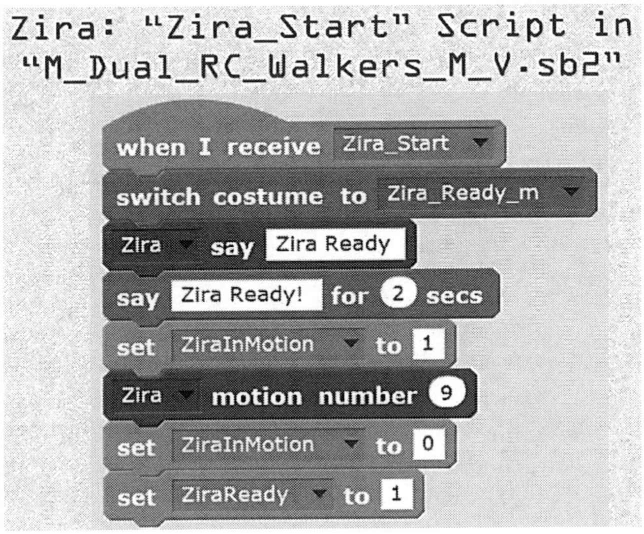

Fig. 8.30 Zira's "Zira_Start" SCRIPT in "M_Dual_RC_Walkers_M_V.sb2".

As an illustration of the "new" Scripts executed by Zira when it receives a Motion Message such as "Advance", Fig. 8.31 displays the "Zira_Advance" Script listing Zira's steps after receiving this message:

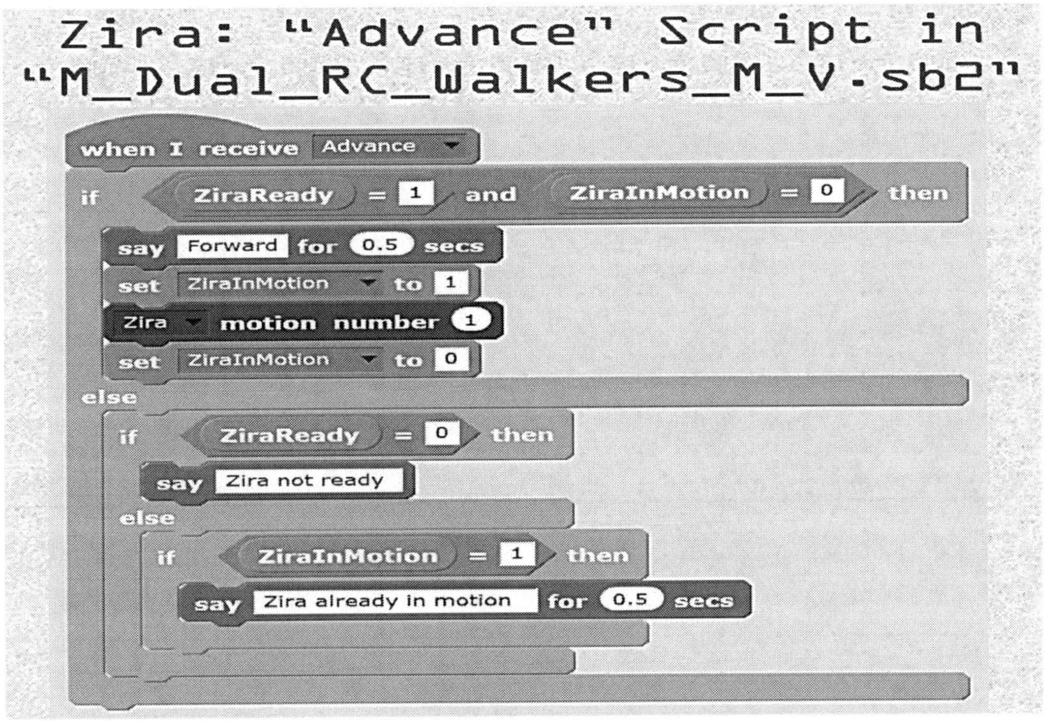

Fig. 8.31 Zira's "Advance" SCRIPT in "M_Dual_RC_Walkers_M_V.sb2".

- First, Parameters "ZiraReady" and "ZiraMotion" are compared respectively to "1" and "0". This is to check if Zira had been connected properly earlier and is not performing any Motion.
- If this condition evaluates to TRUE, then SCRATCH shows the Text Balloon "Forward" for 0.5 second, sets "ZiraInMotion" to "1" and plays MOTION (1) (i.e. (Forward)). When this MOTION is concluded, "ZiraInMotion" is reset to "0".
- However, if this condition evaluates to FALSE, then SCRATCH checks further if the "error" came from "ZiraReady" being set to "0" or from "ZiraInMotion" being set to "1". SCRATCH then displays the appropriate Text Balloon for the actual situation: "Zira not ready" or "Zira already in motion".

The other Zira's MOTION related Scripts are similarly programmed.

Video 8.4 illustrates the runtime performance of "M_Dual_RC_Walkers_M_V.sb2" which was the same as for the "Simplified" version, however the operator was not allowed to "stack up" the keystrokes in this "Verbose" version. The author is not knowledgeable enough in the inner workings of SCRATCH to offer the reader any explanation for this behavior.

8.6 Dual Random Walkers SCRATCH Project

In this Section 8.6, Zira and Zork are programmed as autonomous robots using the overall algorithm developed for Project "M_RandomWalker.tsk3" described in Section 6.6. Same as for the other Dual Robots Projects, 3 SPRITES (Controller, Zira and Zork) are used for this Project "M_Dual_RandomWalkers.sb2" which also uses the Smart Avoider Motion Group.

Fig. 8.32 shows the Controller's FLAG Scripts used separately for Zira and Zork to program their "Getting Ready" behaviors. Again, the author only describes in details Zira's behaviors as a demonstration for Zork's identical behavior:

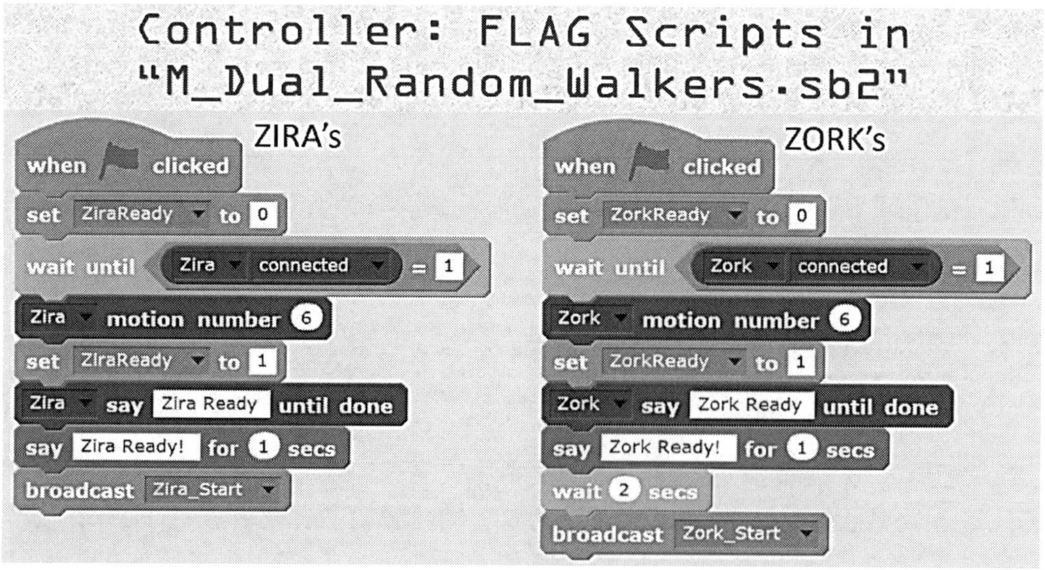

Fig. 8.32 Controller's FLAG SCRIPTS for Zira and Zork in "M_Dual_Random_Walkers.sb2".

- SCRATCH sets Parameter "ZiraReady" to "0" and waits for Zira to be connected via EDBOT.
- Zira plays MOTION (6) which is the usual "Initial Position" and then sets Parameter "ZiraReady" to "1".
- EDBOT issues the audio message "Zira Ready" and SCRATCH displays a Text Balloon with the words "Zira Ready!" for 1 second.
- Lastly, SCRATCH broadcasts the Message "Zira_Start".

The "Zira_Start" Script is designed as the Main Endless Loop for Zira's behaviors and it is displayed in three parts in Figs. 8.33, 8.34 and 8.35. Part 1 is shown in Fig. 8.33:

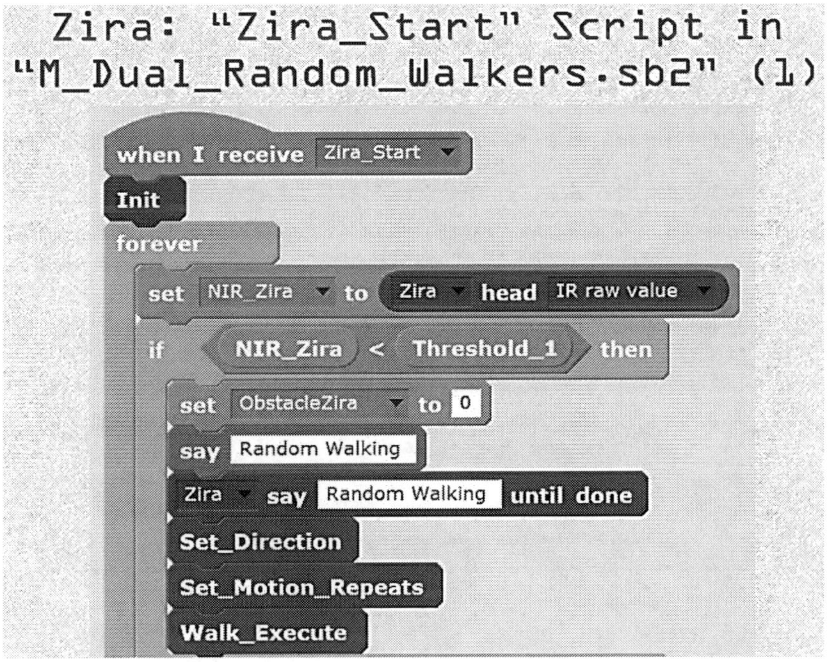

Fig. 8.33 Part 1 of "Zira_Start" Script in "M_Dual_Random_Walkers.sb2".

- First, the User Block/Function "Init" is run to initialize various parameters for Zira (see Fig. 8.36).
- Next, the Main Endless Loop is entered whereas SCRATCH records the latest reading of the Head NIR Sensor into Parameter "NIR_Zira".
- If (NIR_Zira < Threshold_1) is TRUE (i.e. no obstacle found), then Zira executes the "normal" sequence of behaviors:
    - SCRATCH sets Parameter "ObstacleZira" to "0" and displays 'Random Walking".
    - EDBOT voices out "Random Walking" until done.
    - Then the 3 Functions "Set_Direction", "Set_Motion_Repeats" and "Walk_Execute" are called in that order to enable Zira to walk in a random direction with a random number of repeats of its possible moves (Advance, Reverse, Turn Left or Turn Right).

Part 2 of the "Zira_Start' Script is shown in Fig. 8.34 and describes the first 2 steps for Zira to execute when an obstacle is found by its Head NIR Sensor (i.e. the ELSE branch):

- SCRATCH displays "Avoiding Obstacle" and EDBOT voices out "Avoiding Obstacle".
- Then, the First Avoidance Step is taken:
    - Check to see if Parameter "ObstacleZira" is set to 0. If this is TRUE, then set "ObstaclcZira" to "1" and call Function "Reg_Stop". This step is executed only once when an obstacle is first found.
- Next, the Second Avoidance Step is performed:
    - A "Repeat Until" Loop is executed for as long as the latest reading for Parameter "NIR_Zira" is greater or equal than "Threshold_2".
    - For each loop iteration, Zira would perform MOTION (2) (i.e. Reverse) ONCE, then MOTION (3) (i.e. Turn Right) TWICE. Next, SCRATCH waits for 0.1 sec for robot vibrations to dissipate and then picks up the latest Head NIR Sensor input and saves it in Parameter NIR_Zira which is then tested to see if the "Repeat Until" Loop should be repeated once more.
    - Eventually the maneuver "Reverse + 2 Right Turns" should get Zira free of the obstacle (i.e. when (NIR_Zira < Threshold_2)), then the "Repeat Until" Loop is exited and Zira goes on to the Third Avoidance Step (see Fig. 8.35).

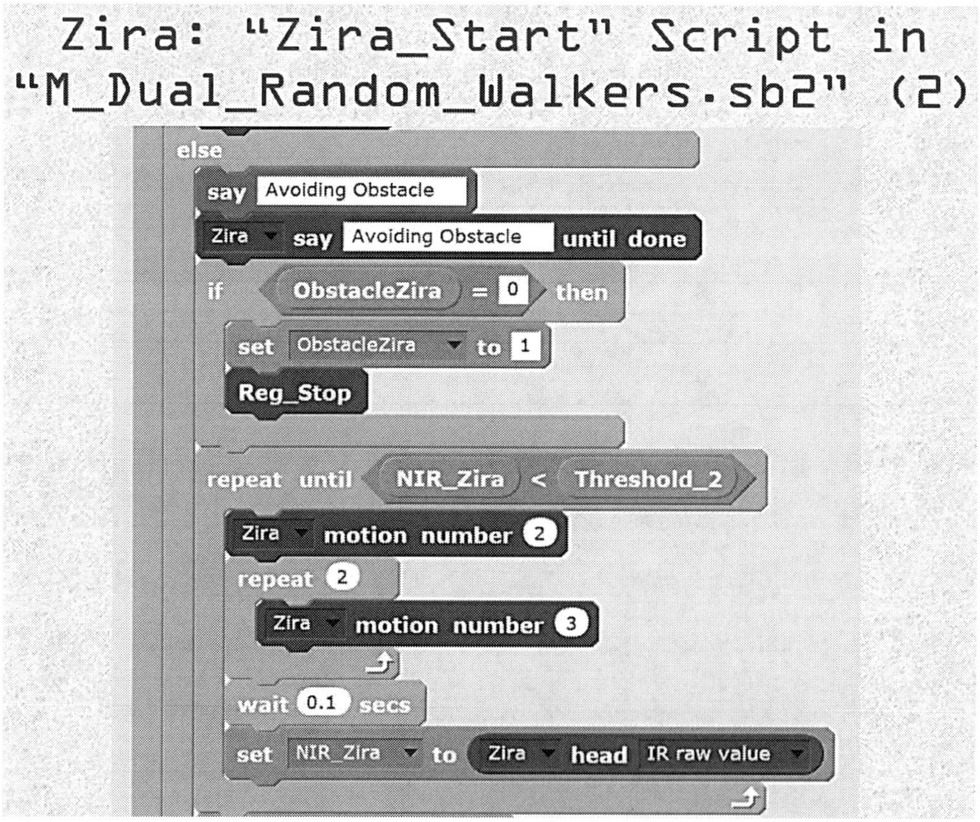

Fig. 8.34 Part 2 of "Zira_Start" Script in "M_Dual_Random_Walkers.sb2".

Part 3 of the "Zira_Start' Script is shown in Fig. 8.35 and describes the steps for Zira to do when it clears the obstacle:

- SCRATCH resets Parameter "ObstacleZira" back to "0".
- EDBOT sets the MOTION LEDs Setting to ON, then sets Servo 17 LED to Green color (i.e. "2").
- Next, Zira performs MOTION (6) (i.e. Initial Position) 4 times to make all LEDs flash 4 times correspondingly.
- SCRATCH then displays "Cleared Obstacle".
- EDBOT voices out "Cleared Obstacle" and resets the MOTION LEDs Setting back to OFF.

At this point in time, Zira reiterates on its Main Endless Loop and keeps on "Random Walking" and "Obstacle Avoiding", until the operator taps on the Space Bar to stop both robots (see Fig. 8.26).

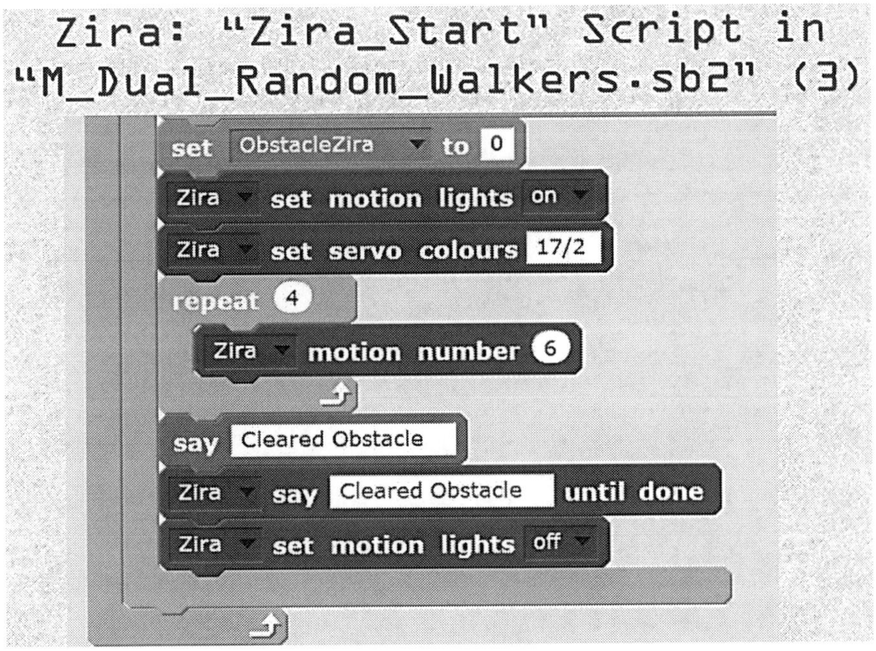

Fig. 8.35 Part 3 of "Zira_Start" Script in "M_Dual_Random_Walkers.sb2".

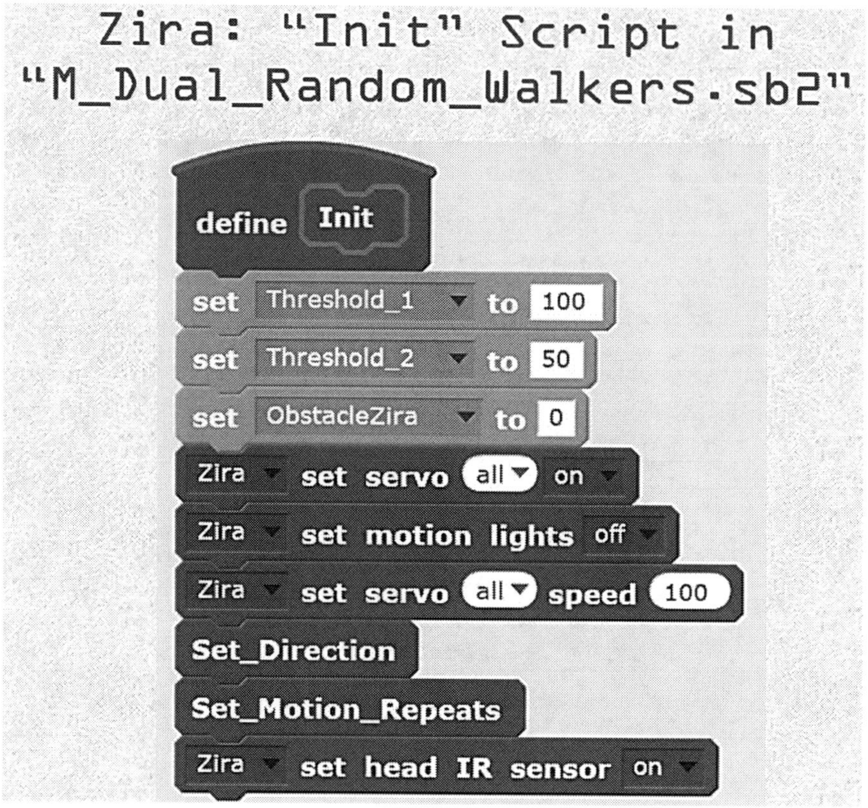

Fig. 8.36 User Block/Function "Init" in "M_Dual_Random_Walkers.sb2".

Fig. 8.36 describes the User Block/Function "Init":

- SCRATCH initializes Parameters "Threshold_1" and "Threshold_2" to their respective values of "100" and "50". SCRATCH also sets Parameter "ObstacleZira" to "0".
- EDBOT sets all servos to ON, turns OFF the MOTION LEDs Setting and sets all servos to Full Speed (100%).
- Function "Set_Direction is called next (see details in Fig. 8.37).
- Then Function "Set_Motion_Repeats" is called (see details in Fig. 8.38).
- Lastly, the "one" Head NIR Sensor on Port 1 is enabled.

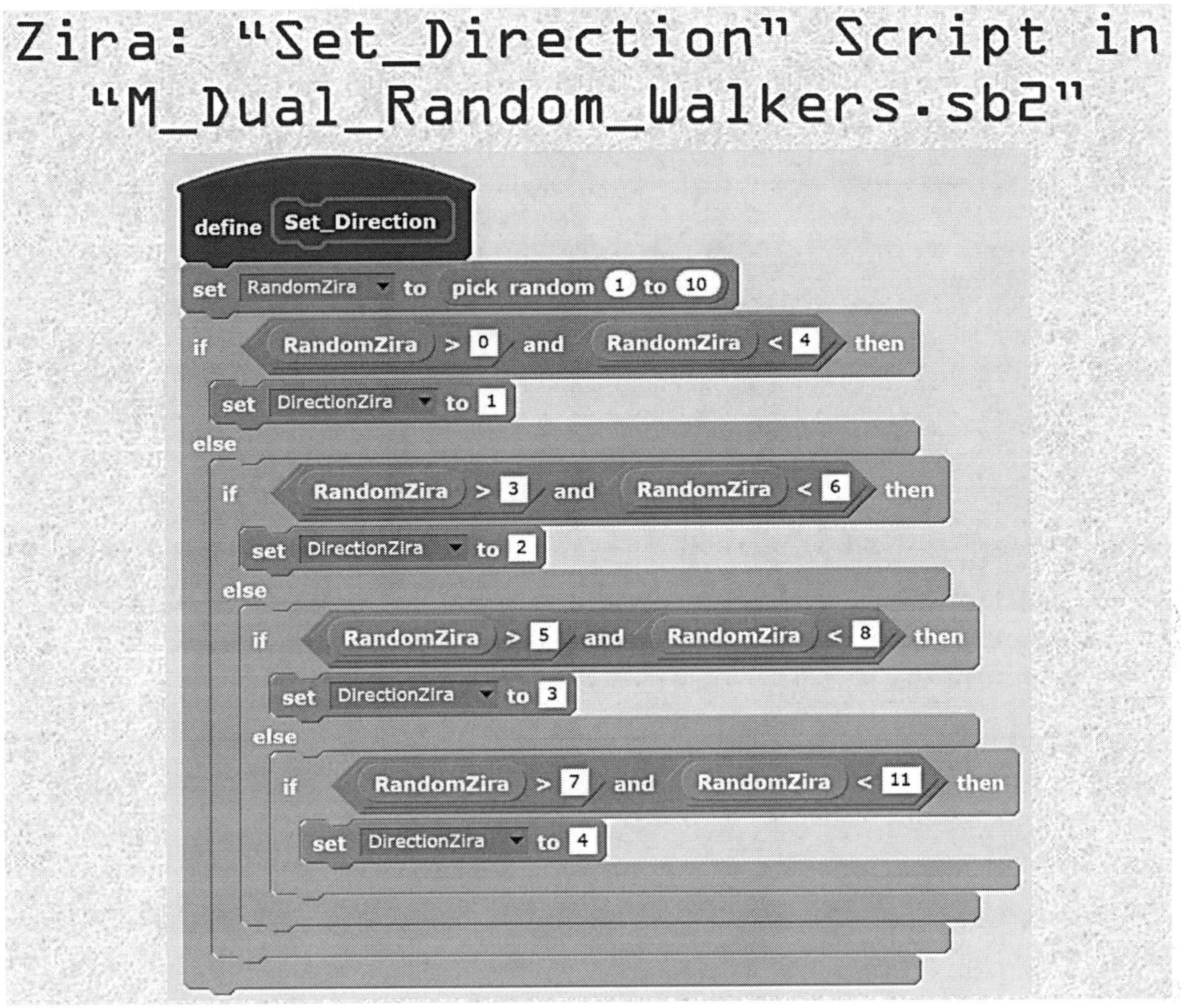

Fig. 8.37 User Block/Function "Set_Direction" in "M_Dual_Random_Walkers.sb2".

Fig. 8.37 provides the details for the User Block/Function "Set_Direction" which is called by Zira for each iteration of Zira's Main Endless Loop:

- SCRATCH is set to generate a Random Integer Number with possible values from 1 to 10:
  - [1,2,3] is chosen to correspond to Direction 1 (i.e. Advance).
  - [4,5] is chosen to correspond to Direction 2 (i.e. Reverse).
  - [6,7] is chosen to correspond to Direction 3 (i.e. Right Turn).
  - [8,10] is chosen to correspond to Direction 4 (i.e. Left Turn).

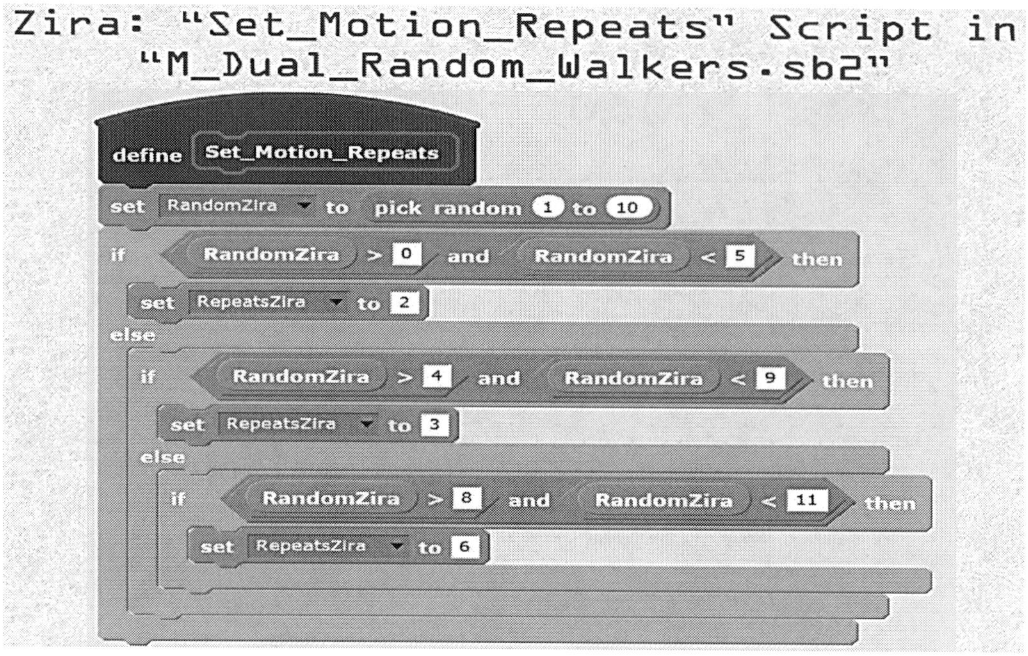

Fig. 8.38 User Block/Function "Set_Motion_Repeats" in "M_Dual_Random_Walkers.sb2".

Fig. 8.38 provides the details for the User Block/Function "Set_Motion_Repeats" which is called by Zira for each iteration of Zira's Main Endless Loop:

- SCRATCH is set to generate a Random Integer Number with possible values from 1 to 10:
    - [1,2,3,4] is chosen to correspond to "RepeatsZira" set to "2".
    - [5,6,7,8] is chosen to correspond to "RepeatsZira" set to "3".
    - [9,10] is chosen to correspond to "RepeatsZira" set to "6".

Fig. 8.39 provides the details for the User Block/Function "Check_NIR" which is called every time that the Use Block/Function "Walk_Execute" is running:

- The latest NIR Sensor reading is saved into Parameter "NIR_Zira".
- Next if the condition (NIR_Zira > Threshold1) evaluates to TRUE, then "ObstacleZira" is set to "1" and Function "Emer_Stop" is called.

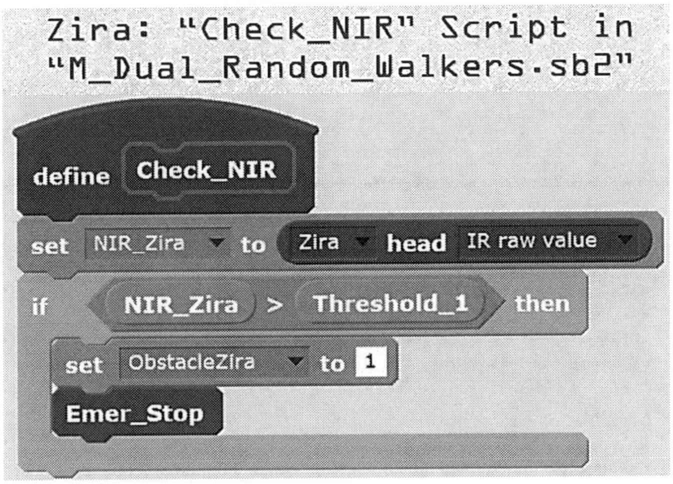

Fig. 8.39 User Block/Function "Check_NIR" in "M_Dual_Random_Walkers.sb2".

Fig. 8.40 provides the details for the User Block/Function "Reg_Stop" which is called by Zira for each iteration of Zira's Main Endless Loop whereas an obstacle is found:

- Zira immediately plays MOTION (-1).
- SCRATCH displays "Regular Stop" and EDBOT voices out "Regular Stop".
- SCRATCH waits for 1 second before displaying "Avoiding Obstacle".

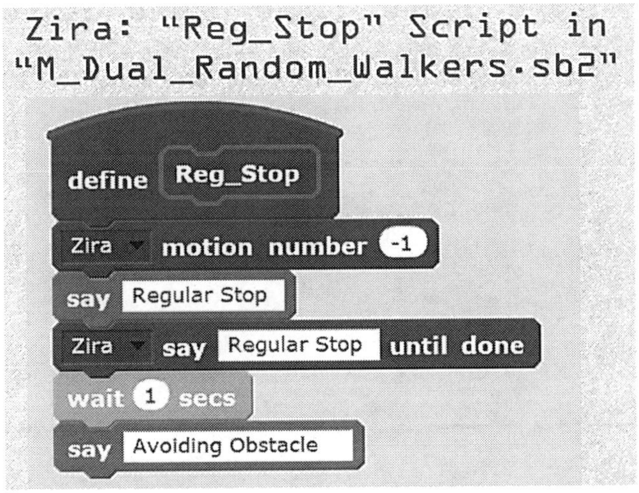

Fig. 8.40 User Block/Function "Reg_Stop" in "M_Dual_Random_Walkers.sb2".

Fig. 8.41 provides the details for the User Block/Function "Emer_Stop" which is called by Function "Check_NIR" whenever an obstacle is found:

- Zira immediately plays MOTION (0).
- SCRATCH displays "Emergency Stop" and EDBOT voices out "Emergency Stop".

- SCRATCH waits for 1 second before displaying "Emergency Obstacle Avoid".

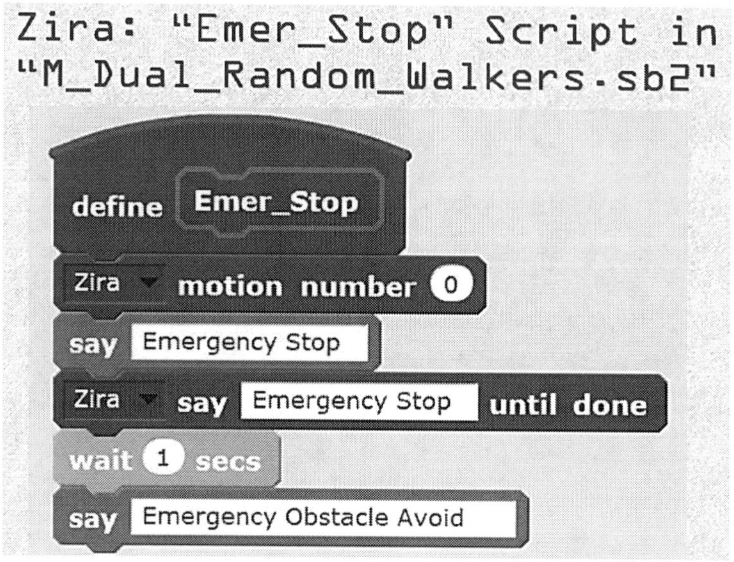

Fig. 8.41 User Block/Function "Emer_Stop" in "M_Dual_Random_Walkers.sb2".

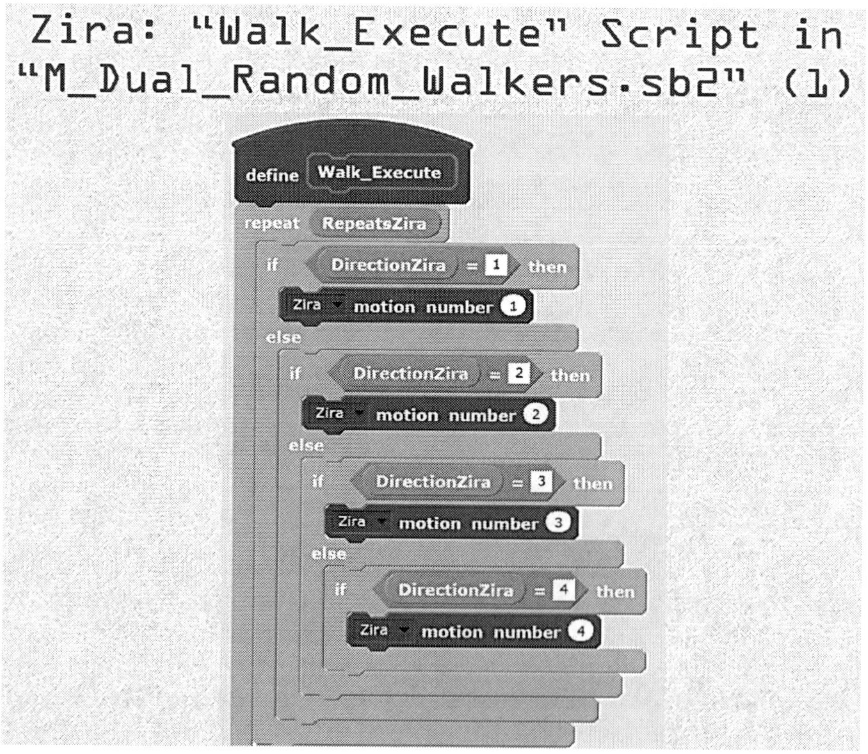

Fig. 8.42 Part 1 of User Block/Function "Walk_Execute" in "M_Dual_Random_Walkers.sb2".

The details for User Block/Function "Walk_Execute" are shown in Figs. 8.42 (Part 1) and 8.43 (Part 2). Function "Walk_Execute" is called by the Main Endless Loop every time that it iterates.

Part 1 of Function "Walk_Execute" is shown in Fig. 8.42:

- A REPEAT Loop based on Parameter "RepeatsZira" is started.
- An IF-ELSE-IF structure is used to activate the proper MOTION-LIST/PAGE among the MOTIONS (1), (2), (3) or (4) depending on the current value of Parameter "DirectionZira".

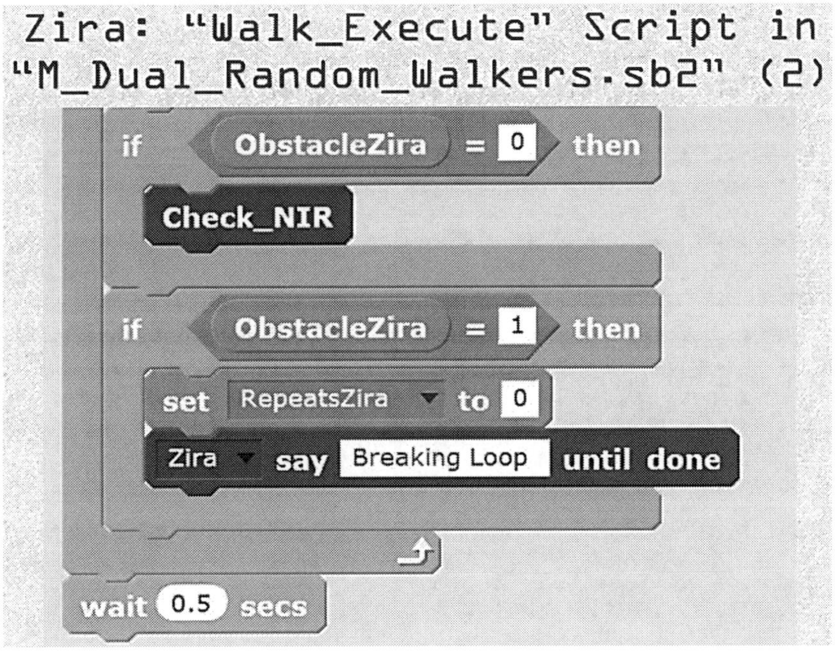

Fig. 8.43 Part 2 of User Block/Function "Walk_Execute" in "M_Dual_Random_Walkers.sb2".

Part 2 of Function "Walk_Execute" continues in Fig. 8.43 with two tasks:

- Task 1: If (ObstacleZira == 0), i.e. "so far" no obstacle is found, then Function "Check_NIR" is called to check for presence of an obstacle. This task is not activated **after** an obstacle is first found by "Check_NIR".
- Task 2: This task was designed to act as a SCRATCH code equivalent to a "BREAK LOOP" command that was used in the TASK version of the corresponding Random Walker Project (see Line 144 of Fig. 6.53). But the author must report that this coding did not work, as once a Repeat Loop is started with a certain number of repeats, nothing can change this number until this loop is concluded.

Video 8.5 illustrates the runtime performance of "M_Dual_Random_Walkers.sb2" which is usually good until EDBOT drops its BT connections with one or both robots for whatever reasons, then the operator would need to recycle the power on the robots (i.e. to reboot the BT-210s as far as EDBOT is concerned) and to restart the whole SCRATCH Project.

In summary for Sections 8.4, 8.5 and 8.6, SCRATCH is a great tool to explore multi-robots control from a single computer program and it offers the beginner programmer an easy and fun introduction to Concurrent Programming using Broadcast Messages, a subject that we can get into a deeper and more detailed level when using PYTHON with EDBOT in the remaining Sections of this Chapter 8.

8.7 Using PYTHON

When using PYTHON with EDBOT, the user can get "under the hood" to learn more details about the interface between EDBOT and the ROBOTIS firmware used on the OpenCM9.04-C. The user also can get to other utilities that were not implemented in the SCRATCH API such as:

- Full access to all 4 I/O ports (i.e. we can use all 3 NIR Sensors IRSS-10 and DMS-80).
- Custom read/write to any address defined in the Control Table of the OpenCM9.04-C (i.e. we can now apply JOINT OFFSETs).
- With V. 5.2.1523, MOTIONs "0" and "-1" do not work properly in PYTHON, although they worked fine for SCRATCH 2.
- By design, EDBOT does not update sensor and servo data while a MOTION is being played or while a set_servo_position() command is running.

A PYTHON program will require more steps than a TASK or SCRATCH program to get the MINI to do a typical action, but its execution speed is much higher and it will allow the user to tap into the vast resources of other Python packages available at the Python Software Foundation (https://pypi.org/ and https://docs.python.org/3/contents.html). The web site Python Central is also a great resource for learning and practicing Python programming (https://www.pythoncentral.io/).

EDBOT V.5.2 was a major upgrade which changed fundamentally its Python API for the MINI, thus all sample codes discussed in this section used the EDBOT V.5.2.0.1523. The reader is assumed to have some basic skills in Python Programming, if not the author is recommending these books to get the reader better prepared for materials presented in this section (web resources abound for Python also):

- "Python for Tweens and Teens: Learn Computational And Algorithmic Thinking" (Bouras and Ainarozidou, 2017).
- "Python Crash Course", 2nd Edition (Matthes, 2019).

For more in-depth treatments of Python, the user is recommended to refer to these works:

- "Programming in Python 3", 2nd Edition (Summerfield, 2010).
- "The Python 3 Standard Library by Example", 1st Edition (Hellmann, 2017).

Documentation for the EDBOT Python API is available at this web link (http://support.ed.bot/edbot-python3.html). This web page also shows how to install Python on the user's PC and provides a short primer on the programming steps needed for controlling the MINI's arms. When the EDBOT tool was installed, it created a few Python Example Codes under the user's "Documents\Edbot\python" folder:

- edbot_arms.py
- edbot_motions.py
- edbot_motions_gui.py
- edbot_twitter.py

The author is using Thonny V.3.2.4 as the Integrated Development Environment (IDE) which is available for download at https://thonny.org/ , where it has a video demonstrating the basic usage of this

IDE. Thonny V.3.2.4 also installs Python 3.7.5 for the user as part of its own installation and Thonny has its own GitHub page at https://github.com/thonny/thonny.

For the best use of the materials in Sections 8.7 and 8.8, it is recommended that the reader opens the featured Python source code into Thonny and uses it while reading the rest of Sections 8.7 and 8.8. In general, single-robot applications will use the robot "Zira", and multiple-robot applications will use the robots "Zira" and "Zork".

8.7.1 Basic Steps for PYTHON programming with EDBOT

The complete EDBOT's PYTHON API is available at this link http://support.ed.bot/edbot-python3.html which has information on how to set up and use EDBOT with PYTHON. The author assumes that the reader has installed EDBOT and configured it properly for the user's robot(s) – please review Section 8.1 and Video 8.1 as needed. The author uses THONNY 3.2.4 as his Integrated Development Environment, and it is bundled with PYTHON V. 3.7.5.

Let's first assume that the user has started EDBOT and connected successfully to a chosen robot such as Zira (see Fig. 8.44).

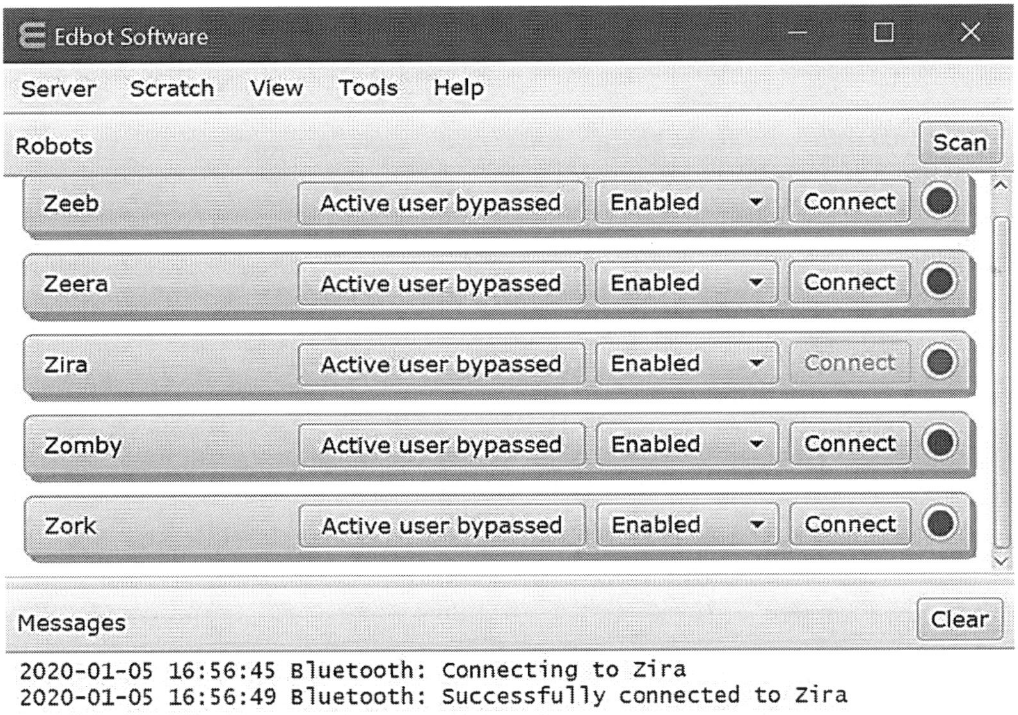

Fig. 8.44 EDBOT connected to robot Zira (author's setup).

Next, there are definite and orderly steps that the user must perform in order to successfully control one or multiple MINI robots in a PYTHON program:

1. Import the "edbot" package from the Python Package Index (PyPI) web site (https://pypi.org/project/edbot/), and "time" and "sys" as needed:
    - import edbot
    - import time   *# if time functions are used*
    - import sys   *# as we'll need to use sys.exit()*
2. Connect to EDBOT as an Edbot Client via the local Port 8080 on your personal computer with the following Python statements:
    - ec = edbot.EdbotClient("localhost", 8080)
    - ec.connect()
3. Create a variable (e.g. "robot1") and assign to it the user's robot name (e.g. "Zira") as previously configured in EDBOT (see Figs. 8.2 and 8.44):
    - robot1 = "Zira"
4. Check the previous robot name with EDBOT and verify its configuration, i.e. whether the user's robot is an "edbot", "dream" or "play" robot. The MINI is an "edbot" type and if there is a configuration error, Thonny ought to do a "system exit":
    - if not robot1 in ec.get_robot_names("edbot"):
    -   print(robot1 + " is not configured")
    -   sys.exit()
5. Wait for actual connection between Thonny (via EDBOT) and the physical robot:
    - print("Waiting to control " + robot1 + "... ", end="")
    - ec.wait_until_active(robot1)
    - print("Got control of " + robot1 + "!")

If the user's PYTHON program gets through these 5 steps with no error, the user can issue, from here on, any legal EDBOT function to control the robot to perform the user's programmed tasks, such as:

- ec.say(robot1, "connected to robot " + robot1) *# have Zira announce "connected to robot Zira"*
- *# Robots initializations*
- ec.set_servo_torque(robot1,"0/1")   *# for "All" servos, set Torque Level to 100%*
- ec.set_servo_speed(robot1,"0/100")   *# for "All" servos, set Speed to 100%*
- ec.set_custom(robot1,"82/1/0") *# write "0" as 1 byte to Address 82, i.e. turn off Motion LEDs*

If the user has another MINI robot to control (e.g. "robot2"), only steps 3, 4 and 5 are needed to be repeated for this "robot2". Later, all EDBOT functions just need the argument "robot1" or "robot2" to distinguish commands for each robot. EDBOT's support folks do not recommend the creation of a second Edbot Client by repeating Step 2 for a new object "ec2" for example, as this approach is not "thread-safe".

In practice, there will be times when, for various reasons, EDBOT drops its BT connection(s) to the robot(s), which would result in PYTHON run-time errors that are not due to the programmer's omissions.

8.7.2 Sensors Throughput Applications

Let's get started with a basic Python program reading and displaying the current readings of two IRSS-10 sensors out of Ports 1 and 4, a DMS-80 out of Port 2, and the Present Positions for Servos #1 and #17 for robot "Zira".

But before we can get into those gory details, we need to examine the Data Structures used by EDBOT to keep track of the real-time values of **all its sensors and actuators connected to each robot** configured in EDBOT's setup.

Each robot's sensors-data structure is a PYTHON dictionary named "reporters" (see link port.ed.bot/edbot-python3.html#reporters), which is a list of paired "keys" and "numerical values" shown below as an example for robot "Zira":

```
"reporters": {
  "port1": 10,
  "port2": 650,
  "port3": 335,
  "port4": 22,
}
```

For example, the current raw value for the "left" IRSS-10 sensor (with "port1" key) is "10" and the current raw value for the "port2" DMS-80 sensor is valued at "650". "So far, so good", the reader must reckon at this point! But it is getting bad quickly as the next paragraph will show.

In order to get to this specific Zira's "reporters" dictionary from inside a Python program, we need to use a method named ***"ec.get_data()"*** (http://support.ed.bot/edbot-python3.html#_get_data) which will return a "huge" dictionary containing sensors data (+ other data structures), i.e. "dictionary within dictionary":

```
{
  "server": {                  # server information
     "version": "5.0.6.1247",
     "platform": "Windows 10, 10.0.17134.523, amd64"
  },
  "auth": "9TBvXvf9",          # private session token
  "initComplete": True,        # true after connect() returns
  "robots": {
    "Zira": {                  # name of the robot
     "enabled": True,          # enabled?
     "connected": True,        # Bluetooth connected?
     "reporters": {
       ...
     },
     "activeUser": "Python...",  # currently active user
     "model": {                # the robot model
        "name": "Edbot",
        "type": "ERM162",
        "key": "edbot"
```

```
      }
    }
  },
  "user": "Python <Username@[192.168.1.24]:51144>",
  "users": [
    "Python <Username@[192.168.1.24]:51144>",
    "Scratch 2.0 <Username@[192.168.1.24]:0>",
  ]
}
```


Unfortunately, the Zira's "reporters" dictionary, that we are interested in, is "buried" inside the "robots" dictionary highlighted in yellow as shown above. This means that we need to use the value "Zira" and apply it to the "robots" dictionary in order to extract out the "reporters" dictionary of interest, which is then can be extracted further out for its individual "sensor" key and its corresponding numerical value! The implementation of this complex procedure will be shown later in Fig. 8.48.

For now, let's open the file "M_SensorsThroughputBasic.py" into THONNY. Figs. 8.45, 8.47 and 8.48 contain only the pertinent statements of this project to be explained further.

```python
import sys
import time
import edbot

# Main program starts here
# Connecting to the Edbot server. The Edbot Software needs to be running.
#
ec = edbot.EdbotClient("localhost", 8080)
ec.connect()

#
# Set the global names for the robots used and verify their configurations
#
robot1 = "Zira"
if not robot1 in ec.get_robot_names("edbot"):
    print(robot1 + " is not configured")
    sys.exit()

#
# Wait for control of the MINI robot(s) used.
#
print("Waiting to control " + robot1 + "... ", end="")
ec.wait_until_active(robot1)
print("Got control of " + robot1 + "!")
```

Fig. 8.45 Steps to connect to Zira and secure its control by THONNY via EDBOT software.

Fig. 8.45 displays the minimal steps to be used to ensure that connection and control have been properly established between THONNY, EDBOT and the MINI's OpenCM9.04-C controller. These statements are present in all example Python codes for this book:

- Line 7 imports the "sys" package so that we can use the methods exit() and print() at a later time.
- Line 8 imports the "time" package so that we can use the method time.sleep() later (Fig. 8.48).
- Line 9 imports the "edbot" package so that we can use its methods to control the MINI robot and to extract its sensors data.
- Line 14 uses the EDBOT's method named "EdbotClient" to set up THONNY as a client to the EDBOT software tool (which should be running already!) and assign this instance to Variable "ec". The argument "localhost" means that we are using the local PC as the EDBOT server, and the argument "8080" specifies that Port 8080 is to be used. These settings need to match the settings used when the user installed and set up the EDBOT package (see Fig. 8.2). Essentially, from the main menu bar of EDBOT, if the user chooses Server >> Setup, then two options are available (see Fig. 8.46):

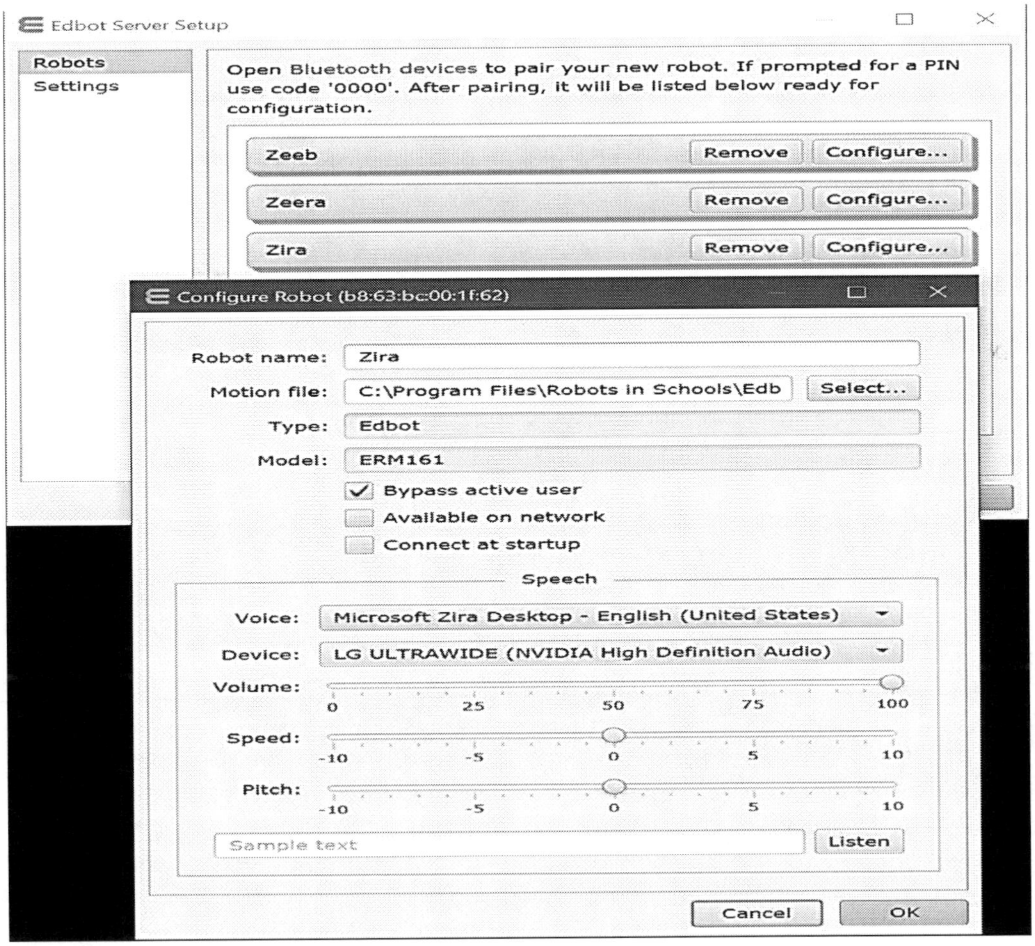

Fig. 8.46 Edbot Server Setup options for "Robots".

  - In “Settings”, the user needs to make sure that the Server Port field is set to “8080”.
  - In “Robots” (see Fig. 8.46), the user needs to note the two options “Bypass active user” and “Available on network”. The option, “Bypass active user”, when checked, allows any connection to control the robot, i.e. both SCRATCH and THONNY can access the robot at the same time, so this option is needed when developing PYTHON applications via THONNY. The option “Available on network” should be unchecked if the user is concerned about other users on the local network accessing inadvertently the robot that the user is working on. The author also prefers to leave the option “Connect at startup” unchecked so that EDBOT won’t waste time trying to connect to robots that the author is not using at that time.
- Line 15 establishes the connection between EDBOT and the user’s PYTHON session, i.e. the handle/object named “***ec***”.
- Line 20 assigns the string “Zira” to Variable “robot1”. Of course, if the user’s program needs to control a second robot, then the user needs to add a similar statement (***robot2 = “Zork”***) for example.
- Next, an IF structure (Lines 21-23) is used to check the **non-existence** of “robot1”, i.e. string “Zira” in the setup configuration of the EDBOT tool for “**edbot**” type of robots, by calling the method ***ec.get_robot_names(“edbot”)***. If this condition happens to be True, the print statement in Line 22 gets executed, then this PYTHON session is exited via method ***sys.exit()*** (Line 23). If this condition is False, meaning that “Zira” is properly configured in the EDBOT tool, then program control flows to Line 28 which prints out ***“Waiting to control Zira…”***. The argument ***end=””*** means that the cursor should stay on the same print line (for the next print command in Line 30). Please note that by default the ***print()*** method goes to a new line after its execution, i.e. ***end=”\n”***.
- Line 29 waits until “robot1” (i.e. Zira) gets active then program control flows to Line 30. Please note that if the option “Bypass active user” was not set properly earlier, THONNY would get stuck at Line 29 indefinitely.
- Line 30 prints out ***“Got control of Zira!”*** - on the same output line as for Line 29 (then it goes to a new line).

Next, Fig. 8.47 shows the initialization steps used for robot Zira:

- Line 35 uses method ***ec.set_servo_torque()*** to turn ALL SERVOS (parameter value “0”) to the ON state (parameter value “1”) via the string “0/1”.
- Line 36 similarly uses method ***ec.set_servo_speed()*** to set ALL SERVO SPEEDS to the 100% level via the string “0/100”.
- Line 37 uses method ***ec.set_custom()*** to set the Motion Play LED Parameter with a “1” byte constant valued with “0” (i.e. not to have the servos’ LED flashing during Motion Play). The Motion Play LED Parameter is located at Address 82 in the Control Table for OpenCM9.04-C (see link http://emanual.robotis.com/docs/en/parts/controller/opencm904/#ram-area).
- Line 38 uses method ***ec.run_motion()*** to make “robot1” perform MOTION-LIST/PAGE 6 (i.e. Initial Position), as we are using the Smart Avoider MOTION-GROUP for this PYTHON application.
- With Line 39, THONNY is then set to “do nothing” for 1 second with method ***time.sleep ()***.
- Line 40 makes robot1 “Zira” perform MOTION-LIST/PAGE 9, i.e. to do a Left Arm Wave.

```
#
# Power up robot and put it in Ready/Initial Position
#
ec.set_servo_torque(robot1,"0/1")
ec.set_servo_speed(robot1,"0/100")
ec.set_custom(robot1,"82/1/0")
ec.run_motion(robot1, 6)
time.sleep (1)
ec.run_motion(robot1, 9)
```

Fig. 8.47 Initialization Steps for robot Zira.

Lastly, Fig. 8.48 shows the main task to be done:

```
41 # Read and display sensors data
42 ec.say(robot1, "Reading sensors data from robot " + robot1)
43 while True:
44     data = ec.get_data()
45     sensors_data = data["robots"][robot1]["reporters"]
46     print("----------------------------------------")
47     if "port1" in sensors_data:
48         distance_port1 = sensors_data["port1"]
49         print("Left IRSS-10 = " + str(distance_port1))
50     if "port2" in sensors_data:
51         distance_port2 = sensors_data["port2"]
52         print("Center DMS-80 = " + str(distance_port2))
53     if "port4" in sensors_data:
54         distance_port4 = sensors_data["port4"]
55         print("Right IRSS-10 = " + str(distance_port4))
56     if "servo-01" in sensors_data:
57         position_servo1 = sensors_data["servo-01"]["position"]
58         print("Servo 1 Current Position = " + str(position_servo1))
59     if "servo-17" in sensors_data:
60         position_servo17 = sensors_data["servo-17"]["position"]
61         print("Servo 17 Current Position = " + str(position_servo17))
62     time.sleep(0.1)
```

Fig. 8.48 Code for reading Sensors and Servos data for robot Zira
in "M_SensorsThroughputBasic.py" – using method ***ec.get_data()***.

- Line 42 makes Zira (i.e. the EDBOT tool) announce the phrase "Reading sensors data from robot Zira".
- Line 43 marks the beginning of an endless loop which includes Lines 44 through 62. Line 43 is the PYTHON equivalent for the start of the familiar Endless/Forever Loop. Thus, the block of commands from Lines 44 to 62 gets repeated forever.

- Line 44 saves the latest robot's informational items into a rather large data structure named "data" using method ***ec.get_data()*** (http://support.ed.bot/edbot-python3.html#_get_data).
- Line 45 uses the value "robot1" for the key "robots" to retrieve the dictionary named "reporters" and save it in "sensors_data".
- Lines 47-61 contains 5 IF structures designed to access specific sensor data as well as servo data, for example:
    - Line 47 checks to see if the key "port1" exists in "sensors_data", and if this evaluates to TRUE, Lines 48-49 are executed next. Line 48 retrieves the actual IRSS-10 raw sensor value (i.e. "sensors_data ["port1]) and saves it into Variable "distance_port1". Line 49 then prints out this value on the computer screen.
    - Line 59 checks to see if the key "servo-17" exists in "sensors_data", and if this is TRUE then Lines 60-61 are executed next. Line 60 retrieves the numerical value that corresponds to the sub-key "position" within the sub-dictionary (sensors_data["servo-17"]) and saves it in Variable "position_servo17". Line 61 prints out this value on the computer screen.

Fig. 8.48 shows a good example of "defensive programming" to avoid "embarrassing" run-time errors for accessing "non-existent" data, but if the programmer is certain of the existence of specific sensors and servos on his/her robot, a more direct access to these data is shown in Fig. 8.49 for the program "M_SensorsThroughputBasic_NoIf.py":

```
while True:
    data = ec.get_data()
    sensors_data = data["robots"][robot1]["reporters"]
    print("-----------------------------------------")
#     if "port1" in sensors_data:
    distance_port1 = sensors_data["port1"]
    print("Left IRSS-10 = " + str(distance_port1))
#     if "port2" in sensors_data:
    distance_port2 = sensors_data["port2"]
    print("Center DMS-80 = " + str(distance_port2))
#     if "port4" in sensors_data:
    distance_port4 = sensors_data["port4"]
    print("Right IRSS-10 = " + str(distance_port4))
#     if "servo-01" in sensors_data:
    position_servo1 = sensors_data["servo-01"]["position"]
    print("Servo 1 Current Position = " + str(position_servo1))
    position_servo17 = sensors_data["servo-17"]["position"]
    print("Servo 17 Current Position = " + str(position_servo17))
    time.sleep(0.1)
```

Fig. 8.49 Alternate code for reading Sensors and Servos data for robot Zira in "M_SensorsThroughputBasic_NoIf.py" – using method ***ec.get_data()***.

For interested readers, a more compact way to code is also shown in Fig. 8.50 (Lines 52 to 58) from program "M_SensorsThroughputBasic_Exceptions.py":

```
48 ec.say(robot1, "Reading sensors data from robot " + robot1)
49 counter = 1
50 while True:
51     try:
52         sensors_data = ec.get_robot(robot1)["reporters"]
53         print("----------------------------------------")
54         print("Left IRSS-10 = " + str(sensors_data["port1"]))
55         print("Center DMS-80 = " + str(sensors_data["port2"]))
56         print("Right IRSS-10 = " + str(sensors_data["port4"]))
57         print("Servo 1 Current Position = " + str(sensors_data["servo-01"]["position"]))
58         print("Servo 17 Current Position = " + str(sensors_data["servo-17"]["position"]))
59         print("Counter = " + str(counter))
60         time.sleep(0.10)
61         counter = counter + 1
62         if (counter == 1001):
63             robot1 = "Spot"
64         ec.set_servo_position(robot1, "17/100")
65         time.sleep (0.5)
66         ec.set_servo_position(robot1, "17/200")
67         time.sleep (0.5)
68         ec.run_motion(robot1, 6)
69 
70     except KeyboardInterrupt:
71         print("Session ended by user")
72         sys.exit()
73     except KeyError:
74         print("Key = " + robot1 + "was not found")
75         sys.exit()
```

Fig. 8.50 Compact code for reading Sensors and Servos data for robot Zira in "M_SensorsThroughputBasic_Exceptions.py" – using method ***ec_get_robot(robot1)***.

Please note that method ***get_robot()*** (http://support.ed.bot/edbot-python3.html#_get_robot) is used in Fig. 8.50 (Line 52), instead of method ***get_data()*** as shown earlier in Figs. 8.48 and 8.49. Furthermore, method ***get_robot()*** requires 1 input argument (i.e. the key "robot1"), while ***get_data()*** requires no input argument. In a way, Line 52 in Fig. 8.50 shows that "***ec.get_robot(robots1)***" is equivalent to the expression "***data[robots][robot1]***" in Line 46 of Fig. 8.49.

In Fig. 8.50, the program "M_SensorsThroughput_Exceptions.py" also illustrates the handling of run-time exceptions - for more information please visit https://docs.python.org/3/tutorial/errors.html?highlight=exception. More details of the built-in exceptions can be found at https://docs.python.org/3/library/exceptions.html#bltin-exceptions.

Two types of exception are considered: 1) "**KeyboardInterrupt**", i.e. when the user pushes "Control-C" on the keyboard; and 2) "**KeyError**", i.e. when a key, used in a dictionary, was not previously configured inside EDBOT:

- Line 49 initializes Variable "counter" to 1 before entering the Endless Loop at Line 50.
- This EndLess Loop spans from Line 51 through Line 75, where a "**try-except**" structure is used to "capture" and "respond" to a specific run-time error:
    - The "**try:**" block contains all the statements needed to perform the needed tasks (Lines 52-68).
    - Lines 52-58 are for the tasks already explained in detail previously.
    - Line 59 prints out the current value of Variable "counter".

- Line 60 makes THONNY wait for 0.1 second.
- Line 61 increments Variable "counter" by 1, and **if (counter == 1001)** (Line 62), updates Variable "robot1" with a new string "Spot" (Lines 62-63), **else** (i.e. counter <=1000) loops back to Line 52. In other words, for the first 1000 loop iterations, "robot1" is set to "Zira" which was well defined in **the author's EDBOT's Setup** – so no runtime errors. But at iteration 1001, "robot1" is set to "Spot" (Line 63), which was not configured in EDBOT's Setup previously, therefore this value change will cause a runtime error when the method ***ec.get_robot(robot1)*** gets called in the next loop iteration (1001) at Line 52, raising the exception "KeyError".
- Line 64 moves the MINI's head to the right and Line 65 waits for 0.5 second.
- Line 66 moves the MINI's head to the left and Line 66 waits for 0.5 second.
- Line 68 plays MOTION-LIST/PAGE 6, i.e. the Initial Pose.
- Next is the first "**except:**" block for the event "**KeyboardInterrupt**", i.e. when the user pushes "Control-C" on the keyboard. For this event, we want to print an appropriate error message before invoking "***sys.exit()***" to quit this program (Lines 70-72).
- Last is the second "**except:**" block for the event "**KeyError**", i.e. when a dictionary is being accessed using a non-existent key value (Lines 73-75), i.e. when "Spot" is set by Line 63 at the end of Iteration 1000.

Video 8.6 illustrates the run-time performances of all three versions of "M_SensorsThroughput".

Next, let's explore what PYTHON can and cannot do with more substantial projects in Section 8.8.

8.8 MINI PYTHON/EDBOT Applications

Using PYTHON with EDBOT will give us the opportunities to compare PYTHON solutions to similar TASK solutions created for the same type of projects such as "RC Walker" and "Random Walker", but it will also allow us to extend the capabilities of the MINI with "Dual MINIs" and "Web Cam" projects using Threading and OpenCV libraries (features that are now not available using ROBOTIS TASK V.3).

Let's first review the key design features and constraints used in EDBOT:

1. In the TASK project "M_RC_Walker_IS.tsk3" (see Section 6.4.3), the parameter "MOTION-STATUS" was key to the logic design of this TASK program, but there is no such equivalent parameter in EDBOT.
2. Previously, in SCRATCH+EDBOT applications (see Section 8.4.2), the Play-Motion Block would wait until the called Motion is finished playing before letting the next Block activate. However, when PYTHON is used with EDBOT, the method ***run_motion()*** can be used with either WAIT or NO-WAIT option (http://support.ed.bot/edbot-python3.html#run_motion).
3. Also, in previous SCRATCH+EDBOT applications, there was no way to set the JOINT-OFFSET of the XL-320 servos. But the PYTHON API has a method called ***set_custom()*** that can now be used to set these JOINT-OFFSET parameters for Servos #1 and #17.
4. The EDBOT PYTHON API does not allow access to sensors and servos data during a MOTION PLAY.

We'll use the Pygame library for the display and keyboard interfaces (https://www.pygame.org/docs/ref/display.html and https://www.pygame.org/docs/ref/key.html) and it turns out that Pygame also implements the same "sequential" and "single-key" input functionalities that

SCRATCH was using. Thus, the combined-buttons approach of TASK programs cannot be reproduced in EDBOT+PYTHON applications also.

8.8.1 RC Walker with Independent Servo Control

In this project "M_RC_Walker_IS.py", the Pygame package is used to monitor the user's keyboard inputs and to provide graphical screen outputs at run time. The web site "https://www.pygame.org/docs/" has all the official PyGame documents that the reader can consult as needed. If the reader prefers a regular textbook, the author recommends the work by Kinsley & McGugan (2015).

The MINI robot named Zira is used in this project and Fig. 8.51 is a typical screen capture at run time.

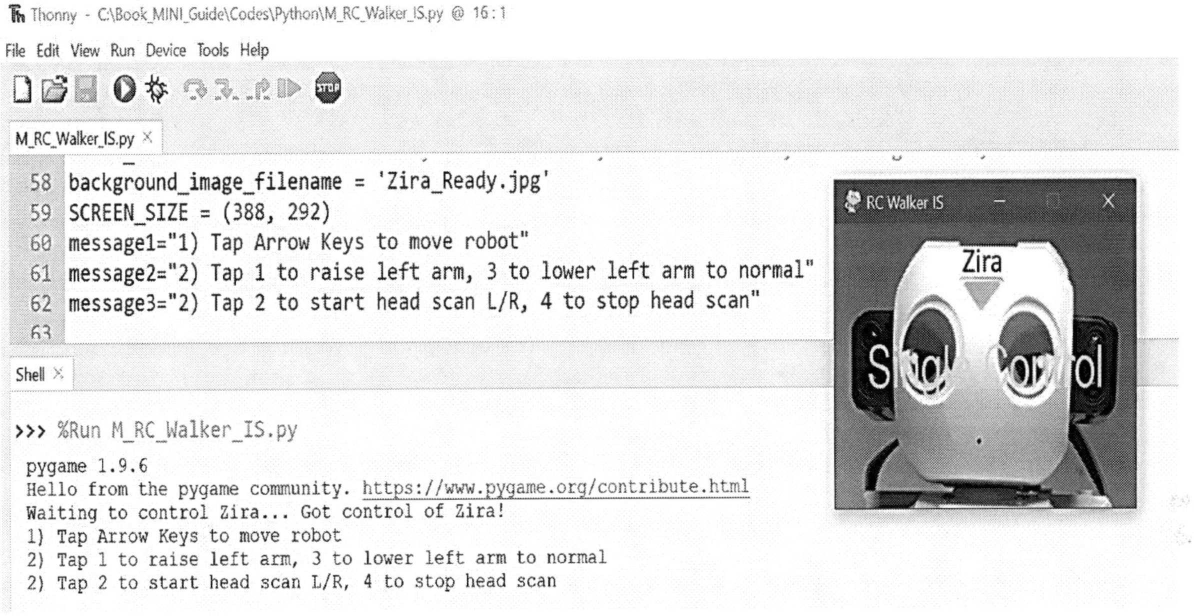

Fig. 8.51 Screen capture at run time for "M_RC_Walker_IS.py".

Figs. 8.52 through 8.60 display the pertinent code sections to be described in the following paragraphs.

"M_RC_Walker_IS.py"
Import Packages &
Selected "Motion" Methods

```
7  import sys
8  import time
9  import edbot
10 import pygame
11 from pygame.locals import *
```

```
14 def init_pose(robot):
15     ec.run_motion(robot, 6, wait=False)
16
17 def forward(robot):
18     ec.run_motion(robot, 1, wait=False)
19
20 def backward(robot):
21     ec.run_motion(robot, 2, wait=False)
```

```
29 def r_arm_wave(robot):
30     ec.run_motion(robot, 8, wait=False)
31
32 def l_arm_wave(robot):
33     ec.run_motion(robot, 9, wait=False)
```

Fig. 8.52 Packages Import and Selected Method Definitions for "M_RC_Walker_IS.py".

Lines 10-11 in Fig. 8.52 show how to import the Pygame package and a limited set of constants and methods that will be needed in this Python project (https://www.pygame.org/docs/tut/ImportInit.html).

In the source code for "M_RC_Walker_IS.py", Lines 14-50 list the definitions of user methods for this project which are: ***init_pose(), forward(), backward(), right(), left(), r_arm_wave(), l_arm_wave(), raise_arm(), lower_arm(), go_scan_head()*** and ***stop_scan_head()***. But for lack of space, Fig. 8.52 contains only the details for ***init_pose(), forward(), backward(), r_arm_wave()*** and ***l_arm_wave()*** which are described further below:

- Lines 14-33 in Fig. 8.52 define the MOTION-related methods whereas each has 1 input argument named "robot" and a single command using the ***ec.run_motion()*** method to play the correct MOTION-LIST/PAGE from the Smart Avoider Motion Group. The option/argument "wait" is set to "False", meaning that this ***run_motion()*** method will not wait for the called motion to finish playing before the next PYTHON statement is executed. The reader can also try using "wait=True" to see if any discernable run-time behavior is exhibited, resulting from this change.

"M_RC_Walker_IS.py"
Methods used for Servos #1 & #17

```
35 def raise_arm(robot):
36     ec.set_custom(robot,"232/2/1024")   # set Servo1 joint-offset to 1024
37     ec.set_servo_position(robot, "1/300")
38
39 def lower_arm(robot):
40     ec.set_custom(robot,"232/2/0")   # set Servo1 joint-offset to 0
42 def go_scan_head(robot):
43     ec.set_servo_position(robot, "17/100")
44     time.sleep (0.5)
45     ec.set_servo_position(robot, "17/200")
46     time.sleep (0.5)
47
48 def stop_scan_head(robot):
49     ec.set_servo_position(robot, "17/150")
50     time.sleep (0.5)
```

Fig. 8.53 Methods used for Servos #1 and #17 in "M_RC_Walker_IS.py".

Fig. 8.53 lists the 4 methods used for Servos #1 and #17 which are to be controlled independently from the moves that were previously assigned by various MOTION-LIST/PAGEs from the Smart Avoider Motion Group:

- ***raise_arm()*** and ***lower_arm()*** (Lines 35-40) – As Servo #1 (right shoulder joint) is involved in the "Forward" and "Backward" motions of the MINI, the programming of these two methods is different from the methods used for Servo #17 (neck joint/head) which stay fixed at Position 150 degrees during **all motions**:
    - For ***raise_arm()***, Line 36 uses ***ec.set_custom()*** to send a string that has 3 components (visit http://support.ed.bot/edbot-python3.html#_set_custom for more details):
        1. "232" is the address that corresponds to the JOINT-OFFSET of Servo #1. This information is found by using the MANAGER Tool (see Fig. 3.7 and use the pull-down menu for Item "Adjust Offset ID" and choose Servo #1).
        2. "2" means that a 2-byte constant is being sent over to the MINI via EDBOT.
        3. "1024" is the actual numerical value of this 2-byte constant.
    - Essentially, Line 36 takes Servo #1 out of the influence of any MOTION being played.
    - Line 37 then sends Servo #1 to a Goal Position of "300" degrees, i.e. raises up the right arm, and hold it there until "reset" by ***lower_arm()***.
    - For ***lower_arm()***, Line 40 uses ***ec.set_custom()*** to send a "0" to Address 232 which effectively puts Servo #1 back under the control of any MOTION that is being played then or at a later time (as long as its JOINT-OFFSET is still set at "0"). Thus, there was no need to send servo #1 back to its initial Goal Position of "150" degrees.
- ***go_scan_head()*** (Lines 42-46) simply sends Servo #17 to Position "100", waits for 0.5 second for it to get there, then sends it to Position "200" and waits for another 0.5 second for it to get this position. This task is possible because the JOINT-OFFSET for Servo #17 is already set to 1024 back in the Main Method (see Fig. 8.56 – Line 94).
- ***stop_scan_head()*** (Lines 48-50) just resets Servo #17 back to its initial Goal Position of "150" degrees.

Fig. 8.54 shows the Variables Initialization section of the Main Method:

"M_RC_Walker_IS.py"
Variables Initialization

```
56 robot1 = "Zira"
57 robot_mode = 'n' # f for Forward, b for Backward, l for Left Turn, r
58 background_image_filename = 'Zira_Ready.jpg'
59 SCREEN_SIZE = (388, 292)
60 message1="1) Tap Arrow Keys to move robot"
61 message2="2) Tap 1 to raise left arm, 3 to lower left arm to normal"
62 message3="2) Tap 2 to start head scan L/R, 4 to stop head scan"
```

Fig. 8.54 Variables Initialization in Main Method for "M_RC_Walker_IS.py".

- Line 56 assigns the string "Zira" to Variable "robot1" which will be used as the "robot name" argument for practically all EDBOT API methods.
- Lines 57-59 initialize constants needed for proper Pygame programming later:
    - Line 57 defines Parameter "robot_mode" with the character value 'n' which stands for the Initial Pose for the MINI.
    - Line 58 declares that the background image file to be used is "Zira_Ready.jpg" and that its SCREEN_SIZE is 388x292 (Line 59) – a list with 2 elements.
    - Lines 60-62 define 3 messages to be used as a simple menu of options for the user at run time.

Fig. 8.55 lists the steps needed to ensure a proper and working instance of an EdbotClient():

"M_RC_Walker_IS.py"
Connecting to EDBOT

```
67 ec = edbot.EdbotClient("localhost", 8080)
68 ec.connect()

74 if not robot1 in ec.get_robot_names("edbot"):
75     print(robot1 + " is not configured")
76     sys.exit()

81 print("Waiting to control " + robot1 + "... ", end="")
82 ec.wait_until_active(robot1)
83 print("Got control of " + robot1 + "!")
84 ec.say(robot1, "connected to robot " + robot1)
```

Fig. 8.55 Creating EdbotClient instance in Main Method for "M_RC_Walker_IS.py".

- Lines 67-68 create an instance of the EdbotClient class named "ec" and connect it to the EDBOT Server software tool via the local PC's port 8080.
- Lines 74-76 check to see if "robot1" (i.e. "Zira") was properly configured in EDBOT tool.
- Lines 81-83 wait for the active control of "robot1".
- Line 84 makes EDBOT voice out "connected to robot Zira".

Fig. 8.56 lists the robot initialization steps:

"M_RC_Walker_IS.py"
Robot Initialization

```
87 ec.set_servo_torque(robot1,"0/1")
88 ec.set_servo_speed(robot1,"0/100")
89 ec.set_custom(robot1,"82/1/0")
90 init_pose(robot1)
91 time.sleep(1.00)
92 l_arm_wave(robot1)
93 ec.say(robot1, robot1 + " is ready for a command")
94 ec.set_custom(robot1,"264/2/1024")  # set Servo17 joint-offset to 1024
95 print(message1)
96 print(message2)
97 print(message3)
```

Fig. 8.56 Robot Initializations in Main Method for "M_RC_Walker_IS.py".

- Line 87 turns power/torque ON (i.e. argument "1") for ALL servos (i.e. argument "0").
- Line 88 sets ALL servos speed to "100%".
- Line 89 turns OFF the "Motion Play LED" option at address "82" with a "1" byte constant equals to "0".
- Line 90 sets the MINI to its Initial Pose.
- Line 91 waits for 1 second.
- Line 92 makes Zira wave the left arm.
- Line 93 announces that "Zira is ready for a command".
- Line 94 sets the JOINT-OFFSET (Address "264") for Servo #17 to "1024" (i.e. a "2" bytes constant), to take it out of the influence of future Motion Plays.
- Lines 95-97 prints out a simple menu of 3 messages for the program user at run-time.

Fig. 8.57 shows how to activate Pygame and use some of its functionalities:

"M_RC_Walker_IS.py"
Pygame Initialization

```
101 pygame.init()
102 screen = pygame.display.set_mode(SCREEN_SIZE, 0, 32)
103 pygame.display.set_caption('RC Walker IS')
104 background = pygame.image.load(background_image_filename).convert()
105 font = pygame.font.SysFont("arial", 60);
106 text_surface1 = font.render("Single Control", True, (255, 255, 0))
107 x1 = ( SCREEN_SIZE[0] - text_surface1.get_width() ) / 2
108 y1 = ( SCREEN_SIZE[1] - text_surface1.get_height() ) / 2
```

Fig. 8.57 Pygame Initializations in Main Method for "M_RC_Walker_IS.py".

- "pygame" needs to be initialized as the first step (Line 101) (https://www.pygame.org/docs/tut/ImportInit.html).
- Line 102 pops up a window using SCREEN_SIZE to know how big to create it for and using 32-bit colors. Line 103 sets the caption "RC Walker IS" as the title for this window (see right area of Fig. 8.51). See more detailed information for these Pygame methods at these links (https://www.pygame.org/docs/ref/display.html?highlight=display%20set_mode#pygame.display.set_mode) and (https://www.pygame.org/docs/ref/display.html#pygame.display.set_caption).
- Line 104 loads the contents of the image file "Zira_ready.jpg" into Variable "background" which is a Surface object (https://www.pygame.org/docs/ref/image.html?highlight=image%20load#pygame.image.load).
- Line 105 sets up Font Arial - Size 60 to be used for the object "text_surface1" to be defined next.
- "text_surface1" is defined by Line 106 and contains the text "Single Control" to be rendered using a Yellow color by specifying its (Red, Green, Blue) components (i.e. 255 for Red, 255 for Green, and 0 for Blue). The argument "True" means that the rendered characters will have smooth edges (https://www.pygame.org/docs/ref/font.html?highlight=font%20render#pygame.font.Font.render).
- Lines 107-108 define the (x1, y1) coordinates where to put "text_surface1" on the previous "RC Walker IS" window – essentially in the middle of the background image (see Fig. 8.51).

Fig. 8.58 describes Part 1 of the Main Endless Loop where Pygame is used to monitor keyboard events:

- For every iteration of the Main Loop (Line 111), Line 113 will get all the messages and remove them from the Pygame events queue via Method event.get() (https://www.pygame.org/docs/ref/event.html?highlight=event%20get#pygame.event.get).

"M_RC_Walker_IS.py"
Main Endless Loop (1)

```
111 while True:
112 # Checking for user's keyboard inputs
113     for event in pygame.event.get():
114         if event.type == QUIT:
115             init_pose(robot1)
116             pygame.quit()
117             sys.exit()
118 
119         if event.type == KEYDOWN:
120             if event.key == K_UP:
121                 robot_mode = 'f'
122             if event.key == K_DOWN:
123                 robot_mode = 'b'
124             if event.key == K_RIGHT:
125                 robot_mode = 'r'
126             if event.key == K_LEFT:
127                robot_mode = 'l'
128             if event.key == K_SPACE:
129                robot_mode = 'n'
130             if event.key == K_1:
131                robot_mode = '1'
132             if event.key == K_3:
133                robot_mode = '3'
134             if event.key == K_2:
135                robot_mode = '2'
136             if event.key == K_4:
137                robot_mode = '4'
```

Fig. 8.58 Part 1 of Main Endless Loop in "M_RC_Walker_IS.py".

- Line 113 also contains a FOR LOOP that will check all the "events" received via Method event.get() for 2 specific event-types: "QUIT" and "KEYDOWN" (the interested reader can search the web page https://www.pygame.org/docs/ref/event.html#pygame.event for a list of other event types).
    - If "QUIT" is found (Line 114), i.e. when the user closes the "RC Walker IS" popup window, then the robot assumes its Initial Pose (Line 115), "pygame" is shut down (Line 116) and then the PYTHON session is also exited (Line 117).
    - If "KEYDOWN" is found (Line 119), meaning that some keys had been pushed down, then the "event.key" variable is examined further (Lines 120-137):
        - If the UP-ARROW key had been pushed (Line 120), then "robot_mode" is set to "f" for going forward (Line 121).
        - If the DOWN-ARROW key had been pushed (Line 122), then "robot_mode" is set to "b" for going backward (Line 123).
        - If the RIGHT-ARROW key had been pushed (Line 124), then "robot_mode" is set to "r" for turning right (Line 125).
        - If the LEFT-ARROW key had been pushed (Line 126), then "robot_mode" is set to "l" for turning left (Line 127).
        - If the SPACE BAR had been pushed (Line 128), then "robot_mode" is set to "n" for Initial Pose (Line 129).

- If the “1” key had been pushed (Line 130), then “robot_mode” is set to “1” for raising the right arm (Line 131).
- If the “3” key had been pushed (Line 132), then “robot_mode” is set to “3” for lowering the right arm (Line 133).
- If the “2” key had been pushed (Line 134), then “robot_mode” is set to “2” for head scanning (Line 135).
- Lastly, if the “4” key had been pushed (Line 136), then “robot_mode” is set to “4” to stop head scanning (Line 137).

“M_RC_Walker_IS.py”
Main Endless Loop (2)

```
139 # Update Pygame Display Window
140     screen = pygame.display.set_mode(SCREEN_SIZE, 0, 32)
141     screen.blit(background, (0,0))
142     screen.blit(text_surface1, (x1, y1))
143     pygame.display.update()
```

Fig. 8.59 Part 2 of Main Endless Loop in “M_RC_Walker_IS.py”.

Fig. 8.59 describes Part 2 of the Main Endless Loop where Pygame is used to display the popup window shown on the right side of Fig 8.51:

- Lines 140-143 fill in the different display components to result in the popup window show on the right in Fig. 8.51:
    - Line 140 initializes a window using SCREEN_SIZE, a “flag” value of 0 (i.e. a software-driven window), and a color depth of 32 bits and assigned this window object to Variable “screen” (https://www.pygame.org/docs/ref/display.html#pygame.display.set_mode).
    - Line 141 “blits” (i.e. draws) the “background” image defined back in Line 104 (Fig. 8.57) into the popup window (i.e. Surface object “screen”).
    - Additionally, Line 142 blits the “text_surface1” object into the middle area of the object “screen” (i.e. coordinates x1 and y1).
    - Finally in Line 143, Method “display.update()” refreshes the popup window so that we can see the final result shown in Fig. 8.51 (https://www.pygame.org/docs/ref/display.html?highlight=display%20update#pygame.display.update).

Fig. 8.60 describes Part 3 of the Main Endless Loop, i.e. the control of various robot’s motions:

- A big IF-ELSE-IF structure (Lines 147-164) is used to “map” each possible value for Variable “robot_mode” (e.g. ‘f’, ‘b’, ‘r’, ‘l’, ‘n’, ‘1’, ‘2’, ‘3’, ‘4’) to the appropriate “motion” for the robot to perform.
- Line 166 carries an important role which is to erase the “old data”, i.e. “robot_mode”, right after the robot performs a commanded motion. This is to ensure that the robot moves only “once”

per keyboard's tap.

"M_RC_Walker_IS.py"
Main Endless Loop (3)

```
#  Robot maneuvers start here
    if (robot_mode == 'f'):
        forward(robot1)
    elif (robot_mode == 'b'):
        backward(robot1)
    elif (robot_mode == 'r'):
        right(robot1)
    elif (robot_mode == 'l'):
        left(robot1)
    elif (robot_mode == 'n'):
        init_pose(robot1)
    elif(robot_mode == '1'):
        raise_arm(robot1)
    elif (robot_mode == '3'):
        lower_arm(robot1)
    elif (robot_mode == '2'):
        go_scan_head(robot1)
    elif (robot_mode == '4'):
        stop_scan_head(robot1)

    robot_mode = ''
```

Fig. 8.60 Part 3 of Main Endless Loop in "M_RC_Walker_IS.py".

Video 8.7 shows the real-time performance of "M_RC_Walker_IS.py" in two scenarios: when no "motion wait" is used (i.e. wait = False) and when motion wait states are implemented (i.e. wait = True).

8.8.2 Dual RC Walkers

The reader may recall the "Dual RC Walkers" project using SCRATCH (see Section 8.5) where the two MINIs' motions "had to be synchronized" due to the constraint on SCRATCH that can handle only one "key-stroke" at any one time. As PYTHON will allow EDBOT programming at a lower level, the author's hope was that independent Remote Control of two MINIs could be achieved when using PYTHON.

Thus, for this project "M_Dual_RC_Walkers.py", the goal is to independently "remote-control" two MINIs (Zira and Zork) from the same PYTHON program. The usual arrow keys U-D-L-R will be used to control Zira (as "robot1") and the keys W-Z-A-S will be the equivalent "arrow" keys for controlling Zork (as "robot2") – see Fig. 8.61.

"M_Dual_RC_Walkers.py"
Variables Definition

```
robot1 = "Zira"
robot2 = "Zork"
robot1_mode = 'n' # f for Forward, b for Backward,
robot2_mode = 'n' # f for Forward, b for Backward,
background_image_filename = 'Zira_Zork_Ready.jpg'
SCREEN_SIZE = (927, 349)
message1="1) Tap Arrow Keys to move Zira"
message2="2) Tap [w,z,a,s] Keys to move Zork"
```

Fig. 8.61 Variables Initializations in "M_Dual_RC_Walkers.py".

Fig. 8.62 lists the user methods defined for this project, they are fewer than for the previous project "M_RC_Walker_IS.py" because the author wanted to show a different way to implement Motion Control for the MINIs whereas the EDBOT method ***run_motion()*** is used directly in the Main Method:

"M_Dual_RC_Walkers.py"
"Motion" Methods Definition

```
14 def init_pose(robot):
15     ec.run_motion(robot, 6, wait=False)
16 
17 def stop(robot):
18 #     ec.run_motion(robot, -1)
19     ec.run_motion(robot, 6, wait=False)
20 
21 def r_arm_wave(robot):
22     ec.run_motion(robot, 8, wait=False)
23 
24 def l_arm_wave(robot):
25     ec.run_motion(robot, 9, wait=False)
```

Fig. 8.62 User Methods in "M_Dual_RC_Walkers.py".

- The only "new" user method is ***stop(robot)*** whereas the author wanted to use MOTION (-1) (Line 18) as applied in previous TASK and SCRATCH projects, unfortunately the EDBOT V. 5.2.0.1523 does not yet support this feature in PYTHON (even though it supported this feature in SCRATCH – see Fig. 8.26). This issue will probably get sorted out in a later version of EDBOT, thus the author left Line 18 in the source code for the reader to try out in the future.
- So for now, method ***stop()*** corresponds simply to run the MOTION (6) to get the robot into Initial Pose.

Fig. 8.63 shows that the same steps used to connect “robot1” to EDBOT are just repeated to connect “robot2” to the same “EdbotClient” which now will route specific commands for each robot via a different BT COM port to either Zira or Zork.

“M_Dual_RC_Walkers.py”
Connecting to EDBOT

```
42 ec = edbot.EdbotClient("localhost", 8080)
43 ec.connect()
44 #
45 # Check robot(s) names and verify their configurations
46 #
47 if not robot1 in ec.get_robot_names("edbot"):
48     print(robot1 + " is not configured")
49     sys.exit()
50 if not robot2 in ec.get_robot_names("edbot"):
51     print(robot2 + " is not configured")
52     sys.exit()
53 #
54 # Wait for control of the MINI robot(s) used.
55 #
56 print("Waiting to control " + robot1 + "... ", end="")
57 ec.wait_until_active(robot1)
58 print("Got control of " + robot1 + "!")
59 ec.say(robot1, "connected to robot " + robot1)
60 print("Waiting to control " + robot2 + "... ", end="")
61 ec.wait_until_active(robot2)
62 print("Got control of " + robot2 + "!")
63 ec.say(robot2, "connected to robot " + robot2)
```

Fig. 8.63 Connecting to EDBOT for “robot1” and “robot2” in “M_Dual_RC_Walkers.py”.

Fig. 8.64 lists the robot initialization steps for “robot1” and “robot2” which are the same as the ones discussed in the previous single-bot project, except that “robot1” would swing its left arm (Line 75) to signal that it is “ready for a command” while “robot2” would swing its right arm (Line 77).

"M_Dual_RC_Walkers.py"
Robots Initialization

```
# Robots initialization
ec.set_servo_torque(robot1,"0/1")
ec.set_servo_torque(robot2,"0/1")
ec.set_servo_speed(robot1,"0/100")
ec.set_servo_speed(robot2,"0/100")
ec.set_custom(robot1,"82/1/0")
ec.set_custom(robot2,"82/1/0")
init_pose(robot1)
init_pose(robot2)
time.sleep(1.00)
l_arm_wave(robot1)
ec.say(robot1, robot1 + " is ready for a command")
r_arm_wave(robot2)
ec.say(robot2, robot2 + " is ready for a command")

print(message1)
print(message2)
```

Fig. 8.64 Robot Initializations for "robot1" and "robot2" in "M_Dual_RC_Walkers.py".

Fig. 8.65 also shows that the Pygame initialization procedure is the same as before, except for the texts used for the Display Caption (Line 87) and "text_surface1" (Line 90).

"M_Dual_RC_Walkers.py"
Pygame Initialization

```
85 pygame.init()
86 screen = pygame.display.set_mode(SCREEN_SIZE, 0, 32)
87 pygame.display.set_caption('Dual RC Walkers')
88 background = pygame.image.load(background_image_filename).convert()
89 font = pygame.font.SysFont("arial", 60);
90 text_surface1 = font.render("Dual Control", True, (255, 255, 0))
91 x1 = ( SCREEN_SIZE[0] - text_surface1.get_width() ) / 2
92 y1 = ( SCREEN_SIZE[1] - text_surface1.get_height() ) / 2
```

Fig. 8.65 Pygame Initializations in "M_Dual_RC_Walkers.py".

Fig. 8.66 shows Part 1 of the Main Endless Loop which follows the same logic as for the single-robot project, but there are more keys to monitor (Lines 113-123). Part 2 of the Main Endless Loop for this dual-robot project is the same as for the previous single-robot project, so it is not repeated here (see Fig. 8.59).

"M_Dual_RC_Walkers.py"
Main Endless Loop (1)

```
while True:
# Checking for user's keyboard inputs
    for event in pygame.event.get():
        if event.type == QUIT:
            stop(robot1)
            stop(robot2)
            pygame.quit()
            sys.exit()

        if event.type == KEYDOWN:
            if event.key == K_UP:
                robot1_mode = 'f'
            if event.key == K_DOWN:
                robot1_mode = 'b'
            if event.key == K_RIGHT:
                robot1_mode = 'r'
            if event.key == K_LEFT:
                robot1_mode = 'l'
            if event.key == K_w:
                robot2_mode = 'f'
            if event.key == K_z:
                robot2_mode = 'b'
            if event.key == K_s:
                robot2_mode = 'r'
            if event.key == K_a:
                robot2_mode = 'l'
            if event.key == K_SPACE:
                robot1_mode = 'n'
                robot2_mode = 'n'
```

Fig. 8.66 Part 1 of Main Endless Loop in "M_Dual_RC_Walkers.py".

Fig. 8.67 describes Part 3 of the Main Endless Loop where the control steps for both robots are listed. In this dual-robot project, the EDBOT API methods ***run_motion()*** are used directly for each robot without wait state, but in separate IF-ELSE-IF structures: Lines 133-143 for "robot1", and Lines 145-155 for "robot2" as the project's goal is to control Zira and Zork **independently** at run-time.

"M_Dual_RC_Walkers.py"
Main Endless Loop (3)

```
133     if (robot1_mode == 'f'):
134         ec.run_motion(robot1, 1, wait=False)
135     elif (robot1_mode == 'b'):
136         ec.run_motion(robot1, 2, wait=False)
137     elif (robot1_mode == 'r'):
138         ec.run_motion(robot1, 3, wait=False)
139     elif (robot1_mode == 'l'):
140         ec.run_motion(robot1, 4, wait=False)
141     elif (robot1_mode == 'n'):
142         stop(robot1)
143     robot1_mode = ''
145     if (robot2_mode == 'f'):
146         ec.run_motion(robot2, 1, wait=False)
147     elif (robot2_mode == 'b'):
148         ec.run_motion(robot2, 2, wait=False)
149     elif (robot2_mode == 'r'):
150         ec.run_motion(robot2, 3, wait=False)
151     elif (robot2_mode == 'l'):
152         ec.run_motion(robot2, 4, wait=False)
153     elif (robot2_mode == 'n'):
154         stop(robot2)
155     robot2_mode = ''
```

Fig. 8.67 Part 3 of Main Endless Loop in "M_Dual_RC_Walkers.py".

Video 8.8 shows the real-time performance of "M_Dual_RC_Walkers.py" in two scenarios: when no "motion wait" is used (i.e. wait = False) and when motion wait states are implemented (i.e. wait = True).

8.8.3 Autonomous Random Walker & Obstacle Avoider

The reader may recall from the Random Walker SCRATCH project (Section 8.6) that SCRATCH would only support the use of **one** IRSS-10 sensor connected to **Port 1**, thus this SCRATCH solution was more limited in its response when an obstacle was found. On the other hand, the PYTHON API allows the use of all sensors on all 4 Ports. Thus, the PYTHON project "M_RandomWalker.py" can use two NIR sensors to respond to a found obstacle in the same way as the TASK-equivalent project "M_RandomWalker.tsk3" in Section 6.6.

The TASK project "M_RandomWalker.tsk3" relied heavily on using MOTION (0) and MOTION (-1), as well as the CALLBACK Function to provide the most recent Sensors Data at any point in this TASK program. Unfortunately, these two features are not available currently in EDBOT/PYTHON V. 5.2.0.1523:

- MOTIONs "0" and "-1" do not work properly in PYTHON, although they worked fine for SCRATCH 2.
- By design, EDBOT does not update sensor and servo data while a MOTION-LIST/PAGE is playing.

Thus, although the overall goals and algorithm are the same between the TASK and PYTHON versions for the Random Walker Project, the reader will see that some implementation details had to be changed in the PYTHON version.

Fig. 8.68 lists the Definition of Variables used in "M_RandomWalker.py":

- Although both robots "Zira" and "Zork" have their respective Variables defined for them (Lines 160-161 and Lines 162-163) this PYTHON code is made to control only 1 robot, but it can be either Zira or Zork. Variables "robot1", "robot2", "obstacle1", "obstacle2" and "threshold1" are Global.

"M_RandomWalker.py"
Variables Definition

```
160 robot1 = "Zira"
161 robot2 = "Zork"
162 obstacle1 = 0
163 obstacle2 = 0
164 threshold1 = 100
165 threshold2 = 50
166 threshold3 = 30
167 w_direction = 0
168 w_repeats = 0
169 ir_left = 0
170 ir_right = 0
```

Fig. 8.68 Variables Definition in "M_RandomWalker.py".

- Please note that "obstacle1" and "obstacle2" are designed to be set to "0" when no obstacle is found, and when an obstacle is found by the respective robot ("robot1" or "robot2"), they will be set to "1".

There are many methods used in the program "M_RandomWalker.py":

- The various MOTION related methods already described in previous PYTHON projects, such as ***init_pose(), forward()*** and ***l_arm_raise()***, won't be repeated here.
- Two methods of robot stoppage are defined - ***reg_stop()*** and ***emer_stop()***. Method ***reg_stop()*** is to be used when the robot encounters an obstacle when it **has just concluded** a set of randomized moves. Method ***emer_stop()*** is to be used when the robot encounters an obstacle **while still performing** a set of randomized moves (see Fig. 8.69).
- Two "random-based" methods are used to set the robot's move direction and the number of repeats for such move – ***set_direction()*** and ***set_repeats()*** (see Fig. 8.70).
- Method ***walk_execute()*** takes the previous randomized move-direction and move-repeat values and commands the robot to perform such-directed moves (Figs. 8.73 and 8.74).
- Method ***set_obstacle()*** sets Parameter "obstacle" to either "obstacle1" or "obstacle2" depending on which robot among "robot1" or "robot2" is being tested (Fig. 8.71).
- Method ***check_NIR()*** is called by Method ***walk_execute()*** at the end each robot move to check if an obstacle is found, and it issues the start of an "emergency stop" when needed (Fig. 8.72).
- Method ***flash_LED()*** is called at the conclusion of a **successful** obstacle avoidance procedure (Fig. 8.75).

Fig. 8.69 illustrates the details for Methods ***reg_stop()*** and ***emer_stop()***:

"M_RandomWalker.py"
"Motion" Methods Definition

```
35 def reg_stop(robot):
36 #     ec.run_motion(robot, -1)
37     ec.run_motion(robot, 6)
38     ec.say(robot, robot + " regular stop")
39
40 def emer_stop(robot):
41 #     ec.run_motion(robot, 0)
42     ec.run_motion(robot, 6)
43     ec.say(robot, robot + " emergency stop end")
```

Fig. 8.69 Methods ***reg_stop()*** and ***emer_stop()*** in "M_RandomWalker.py".

- Ideally, MOTION (-1) should be used for ***reg_stop()*** but that option is not currently implementable with EDBOT V. 5.2.0.1523 (may be OK in a later version). Thus, MOTION (6) is used instead (Line 37). Next, the robot announces that it is doing a "regular stop" (Line 38).

- Similarly, MOTION (0) should be used for ***emer_stop()*** but that option is not currently available with EDBOT V. 5.2.0.1523 (may be OK in a later version). Thus, MOTION (6) is used instead (Line 42). Next, the robot announces that it is ending an "emergency stop" (Line 43).

Fig. 8.70 lists the details for Methods ***set_direction()*** and ***set_repeats()***:

"M_RandomWalker.py"
"Random" Methods Definition

```
45 def set_direction():
46     rand_no = random.randint(1,4)
47     if (rand_no == 1):
48         walk_direction = 1
49     elif (rand_no == 2):
50         walk_direction = 2
51     elif (rand_no == 3):
52         walk_direction = 3
53     elif (rand_no == 4):
54         walk_direction = 4
55     return walk_direction

57 def set_repeats():
58     rand_no = random.randint(1,10)
59     if ((rand_no >= 1) and (rand_no <= 4)):
60         walk_repeats = 2
61     elif ((rand_no >= 5) and (rand_no <= 8)):
62         walk_repeats = 3
63     elif ((rand_no >= 9) and (rand_no <= 10)):
64         walk_repeats = 6
65     return walk_repeats
```

Fig. 8.70 Methods ***set_direction()*** and ***set_repeats()*** in "M_RandomWalker.py".

- For ***set_direction()*** - Line 46 throws the "dice" and obtains either 1, 2, 3 or 4 to save in Parameter "rand_no". Next, based on the value of "rand_no", an IF-ELSE-IF structure is used to set the appropriate value to Parameter "walk_direction" which is then returned to the calling statement elsewhere in the PYTHON code.
- For ***set_repeats()*** - Line 46 throws the "dice" and obtains an integer between 1 and 10 which is recorded into Parameter "rand_no". Next, based on the value of "rand_no", an IF-ELSE-IF structure is used to set the appropriate value (2, 3 or 6) to Parameter "walk_repeats" which is then returned to the calling statement elsewhere in the PYTHON code. On the average, a "walk_repeats" value of 2 would be returned 40% of the time, while a value of "3" would be returned also for 40% of the time, with a value of "6" for the remaining 20%. Later, in the Main Method a special value of "1" for "walk_repeats" is used when an obstacle has been determined.

Method ***set_obstacle()*** is created because there is a need to let some Methods know about the status of the Global Parameter "obstacle1" and "obstacle2". It is also designed so that either "robot1" or "robot2" can be used with it via the string input "robot". Fig. 8.71 list details of Method ***set_obstacle()*** which are rather self-explanatory. Please note that "obstacle1" and "obstacle2" are Global Parameters and thus they can be set anywhere else in the program "M_RandomWalker.py".

"M_RandomWalker.py"
"set_obstacle" Method

```
def set_obstacle(robot):
    global robot1, robot2, obstacle1, obstacle2
    if (robot == robot1):
        obstacle = obstacle1
    elif (robot == robot2):
        obstacle = obstacle2
    return obstacle
```

Fig. 8.71 Method ***set_obstacle()*** in "M_RandomWalker.py".

Method ***check_NIR()*** (see Fig. 8.72) is designed to work with any robot thus it has a string input "robot" which can be set to either "robot1" or "robot2" at run-time. It is created to check on the current values of the NIR Sensors on Port 1 and Port 4 and to determine whether there is an obstacle being detected:

"M_RandomWalker.py"
"check_NIR" Method

```
def check_NIR(robot):
    global robot1, robot2, obstacle1, obstacle2, threshold1
    sensors = ec.get_robot(robot)["reporters"]
    ir_l = sensors["port1"]
    ir_r = sensors["port4"]
    if ((ir_l >= threshold1) or (ir_r >= threshold1)):
        ec.say(robot, robot + " emergency stop starts")
        emer_stop(robot)
        if (robot == robot1):
            obstacle1 = 1
            obstacle = obstacle1
        elif (robot == robot2):
            obstacle2 = 1
            obstacle = obstacle2
    else:
        if (robot == robot1):
            obstacle1 = 0
            obstacle = obstacle1
        elif (robot == robot2):
            obstacle2 = 0
            obstacle = obstacle2
    return obstacle
```

Fig. 8.72 Method ***check_NIR()*** in "M_RandomWalker.py".

- First, Method ***get_robot()*** is used to receive the sensors data from a specified "robot" into a dictionary named "sensors" (Line 115).
- Lines 116 and 117 extract the sensor data respectively from "port1" and "port4" and save them in Parameters "ir_l" and "ir_r"
- Next, if either (ir_l >= threshold1) or (ir_r >= threshold1) is TRUE (Line 118) then an obstacle is detected and this robot is entering an "emergency stop" procedure, and statements from Line 120 to Line 126 are executed:
  - EDBOT first announces that for this "robot", an "emergency stop starts" (Line 119)
  - Method ***emer_stop()*** is called to stop the robot, i.e. to put it into its Initial Pose (see Fig. 8.69).
  - The IF-ELSE-IF structure (Lines 121-126) is executed next to set the appropriate Global Parameter "obstacle1" or "obstacle2" to "1" depending on the run-time value of the string "robot" (Line 122 or Line 125). The local variable "obstacle" is also set appropriately (Line 123 or 126).
- When there is no obstacle detected, the ELSE branch on Line 127 is taken instead. The IF-ELSE-IF structure (Lines 128-133) is executed next to set the appropriate Global Parameter "obstacle1" or "obstacle2" to "0" depending on the run-time value of the string "robot" (Line 129 or Line 132). The local variable "obstacle" is also set appropriately (Line 130 or 133).
- Lastly, the value of Local Parameter "obstacle" is returned to the statement/method that calls it previously. Method ***check_NIR()*** is only called by Method ***walk_execute()*** (see Figs. 8.73 and 8.74).

Figs. 8.73 (Part 1) and 8.74 (Part 2) lay out the programming details for Method ***walk_execute()*** which is **more complex in its design** that its previous TASK version (see Figs. 6.52 and 6.53) because MOTION (0), i.e. emergency stop, cannot be used and CALLBACK is also not available to provide updated sensors data on demand:

- **Fig. 8.73** - Part 1 of Method ***walk_execute()*** shows that an overall FOR LOOP is used to iterate over the run-time value of Parameter "walk_repeats" which can be either "2", "3", "6" or "1". The value of "1" is set when an obstacle is found by other codes in the Main Method, then the IF-ELSE-IF structure from Line 70 to Line 78 is executed, i.e. to make the robot perform the commanded move among F/B/R/L for only "**one time**" per call of Method ***walk_execute()***.

"M_RandomWalker.py"
"walk_execute" Method (1)

```
67 def walk_execute(robot, walk_direction, walk_repeats):
68     global robot1, robot2, obstacle1, obstacle2
69     for i in range(walk_repeats):
70         if (walk_repeats == 1):    # after obstacle found
71             if (walk_direction == 1):
72                 forward(robot)
73             elif (walk_direction == 2):
74                 backward(robot)
75             elif (walk_direction == 3):
76                 right(robot)
77             elif (walk_direction == 4):
78                 left(robot)
```

Fig. 8.73 Part 1 of Method ***walk_execute()*** in "M_RandomWalker.py".

"M_RandomWalker.py"
"walk_execute" Method (2)

```
79         else:                  # when repeats set to 2, 3, or 6
80             obstacle = set_obstacle(robot)
81             if (obstacle == 0):
82                 if (walk_direction == 1):
83                     forward(robot)
84                 elif (walk_direction == 2):
85                     backward(robot)
86                 elif (walk_direction == 3):
87                     right(robot)
88                 elif (walk_direction == 4):
89                     left(robot)
90                 time.sleep(0.1)
91                 obstacle = check_NIR(robot)
92             elif (obstacle == 1):
93                 init_pose(robot)
```

Fig. 8.74 Part 2 of Method ***walk_execute()*** in "M_RandomWalker.py".

- **Fig. 8.74** – Part 2 of Method ***walk_execute()*** is executed when the run-time value of Parameter "walk_repeats" turns out to be "2", "3" or "6" (starting at the "ELSE" statement at Line 79):
    - First, Local Parameter "obstacle" is set to the current value of either Global Parameters "obstacle1" or "obstacle2", depending on the robot tested ("robot1" or "robot2"), by using Method ***set_obstacle()*** (Line 80). Please refer to Fig. 8.71 for details. The value for Parameter "obstacle" could then be either "1" or "0" depending on whether an obstacle was found, or not, by other pieces of code in the Main Method or in Method ***check_NIR()*** (see Fig. 8.72).

- Next, **if no obstacle is found (Line 81)**, then the IF-ELSE-IF structure based on the run-time value of "walk_direction" (Lines 82-89) is executed to make the robot perform the motion commanded (until it is finished). Then a wait of 0.1 second occurs (Line 90), before calling Method ***check_NIR()*** at Line 91. This 0.1 second pause is **crucial** to allow EDBOT time to resume collecting sensors data that the Method ***check_NIR()*** will need to use – see Fig. 8.72 - (*remember that EDBOT halts its sensors and servos data collecting task when a MOTION-LIST/PAGE is being performed in the previous IF-ELSE-IF structure*). Method ***check_NIR()*** returns a "0" to Local Parameter "obstacle" if no obstacle is found, or a "1" if an obstacle is found. Then the IF structure (Lines 81-91) is exited and "program logic" flows back to the overall FOR-LOOP structure at Line 69 of Part 1 in Fig. 8.73 which would iterate again if (counter "i" < "walk_repeats"). Here, "program logic" gets more "subtle" because this new iteration happens with "obstacle" being **set to "0" or "1"**:
    - **If (obstacle == 0)**, then Lines 80-91 are executed again, meaning that the robot performs the previous random walk direction one more time and ***check_NIR()*** once again determines whether an obstacle is found or not. Then the FOR-LOOP continues to iterate once more, if its counter "i" has not exhausted its repeat limit.
    - **Else If (obstacle == 1)**, then Lines 92-93 are executed, meaning that the robot assumes its Initial Pose, i.e. it **stops moving**. From then on, if there are some remaining iterations for the FOR-LOOP to do, it would just assume again its Initial Pose for each of those iterations, meaning that from the outside, the robot seems to stop and ponder how it can get itself "clear from the obstacle"! (that job will be performed by codes in the Main Method – see Figs. 8.77 to 8.81).
- However, if an obstacle is found from/because of Line 80 when the FOR-LOOP is first entered (because "obstacle1" was set to 1 from Line 216 in the Main Method – see Fig. 8.77), the ELSE-IF at Line 92 is therefore TRUE, then the robot is commanded to go to its Initial Pose (Line 93) and essentially stops. Then the FOR-LOOP remaining iterations work as described the just-preceding paragraph.

Fig. 8.75 lists the programming details for Method ***flash_LED()*** which is used when the robot just clears an obstacle:

- Lines 131-141 essentially turn ON the Motion Play LED option for either "robot1" or "robot2" depending on which one is being tested, by using ***ec.set_custom()*** to send a 1-byte constant (= 1) to Address 82.
- Lines 143-144 play the MOTION (6), i.e. "Initial Pose", 4 times. During this time period, the user would see the robot's servo LEDs flashing.
- Lines 146-149 turn OFF the Motion Play LED option for "robot1" or "robot2" appropriately also.
- Last, but very importantly, Lines 151-154 resets either Global Parameter "obstacle1" or "obstacle2" (as appropriate) back to the value of "0" as the robot is truly clear of the obstacle at this point.

"M_RandomWalker.py"
"flash_LED" Method

```
def flash_LED(robot):
    global robot1, robot2, obstacle1, obstacle2
    if (robot == robot1):
        ec.set_custom(robot1,"82/1/1")
    elif (robot == robot2):
        ec.set_custom(robot2,"82/1/1")

    for i in range(4):
        ec.run_motion(robot, 6)

    if (robot == robot1):
        ec.set_custom(robot1,"82/1/0")
    elif (robot == robot2):
        ec.set_custom(robot2,"82/1/0")

    if (robot == robot1):
        obstacle1 = 0
    elif (robot == robot2):
        obstacle2 = 0
```

Fig. 8.75 Method ***flash_LED()*** in "M_RandomWalker.py".

We are now ready to look at the details of the Main Method showing how all the previous Methods work together to create a RANDOM WALKER. The Main Method is described via Fig. 8.76 through Fig. 8.80.

Fig. 8.76 describes Part 1 of the Main Method:

- The familiar Endless Loop starts at Line 200.
- Lines 202-204 record sensors data from "port1" and "port4" into Parameters "ir_left" and "ir_right" respectively.
- Please first note that Line 206 uses "threshold1" (=100) to check "ir_left" and "ir_right" against itself, this is the first level of "obstacle detection" and the "100" value would set the robot quite close to the obstacle. There are two other thresholds to be used in later code sections: "threshold2" set to "50" and "threshold3" set to "30" (see Fig. 8.68).
- If the IF condition in Line 206 evaluates to TRUE, meaning that the robot has not detected any obstacle, Lines 207-211 are executed next:
    - Line 207 sets Parameter "obstacle1" to 0.
    - Line 208 announces that "robot1" is "random walking".
    - Line 209 uses Method ***set_direction()*** to assign a randomized integer value [1-4] to Parameter "w_direction".
    - Line 210 uses Method ***set_repeats()*** to assign a randomized integer value [2, 3, 6] to Parameter "w_repeats".
    - Line 211 calls Method ***walk_execute()*** using input arguments "robot1", "w_direction" and "w_repeats".

- If the IF condition in Line 206 evaluates to FALSE, meaning the robot has detected an obstacle via its "port1" sensor or via its "port4" sensor, or both, then Part 2 of the Main Method is activated next (see Fig. 8.77).

"M_RandomWalker.py"
Main Endless Loop (1)

```
200 while True:
201
202     sensors_data = ec.get_robot(robot1)["reporters"]
203     ir_left = sensors_data["port1"]
204     ir_right = sensors_data["port4"]
205     # if no obstacle
206     if ((ir_left < threshold1) and (ir_right < threshold1)):
207         obstacle1 = 0
208         ec.say(robot1, robot1 + " random walking")
209         w_direction = set_direction()
210         w_repeats = set_repeats()
211         walk_execute(robot1, w_direction, w_repeats)
```

Fig. 8.76 Part 1 of Main Method in "M_RandomWalker.py".

"M_RandomWalker.py"
Main Endless Loop (2)

```
212     # if obstacle present
213     else:
214         ec.say(robot1, robot1 + " avoiding obstacle")
215         if (obstacle1 == 0):
216             obstacle1 = 1
217             reg_stop(robot1)
218         time.sleep (0.05)
219         sensors_data = ec.get_robot(robot1)["reporters"]
220         ir_left = sensors_data["port1"]
221         ir_right = sensors_data["port4"]
```

Fig. 8.77 Part 2 of Main Method in "M_RandomWalker.py".

Fig. 8.77 describes Part 2 of the Main Method showing the First Phase of actions taken to deal with the situation when an obstacle is detected:

- First, "robot1" announces that it is "avoiding obstacle" with Line 214.
- The IF structure at Line 215 checks to see if Parameter "obstacle1" is set to "0":
    - If it evaluates to TRUE, this means that Method ***check_NIR()*** did not find any obstacle during the execution of Method ***walk_execute()*** during Part 1 of Main Method (i.e. Line

211 in Fig. 8.76). Then it is "logical" to set "obstacle1" to "1" and performs a "regular stop" by calling Method ***reg_stop()*** (Line 217).
    - If it evaluates to FALSE, this means that Method ***check_NIR()*** did find an obstacle during the execution of Method ***walk_execute()*** during Part 1 of Main Method. In that case, Method ***check_NIR()*** already sets "obstacle1" to "1" (see Line 122 or Line 125 in Fig. 8.72) and it has already issued an "emergency stop" (see Line 120 in Fig. 8.72). Thus, there is really nothing to do in this case, so there was **no** ELSE branch used with this IF structure.
- Line 218 does nothing for 0.5 second (just to be on the safe side) so to give EDBOT time to **turn back on** its sensor/servo data collecting task after Method ***walk_execute()*** is used in Part 1 of Main Method.
- Lines 219-221 then collect a new set of input data for "ir_left" and "ir_right", to be used in the next Part 3 of the Main Method.

Fig. 8.78 describes Part 3 of the Main Method showing the Second Phase of actions taken to deal with the situation when an obstacle is detected:

"M_RandomWalker.py"
Main Endless Loop (3)

```
222     while ((ir_left > threshold2) or (ir_right > threshold2)):
223         # if obstacle in front then go backward one step
224         ir_diff = ir_left - ir_right
225         if ((ir_diff <= 10) and (ir_diff >= -10)):
226             walk_execute(robot1, 2, 1)
227
228         # get current IR data
229         time.sleep(0.10)
230         sensors_data = ec.get_robot(robot1)["reporters"]
231         ir_left = sensors_data["port1"]
232         ir_right = sensors_data["port4"]
```

Fig. 8.78 Part 3 of Main Method in "M_RandomWalker.py".

- Please note that the WHILE LOOP starting at Line 222 uses "threshold2" (=50) to test "ir_left" and "ir_right" against itself. This would result in the robot being **"further away"** from the obstacle than when using "threshold1" (=100) as in Part 1 of Main Method.
- Furthermore, this WHILE LOOP is using an **OR** in its looping condition, because at this point, there are 3 possible situations:
    a) The robot was approaching squarely into the obstacle, then it should make 1 step backward first, then turn left or right depending on which side of the robot is closer to the obstacle at that time (i.e. situations b) or c) to be described next).
    b) The robot was approaching the obstacle on its left side, i.e. ("ir_left" > "ir_right"), so it should keep on turning right until clear of the obstacle (see Fig. 8.79).
    c) The robot was approaching the obstacle on its right side, i.e. ("ir_right" > "ir_left"), so it should keep on turning left until clear of the obstacle (see Fig. 8.79).

- Lines 224 and 225 are used to figure out if "situation a)" is encountered at run-time (i.e. when "ir_left" and "ir_right" are off from each other within +/- 10 units). If this turns out to be TRUE, then the robot is commanded to make one step backward by calling Method ***walk_execute()*** with "2" for "w_direction" and "1" for "w_repeats" (see Line 226). If this turns out not to be an instance of "situation a)", then just collect a new set of sensors data (Lines 229-232) and off to Part 4 of the Main Method (see Fig. 8.79).

Fig. 8.79 describes Part 4 of the Main Method showing the Fourth Phase of actions taken to deal with the situation when an obstacle is detected by the robot either by its left side or by its right side:

"M_RandomWalker.py"
Main Endless Loop (4)

```
233         # case when obstacle on left, turn right until obstacle cleared
234         while ((ir_left > threshold3) and (ir_left > ir_right)):
235             walk_execute (robot1, 3, 1)
236             time.sleep(0.10)
237             sensors_data = ec.get_robot(robot1)["reporters"]
238             ir_left = sensors_data["port1"]
239             ir_right = sensors_data["port4"]
240         # case when obstacle on right, turn left until obstacle cleared
241         while ((ir_right > threshold3) and (ir_right > ir_left)):
242             walk_execute (robot1, 4, 1)
243             time.sleep(0.10)
244             sensors_data = ec.get_robot(robot1)["reporters"]
245             ir_left = sensors_data["port1"]
246             ir_right = sensors_data["port4"]
```

Fig. 8.79 Part 4 of Main Method in "M_RandomWalker.py".

- The reader may recall from Part 3 (Fig. 8.78) that if the obstacle happens to be right in front of the robot then it just finished its one-step backward (Line 226 in Fig. 8.78), when Part 4 gets started on Line 234. Part 4 uses two independent WHILE LOOPs (Lines 234-239 and Lines 241-246) to respond, respectively, to the cases when the obstacle is on the robot's left or when the obstacle is on the robot's right. Please also note that these two WHILE LOOPs use "threshold3" (=30) in their loop condition and that they are inside the WHILE LOOP construct starting at Line 222 (Fig. 8.78) which uses "threshold2" (=50) in its loop condition.
- As an example, let's look more closely at Lines 234 to 239:
    - Using the latest values for "ir_left" and "ir_right" obtained from Lines 231-232, Line 234 checks to see if the obstacle is closer to the Left NIR sensor than to the Right NIR sensor. If this is TRUE, the robot is commanded to **turn right once** (Line 235). Then a short wait time of 0.1 second is activated (Line 236) for new sensors data to be retrieved properly (Lines 238-239). Then, this code segment goes back to Line 234 for another iteration of this WHILE LOOP. Eventually, the robot should turn away enough from the obstacle to make the LOOP condition at Line 234 evaluate to FALSE, then this WHILE LOOP is terminated and the program flows to the second WHILE LOOP at Line 241 which will for sure be skipped because at that time, both NIR sensors should

respond with very low values less than threshold3 (=30), as the robot should be clear of the obstacle at that point.

- Similar behavior for the robot with Lines 242-246, but the robot turns to the right instead.
- Thus, after Line 246, the robot has concluded its obstacle avoidance procedure and both "ir_left" and "ir_right" should be less than "threshold3" (=30). Part 5 of the Main Method can now start.

Fig. 8.80 shows the details for Part 5 of Main Method which is designed to perform a proper exit for the OUTER WHILE LOOP started at Line 222 in Fig. 8.78 and light up the robot LEDs to signal that it is ready to resume "random walking":

- Similar as before, Lines 248-251 retrieve the latest sensor data for "ir_left" and "ir_right", and program logic flows back to Line 222 (OUTER WHILE LOOP). But at this point, both Parameters "ir_left" and "ir_right" should be less than "threshold3" (=30) and therefore they should be less than "threshold2" (=50), the OUTER WHILE LOOP's condition would then evaluate to FALSE, and this loop terminates.
- Line 253 is activated next, and Method ***flash_LED()*** is called to make the robot flash its LEDs four times. Line 253 also corresponds to the ending of the big ELSE branch that started out on Line 213 of Fig. 8.77.
- Then the Main Endless Loop reiterates, and the robot goes on another round of "random walking" or "obstacle avoiding" if needed.

"M_RandomWalker.py"
Main Endless Loop (5)

```
247             # back to the outer while
248             time.sleep(0.20)
249             sensors_data = ec.get_robot(robot1)["reporters"]
250             ir_left = sensors_data["port1"]
251             ir_right = sensors_data["port4"]
252         # end of outer while
253         flash_LED(robot1)
254     # end of else
255     time.sleep(0.10)
256 # end of main endless loop
```

Fig. 8.80 Part 5 of Main Method in "M_RandomWalker.py".

Video 8.9 showcases the run-time performance of "M_RandomWalker.py". Although its obstacle avoidance response is not quite "snappy" as compared to the TASK version (Video 6.10), it does the job. But in the next project we'll need to use Thread Programming which is a feature beyond the current capabilities of the ROBOTIS TASK V.3 package.

8.8.4 Dual Random Walkers with Thread Programming

This project was described in Section 8.6 for its SCRATCH implementation which was using the concepts of Event Programming via "Message Broadcast" and "Message Received" between 3 independent Sprites (Zira, Zork and Duck as the Game Controller). The same Event Programming approach will be used in this PYTHON project using PYTHON-equivalent constructs which are "Thread" and "Event" objects from the "**threading**" module of the PYTHON Standard Library (https://docs.python.org/3/library/threading.html#). The following web page was very helpful to the author by providing working code snippets used in this project: https://pymotw.com/3/threading/index.html#module-threading – it is maintained by Doug Hellmann.

This PYTHON project "M_Dual_RandomWalkers.py" uses the same approaches as previous projects regarding EDBOT robot control and PYGAME keyboard/graphics interfacing so only the newer concepts of threading and event signaling will be described further in the following paragraphs. If the reader is interested in textbooks to learn more about "threads" or "concurrency" in general, the author would suggest Hellmann (2017) and Nguyen (2018).

The author was delighted to find that all Methods created for the Single Random Walker project (with Zira) in Section 8.8.3 were re-usable in the Dual Random Walkers project (with Zira and Zork). The Main Endless Loop used in the Single Random Walker project needed to be recast as a Thread for Zira and duplicated as another Thread for Zork in the Dual Random Walkers project, with minor code changes to adjust for the robot names, and some code additions at the beginning and at the end of each of these Threads to handle how they would start and end. A third new Thread had to be created to act as the Game Closer, while the Main Method acts as the Game Starter.

Fig. 8.81 shows all the import modules used in "M_Dual_RandomWalkers.py".

"M_Dual_RandomWalkers.py"
Import Packages

```
import sys
import time
import edbot
import random
import pygame
from pygame.locals import *
import threading
```

Fig. 8.81 Modules imported in "M_Dual_RandomWalkers.py".

In PYTHON, a thread can be simply created as a regular function like the ones previously created for various robot maneuvers (https://en.wikibooks.org/wiki/Python_Programming/Threading). As we plan to use the Event Object formalism (https://docs.python.org/3/library/threading.html#event-objects), each thread will also need a special command called ***message.wait()*** which indefinitely blocks the execution of **all codes that follow** the ***message.wait()*** command. When the ***message.set()*** command is asserted in the main or in other threads, then this thread's code execution restarts and proceeds as normal for the rest of the thread/method. Thus the ***message.wait()*** command is the equivalent of the SCRATCH's HAT command "**When I receive Message**".

As an example, let's look at the details of Thread ***go_Zira()*** (see Fig. 8.82):

- Line 164 lists the Global Variables used for this thread. "game_start" and "game_over" are Event objects, while (reset, winner, robot1, obstacle1, threshold1, threshold2, threshold3) are standard Global Variables defined in the Main Method.
- Line 166 uses the command ***game_start.wait()*** which effectively blocks the execution of the remaining code for this thread which then can only wait for a signal ***game_start.set()*** from another thread/method (which will be the Main Method for this case – see Fig. 8.86). Thus the signal ***game_start.set()*** is equivalent to the SCRATCH's **Broadcast Message** block.
- Once the signal ***game_start.set()*** is received by this thread, Line 167 prints out, on THONNY's output console, the message "Thread go_Zira starts".

"M_Dual_RandomWalkers.py"
Thread 1 "go_Zira()": Start & End

```
162 # Thread 1 - Go Zira!
163 def go_Zira():
164     global game_start, game_over, reset, robot1, obstacle1, threshold1, threshold2, threshold3
165 
166     game_start.wait()   # blocking until game_start is set in Main program
167     print("\nThread go_Zira() starts")
168 
169     while True:
170     # Codes for Robot1
171         sensors_data1 = ec.get_robot(robot1)["reporters"]
172         ir_left1 = sensors_data1["port1"]
173         ir_right1 = sensors_data1["port4"]
174     # if no obstacle
175         if ((ir_left1 < threshold1) and (ir_right1 < threshold1)):
176             obstacle1 = 0
177             ec.say(robot1, robot1 + " random walking")
221             flash_LED(robot1)
222             # end of else Robot1
223 
224         if (reset == True):
225             init_pose(robot1)
226             break   # to break out of endless loop
227         time.sleep(0.10)
228     # end of endless loop
229 
230     print("\nGetting out of go_Zira()")
231     return
```

Fig. 8.82 Thread/Method ***go_Zira()*** in "M_Dual_RandomWalkers.py".

- The Main Endless Loop from the Single Random Walker project "M_RandomWalker.py" is inserted "unchanged" at Line 169 (see Figs. 8.76 to 8.80), as "robot1" is properly set to "Zira" in the Main Method's Variables Initializations (see Fig. 8.68). As this code section was already described in Section 8.8.3, Fig. 8.82 skips to Line 221 which is the last line from the "Single Walker" Main Endless Loop.

- Lines 224 to 231 are new lines of code needed to "wrap up" the Thread ***go_Zira()*** at termination time:
    - In Line 224, "reset" is a Global Variable set to TRUE in the Main Method when the user pushes on the key "s" to stop this game and all Threads. This is the author's approach to implement the SCRATCH Block "Stop All".
    - When "reset" is found to be set to TRUE, "robot1" is set to its Initial Pose (Line 225), (i.e. it stops). Next, Line 226 issues a "break" command to get out of this Thread's Endless Loop that started on Line 169. Then, Line 230 prints out the message "Getting out of go_Zira()" and this **Thread terminates** with a Return at Line 231.
    - If (reset == FALSE) instead, then Line 227 waits 0.10 second for the EDBOT's sensors data collection process to restart after Motion Plays, and this Thread's Endless Loop reiterated for another round from Line 169.

Fig. 8.83 lists the details for Thread ***go_Zork()*** which is virtually identical in design to Thread ***go_Zira()***, except that the Global Variable "robot2" is used.

"M_Dual_RandomWalkers.py"
Thread 2 "go_Zork()": Start & End

```
233 # Thread 2 - Go Zork!
234 def go_Zork():
235     global game_start, game_over, reset, robot2, obstacle2, threshold1, threshold2, threshold3
236 
237     game_start.wait()   # blocking until game_start is set in Main program
238     print("\nThread go_Zork() starts")
239 
240     while True:
241     # Codes for Robot2
242         sensors_data2 = ec.get_robot(robot2)["reporters"]
243         ir_left2 = sensors_data2["port1"]
244         ir_right2 = sensors_data2["port4"]
245     # if no obstacle
246         if ((ir_left2 < threshold1) and (ir_right2 < threshold1)):
247             obstacle2 = 0
248             ec.say(robot2, robot2 + " random walking")
```

```
292             flash_LED(robot2)
293             # end of else Robot2
294 
295         if (reset == True):
296             init_pose(robot2)
297             break   # to break out of endless loop
298         time.sleep(0.10)
299 # end of endless loop
300 
301     print("\nGetting out of go_Zork()")
302     return
```

Fig. 8.83 Thread/Method ***go_Zork()***" in "M_Dual_RandomWalkers.py".

Fig. 8.84 lists the details for the third Thread ***game_end()*** which is designed to wrap up a game session once it receives the signal ***game_over.set()*** which is sent by the Main Method (Line 431 of Fig. 8.86):

"M_Dual_RandomWalkers.py"
Thread 3 "game_end()"

```
304 # Thread 3 - game_end()
305 def game_end():
306     global game_over, reset, robot1, robot2
307 
308     game_over.wait()  # blocking until game_over is set
309     init_pose(robot1)
310     init_pose(robot2)
311     ec.say(robot1, "Game is over")
312     ec.say(robot2, "Game is over")
313     reset = True
314     print("\nGetting out of game_end()")
315     return
```

Fig. 8.84 Thread/Method ***game_end()*** in "M_Dual_RandomWalkers.py".

- This thread uses the Event object ***game_over.wait()*** to block further execution of its remaining code (Line 308).
- Once the signal ***game_over.set()*** is sent from the Main Method, this thread immediately proceeds to stop robot1 and robot2 (Lines 309-310) and makes them "say" "Game is over" (Lines 311-312).
- This thread's final actions are to set the Global Variable "reset" to TRUE (Line 313) and to print out an informational statement (Line 314).

In short, creating a Python thread is no different than creating a regular Python function except for the execution control aspect which is provided by the signaling service of an Event object.

Next, let's look at how the various Thread and Event objects are defined and started, as shown in Fig. 8.85 (which are codes residing in the Main Method):

- Lines 396 and 397 define the "**threading.Event**" objects used in this project which are "**game_start**" and "**game_over**". These Event objects manage an internal flag that is initially set to FALSE, which makes its corresponding method ***wait()*** block all code execution flow wherever they are used. Code execution would resume as normal when this flag is set to TRUE somewhere else by using its corresponding method ***set()***. This web page https://docs.python.org/3/library/threading.html#event-objects documents other parameters that can be applied depending on the user's needs.
- Line 400 associates Thread "t1" with the Target Method "**go_Zira**" and Line 401 starts "t1", i.e. Thread/Method ***go_Zira()*** is called, but only Line 166 of Fig. 8.82 really gets executed. Next, Line 167, with its ***game_start.wait()*** method, effectively holds this thread in hiatus until ***game_start.set()*** is used somewhere else at a later time.

"M_Dual_RandomWalkers.py"
Threads & Events Definition

```
395 # Define the two thread controling events used
396 game_start = threading.Event()
397 game_over = threading.Event()
398
399 # Define the three threads used
400 t1 = threading.Thread(name='Go Zira', target=go_Zira, args=() )
401 t1.start()
402
403 t2 = threading.Thread(name='Go Zork', target=go_Zork, args=() )
404 t2.start()
405
406 t3 = threading.Thread(name='End Game', target=game_end, args=() )
407 t3.start()
```

Fig. 8.85 Thread and Event objects definitions and starting actions in "M_Dual_RandomWalkers.py".

- Lines 403-404 do a similar job of associating "t2" to Target Thread ***go_Zork()*** and for starting "t2". And in the same way as for "t1", "t2" halts its execution flow at Line 237 of Fig. 8.83 and waits for the signal ***game_start.set()*** to be activated from Main Method at a later time.
- Lastly, Lines 406-407 start the third thread "t3", i.e. Thread ***game_end()***, which also halts its execution flow at Line 308 of Fig. 8.84 and waits for the signal ***game_over.set()*** to be activated elsewhere.

If the reader goes back and pulls up the SCRATCH code "M_Dual_Random_Walkers.sb2", the reader can see that so far for PYTHON the author has kept the same code design approach as in SCRATCH: with Sprite "Zira" corresponding to Thread ***go_Zira***() and Sprite "Zork" corresponding to Thread ***go_Zork()***. The Thread ***game_end()*** corresponds to the "Space Key" Script which is located in the "Duck" Sprite, but it could have been put in either the "Zira" or "Zork" Sprites and the run-time functionality would be the same. Let's next see how we can implement the "Duck" Sprite to act as the Game Controller in PYTHON.

Fig. 8.86 shows Part 1 of the Endless Loop that acts as the Game Controller (which resides in the Main Method):

- Line 410 initializes Variable "time_game_start" to 0, but it won't actually be set until the user pushes on the key "p" during run-time.
- The Endless Loop starts at Line 411, and the first task is to process the Pygame Event Queue with a For Loop defined by Lines 413-436, looking for 2 event types – QUIT or KEYDOWN:
  - If the event type is QUIT (Lines 410-418), i.e. the user chooses to close the "Dual Random Walkers Game" window as shown in Fig. 8.87, then both robots are stopped (Lines 415-416), the Pygame interface closes (Line 417) and the PYTHON session is also exited (Line 418).
  - If the user pushes down on key "p", then the "Game" starts with the sending of a TRUE signal by ***game_start.set()*** (Line 422) to release the blocks set by the methods ***game_start.wait()*** used in the Threads ***go_Zira()*** (Line 166 of Fig. 8.82) and ***go_Zork()*** (Line 237 of Fig. 8.83). At this time the user would see the robots performing their random walks as the full codes for ***go_Zira()*** and ***go_Zork()*** are finally doing their jobs! Line 423 marks the effective "time_game_start" using Method ***process_time()*** from the time module. This parameter is also printed out on THONNY's Shell for informational

purposes (Line 424). Lines 425-426 “screen.blit” “text_surface1” which is “Start” in the middle of the Pygame window named “Dual Random Walkers Game”. Next, ***time.sleep()*** of 1 second is used to display the “Start” message hopefully long enough for the user to notice it.

“M_Dual_RandomWalkers.py” Main Controller Loop (1)

```
409 # Main controller loop starts here
410 time_game_start = 0.0  # initial value, actual value is set by event.key "p"
411 while True:
412 # Checking for user's keyboard inputs
413     for event in pygame.event.get():
414         if event.type == QUIT:
415             init_pose(robot1)
416             init_pose(robot2)
417             pygame.quit()
418             sys.exit()  # sys.exit() when user closes Pygame window
419 
420         if event.type == KEYDOWN:
421             if event.key == K_p:
422                 game_start.set()  # signals go_Zira and go_Zork to run
423                 time_game_start = time.process_time()   # the effective time_game_start
424                 print("time_game_start = " + str(time_game_start))
425                 screen.blit(text_surface1, (x1, y1))
426                 pygame.display.update()
427                 time.sleep(1)
428             if event.key == K_s:
429                 init_pose(robot1)
430                 init_pose(robot2)
431                 game_over.set()
432                 reset = True  # this flag is also used to control threads behavior
433                 screen.blit(text_surface5, (x5, y5))
434                 pygame.display.update()
435                 time.sleep(5.0)
436                 sys.exit()  # sys.exit() when user wants to stop the game
```

Fig. 8.86 Part 1 of Main Controller Loop in “M_Dual_RandomWalkers.py”.

- If the user pushes down on key “s”, then the “Game” exits by first stopping both robots (Lines 429-430). Next, Line 431 send the signal ***game_over.set()*** to let Thread ***game_end()*** do its game wrapping-up job. With Line 432, the Global Parameter “reset” is set to TRUE to signal the Threads ***go_Zira()*** and ***go_Zork()*** to stop their respective robots (once more – just to be sure!) and also to force a break from the Endless Loops that both threads were in at this point in time (see Lines 224-226 of Fig. 8.82 and Lines 295-297 of Fig. 8.83) and thus these threads/methods are terminated properly and cleanly. This is an important step to remember to do when ones use multiple threads in an application. Lines 433-435 display “text_surface5”, i.e. “Game Stopped” for 2 seconds, before a system exit is executed at Line 436.

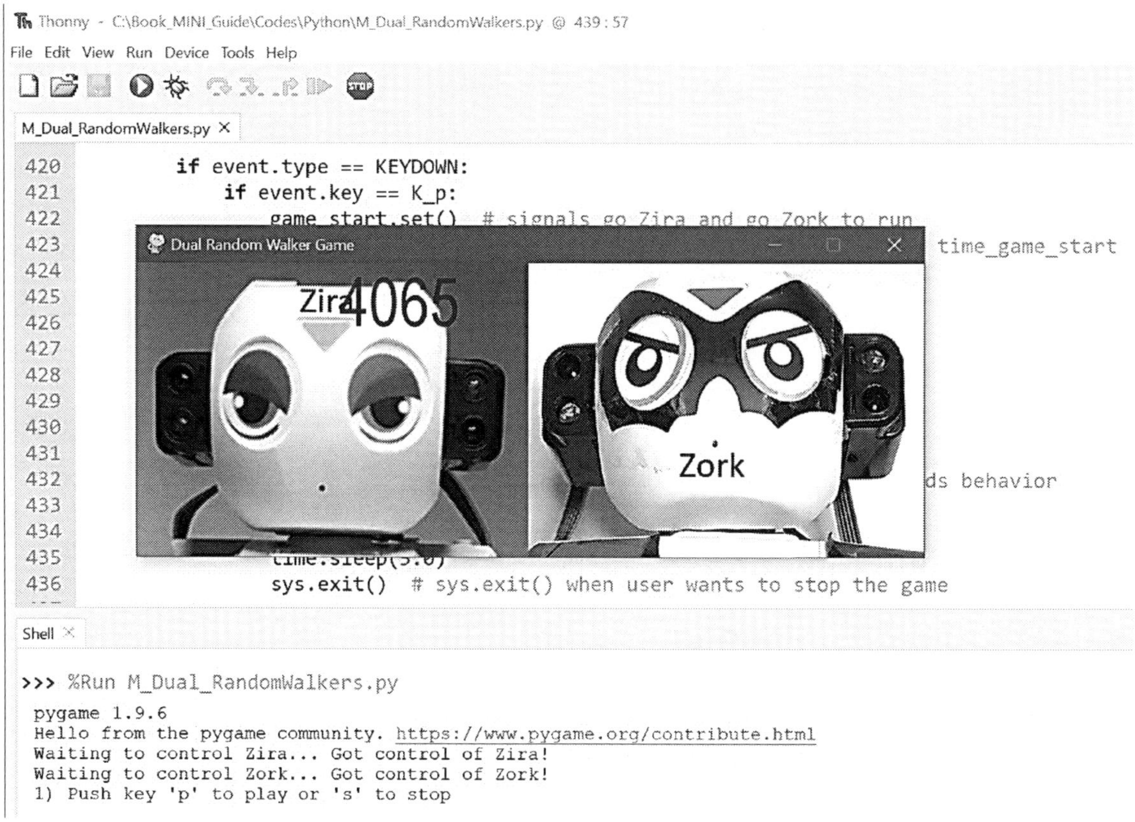

Fig. 8.87 Runtime Display for "M_Dual_RandomWalkers.py".

Fig. 8.88 describes Part 2 of the Endless Loop that acts as the Game Controller (which resides in the Main Method):

"M_Dual_RandomWalkers.py"
Main Controller Loop (2)

```
438     counter += 1
439     screen = pygame.display.set_mode(SCREEN_SIZE, 0, 32)
440     screen.blit(background, (0,0))
441     text_surface7 = font.render(str(counter), True, (0, 0, 255))
442     screen.blit(text_surface7, (240, 0))
443     pygame.display.update()
```

Fig. 8.88 Part 2 of Main Controller Loop in "M_Dual_RandomWalkers.py".

- Line 438 increments Variable "counter" by 1.
- Lines 439-440 put up the background image with Zira and Zork (see Fig. 8.87).

- Line 441 prepares "text_surface_7" with the current value of Variable "counter" (converted to a String, of course).
- Line 442 "blits" "text_surface7" to the background image at position (240,0).
- Line 443 updates all the Pygame display objects to show in the final result such like the one shown in Fig. 8.87.

Video 8.10 shows the run-time performance of "M_Dual_RandomWalkers.py" where the viewer can see that the Threads ***go_Zira()*** and ***go_Zork()*** alternate their execution processes. The author so far cannot ascertain if this effect comes from the way EDBOT is sending its commands to Zira and Zork, or because the time-slice approach used by the PYTHON Threading package. This video also shows that, at unexpected times, EDBOT may decide to drop its connection with one of the robots, so the affected one would just stop "random walking", but the other robot would still perform its job – showing the power of Threads!

8.8.5 Color Tracker with Web Cam using OpenCV

The reader may recall the TASK project "M_CameraTracker.tsk3" in Section 7.6, leveraging the R+m.PLAY700 App to use the camera from a smartphone. This project was limited because of the R+m.PLAY700 App itself and because of the weight of the smartphone as the MINI could only turn left or right in a stable manner.

But now that we are using PYTHON from a Desktop Computer, inexpensive and light-weight USB webcams are available to be mounted of the MINI's head. The author uses a Logitech C270 with its stand removed for Zira and Zork (see Fig. 8.89). Furthermore, USB cables can be up to 25 ft long, so the MINI should have good freedom of movement, although it does have to drag that USB cable around which may influence its walking moves.

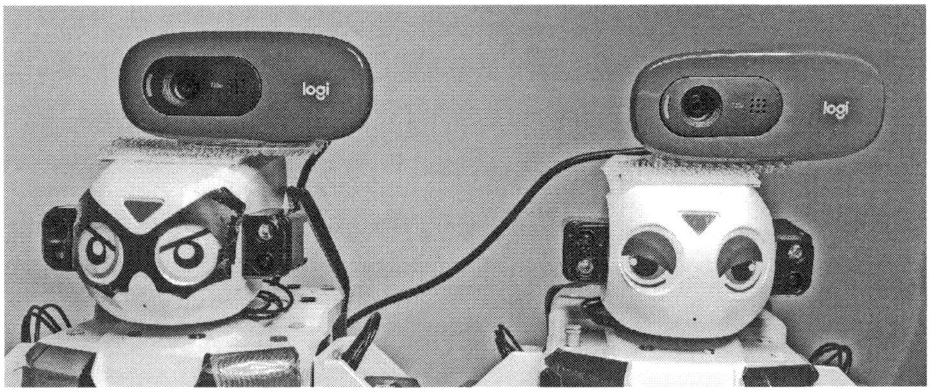

Fig. 8.89 Zira and Zork equipped with webcams mounted on their head pieces.

Another advantage when using PYTHON which is a widely adopted language is the large repository of modules/tools found at the Python Software Foundation https://pypi.org/. This section 8.8.5 would illustrate the use of the OpenCV module and a related module called NumPy (https://pypi.org/project/opencv-python/). If the reader is using THONNY, it is recommended to use the "Manage Package" tool to search for the module "opencv-python" and install it. This process will install all related modules for the reader in one step. The author is using OpenCV-Python V.4.1.2.30 and NumPy V.1.18.1.

The field of Computer Vision or Machine Vision is large and complex, thus in this section the author is only trying to show that, with a bit of work, the reader can incorporate Machine Vision capabilities to a MINI robot. References for Machine Vision, OpenCV and NumPy abound, so the author is only recommending a few that he is familiar with:

- "Learning OpenCV 3" by Kaehler and Bradski (2017), this book is written for the C/C++ programming language but it is a thorough reference to keep.
- For Cookbook types of OpenCV book references in Python, the author has used "Learn OpenCV 4 by Building Projects" by Escrivá and Mendonça (2018) and "OpenCV 3 Computer Vision with Python Cookbook" by Spizhevoy and Rybnikov (2018) and had found them useful.
- Web resources (free and at cost) for OpenCV-Python and NumPy can be found at many places and here are a few:
    - https://opencv-python-tutroals.readthedocs.io/en/latest/index.html
    - https://www.learnopencv.com/about/
    - https://www.pyimagesearch.com/
    - http://www.numpy.org/ and https://docs.scipy.org/doc/numpy/.
    - http://cs231n.github.io/python-numpy-tutorial/

The author assumes that most readers are aware that a color digital image consists of physical pixels in a 2-dimensional array such as 1280 x 1024 or 3840 x 1600 (https://en.wikipedia.org/wiki/Color_image). Each of these physical pixels is in fact a triplet of pixels in the primary colors Red, Green, and Blue (also known as the RGB Color Model - https://en.wikipedia.org/wiki/RGB_color_model). The RGB Color Model is needed to capture a color scene from the real world into a video camera sensor and to display the same color scene on a computer monitor (https://www.visiononline.org/userassets/aiauploads/file/cvp_the-fundamentals-of-camera-and-image-sensor-technology_jon-chouinard.pdf). When these triplet light sources from a color display reach the human eye, a very complex bio-physical process occurs to give us the sensation of color (https://www.olympus-lifescience.com/en/microscope-resource/primer/lightandcolor/humanvisionintro/). In turn, we usually describe those color sensations as red, green and blue (the primary colors), but also as yellow or brown or violet (generally called "hue"). For example, to create the "yellow" sensation we would have to turn on the "red" and "green" light sources while providing no "blue" light. Thus, from the point of view of color image processing, we would have to use at least two parameters (Red & Green) to define a single hue (Yellow). For this reason, other color models are created (derived from the RGB model) to more closely align with how humans classify perceived colors (https://en.wikipedia.org/wiki/HSL_and_HSV), where we would need only one parameter to describe Hue (with the other two parameters S or L/V used to describe more subtle color perceptions such as vividness or brightness of a particular color perceived). We will be using the HSV Color Model in the Python projects described in Sections 8.8.5 and 8.8.6.

First let's look at the project "M_RC_Vision_HSV.py" where the user can remotely control the movement of Zork (therefore the camera – see Fig. 8.90) to position a central Region Of Interest (ROI) onto an object of interest to display its HSV data. This project also gives the user the option of recording and pausing multiple video clips into a single video file.

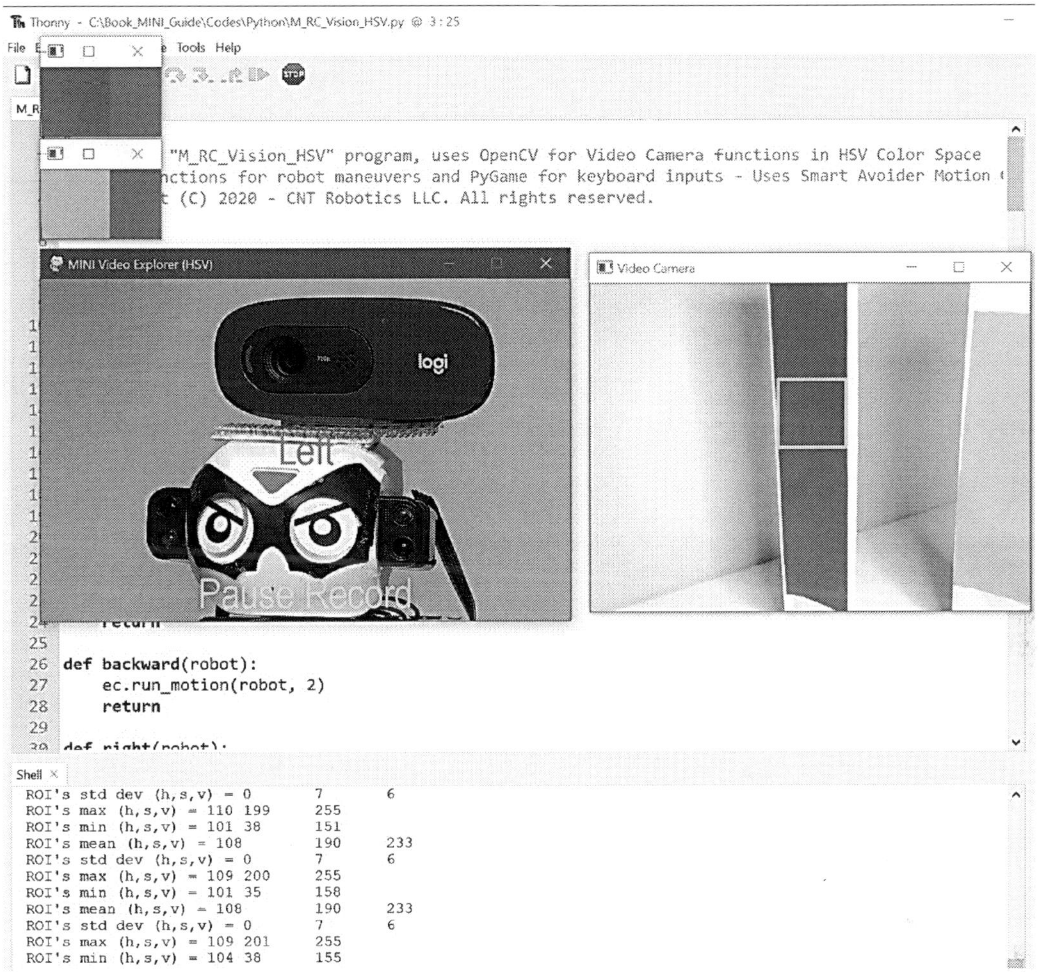

Fig. 8.90 Run-time Displays for "M_RC_Vision_HSV.py".

Fig. 8.91 shows the modules imported for this project, please note "cv2" and "numpy".

"M_RC_Vision_HSV.py"
Import Packages

```
import edbot
import time
import sys
import pygame
from pygame.locals import *
import cv2
import numpy as np
```

Fig. 8.91 Imported modules in "M_RC_Vision_HSV.py".

Fig. 8.92 shows the main Variables used and the Messages for the run-time options menu:

"M_RC_Vision_HSV.py"
Variables & Constants Init.

```
robot1 = "Zork"
robot_mode = 's' # Initial mode is 'stop'
text_mode = 's' # Initial Text display is "stopped"
background_image_filename = 'Zork_Cam_Ready.jpg'
SCREEN_SIZE = (764, 501)
message1 = "1) Use Arrow keys to move robot around or 's' to stop the robot"
message2 = "2) Tap key 'r' to toggle start/stop video and 'Esc' to quit this app"
```

Fig. 8.92 Initialization of Variables and Constants used in "M_RC_Vision_HSV.py".

Fig. 8.93 illustrates the various initialization and checking steps that need to be done in order to use a video camera as a VideoCapture object and a VideoWriter object to record video clips selected from the video stream:

"M_RC_Vision_HSV.py"
Video Objects Settings

```
vid_cam = cv2.VideoCapture(0)
if(not vid_cam.isOpened()):
    print("Error connecting to camera - Exiting program")
    vid_cam.release()
    sys.exit()
f_width = int(vid_cam.get(cv2.CAP_PROP_FRAME_WIDTH))
f_height = int(vid_cam.get(cv2.CAP_PROP_FRAME_HEIGHT))
roi_size = 100
w1 = int((f_width - roi_size)/2)
h1 = int((f_height - roi_size)/2 - 50)
w2 = int((f_width + roi_size)/2)
h2 = int((f_height + roi_size)/2 - 50)
# in this version we use HSV Color Space for the ROI
min_hsv = [None] * 3
max_hsv = [None] * 3
recording = False

vid_filename = "video.avi"
vid_recorder = cv2.VideoWriter(vid_filename, cv2.VideoWriter_fourcc('M','J','P','G'),
                               25, (f_width, f_height))
if(not vid_recorder.isOpened()):
    print("Error opening video recording file - Exiting program")
    vid_recorder.release()
    sys.exit()
```

Fig. 8.93 Setting up VideoCapture and VideoWriter objects in "M_RC_Vision_HSV.py".

- Line 93 associates VideoCapture's Device 0 with the object named "vid_cam" (https://docs.opencv.org/4.1.0/d8/dfe/classcv_1_1VideoCapture.html).
- Line 94-97 check for proper connection to the camera by checking on Method ***isOpened()***, and if there is a connection error, the program releases the "vid_cam" resource (Line 96) and exits promptly at Line 97. If not, the program proceeds on to Line 98.
- Lines 98-108 list the initializations of various Variables and Constants used in the program:
  - Please note that the ROI is a square of 100x100 pixels and (w1, h1) and (w2, h2) denote respectively its top left and bottom right corners in relation to the video frame coordinates system (f_width, f_height). The ROI is centered width wise with respect to the video frame's width, but its height is shifted up by 50 pixels, due to the mounting position chosen for the webcam on top of the MINI's head piece (see yellow square in Fig. 8.90).
  - "min_hsv" and "max_hsv" are 1-dimensional arrays with 3 elements having zero for initial values (Lines 106-107).
  - Variable "recording" is initialized to FALSE (Line 108).
- Line 110 assigns a local file name "video.avi" to Variable "vid_filename".
- In Line 111, the class "VideoWriter()" creates a "vid_recorder" object using 4 arguments:
  - "vid_filename", i.e. the string "video.avi".
  - a FOURCC value of "MJPG", specifying the video codec to use when saving the video frames (see http://www.fourcc.org/codecs.php for other codecs that the reader can try out). At this point, some readers may ask why did the author use the file extension of "avi" for "vid_filename" in Line 111, instead of "mp4"? The reason was that so that most Windows video utilities like "Windows Media Player" or "Movies & TV" can play this video file. When the author changed the file extension to "mp4", then only "VLC Media Player" or "5KPlayer" can play such file, and the author so far could not find any technical reason for this effect!
  - "25" means 25 video frames per second.
  - The pair (f_width, f_height) means that the original video frame size streaming in from the camera 0 is kept (see Lines 98-99).
- Lines 113-116 check to see if the "vid_recorder" object can be opened properly, if not the whole program exits at Line 116. If "vid_recorder" opens properly, then the program flows into the code segment displayed in Fig. 8.94.

Fig. 8.94 defines the code used to display the main Pygame window named "MINI Video Explorer (HSV)" with its background image "Zork_Cam_Ready.jpg" and various Text Surfaces to be used in later parts of the program.

"M_RC_Vision_HSV.py" Pygame Set Up

```
121 pygame.init()
122 screen = pygame.display.set_mode(SCREEN_SIZE, 0, 32)
123 pygame.display.set_caption('MINI Video Explorer (HSV)')
124 background = pygame.image.load(background_image_filename).convert()
125 font = pygame.font.SysFont("arial", 60);
126 text_surface1 = font.render("Forward", True, (0, 255, 0))   # PyGame uses RGB
127 x1 = ( SCREEN_SIZE[0] - text_surface1.get_width() ) / 2
128 y1 = ( SCREEN_SIZE[1] - text_surface1.get_height() ) / 2
129 text_surface2 = font.render("Left", True, (0, 255, 0))
130 x2 = ( SCREEN_SIZE[0] - text_surface2.get_width() ) / 2
131 y2 = ( SCREEN_SIZE[1] - text_surface2.get_height() ) / 2
132 text_surface3 = font.render("Right", True, (0, 255, 0))
133 x3 = ( SCREEN_SIZE[0] - text_surface3.get_width() ) / 2
134 y3 = ( SCREEN_SIZE[1] - text_surface3.get_height() ) / 2
135 text_surface4 = font.render("Backward", True, (0, 255, 0))
136 x4 = ( SCREEN_SIZE[0] - text_surface4.get_width() ) / 2
137 y4 = ( SCREEN_SIZE[1] - text_surface4.get_height() ) / 2
138 text_surface5 = font.render("Stopped", True, (255, 0, 0))
139 x5 = ( SCREEN_SIZE[0] - text_surface5.get_width() ) / 2
140 y5 = ( SCREEN_SIZE[1] - text_surface5.get_height() ) / 2
141 text_surface6 = font.render("Recording", True, (255, 255, 0))
142 x6 = ( SCREEN_SIZE[0] - text_surface6.get_width() ) / 2
143 y6 = 425
144 text_surface7 = font.render("Pause Record", True, (255, 255, 0))
145 x7 = ( SCREEN_SIZE[0] - text_surface7.get_width() ) / 2
146 y7 = 425
```

Fig. 8.94 Setting up various Pygame objects in "M_RC_Vision_HSV.py".

Fig. 8.95 displays Part 1 of the Main Endless Loop (starting at Line 150) which checks for the Pygame event type "QUIT" (Line 153), i.e. when the user closes the window named "MINI Video Explorer (HSV)" in Fig. 8.90, then:

- Line 154 stops the MINI robot.
- Lines 155-157 close all OpenCV devices and windows used in the program.
- Line 158 quits the Pygame module and Line 159 exits the Python shell.

"M_RC_Vision_HSV.py"
Part 1 of Main Endless Loop

```python
while True:
# Checking for user's keyboard inputs
    for event in pygame.event.get():
        if event.type == QUIT:  # when user closes Pygame window
            init_pose(robot1)
            vid_cam.release()
            vid_recorder.release()
            cv2.destroyAllWindows()
            pygame.quit()
            sys.exit()
```

Fig. 8.95 Part 1 of Main Endless Loop in "M_RC_Vision_HSV.py".

Fig. 8.96 displays Part 2 of the Main Endless Loop which checks for the Pygame event type "KEYDOWN", i.e. when the user pushes on specific keys:

"M_RC_Vision_HSV.py"
Part 2 of Main Endless Loop

```python
        if event.type == KEYDOWN:
            if (event.key == K_r):  # toggling start/stop video recording
                if (recording == False):
                    recording = True
                    screen.blit(text_surface6, (x6, y6))
                    pygame.display.update()
                else:
                    recording = False
                    screen.blit(text_surface7, (x7, y7))
                    pygame.display.update()
            if event.key == K_s:
                robot_mode = 's'
                text_mode = 's'
            if event.key == K_UP:
                robot_mode = 'f'
                text_mode = 'f'
            if event.key == K_DOWN:
                robot_mode = 'b'
                text_mode = 'b'
            if event.key == K_LEFT:
                robot_mode = 'l'
                text_mode = 'l'
            if event.key == K_RIGHT:
                robot_mode = 'r'
                text_mode = 'r'
```

Fig. 8.96 Part 2 of Main Endless Loop in "M_RC_Vision_HSV.py".

- Lines 161-169 illustrate when the user toggles on key "r". Essentially, if "recording" was not in effect (Line 162) then "recording" is set to TRUE (Line 163) and "text_surface6", i.e. "Recording", is displayed (Lines 164-165). Otherwise "recording" is set to FALSE (Line 167) and "text_surface7", i.e. "Pause Record", is displayed (Lines 168-169).
- Lines 170-184 set the proper values to "robot_mode" and "text_mode" depending on the robot motions wanted by the user: "stop", "forward", "backward", "left" and "right" (using key "s" and the respective arrow keys).

Fig 8.97 displays Part 3 of the Main Endless Loop where Pygame blits the proper text surfaces depending on the current values of Parameters "text_mode" and "recording":

"M_RC_Vision_HSV.py"
Part 3 of Main Endless Loop

```
# Updating Pygame screen
    screen = pygame.display.set_mode(SCREEN_SIZE, 0, 32)
    screen.blit(background, (0,0))
    if (text_mode == 'f'):          # robot forward
        screen.blit(text_surface1, (x1, y1))
    elif (text_mode == 'l'):        # robot left
        screen.blit(text_surface2, (x2, y2))
    elif (text_mode == 'r'):        # robot right
        screen.blit(text_surface3, (x3, y3))
    elif (text_mode == 'b'):        # robot backward
        screen.blit(text_surface4, (x4, y4))
    elif (text_mode == 's'):        # stop the robot
        screen.blit(text_surface5, (x5, y5))
    # Updating video recording notifications
    if (recording == True):
        screen.blit(text_surface6, (x6, y6))
    elif (recording == False):
        screen.blit(text_surface7, (x7, y7))
    pygame.display.update()
```

Fig. 8.97 Part 3 of Main Endless Loop in "M_RC_Vision_HSV.py".

Fig. 8.98 shows Part 4 of the Main Endless Loop where the appropriate robot MOTIONs are activated depending on the current value of Variable "robot_mode" (Lines 207-216):

- Please note Line 217 where "robot_mode" is cleared of its previous value to ensure that only ONE robot move is performed per key input.

"M_RC_Vision_HSV.py"
Part 4 of Main Endless Loop

```
# Updating robot maneuvers
    if (robot_mode == 'f'):         # robot forward
        forward(robot1)
    elif (robot_mode == 'l'):       # robot left
        left(robot1)
    elif (robot_mode == 'r'):       # robot right
        right(robot1)
    elif (robot_mode == 'b'):       # robot backward
        backward(robot1)
    elif (robot_mode == 's'):       # stop the robot
        init_pose(robot1)
    robot_mode = ''  # Motion to be played only once per key input
```

Fig. 8.98 Part 4 of Main Endless Loop in "M_RC_Vision_HSV.py".

Fig. 8.99 shows Part 5 where for each iteration of the Main Endless Loop a video frame is captured (Line 220) and checked for proper capture (Line 221), if an error occurred then break out of the Main Endless Loop (Lines 222-223). Otherwise check to see if "recording" is wanted (Line 224) and if "recording" is TRUE save the current video frame using Method ***vid_recorder.write()*** (Line 225).

"M_RC_Vision_HSV.py"
Part 5 of Main Endless Loop

```
219 # Operating video camera
220     got_frame, vid_frame = vid_cam.read()  # picking up a new video frame
221     if not got_frame:   # error capturing video frame
222         print("No video frame captured!")
223         break   # get out of Main Loop
224     if (recording == True):  # saving video frame per user's toggling the 'r' key
225         vid_recorder.write(vid_frame)
```

Fig. 8.99 Part 5 of Main Endless Loop in "M_RC_Vision_HSV.py".

A special note should be mentioned for Line 220 which shows the compactness of the PYTHON language where two items get returned whenever Method ***vid_cam.read()*** is invoked: "**got_frame**" is a logical flag while "**vid_frame**" is a 3-dimensional array representing the image captured in RGB Color Model (or to be precise in BGR format – a peculiar quirk of OpenCV). "**vid_frame**" will be further processed in Part 6.

Fig. 8.100 shows Part 6 where the true powers of OpenCV and NumPy are displayed:

- Line 227 is quite a "nifty" NumPy operation combining indexing and slicing operations onto the 3-channel array "vid_frame" to extract out a smaller 3-channel array called "roi_bgr", using "coordinates" (w1, h1) and (w2, h2) which define the top-left and bottom-right corners of a ROI previously defined in the video frame captured (see Lines 101-104 in Fig. 8.93 and the yellow box in the window labeled "Video Camera" shown in Fig. 8.90) (see https://docs.scipy.org/doc/numpy/reference/arrays.indexing.html for more details).

- Lines 228-229 use the OpenCV Methods ***imshow()*** and ***moveWindow()*** to display this "roi_bgr" in a small window located at position (50, 50) i.e. near the top left area of the computer display (see Fig. 8.90). Please note that ***imshow()*** uses a window wider that the ROI's width, thus creating the extra "blank-grey" area on the right of the displayed window. For more details regarding the use of ***imshow()*** and ***moveWindow()*** please visit this web link https://docs.opencv.org/4.0.1/d7/dfc/group__highgui.html#ga453d42fe4cb60e5723281a89973ee563)

"M_RC_Vision_HSV.py"
Part 6 of Main Endless Loop

```
226    # working on the ROI image/window
227    roi_bgr = vid_frame[h1:h2, w1:w2] # defining the ROI - numpy function
228    cv2.imshow('ROI_BGR', roi_bgr) # first ROI video window (BGR)
229    cv2.moveWindow('ROI_BGR', 50, 50)
230    roi_hsv = cv2.cvtColor(roi_bgr,cv2.COLOR_BGR2HSV)
231    cv2.imshow('ROI_HSV', roi_hsv) # second ROI video window (HSV)
232    cv2.moveWindow('ROI_HSV', 50, 200)
233    roi_mean, roi_std_dev = cv2.meanStdDev(roi_hsv)  # computing mean & std. dev.values of ROI
234    print("ROI's mean (h,s,v) = " + str(int(roi_mean[0])) +
235          "\t" + str(int(roi_mean[1])) + "\t" + str(int(roi_mean[2])))
236    print("ROI's std dev (h,s,v) = " + str(int(roi_std_dev[0])) +
237          "\t" + str(int(roi_std_dev[1])) + "\t" + str(int(roi_std_dev[2])))
238    (h, s, v) = cv2.split(roi_hsv)    # split the ROI_HSV into its H-S-V image components
239    max_hsv[0] = np.max(h)
240    max_hsv[1] = np.max(s)
241    max_hsv[2] = np.max(v)
242    print("ROI's max (h,s,v) = " + str(max_hsv[0]) +
243          "\t" + str(max_hsv[1]) + "\t" + str(max_hsv[2]))
244    min_hsv[0] = np.min(h)
245    min_hsv[1] = np.min(s)
246    min_hsv[2] = np.min(v)
247    print("ROI's min (h,s,v) = " + str(min_hsv[0]) +
248          "\t" + str(min_hsv[1]) + "\t" + str(min_hsv[2]))
```

Fig. 8.100 Part 6 of Main Endless Loop in "M_RC_Vision_HSV.py".

- Line 230 converts the ROI's BGR image into an HSV image named "roi_hsv" (see more details at https://docs.opencv.org/4.0.1/d8/d01/group__imgproc__color__conversions.html#ga397ae87e1288a81d2363b61574eb8cab and https://docs.opencv.org/master/df/d9d/tutorial_py_colorspaces.html).

- Lines 231-232 display “roi_hsv” in a window located at coordinates (50, 200) with respect to the computer display, i.e. right below the window for “roi_bgr” (see Fig. 8.90).
- Line 233 uses the OpenCV’s Method ***meanStdDev()*** to compute the respective means and standard deviations of all respective pixel values in each ROI image channel H-S-V, and save them in two 1-D arrays named “roi_mean” and “roi_std_dev” which have 3 elements each (in the HSV order). For more details please see web link at https://docs.opencv.org/4.0.1/d2/de8/group__core__array.html#ga846c858f4004d59493d7c6a4354b301d).
- Lines 234-237 print out the ROI’s mean and standard deviation values found for each image channel Hue (element [0]), Saturation (element [1]) and Value (element [2]).
- Line 238 uses the OpenCV’s Method ***split()*** to unpack the 3-channel array “roi_hsv” into three single-channel arrays named “h”, “s” and “v”. See more details at web link (https://docs.opencv.org/master/d2/de8/group__core__array.html#ga8027f9deee1e42716be8039e5863fbd9).
- Lines 239-241 use the NumPy’s Method ***max()*** to find the maximum pixel values found in each respective single-channel array “h”, “s” and “v” and save them in the respective element max_hsv[0], max_hsv[1] and max_hsv[2]. Lines 242-243 print out the current results found.
- Lines 244-248 perform a similar job to find the respective minimum pixel values found in each (h, s, v) channel using the NumPy’s Method ***min()***.

Part 7 (see Fig. 8.101) wraps up operations for the Main Endless Loop:

- Line 250 uses the OpenCV Method ***rectangle()*** to draw a “yellow square ROI” inside the current “vid_frame”, at locations defined by the two coordinates (w1, h1) and (w2, h2), using the color triplet (0, 255, 255) (remember OpenCV’s BGR order here!), and the last argument “3” means a line thickness of 3. See link https://docs.opencv.org/master/d6/d6e/group__imgproc__draw.html#ga07d2f74cadcf8e305e810ce8eed13bc9 for more details about its usage.
- Line 251 then shows the current “vid_frame” with the yellow square embedded in its higher center area. Please note that the actual video frame was recorded (Line 225 of Fig. 8.99) **before** the ***rectangle()*** operation so the video file “video.avi” does not contain the yellow ROI.

“M_RC_Vision_HSV.py”
Part 7 of Main Endless Loop

```
249 # Adding the ROI rectangle in YELLOW onto the original video frame
250     cv2.rectangle(vid_frame, (w1, h1), (w2, h2), (0, 255, 255), 3)  # OpenCV uses BGR
251     cv2.imshow('Video Camera', vid_frame) # display modified captured frame
252 
253 # End of MAIN LOOP
254 
255 vid_cam.release()
256 vid_recorder.release()
257 cv2.destroyAllWindows()
258 sys.exit()
259 # End of program
```

Fig. 8.101 Part 7 of Main Endless Loop in “M_RC_Vision_HSV.py”.

If a **break-out** of the Main Endless Loop happens at run time, the program would flow to Lines 255-258 in Fig. 8.101 to terminate all OpenCV devices and windows, and the PYTHON session terminates.

Video 8.11 shows the run time performance of "M_RC_Vision_HSV.py" and illustrates how it can be used to record the various HSV data for Blue and Yellow objects of interest to be used in the next PYTHON project "M_HSV_Tracker_Walker.py".

In the project "M_HSV_Tracker_Walker.py", the user can specify either "Blue" or "Yellow" objects for the robot Zork to autonomously search for and then, once found, approach it within a given distance controlled by the Left and Right IRSS-10 sensors on the MINI.

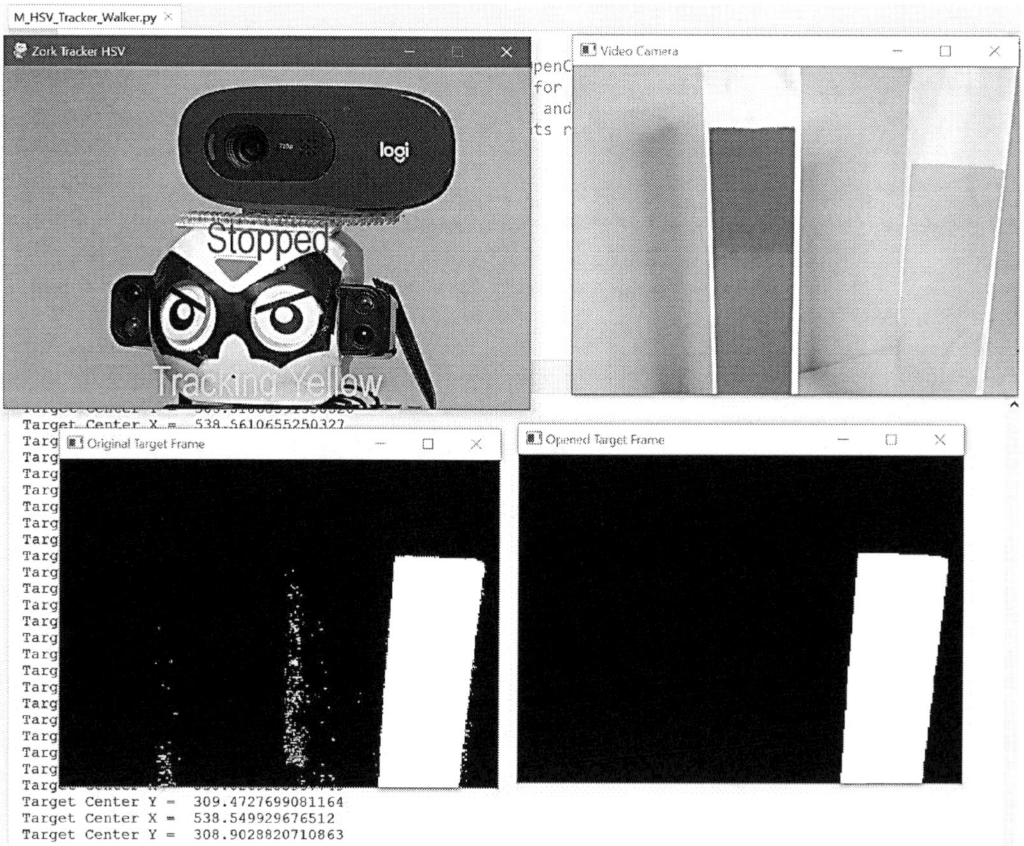

Fig. 8.102 "M_HSV_Tracker_Walker.py" can be used to track "Blue" or "Yellow" objects.

Fig. 8.103 lists the beginning code for "M_HSV_Tracker_Walker.py" where various Variables and Constants are initialized:

- Lines 118-127 are for the usual Variables and Messages.
- Lines 130-137 contain the Constants that were collected using "M_RC_Vision_HSV.py" on the actual physical set up of the scene and objects to be tracked, it is recommended that the user

keeps the same lighting setup between these two projects for best transfer of data and runtime performances.

"M_HSV_Tracker_Walker.py" Variables & Constants Init.

```
118 robot1 = "Zork"
119 robot_mode = 's' # Initial mode is 'stop'
120 text_mode ='s'
121 ir_left = 0
122 ir_right = 0
123 background_image_filename = 'Zork_Cam_Ready.jpg'
124 SCREEN_SIZE = (764, 501)
125 message1 = "1) Use Arrow keys to move robot around or 's' to stop the robot"
126 message2 = "2) Tap key 'b' to track blue target or 'y' to track yellow target"
127 message3 = "3) AFTER target's color is chosen, tap key 'a' for autonomous search & approach"
128
129 # HSV Data previously collected from M_RC_Vision_HSV.py
130 BLUE_OBJECT_MEAN = (106, 161, 254)
131 BLUE_OBJECT_STDDEV = (9, 22, 5)
132 BLUE_OBJECT_MIN = (1, 1, 181)
133 BLUE_OBJECT_MAX = (111, 190, 255)
134 YELLOW_OBJECT_MEAN = (26, 237, 242)
135 YELLOW_OBJECT_STDDEV = (1, 21, 3)
136 YELLOW_OBJECT_MIN = (26, 49, 209)
137 YELLOW_OBJECT_MAX = (31, 245, 249)
147 # Pixel values limits to be set for object to be tracked
148 obj_hsv_lo = [None] * 3
149 obj_hsv_hi = [None] * 3
150
151 # user needs to set up kernel size as needed for image processing
152 kernel = np.ones((20,20),np.uint8)
```

Fig. 8.103 Initializations for Variables and Constants used in "M_HSV_Tracker_Walker.py".

- Lines 148 and 149 define and initialize (to 0) two 1-D arrays for the Low and High HSV Limits which will be used for tracking the wanted object. Their actual values will be set by either Method ***set_blue_object()*** or ***set_yellow_object()*** depending on the user's choice to be made at a later time (see Fig. 8.104).
- Line 152 defines a 2-D kernel or mask array of the size 20x20 filled with "1" for all elements. This kernel will be used in various image processing operations described in later code segments.

Fig. 8.104 lists the two Methods ***set_blue_object()*** and ***set_yellow_object()*** where the reader can see that low and high limits are set to the respective values "**mean ± 2 * Std Dev**".

"M_HSV_Tracker_Walker.py" Methods: set_blue_object() & set_yellow_object()

```
def set_blue_object():
    global obj_hsv_lo, obj_hsv_hi
    for i in range(3):
        obj_hsv_lo[i] = BLUE_OBJECT_MEAN[i] - 2 * BLUE_OBJECT_STDDEV[i]
        obj_hsv_hi[i] = BLUE_OBJECT_MEAN[i] + 2 * BLUE_OBJECT_STDDEV[i]
        if obj_hsv_lo[i] < 0:
            obj_hsv_lo[i] = 0
        if obj_hsv_hi[i] > 255:
            obj_hsv_hi[i] = 255

def set_yellow_object():
    global obj_hsv_lo, obj_hsv_hi
    for i in range(3):
        obj_hsv_lo[i] = YELLOW_OBJECT_MEAN[i] - 2 * YELLOW_OBJECT_STDDEV[i]
        obj_hsv_hi[i] = YELLOW_OBJECT_MEAN[i] + 2 * YELLOW_OBJECT_STDDEV[i]
        if obj_hsv_lo[i] < 0:
            obj_hsv_lo[i] = 0
        if obj_hsv_hi[i] > 255:
            obj_hsv_hi[i] = 255
```

Fig. 8.104 Methods ***set_blue_object()*** and ***set_yellow_object()*** in "M_HSV_Tracker_Walker.py".

Fig. 8.105 describes Method ***check_NIR()*** which is used to estimate the distance between the robot and selected target during the final approach, using the IRSS-10 sensors mounted on both sides of the robot's head piece (see Fig. 8.102). Ideally, the author wanted to access the sensors data during the robot's swing of its head from the right (i.e. Position 120 deg.) to the left (i.e. Position 180 deg.), similarly to the way he used the DMS-80 in the TASK Project "Dowel Scanner" (see Section 6.9). Unfortunately, the current software design in EDBOT does not allow sensors and servos data collection during any Servo Motion. Thus the author had to use a less precise way to estimate the distance between the robot and the target as shown in Method ***check_NIR()***:

- Line 48 puts the robot into the Initial Pose.
- Parameter "max_ir" is initialized to 0 at Line 49.
- Line 50 sends Servo #17 to Position 120 and Line 51 waits for 0.5 second for it to get to Position 120 and for the sensor data collection process to restart.
- Line 52 collects all input data and save them in the dictionary named "sensors".
- Line 53 extracts "port 1" data and saves it into Parameter "ir_l".
- Lines 54-55 essentially compares Parameters "ir_l" and "max_ir" with each other and saves the maximum of these two into "max_ir".
- Line 56 extracts "port 4" data and saves it into Parameter "ir_r".
- Lines 57-58 perform the same "maximum-seeking" procedure between the current "max_ir" and the new "ir_r" and save the result in "max_ir". Thus, at this point, "max_ir" contains the maximum of the two NIR sensors that the robot could collect at Servo17's Position 120.

"M_HSV_Tracker_Walker.py"
Method check_NIR()

```
47 def check_NIR(robot):
48     init_pose(robot)
49     max_ir = 0
50     ec.set_servo_position(robot, "17/120")   # turn head to the right
51     time.sleep (0.5)
52     sensors = ec.get_robot(robot)["reporters"]
53     ir_l = sensors["port1"]
54     if (ir_l > max_ir):
55         max_ir = ir_l
56     ir_r = sensors["port4"]
57     if (ir_r > max_ir):
58         max_ir = ir_r
59
60     ec.set_servo_position(robot, "17/180")   # turn head to the left
61     time.sleep (0.5)
62     sensors = ec.get_robot(robot)["reporters"]
63     ir_l = sensors["port1"]
64     if (ir_l > max_ir):
65         max_ir = ir_l
66     ir_r = sensors["port4"]
67     if (ir_r > max_ir):
68         max_ir = ir_r
69
70     ec.set_servo_position(robot, "17/150")  # turn head forward before exiting check_NIR
71     time.sleep (0.5)
72     return max_ir
```

Fig. 8.105 Method ***check_NIR()*** in "M_HSV_Tracker_Walker.py".

- Lines 60-68 perform the same "maximum-seeking" procedure between the current "max_ir" and the new "port 1" and "port 4" data collected when Servo17 gets to its new Position 180. Thus, after Line 68, "max_ir" contains the maximum of all 4 NIR sensors that were collected during the robot's head-swing maneuver.
- Line 70 sends Servo17 to Position 150, so that the webcam is back to facing forward, and Line 71 waits for 0.5 second for it to get there.
- Line 72 returns the final value obtained for "max_ir" back to the code that called this method. It will be used to decide whether the robot has approached close enough to the selected target.

Fig. 8.106 displays the usual preparation steps to get the camera ready for scene capture and video frame input check (Lines 193-204). Lines 204-205 define two new Variables needed for this project: "search_left_limit" and "search_right_limit" in the video frame coordinates system and please notice that only the width coordinate "f_width" is of use to this project as the robot travels on a flat surface for this application. Please note that Line 192 initializes Variable "camera_mode" to NULL, it will be set with a proper value at a later time in the Main Method.

"M_HSV_Tracker_Walker.py"
Camera Set Up

```
camera_mode = ''
vid_cam = cv2.VideoCapture(0)
if(not vid_cam.isOpened()):
    print("Error connecting to camera - Exiting program")
    vid_cam.release()
    sys.exit()   # terminate program

got_frame, vid_frame = vid_cam.read()  # picking up a new video frame
if not got_frame:   # error capturing video frame
    print("No video frame captured - Exiting program")
    sys.exit()  # terminate program
f_width = int(vid_cam.get(cv2.CAP_PROP_FRAME_WIDTH))
f_height = int(vid_cam.get(cv2.CAP_PROP_FRAME_HEIGHT))
search_left_limit = f_width/2 - 20
search_right_limit = f_width/2 + 20
print ("left_limit = " + str(search_left_limit))
print ("right_limit = " + str(search_right_limit))
```

Fig. 8.106 Preparation steps for camera usage in "M_HSV_Tracker_Walker.py".

Fig. 8.107 defines the usual "motion-related" text messages and 3 "new" text messages for this project which are "Tracking Blue", "Tracking Yellow" and "Autonomous S&A" (i.e. search and approach).

"M_HSV_Tracker_Walker.py"
Pygame Set Up

```
pygame.init()
screen = pygame.display.set_mode(SCREEN_SIZE, 0, 32)
pygame.display.set_caption('Zork Tracker HSV')
background = pygame.image.load(background_image_filename).convert()
font = pygame.font.SysFont("arial", 60);
text_surface1 = font.render("Forward", True, (0, 255, 0))
x1 = ( SCREEN_SIZE[0] - text_surface1.get_width() ) / 2
y1 = ( SCREEN_SIZE[1] - text_surface1.get_height() ) / 2
text_surface2 = font.render("Left", True, (0, 255, 0))
x2 = ( SCREEN_SIZE[0] - text_surface2.get_width() ) / 2
y2 = ( SCREEN_SIZE[1] - text_surface2.get_height() ) / 2

text_surface6 = font.render("Tracking Blue", True, (0, 0, 255))
x6 = ( SCREEN_SIZE[0] - text_surface6.get_width() ) / 2
y6 = 425
text_surface7 = font.render("Tracking Yellow", True, (255, 255, 0))
x7 = ( SCREEN_SIZE[0] - text_surface7.get_width() ) / 2
y7 = 425
text_surface8 = font.render("Autonomous S&A", True, (0, 255, 0))
x8 = ( SCREEN_SIZE[0] - text_surface8.get_width() ) / 2
y8 = ( SCREEN_SIZE[1] - text_surface8.get_height() ) / 2
```

Fig. 8.107 Relevant Pygame text messages in "M_HSV_Tracker_Walker.py".

Fig. 8.108 shows Part 1 of the Main Endless Loop:

"M_HSV_Tracker_Walker.py"
Part 1 of Main Endless Loop

```
246 while True:
247 # Checking for user's keyboard inputs
248     for event in pygame.event.get():
249         if event.type == QUIT:  # when user closes Pygame window
250             init_pose(robot1)
251             vid_cam.release()
252             cv2.destroyAllWindows()
253             pygame.quit()
254             sys.exit()
```

Fig. 8.108 Part 1 of Main Algorithm in "M_HSV_Tracker_Walker.py".

- Lines 249-254 check if the user closes the Pygame window (labeled "Zork Tracker HSV" in Fig. 8.102) and act accordingly by stopping the robot (Line 250), by closing all OpenCV resources (Lines 251-252) and by quitting the Pygame module (Line 253) before making a call to ***sys.exit()*** (Line 154) to close the PYTHON session.

Fig. 8.109 lists Part 2 of the Main Endless Loop which is a code segment used to check for a KEYDOWN event:

- If Key "b" is pushed, "camera_mode" is set to 'b' and Method ***set_blue_object()*** is called (Lines 256-258).
- If Key "y" is pushed, "camera_mode" is set to 'y' and Method ***set_yellow_object()*** is called (Lines 259-261).
- If Key "s" is pushed, both Parameters "robot_mode" and "text_mode" are set to 's' (Lines 262-264).
- If the 4 Arrow Keys are pushed, the appropriate values 'f', 'b', 'l' or 'r' are assigned to "robot_mode" and "text_mode" (Lines 265-276).
- Lines 277-279 are activated when the user pushes on Key "a". This is the option for the robot to start its autonomous "search and approach" procedure.

"M_HSV_Tracker_Walker.py"
Part 2 of Main Endless Loop

```
255         if event.type == KEYDOWN:
256             if event.key == K_b:  # tracking blue
257                 camera_mode ='b'
258                 set_blue_object()
259             if event.key == K_y:  # tracking yellow
260                 camera_mode ='y'
261                 set_yellow_object()
262             if event.key == K_s:  # stop the robot
263                 robot_mode = 's'
264                 text_mode = 's'
265             if event.key == K_UP:  # robot forward
266                 robot_mode = 'f'
267                 text_mode = 'f'
268             if event.key == K_DOWN:  # robot backward
269                 robot_mode = 'b'
270                 text_mode = 'b'
271             if event.key == K_LEFT:  # robot left
272                 robot_mode = 'l'
273                 text_mode = 'l'
274             if event.key == K_RIGHT:  # robot right
275                 robot_mode = 'r'
276                 text_mode = 'r'
277             if event.key == K_a:  # robot autonomous search & approach
278                 robot_mode = 'a'
279                 text_mode = 'a'
```

Fig. 8.109 Part 2 of Main Endless Loop in "M_HSV_Tracker_Walker.py".

Part 3 of the Main Endless Loop (Fig. 8.110) lists the various "blitting" tasks needed for this project based on the current value of Variable "text_mode" which could be "f", "l", "r", "b", "s" or "a" (Lines 285-296). Similar tasks are also programmed for a possible value for Variable "camera_mode": "b" or "y" (Lines 298-301). Lastly, the Pygame display engine is "refreshed" by Line 303.

"M_HSV_Tracker_Walker.py"
Part 3 of Main Endless Loop

```
    screen = pygame.display.set_mode(SCREEN_SIZE, 0, 32)
    screen.blit(background, (0,0))

    if (text_mode == 'f'):   # updating robot motion directions
        screen.blit(text_surface1, (x1, y1))
    elif (text_mode == 'l'):
        screen.blit(text_surface2, (x2, y2))
    elif (text_mode == 'r'):
        screen.blit(text_surface3, (x3, y3))
    elif (text_mode == 'b'):
        screen.blit(text_surface4, (x4, y4))
    elif (text_mode == 's'):
        screen.blit(text_surface5, (x5, y5))
    elif (text_mode == 'a'):  # search & approach
        screen.blit(text_surface8, (x8, y8))

    if (camera_mode == 'b'):   # when tracking blue
        screen.blit(text_surface6, (x6, y6))
    elif (camera_mode == 'y'):  # when tracking yellow
        screen.blit(text_surface7, (x7, y7))

    pygame.display.update()
```

Fig. 8.110 Part 3 of Main Endless Loop in "M_HSV_Tracker_Walker.py".

Part 4 (Fig. 8.111) shows the activation of the appropriate MOTION-LIST/PAGE depending on the run-time value of "robot_mode" (i.e. when the user wants to control the robot directly via the Arrow Keys). Please note the IF construct used in Lines 316-317 to ensure that these RC motions are played only once per keyboard input by clearing out "robot_mode" per iteration of the Main Endless Loop (but not when "Search & Approach" mode 'a' is active).

"M_HSV_Tracker_Walker.py"
Part 4 of Main Endless Loop

```
306     if (robot_mode == 'f'):          # robot forward
307         forward(robot1)
308     elif (robot_mode == 'l'):        # robot left
309         left(robot1)
310     elif (robot_mode == 'r'):        # robot right
311         right(robot1)
312     elif (robot_mode == 'b'):        # robot backward
313         backward(robot1)
314     elif (robot_mode == 's'):        # stop the robot
315         init_pose(robot1)
316     if (robot_mode != 'a'):   # except for autonomous search option
317         robot_mode = ''   # RC Motion to be played only once per key input
```

Fig. 8.111 Part 4 of Main Endless Loop in "M_HSV_Tracker_Walker.py".

Part 5 (Fig. 8.112) shows the capture and display of a new video frame for each Main Loop iteration.

"M_HSV_Tracker_Walker.py"
Part 5 of Main Endless Loop

```
319 # Operating video camera
320     got_frame, vid_frame = vid_cam.read()  # picking up a new video frame
321     if not got_frame:    # error capturing video frame
322         print("No video frame captured!")
323         break  # get out of Main Loop
324     cv2.imshow('Video Camera', vid_frame) # display original captured frame
```

Fig. 8.112 Part 5 of Main Endless Loop in "M_HSV_Tracker_Walker.py".

Part 6 (Fig. 8.113) lists the more interesting steps of processing the current video frame to look for the object of interest and to autonomously approach the found object:

- Part 6 is only activated when "camera_mode" is set to "b" or "y" (Line 327).

"M_HSV_Tracker_Walker.py"
Part 6 of Main Endless Loop

```
326 # Process image to track blue or yellow object
327     if (camera_mode == 'b' or camera_mode == 'y'):
328 # Convert vid_frame (BGR) to hsv_frame
329         hsv_frame = cv2.cvtColor(vid_frame,cv2.COLOR_BGR2HSV)
330 # working on the TARGET image/window
331         (h, s, v) = cv2.split(hsv_frame)    # split hsv_frame into its H-S-V image components
332         target_frame = cv2.inRange (h, obj_hsv_lo[0], obj_hsv_hi[0]) # using only Hue frame
333         cv2.imshow('Original Target Frame', target_frame) # display original target frame
334         target_opened = cv2.morphologyEx(target_frame, cv2.MORPH_OPEN, kernel)
335         cv2.imshow('Opened Target Frame', target_opened) # display opened target frame
```

Fig. 8.113 Part 6 of Main Endless Loop in "M_HSV_Tracker_Walker.py".

- The first step is to convert the original video frame from BGR to HSV color system (Line 329).
- The second step is to split the HSV video frame into its individual component "h", "s", and "v" and only "h" will be needed in this project (Line 331).
- The third step is to use the OpenCV Method ***inRange()*** on the "h" frame (Line 332) to get a **new** "target_frame" which is a "binary" (black & white) version of the "h" frame where:
    - H pixels that happen to have intensity values between the limits "obj_hsv_lo[0]" and "obj_hsv_hi[0]" are set to "255", i.e. they become "white".
    - For H pixels that fall outside of the range of those two limits they will be set to "0", i.e. they become "black".

- See this web page for more details on Method ***inRange()*** (https://docs.opencv.org/master/d2/de8/group__core__array.html#ga48af0ab51e36436c5d04340e036ce981).

- Line 333 displays a typical "target_frame" obtained after the "inRange" step had been done for a Yellow object (see window labeled "Video Camera" in Fig. 8.102). The reader can see that the main Yellow object had been identified, but other smaller blobs were also identified by the "inRange" method (see window labeled "Original Target Frame" in Fig. 8.102). This was because we did not use the other limits on S and V with the "inRange" method (which would cost us more computer processing time). We were trading process speed for process accuracy because we know that the MINI would have to track quickly enough to provide real-time commands to its servos.
- Fortunately, there is another speedy method that can be used on "target_frame" to get rid of most of the smaller blobs. It is an OpenCV morphological operator named "OPEN" (Line 334) (see link https://docs.opencv.org/master/d4/d86/group__imgproc__filter.html#gga7be549266bad7b2e6a04db49827f9f32a08d3cc3a2ace00cec488966d31fa29ea) which actually is a ERODE operation followed by a DILATE operation (https://docs.opencv.org/master/d4/d86/group__imgproc__filter.html#gaeb1e0c1033e3f6b891a25d0511362aeb). The result frame is displayed by Line 335 (see window labeled "Opened Target Frame" in Fig. 8.102). This result depends on the size of the kernel used (Line 152 of Fig. 8.103), thus readers may have to adjust this parameter for their own situations. Fig. 8.114 shows another result of the above image processing steps for a Blue object.

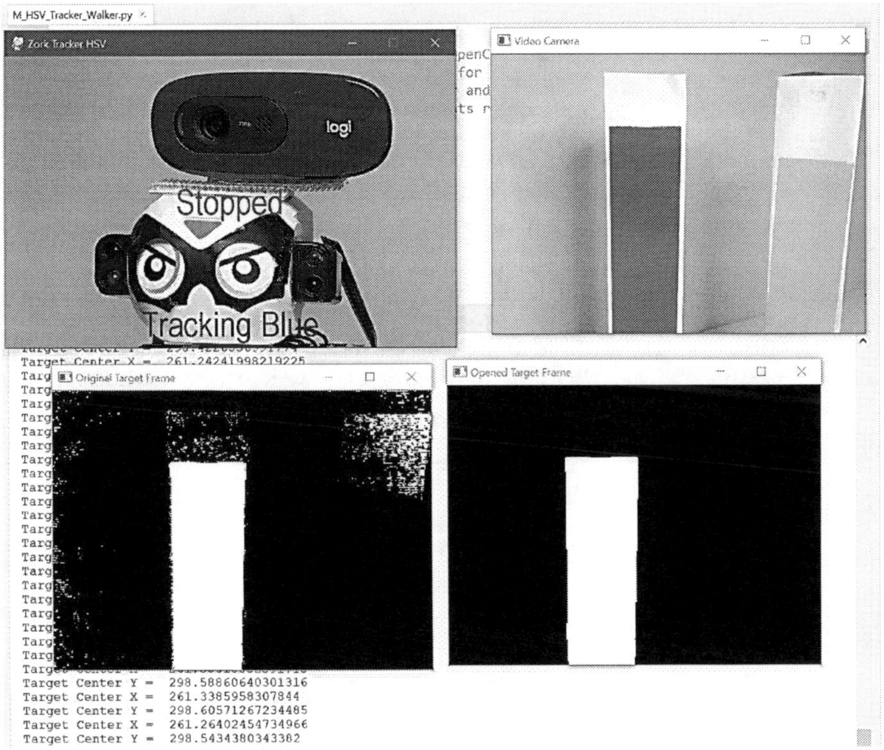

Fig. 8.114 Result of Part 6 code, searching for a Blue object, in "M_HSV_Tracker_Walker.py".

Fig. 8.115 describes Part 7 of the Main Endless Loop where more information is extracted from the "Opened Target Frame" such as the one shown on the bottom right of Fig. 8.114:

"M_HSV_Tracker_Walker.py"
Part 7 of Main Endless Loop

```
337         target_moments = cv2.moments(target_opened)
338         if (target_moments['m00'] >= 50):
339             target_x = target_moments['m10']/target_moments['m00']
340             target_y = target_moments['m01']/target_moments['m00']
341             print('Target Center X = ', target_x)
342             print('Target Center Y = ', target_y)
343             if (robot_mode == 'a' and target_x > 30):
344                 ir_max = 0
345                 if (target_x < search_left_limit):
346                     left(robot1)
347                 elif (target_x > search_right_limit):
348                     right(robot1)
349                 elif ((target_x >= search_left_limit) and (target_x <= search_right_limit)):
350                     forward(robot1)
351                     ir_max = check_NIR(robot1)
352                     print('ir_max = ', ir_max)
353                     if (ir_max > 15):
354                         ec.say(robot1, robot1 + " has found the target")
355                         print(robot1 + ' has found the target')
356                         init_pose(robot1)
357                         break
```

Fig. 8.115 Part 7 of Main Endless Loop in "M_HSV_Tracker_Walker.py".

- As we only need to know the width coordinate of the CG of the white "column" in such "Opened Target Frame" in Fig. 8.114 in order to steer the robot left or right to align it with the selected object in the physical realm, Fig. 8.114 tells us that we are on the right "track"!
- The width coordinate of the CG of the white "column" in the "Open Target Frame" of Fig. 8.114 (i.e. Variable "target_x") can be determined with the use of the OpenCV Class called "moments()" (https://docs.opencv.org/master/d8/d23/classcv_1_1Moments.html) on the binary frame "target_opened" (Line 337). The "moments()" class returns many parameters stored in the dictionary named "target_moments", and we will only need to use two type of parameters from it:
    - The "m00" item represents the total number of white pixels occupied by the white blob in the "target_opened" frame (or so called zeroth order moment). If "m00" is zero or smaller than 50, we shall assume that no valid target had been found (Line 338) and this control algorithm is by-passed back out to the Main Endless Loop.
    - The "m10" item represents the first moment of those white pixels with respect to the width coordinate wherever a given white pixel is found within the "target_opened" frame.

- When (m00 >= 50), the expression (m10 / m00) can be computed and would give us the width coordinate of the CG (or geometrical barycenter) of the white blob (i.e. "target_x" in Line 339), wherever it may appear as the webcam+robot pans across the physical scene.
- In Fig. 8.115, the height-based barycenter coordinate "target_y" is also computed and printed, but it is just for informational use only.

- Lines 343-357 describe the target alignment and approach algorithm for the robot:
  - The IF condition used in Line 343 implies that a little bit of the user-designated object needs to be within the camera viewport at the start of this algorithm, i.e. target_x > 30 (*the reader could improve on this algorithm by making the robot turning left or right automatically when no target is within the camera's viewport, but for this project the operator can just use the left and right arrow keys to get this task done*). When this condition is TRUE, "ir_max" is set to 0 (Line 344) and "target_x" is further checked to see if "target_x" is less than "search_left_limit" (Line 345). If this latter condition is TRUE, then the target is on the physical left of the robot, so it should turn left to bring more of the target into the camera's viewpoint (Line 346). However, if Line 345 turns out to be FALSE, then Line 347 needs to be checked next, to see if (target_x > search_right_limit). If this condition is TRUE, then the target is on the physical right of the robot, then it should turn right to get more of the target into the camera viewpoint (Line 348).
  - The two robot actions described above should turn, in place, the robot right and left as needed until the condition in Line 349 becomes TRUE, i.e. the robot is "acceptably" aligned with the target (i.e. target's CG is within ± 20 pixels on either side of the vertical middle line across the video frame), then the robot is commanded to step forward (Line 350).
  - Next, Method ***check_NIR()*** is called to return the current value of "ir_max" which is then printed for user information (Lines 351-352). If "ir_max" greater than "15" (Line 353), the physical distance between the physical target and the robot is considered "close enough", and the algorithm announces (Line 354) and prints out an informational message (Line 355), then stops the robot (Lines 356) and finally breaks out of the Main Endless Loop (Line 357). Next, the PYTHON session would terminate as normal.

Video 8.12 showcases a typical run-time performance of this project "M_HSV_Tracker_Walker.py".

8.8.6 Dual Color Trackers with Web Cam using OpenCV & Threading

For this last PYTHON project, the code developed for "M_HSV_Tracker_Walker.py" is adapted into suitable Threads for Zira and Zork and the author chose a "game" orientation for this Thread Programming application project "M_Dual_HSV_Trackers.py".

Fig. 8.116 lists the Variables and Constants used in "M_Dual_HSV_Trackers.py":

- In general, two copies of each type of Variables are used as we are now dealing with 2 robots ("robot1" and "robot2") as compared to the previous Single Robot Project (see Fig. 8.103).
- The last character "1" or "2" on these Variables designates to which robot they belong to.

- Variable "winner" (Line 263) is initialized to "0", its final value will be determined by Threads ***go_Zira()*** and ***go_Zork()***.

"M_Dual_HSV_Trackers.py"
Variables & Constants Init.

```
252 robot1 = "Zira"
253 robot2 = "Zork"
254 ir_l1 = 0
255 ir_r1 = 0
256 ir_l2 = 0
257 ir_r2 = 0
258 ir_max1 = 0
259 ir_max2 = 0
260 background_image_filename = 'Zira_Zork_Cam_Ready.jpg'
261 SCREEN_SIZE = (1119, 416)
262 reset = False
263 winner = 0 # '1' if Zira won, '2' if Zork won
264 message1="1) Push key 'p' to play or 's' to stop"
284 # Pixel values limits to be set for object to be tracked
285 obj_hsv_lo1 = [None] * 3
286 obj_hsv_hi1 = [None] * 3
287 obj_hsv_lo2 = [None] * 3
288 obj_hsv_hi2 = [None] * 3
289 # user needs to set up kernel size as needed for image processing
290 kernel1 = np.ones((20,20),np.uint8)
291 kernel2 = np.ones((20,20),np.uint8)
```

Fig. 8.116 Variables and Constants used in "M_Dual_HSV_Trackers.py".

Fig. 8.117 shows that robot1 "Zira" is hard-coded to track a Yellow object while robot2 "Zork" is hard-coded to track a Blue object in this Dual Color Trackers project.

"M_Dual_HSV_Trackers.py"
Methods *set_yellow_object()* & *set_blue_object()*

```
def set_yellow_object():    # Zira is set to search for YELLOW
    global obj_hsv_lo1, obj_hsv_hi1
    for i in range(3):
        obj_hsv_lo1[i] = YELLOW_OBJECT_MEAN[i] - 2 * YELLOW_OBJECT_STDDEV[i]
        obj_hsv_hi1[i] = YELLOW_OBJECT_MEAN[i] + 2 * YELLOW_OBJECT_STDDEV[i]
        if obj_hsv_lo1[i] < 0:
            obj_hsv_lo1[i] = 0
        if obj_hsv_hi1[i] > 255:
            obj_hsv_hi1[i] = 255

def set_blue_object():  # Zork is set to search for BLUE
    global obj_hsv_lo2, obj_hsv_hi2
    for i in range(3):
        obj_hsv_lo2[i] = BLUE_OBJECT_MEAN[i] - 2 * BLUE_OBJECT_STDDEV[i]
        obj_hsv_hi2[i] = BLUE_OBJECT_MEAN[i] + 2 * BLUE_OBJECT_STDDEV[i]
        if obj_hsv_lo2[i] < 0:
            obj_hsv_lo2[i] = 0
        if obj_hsv_hi2[i] > 255:
            obj_hsv_hi2[i] = 255
```

Fig. 8.117 Methods ***set_yellow_object()*** and ***set_blue_object()*** in "M_Dual_HSV_Trackers.py".

The same Thread Design used in the "Dual Random Walkers" Project (see Section 8.8.4) was found to be applicable for this "Dual Color Trackers" project also, i.e. 3 threads are needed:

1. Thread ***go_Zira()*** (see Figs. 8.118 and 8.119) is set for "robot1".
2. Thread ***go_Zork()*** is identical in program logic and design, but is set for "robot2", so no figure for this thread is shown.
3. Thread ***game_end()*** is shown in Fig. 8.120.

Fig. 8.118 shows the beginning of Thread ***go_Zira()*** which is of course adapted from the Main Endless Loop of the previous project "M_HSV_Tracker_Walker.py":

"M_Dual_HSV_Trackers.py"
Thread *go_Zira()* (1)

```
116 def go_Zira():
117     global game_start, game_over, reset, winner, robot1, vid_cam1, search_left_limit1, search_right_limit1
118 
119     game_start.wait()  # blocking until game_start is set in Main program
120 
121     while True:
122     # Codes for Robot1
123     # Operating video camera 1
124         got_frame1, vid_frame1 = vid_cam1.read() # picking up a new video frame
125         if not got_frame1:  # error capturing video frame
126             print("No video frame 1 captured!")
127             break  # get out of endless Loop
128         cv2.imshow('Video Camera 1', vid_frame1) # display original captured frame
129 
130     # Process image to track yellow object
131     # Convert vid_frame (BGR) to hsv_frame
132         hsv_frame1 = cv2.cvtColor(vid_frame1,cv2.COLOR_BGR2HSV)
133     # working on the TARGET image/window
134         (h1, s1, v1) = cv2.split(hsv_frame1)    # split hsv_frame into its H-S-V image components
135         target_frame1 = cv2.inRange (h1, obj_hsv_lo1[0], obj_hsv_hi1[0]) # using only Hue frame
136 #        cv2.imshow('Original Target Frame 1', target_frame1) # display original target frame
137         target_opened1 = cv2.morphologyEx(target_frame1, cv2.MORPH_OPEN, kernel1)
138 #        cv2.imshow('Opened Target Frame 1', target_opened1) # display opened target frame
```

Fig. 8.118 Beginning of Method ***go_Zira()*** in "M_Dual_HSV_Trackers.py".

- Line 119 has the usual Threading Method "game_start.wait()" to wait for the signal "game_start.set()" to be sent by the Main Method when the user pushes on the Key "p" at run-time (Line 445 of Fig. 8.124).

- During the development of this code, the author had noticed that all the OpenCV ***imshow()*** methods were using lots of computer resources, thus the use of these methods are minimized as compared to the single-robot version "M_HSV_Tracker_Walker.py". Please take a note of the commented-out Lines 136 and 138.

Fig. 8.119 shows the ending of Method ***go_Zira()***:

"M_Dual_HSV_Trackers.py"
Thread *go_Zira()* (2)

```
145             if (target_x1 > 30):
146                 ir_max1 = 0
147                 if (target_x1 < search_left_limit1):
148                     left(robot1)
149                 elif (target_x1 > search_right_limit1):
150                     right(robot1)
151                 elif ((target_x1 >= search_left_limit1) and (target_x1 <= search_right_limit1)):
152                     forward(robot1)
153                     ir_max1 = check_NIR(robot1)
154                     print('ir_max1 = ', ir_max1)
155                     if (ir_max1 > 15):
156                         if (winner == 0):
157                             winner = 1 # Zira won!
158                         game_over.set()
159                         break   # to break out of endless loop
160         if (reset == True):
161             init_pose(robot1)
162             break   # to break out of endless loop
163         time.sleep(0.01)  # to prevent Windows error 10048
164     # end of endless loop
165 
166     print("\nGetting out of go_Zira()")
167     print("winner is " + str(winner))
168     return
```

Fig. 8.119 Ending of Method ***go_Zira()*** in "M_Dual_HSV_Trackers.py".

- Lines 145-155 are the same statements used in the Main Endless Loop for "M_HSV_Tracker_Walker.py" but adapted for "robot1" while Lines 156-168 are added for this project "M_Dual_HSV_Trackers.py".
- At run-time, when Line 155 is executed, it means that Zira had located the Yellow object and had approached it close enough to declare "mission accomplished". But as Thread ***go_Zork()*** is also working at the same time, the code needs to check if Zork had not been declared a winner yet.
- Lines 156-157 check to see if no winner had been declared (i.e. winner == 0), then Zira is the true winner (Line 157).

- Next, Line 158 sends the signal ***game_over.set()*** to Thread ***game_end()*** (see Fig. 8.120) to let it wrap up the current game.
- Then, Line 161 breaks out of the Endless Loop that was started at Line 121 in Fig. 8.118. Next, Lines 166-168 are be executed and Thread ***go_Zira()*** ends.
- Lines 160-162 are executed if-and-only-if Parameter "reset" is set to TRUE by Thread ***game_end()*** meaning that Zork was the winner already, or that the user had pushed Key "s" (Line 458 in Fig. 8.124), or that the time allotted to the game (100 seconds) had run out (Line 482 of Fig. 8.126).
- Line 163 is used to prevent a Windows error message when it finds that this endless loop is zipping along too fast from previous similar programming projects.

Fig. 8.120 displays the programming steps used in Thread ***game_end()***:

- Line 229 performs the same blocking as explained before.
- Line 230 immediately set Global Parameter "reset" to TRUE so that a clear winner can be declared ASAP.
- Lines 231-232 stop the two robots and Lines 233-234 make them say "Game is over" one after the other.
- If Line 235 evaluates to TRUE, Zira is the winner and it claims so with Line 236, and then waves its left arm with Line 237.

"M_Dual_HSV_Trackers.py"
Thread *game_end()*

```
226 def game_end():
227     global game_over, reset, robot1, robot2, winner, draw
228 
229     game_over.wait()  # blocking until game_over is set in Main program or in the other 2 threads
230     reset = True   # set reset to True ASAP
231     init_pose(robot1)
232     init_pose(robot2)
233     ec.say(robot1, "Game is over")
234     ec.say(robot2, "Game is over")
235     if (winner == 1):
236         ec.say(robot1,"and Zira won")
237         l_arm_wave(robot1)
238     elif (winner == 2):
239         ec.say(robot2,"and Zork won")
240         r_arm_wave(robot2)
241     elif (winner == 0):
242         ec.say(robot1,"because game")
243         ec.say(robot2,"time elapsed")
244 
245     print("\nGetting out of game_end()")
246     return
```

Fig. 8.120 Method ***game_end()*** in "M_Dual_HSV_Trackers.py".

- But if Line 238 evaluates to TRUE instead, Zork is the winner and it claims so with Line 239, and then waves its right arm with Line 240.
- Otherwise, if Line 241 evaluates TRUE, then we got a draw situation because the allotted game time of 100 seconds had elapsed.
- Line 245 is just an informational message for the programmer to help with debugging at run-time.

Fig. 8.121 displays Robots Initializations steps used in Main Method of "M_Dual_HSV_Trackers.py":

"M_Dual_HSV_Trackers.py"
Robots Initializations

```
321 # Robots initialization
322 ec.set_servo_torque(robot1,"0/1")
323 ec.set_servo_speed(robot1,"0/100")
324 ec.set_custom(robot1,"82/1/0")
325
326 ec.set_servo_torque(robot2,"0/1")
327 ec.set_servo_speed(robot2,"0/100")
328 ec.set_custom(robot2,"82/1/0")
329
330 init_pose(robot1)
331 init_pose(robot2)
332
333 time.sleep(1.00)
334
335 l_arm_wave(robot1)
336 ec.set_custom(robot1,"264/2/1024")  # set Servo17 joint-offset to 1024 on robot1
337 ec.say(robot1, robot1 + " is ready for a game")
338
339 r_arm_wave(robot2)
340 ec.set_custom(robot2,"264/2/1024")  # set Servo17 joint-offset to 1024 on robot2
341 ec.say(robot2, robot2 + " is ready for a game")
```

Fig. 8.121 Robots Initializations in Main Method of "M_Dual_HSV_Trackers.py".

- All lines of code had been described before in the previous Single Robot project.
- Only Lines 336 and 340 are added for this Dual Color Trackers project and they are for setting the JOINT-OFFSET of Servo17 to 1024 in each robot "Zira" and "Zork". This "1024" value will take these Servo17 out of the effects due to MOTION-LIST/PAGE being played in this project so that we can set the Goal Position of these servos directly in Method ***check-_NIR(robot)***.

Fig. 8.122 shows the Pygame Initialization steps in the Main Method of "M_Dual_HSV_Trackers.py":

"M_Dual_HSV_Trackers.py"
Pygame Initialization

```
pygame.init()
screen = pygame.display.set_mode(SCREEN_SIZE, 0, 32)
pygame.display.set_caption('Dual Color Search Walkers Game')
background = pygame.image.load(background_image_filename).convert()
font = pygame.font.SysFont("arial", 80);
text_surface1 = font.render("Start", True, (255, 0, 255))
x1 = ( SCREEN_SIZE[0] - text_surface1.get_width() ) / 2
y1 = ( SCREEN_SIZE[1] - text_surface1.get_height() ) / 2
text_surface2 = font.render("Zira won", True, (255, 0, 0))
x2 = ( SCREEN_SIZE[0] - text_surface2.get_width() ) / 2
y2 = ( SCREEN_SIZE[1] - text_surface2.get_height() ) / 2
text_surface3 = font.render("Zork won", True, (255, 0, 0))
x3 = ( SCREEN_SIZE[0] - text_surface3.get_width() ) / 2
y3 = ( SCREEN_SIZE[1] - text_surface3.get_height() ) / 2
text_surface4 = font.render("It's a draw", True, (0, 255, 0))
x4 = ( SCREEN_SIZE[0] - text_surface4.get_width() ) / 2
y4 = ( SCREEN_SIZE[1] - text_surface4.get_height() ) / 2
text_surface5 = font.render("Game stopped", True, (255, 0, 0))
x5 = ( SCREEN_SIZE[0] - text_surface5.get_width() ) / 2
y5 = ( SCREEN_SIZE[1] - text_surface5.get_height() ) / 2
text_surface6 = font.render("Over Time Limit", True, (255, 0, 255))
x6 = ( SCREEN_SIZE[0] - text_surface6.get_width() ) / 2
y6 = ( SCREEN_SIZE[1] - text_surface6.get_height() ) / 2
```

Fig. 8.122 Pygame Initializations in Main Method of "M_Dual_HSV_Trackers.py".

Fig. 8.123 shows Part 1 of the Main Endless Loop which takes care of the case when the user closes the Pygame Main Window:

"M_Dual_HSV_Trackers.py"
Part 1 of Main Endless Loop

```
time_game_start = 0.0  # initial value
while True:
# Checking for user's keyboard inputs
    for event in pygame.event.get():
        if event.type == QUIT:  # when
            init_pose(robot1)
            init_pose(robot2)
            vid_cam1.release()
            vid_cam2.release()
            cv2.destroyAllWindows()
            pygame.quit()
            sys.exit()
```

Fig. 8.123 Part 1 of Main Endless Loop in "M_Dual_HSV_Trackers.py".

Fig. 8.124 shows Part 2 of the Main Endless Loop which takes care of the cases when the user pushes on Key ‘p’ to start playing the robot game or Key ‘s’ when the user wants to stop the game:

"M_Dual_HSV_Trackers.py"
Part 2 of Main Endless Loop

```
443             if event.type == KEYDOWN:
444                 if event.key == K_p:
445                     game_start.set()  # signals go_Zira and go_Zork to
446                     time_game_start = time.process_time()   # the effec
447                     print("time_game_start = " + str(time_game_start))
448                     camera_mode1 ='y'
449                     set_yellow_object()
450                     camera_mode2 ='b'
451                     set_blue_object()
452                     screen.blit(text_surface1, (x1, y1))
453                     pygame.display.update()
454                     time.sleep(1)
455                 if event.key == K_s:
456                     init_pose(robot1)
457                     init_pose(robot2)
458                     reset = True  # this flag is also used to control t
459                     screen.blit(text_surface5, (x5, y5))
460                     pygame.display.update()
461                     time.sleep(2.0)
462                     sys.exit()  # sys.exit() when user wants to stop th
```

Fig. 8.124 Part 2 of Main Endless Loop in “M_Dual_HSV_Trackers.py”.

- Lines 444-454 are used for the case when Key ‘p’ is pressed by the user:
    - Line 445 sends the signal ***game_start.set()*** to get Threads ***go_Zira()*** and ***go_Zork()*** to perform their programmed tasks of locating and approaching their assigned targets.
    - Parameter “time_game_start” is assigned the current value of Method ***time.process_time()***
    - Lines 448-451 set the Yellow object to robot1 (Zira) and the Blue object to robot2 (Zork).
    - Lines 452-453 display “text_surface1” “Start” in the Pygame Main Window.
    - Line 454 waits for 1 second (for the user to notice the change on the Desktop Display.
- Lines 455-462 are used for the case when Key ‘s’ is pressed by the user:
    - Lines 456-457 stop the two robots.
    - Line 458 sets the Global Parameter “reset” to TRUE.
    - Lines 459-460 display “text_surface5” “Game Stopped” in the Pygame Main Window.
    - Line 461 waits for 2 seconds (for the user to notice the change on the Desktop Display before performing a ***sys.exit()*** at Line 462.

Fig. 8.125 shows Part 3 of the Main Endless Loop which takes care of the proper "winner" notification at the end of this game/project:

"M_Dual_HSV_Trackers.py"
Part 3 of Main Endless Loop

```
464 # Updating Pygame screen
465     screen = pygame.display.set_mode(SCREEN_SIZE, 0, 32)
466     screen.blit(background, (0,0))
467     if (winner == 1):
468         screen.blit(text_surface2, (x2, y2))
469         time.sleep(5.0)
470     elif (winner == 2):
471         screen.blit(text_surface3, (x3, y3))
472         time.sleep(5.0)
473     elif (winner == 0):
474         screen.blit(text_surface4, (x4, y4))
475         time.sleep(5.0)
476
477     pygame.display.update()
```

Fig. 8.125 Part 3 of Main Endless Loop in "M_Dual_HSV_Trackers.py".

Fig. 8.126 shows Part 4 of the Main Endless Loop which takes care of the case when the game time limit of 100 second is exceeded at run-time (Line 479), and no robot had achieved their tracking objectives:

"M_Dual_HSV_Trackers.py"
Part 4 of Main Endless Loop

```
479     if ( (time_game_start > 0.0) and (time.process_time() - time_game_start) >= 100.0 ):
480         print("time_game_elapsed = " + str((time.process_time() - time_game_start)))
481         game_over.set()
482         reset = True
483         screen.blit(background, (0,0))
484         screen.blit(text_surface6, (x6, y6))
485         pygame.display.update()
486         break   # to break out of main endless loop
487 # end of Main endless loop
```

Fig. 8.126 Part 4 of Main Endless Loop in "M_Dual_HSV_Trackers.py".

- Line 480 prints the process time elapsed (which is not the "physical" time elapsed).
- Line 481 sends out the signal ***game_over.set()*** to get Thread ***game_end()*** to finish its programmed job, i.e. to stop all robots and to close out all threads.

- Line 482 sets Global Parameter "reset" to TRUE (just to be double sure that all Threads are stopped).
- Lines 483-485 display "text_surface6" "Over Time Limit".
- Line 486 breaks out of the Main Endless Loop.

Fig. 8.127 shows the Program Closure tasks to be executed when the Main Endless Loop is exited:

"M_Dual_HSV_Trackers.py"
Program Closure

```
time.sleep(5)  # so that user
vid_cam1.release()
vid_cam2.release()
cv2.destroyAllWindows()
sys.exit()   # Main sys.exit()
# End of program
```

Fig. 8.127 Program Closure for "M_Dual_HSV_Trackers.py".

Video 8.13 illustrates the runtime behavior for "M_Dual_HSV_Trackers.py" and although this video shows that the robots behaved as programmed, the two live streams of the webcams used did not appear as they did for the Single Tracker Project, so there was no independent verification of what the robots were "seeing" via their webcams during this video recording. Fig. 8.128 shows three screen-captures during the smart phone video recording:

- Top Left Image: this is when the main Initial Pygame Display was drawn but before the user pushes on Key 'p'.
- Top Right Image: this is when the user just pushed on Key 'p' to play the robots game. Ones can see the Text Message "Start" appearing in the middle of the Pygame Display and behind it, the two windows for "Video Camera 1" and "Video Camera 2" popped up, but they were greyed out and would disappear a few seconds later. However, Threads ***go_Zira()*** and ***go_Zork()*** were working as they should in the physical world. So far, the author has not found the reason(s) for this issue yet.
- Bottom Center Image: this is at the conclusion of the game showing that "Zira won".

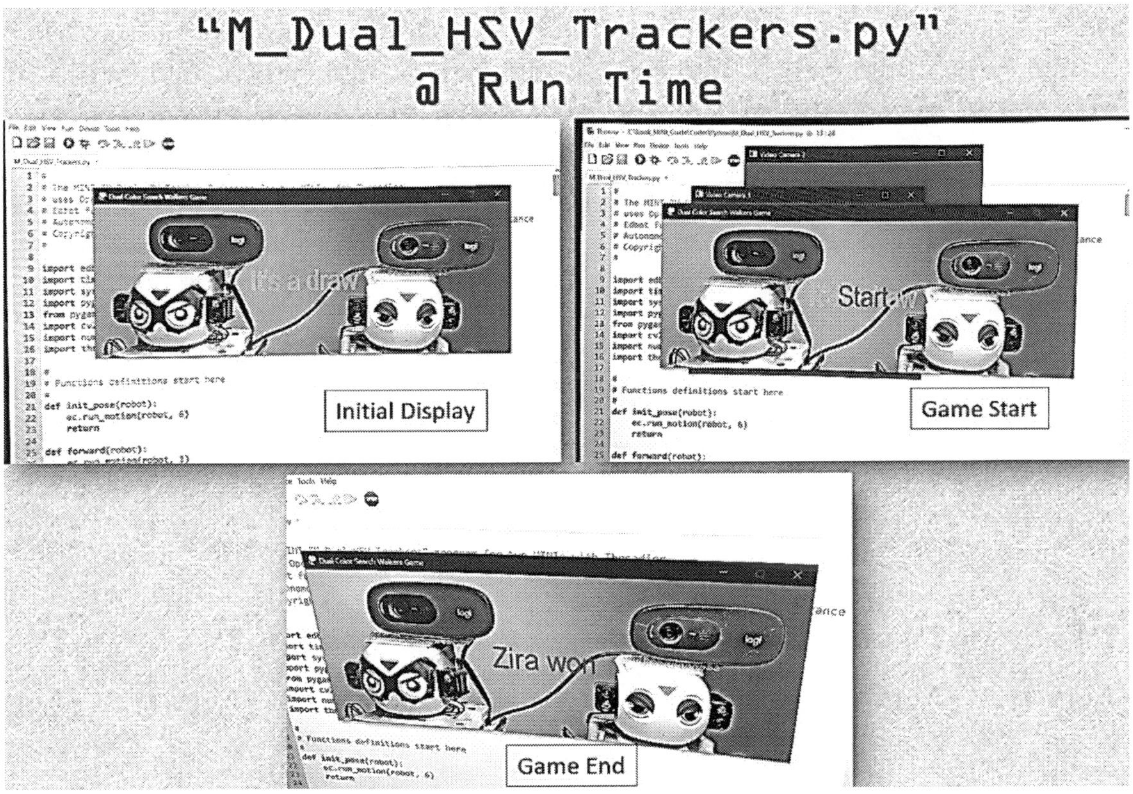

Fig. 8.128 Three Screen Captures during Video 8.13 for "M_Dual_HSV_Trackers.py".

However, THONNY's output shell kept a record of the various print commands that were interspersed within the two Threads ***go_Zira()*** and ***go_Zork()*** and this record can be used as an alternate view of the runtime progress of these two threads:

>>> %Run M_Dual_HSV_Trackers.py
pygame 1.9.6
Hello from the pygame community. https://www.pygame.org/contribute.html
Waiting to control Zira... Got control of Zira!
Waiting to control Zork... Got control of Zork!
1) Push key 'p' to play or 's' to stop
left_limit 1 = 300.0 *-> Please note Left and Right Limits used for "Target Center X1 & X2"*
right_limit 1 = 340.0
left_limit 2 = 300.0
right_limit 2 = 340.0
time_game_start = 3.609375
Target Center X2 = **232.66091749156817** *-> Zork's target is on its left*
Target Center Y2 = 175.18072883819133
Target Center X1 = **224.44913497736982** *-> Zira's target is on its left*
Target Center Y1 = 266.49892136542445
Target Center X2 = 279.5116856521183

Target Center Y2 = 180.32410085754512
Target Center X1 = **328.46608493310066** *-> Zira's target is in front of it, so Zira moves forward*
Target Center Y1 = 341.86561954624784
Target Center X2 = **333.2049716528565** *-> Zork's target is in front of it, so Zork moves forward*
Target Center Y2 = 168.34054731792412
ir_max1 = 13 *-> Zira just used check_NIR() but target not close enough*
Target Center X1 = **322.2219293072947** *-> Zira's target still in front of it, so 2nd step forward*
Target Center Y1 = 317.9612889798467
ir_max2 = 8 *-> Zork just used check_NIR() but target not close enough*
Target Center X2 = 188.32026281825648 *-> Zork's target is now further to its left!*
Target Center Y2 = 191.53612577730846
Target Center X2 = 147.1814021357034
Target Center Y2 = 181.1810041785501
Target Center X2 = 204.6444553667556
Target Center Y2 = 180.72630248560594
Target Center X2 = 226.83274746906636
Target Center Y2 = 188.16872890888638
Target Center X2 = 227.14149570442703
Target Center Y2 = 184.3449838035773
Target Center X2 = 258.1759722317371 ***-> After 6 Left Turns Zork's target is still to its left!***
Target Center Y2 = 190.87378685117946
ir_max1 = 13 *-> Zira distance to target unchanged*
Target Center X1 = **237.4497363262376** *-> Zira's target is on its left*
Target Center Y1 = 304.1273364808526
Target Center X2 = 278.59021316954204
Target Center Y2 = 192.38325354529852
Target Center X1 = 296.52230237756555
Target Center Y1 = 293.7833265596424
Target Center X2 = 225.39838513101768
Target Center Y2 = 197.50073996691913
Target Center X1 = **387.6372836218375** *-> Zira's target is on its right*
Target Center Y1 = 299.10021304926767
Target Center X2 = 194.14525856537918
Target Center Y2 = 198.92995850977374
Target Center X1 = 242.60988250617316
Target Center Y1 = 303.72167801807177
Target Center X2 = 178.00874843030556
Target Center Y2 = 195.37371285056508
Target Center X1 = **330.8040126096625** *-> Zira's target is in front of it, so Zira moves forward*
Target Center Y1 = 283.95129640038124
Target Center X2 = 212.9436761377923
Target Center Y2 = 187.43020283295616
Target Center X2 = 224.46241307223005
Target Center Y2 = 195.6824949869875
Target Center X2 = 250.5483620653496

Target Center Y2 = 196.37552954993498
Target Center X2 = 254.544158243218
Target Center Y2 = 198.77886631104838
Target Center X2 = 253.79963257318582
Target Center Y2 = 194.2537441591118
Target Center X2 = 296.0738553600252
Target Center Y2 = 195.52763670721626
ir_max1 = 13 *-> Zira's distance to target unchanged*
Target Center X1 = 278.2718548669704
Target Center Y1 = 307.84473137989596
Target Center X2 = **353.74803653499333**
Target Center Y2 = 198.0155331898307
Target Center X1 = 322.33679784570325
Target Center Y1 = 283.8706240846601
Target Center X2 = **144.97689028494503**
Target Center Y2 = 227.9936141937488
Target Center X2 = 219.8409074115126
Target Center Y2 = 207.56264316855095
Target Center X2 = 224.12004314218947
Target Center Y2 = 203.53248247348552
Target Center X2 = 290.0529147141518
Target Center Y2 = 210.71441424554826
Target Center X2 = **322.102551976281** ***-> Zork's target is in front of it, so Zork moves forward***
Target Center Y2 = 204.7229023046001
ir_max1 = 13 *-> Zira's distance to target unchanged*
Target Center X1 = 311.85365221000205
Target Center Y1 = 307.2974683544304
ir_max2 = 10 *-> Zork's distance to target slighty "closer"*
Target Center X2 = 303.7496503496503
Target Center Y2 = 228.6042780748663
ir_max1 = 14 *-> Zira's distance to target slighty "closer"*
Target Center X1 = 302.8389705132129
Target Center Y1 = 313.095423936333
ir_max2 = 12 *-> Zork's distance to target slighty "closer"*
Target Center X2 = 304.49174998844535
Target Center Y2 = 229.12133910551697
ir_max1 = 21 ***Zira's distance to target (> 15) – Zira won this game!***

Getting out of go_Zira() ***Exiting go_Zira()***
winner is 1
ir_max2 = 11 ***Zork's distance to target (< 15) – Zork loses this game!***
Getting out of go_Zork() ***Exiting go_Zork() – forced to exit by go_Zira() reset = True***
winner is 1
Getting out of game_end() ***All 3 threads are now stopped!***

The biggest issue for the author was when EDBOT chose to close BT connection to one robot (or both robots) for whatever reasons, then the Thread for that robot will terminate, but in the shell log below, ones can see that the other robot's Thread is still active until this robot achieve its goal of getting close enough to the target:

```
>>> %Run M_Dual_HSV_Trackers.py
pygame 1.9.6
Hello from the pygame community. https://www.pygame.org/contribute.html
Waiting to control Zira... Got control of Zira!
Waiting to control Zork... Got control of Zork!
1) Push key 'p' to play or 's' to stop
left_limit 1 = 300.0
right_limit 1 = 340.0
left_limit 2 = 300.0
right_limit 2 = 340.0
time_game_start = 4.515625
Target Center X1 =  343.9450314465409
Target Center Y1 =  329.614465408805
Target Center X2 =  304.84401002057797   -> Zork is in range already!
Target Center Y2 =  242.97785631206943
Target Center X1 =  344.42884533446954   -> go_Zira() is the only thread running from here
Target Center Y1 =  329.87925010651895   -> for more than 40 cycles
Target Center X1 =  344.76897349079195
Target Center Y1 =  329.23978032868115
Target Center X1 =  343.5503614266423
Target Center Y1 =  326.68865676384144
Target Center X1 =  344.9101957323921
-------      -------   -------   -------   -------   -> Log deleted to save space
Target Center X1 =  344.51470906806463
Target Center Y1 =  331.93522810970063
Target Center X1 =  344.7274593338497
Target Center Y1 =  323.01568551510456
Target Center X1 =  344.3537401257456
Target Center Y1 =  326.3069885539255
ir_max2 =  9                             -> Finally go_Zork() got some actions done
Target Center X2 =  506.7581559717286
Target Center Y2 =  256.70438845186277
Target Center X1 =  344.2338329582608
Target Center Y1 =  332.2705192985436
Target Center X1 =  344.2338329582608    -> Then another round of go_Zira() exclusive actions
Target Center Y1 =  332.2705192985436
Target Center X1 =  345.71664409075515
Target Center Y1 =  328.1131053166272
Target Center X1 =  344.2231230324723
-------      ---------    -------------      ------------
Target Center X1 =  344.11615345187874
```

```
Target Center Y1 =  325.4114655206403
Target Center X1 =  345.08530252405296
Target Center Y1 =  327.2630328891752
Target Center X1 =  344.19409660107334
Target Center Y1 =  320.5073792486583
Target Center X1 =  344.512628176994
Target Center Y1 =  325.76415433108565
Target Center X1 =  343.0883233532934
Target Center Y1 =  321.34557807462
Target Center X2 =  261.2857017889166     -> Once more go_Zork() got ONE action in
Target Center Y2 =  264.0588942833399
Target Center X1 =  344.34853097884       -> Then another round of go_Zira()-only actions
Target Center Y1 =  321.67696757497333
Target Center X1 =  343.83310989234207
Target Center Y1 =  329.9173674588665
Target Center X1 =  344.31481553022985
Target Center Y1 =  323.9874830983195
--------        --------  ------------       ----------
Target Center X1 =  341.2858523205436
Target Center Y1 =  303.5674344668516
Target Center X1 =  339.6059267493564
Target Center Y1 =  295.972531008659
Exception in thread Go Zira:              -> Until here, when EDBOT disconnected Zira from itself
Traceback (most recent call last):
  File "C:\Users\cc101\AppData\Local\Programs\Thonny\lib\threading.py", line 926, in
_bootstrap_inner
    self.run()
  File "C:\Users\cc101\AppData\Local\Programs\Thonny\lib\threading.py", line 870, in run
    self._target(*self._args, **self._kwargs)
  File "C:\Book_MINI_Guide\Codes\Python\M_Dual_HSV_Trackers.py", line 153, in go_Zira
    ir_max1 = check_NIR(robot1)
  File "C:\Book_MINI_Guide\Codes\Python\M_Dual_HSV_Trackers.py", line 55, in check_NIR
    ir_l = sensors["port1"]
TypeError: 'NoneType' object is not subscriptable

ir_max2 =  6                              -> From here on, go_Zork() still runs
Target Center X2 =  403.3373377337734
Target Center Y2 =  276.06213421342136
Target Center X2 =  310.27142083652933
Target Center Y2 =  282.5570094301313
ir_max2 =  7
Target Center X2 =  424.12919013564544
Target Center Y2 =  276.6968968208763
Target Center X2 =  317.21379203402597
Target Center Y2 =  283.5820743431837
```

```
---  ----      -----     ------          ----
Target Center X2 =  345.81736230352664
Target Center Y2 =  329.9378384625545
```

Target Center X2 = 307.0924590594767
Target Center Y2 = 329.0696933361804
ir_max2 = 18 -> ***Until Zork won the game!***
Getting out of go_Zork()
winner is 2
Getting out of game_end()

8.9 Review Questions

1. EDBOT only works with BT-410 (True/False).
2. EDBOT supports MOTION (0) and MOTION (-1) (True/False).
3. How many SCRATCH Blocks does EDBOT support?
4. What is the difference between SINGLE-SERVO and MULTI-SERVO EDBOT Blocks?
5. How many IRSS-10 Sensor(s) does EDBOT support in SCRATCH or PYTHON?
6. Can JOINT-OFFSET be set in EDBOT+SCRATCH?
7. Can JOINT-OFFSET be set in EDBOT+PYTHON?
8. What are the two modes of execution allowed in PYTHON Method run_motion()?
9. Which SCRATCH Block(s) does the programmer need to use to apply Event Programming techniques?
10. Which PYTHON Method(s) does the programmer need to use to apply Event Programming/Threading techniques?
11. PYTHON is a compiled language (True/False).
12. TASK is a compiled language (True/False).
13. What is the name of the PYTHON module that this chapter uses for Machine Vision applications?
14. What us the name of the PYTHON module that this chapter uses for Array Processing applications?
15. List the steps necessary to prepare for a proper connection between a PYTHON program and a given robot registered with EDBOT.

8.10 References

Bouras AS, Ainarozidou LV (2017) Python for Tweens and Teens: Learn Computational And Algorithmic Thinking. Published by Amazon Digital Services LLC.

Escrivá DM and Mendonça VG (2018) Learn OpenCV 4 by Building Projects. Packt, Birmingham, UK.

Ford JL Junior (2014) Scratch 2.0 programming for teens. CENGAGE Learning PTR, Boston USA.

Hellmann D (2017) The Python 3 Standard Library by Example. Addison-Wesley Professional, Boston USA.

Kaehler A and Bradski G (2017) Learning OpenCV 3: Computer Vision in C++ with the OpenCV Library, 1st Edition. O'Reilly Media, Sebastopol USA.

Kinsley H and McGugan W (2015) Beginning Python Games Development with PyGame, 2nd Edition. Apress, New York.

Matthes E (2019) Python Crash Course, 2nd Edition. No Starch Press, San Francisco USA.

Nguyen Q (2018) Mastering Concurrency in Python. Packt, Birmingham, UK.

Spizhevoy A and Rybnikov A (2018) OpenCV 3 Computer Vision with Python Cookbook: Leverage the power of OpenCV 3 and Python to build computer vision applications. Packt, Birmingham, UK.

Summerfield M (2010) Programming in Python 3, 2nd Edition. Addison-Wesley Professional, Boston USA.

Vlieg EA (2016) Scratch by Example: Programming for All Ages. Apress Media, LLC, New York USA

Warner TL (2015) Teach yourself Scratch 2.0 in 24 hours. SAMS, Indianapolis USA.

Made in the USA
Coppell, TX
05 November 2026

11365024R00205